An Introduction to Clouds

From the Microscale to Climate

Clouds, in their various forms, are a vital part of our lives. Their effects on the Earth's energy budget and the hydrological cycle depend on processes on the microphysical scale, encompassing the formation of cloud droplets, ice crystals and precipitation. Cloud formation, in turn, depends on the large-scale environment as well as the characteristics and availability of aerosol particles. An integrated approach drawing on information from all these scales is essential to gain a complete picture of the behavior of clouds in the atmosphere.

An Introduction to Clouds provides a fundamental understanding of clouds, ranging from cloud microphysics to the large-scale impacts of clouds on climate. On the microscale, phase changes and ice nucleation are covered comprehensively, including aerosol particles and the thermodynamics relevant for the formation of clouds and precipitation. At larger scales, cloud dynamics, mid-latitude storms and tropical cyclones are discussed, leading to the role of clouds in the hydrological cycle and their effect on climate.

Each chapter ends with problem sets and multiple-choice questions that can be completed online; important equations are highlighted in boxes for ease of reference. Combining mathematical formulations with qualitative explanations of the underlying concepts, this accessible book requires relatively little previous knowledge, making it ideal for advanced undergraduate and graduate students in atmospheric science, environmental sciences and related disciplines.

Ulrike Lohmann is a professor at the Institute for Atmospheric and Climate Science, ETH Zurich. She obtained her Ph.D. in climate modeling and her research now focuses on the role of clouds and aerosol particles in the climate system, with an emphasis on clouds containing ice. Professor Lohmann has published more than 200 peer-reviewed articles and several book chapters, and was a lead author of the Fourth and Fifth IPCC Assessment Reports. She was awarded the Canada Research Chair in 2002 and was the recipient of the AMS Henry G. Houghton Award in 2007. She is a fellow of the American Geophysical Union and the German Academy of Sciences, Leopoldina. Ulrike Lohmann has been teaching classes in cloud microphysics and cloud dynamics for almost 20 years at both undergraduate and graduate levels.

Felix Lüönd is a researcher at the Swiss Federal Institute of Metrology, METAS. He obtained his Ph.D. in atmospheric ice nucleation, for which he was awarded the ETH medal. His experimental work focused on cloud microphysics. He specialized in the development of dedicated instrumentation to study the aerosol-induced freezing of cloud droplets and the interpretation of the resulting experimental data in the framework of nucleation theory and its advancements. Currently, Dr. Lüönd's research activities are concentrated on aerosol metrology, particularly in the generation of ambient-like aerosols dedicated to establish traceability in measurements of ambient particulate matter and particle number concentration.

Fabian Mahrt is a Ph.D. student at the Institute for Atmospheric and Climate Science, ETH Zurich. He obtained a Master's degree in Atmospheric and Climate Sciences from ETH. Early in his career he developed a passion for cloud microphysics. He is particularly interested in aerosol particles and their role in cloud droplets and ice crystal formation. Fabian Mahrt's work is experimental in nature, measuring and understanding aerosol–cloud interactions in both the laboratory and the field.

An Introduction to Clouds

From the Microscale to Climate

ULRIKE LOHMANN, FELIX LÜÖND AND FABIAN MAHRT

ETH Zurich, Institute for Atmospheric and Climate Science, Switzerland

CAMBRIDGE
UNIVERSITY PRESS

CAMBRIDGE
UNIVERSITY PRESS

Shaftesbury Road, Cambridge CB2 8EA, United Kingdom

One Liberty Plaza, 20th Floor, New York, NY 10006, USA

477 Williamstown Road, Port Melbourne, VIC 3207, Australia

314–321, 3rd Floor, Plot 3, Splendor Forum, Jasola District Centre, New Delhi – 110025, India

103 Penang Road, #05–06/07, Visioncrest Commercial, Singapore 238467

Cambridge University Press is part of Cambridge University Press & Assessment, a department of the University of Cambridge.

We share the University's mission to contribute to society through the pursuit of education, learning and research at the highest international levels of excellence.

www.cambridge.org
Information on this title: www.cambridge.org/9781107018228

First published 2016 (version 4, May 2024)

Printed in Great Britain by CPI Group (UK) Ltd, Croydon CR0 4YY, May 2024

A catalogue record for this publication is available from the British Library

Library of Congress Cataloging-in-Publication data

ISBN 978-1-107-01822-8 Hardback

Additional resources for this publication at www.cambridge.org/clouds

To our families
Kassiem, Stefanie, Claudia, Jana and Rainer

Contents

The plate section is to be found between pages 214 and 215.

Preface

Clouds, in their various forms, are a vital part of our lives. They are a crucial part of the global hydrological cycle, redistributing water to Earth's surface in the form of precipitation. In addition, they are a key element for the global energy budget since they interact with both shortwave (solar) and longwave (terrestrial) radiation. These so-called cloud–radiation interactions depend strongly on the type of cloud. Clearly clouds affect the global climate and thus understanding clouds is an important factor for future climate projections. The effects on Earth's energy budget and on the hydrological cycle both depend on processes on the microphysical scale, encompassing the formation of cloud droplets, ice crystals, raindrops, snowflakes, graupel and hailstones.

Establishing an understanding of clouds and precipitation requires a knowledge of the environment in which they form, i.e. the atmosphere, with all the gases and airborne particles present there. The latter are usually referred to as aerosol particles and encompass a wide range of solid and liquid particles suspended in air. Some aerosol particles can act as nuclei to form cloud droplets or ice crystals and thus initiate the formation of clouds or change their phase from liquid to solid. Thus they influence the microphysical properties of clouds. In turn aerosol particles are removed from the atmosphere when clouds precipitate. In order to gain a complete picture of the behavior of clouds in the atmosphere, the strong interplay between aerosol particles and clouds requires one to tackle the subject in an integrated approach.

This book is intended to offer a fundamental understanding of clouds in the atmosphere. It is primarily written for students at an advanced undergraduate level who are new to the field of atmospheric sciences. The content of this book evolved from the atmospheric physics lectures held at ETH Zurich. This book is intended to serve students with a multidisciplinary background as an introduction to cloud physics, assuming that most readers will have a basic understanding of physics.

The book is organized into 12 chapters, each focusing on a particular topic. Chapter 1 introduces the major cloud types found in the atmosphere and discusses them from a macroscopic point of view. Chapters 2–4 focus on the meteorological conditions and atmospheric dynamics needed for cloud formation and the thermodynamic principles needed to describe atmospheric processes, including phase transitions.

Chapter 5 treats atmospheric aerosol particles and their physical characteristics. The sources and sinks of aerosol particles are discussed at the process level as well as in terms of their global distributions and lifetimes.

Chapters 6–8 cover cloud microphysics. Chapter 6 discusses the fundamental equations that describe the formation of cloud droplets. Chapter 7 introduces the processes which ultimately lead to the formation of rain drops. Ice formation and other microphysical processes occurring in cold clouds are presented in Chapter 8.

Chapter 9 combines the macroscopic view of Chapter 1 with the microscopic view needed to understand the physics of precipitation as well as the differences between stratiform and convective precipitation. Also, the change in precipitation since pre-industrial times and projections into the future are included.

To understand convective clouds, knowledge about cloud dynamics is needed. This is provided in Chapter 10, where convective clouds at all scales, from isolated thunderstorms with lightning and thunder to multicells, supercells and mesoscale convective systems, including tropical cyclones, are discussed.

Finally, Chapters 11 and 12 bring the reader to the global scale. Chapter 11 outlines the physical principles of the global energy budget and discusses the effects of clouds on it. On the basis of the information in Chapter 11 the impact of aerosols and clouds on the climate since pre-industrial times and in future climate projections is considered in Chapter 12.

To strengthen concepts and test the reader's understanding, qualitative exercises and mathematical problems are provided at the end of each chapter. This allows the reader to apply directly the material of the text and provides an opportunity for further learning. To this end, online solutions are provided and can be accessed at www.cambridge.org/clouds. For some of the problem sets the usage of a tephigram will be helpful. This, along with some other material can be accessed from: www.cambridge.org/clouds. Some useful online information about atmospheric science includes the following links:

- Glossary of Meteorology: http://glossary.ametsoc.org/wiki
- Encyclopedia of Atmospheric Sciences: http://app.knovel.com/web/toc.v/cid:kpEASV0002/viewerType:toc/root_slug: encyclopedia-atmospheric/url_slug:encyclopedia-atmospheric/?
- NOAA glossary: http://w1.weather.gov/glossary/
- Fifth Assessment Report of the Intergovernmental Panel on Climate Change: http://www.climatechange2013.org

Throughout the book, important equations are underlaid in gray. All quantities are given in SI units unless stated otherwise. However, as we often refer to processes occurring above or below 0 °C, we will use degrees celsius whenever convenient, keeping in mind that temperatures need to be in kelvins in the equations given (if not noted otherwise).

The outline of the book follows a similar structure to the classic book *A Short Course in Cloud Physics* by Rogers and Yau (1989), which served the present authors not only for their own studies but also for over a decade of teaching at undergraduate level in the atmospheric physics course. Inspired by the straightforwardness of Rogers and Yau (1989) in explaining complex concepts of cloud physics and their style of imparting knowledge to readers new to the atmospheric sciences, paired with the enormous developments in this field over recent years, the authors decided to come up with this new introductory textbook, which places a stronger focus on ice clouds, cloud dynamics and climate change.

We felt that, although there are many excellent textbooks at the graduate level, a textbook introducing the physics of clouds, aerosols and precipitation in an integrated manner combining quantitative discussions at the undergraduate level was lacking. We believe that this book fills this niche in giving intuitive interpretations of the physical processes discussed. Through this approach we hope to present the fascination of clouds that has captured us and thus to stimulate the interest of the readers in this diverse field. The book provides a fundamental understanding, which can be deepened by the excellent further literature that is available.

Writing this book would not have been possible without the knowledge we received from many pioneers of the field of atmospheric sciences; these are named in the appropriate context throughout the book. Equally important, the development of this book relied on the help and support from many colleagues and we are very grateful for help in different aspects. We owe a great debt to Anina Gilgen for her invaluable contribution in putting together the exercises. Chief among those who provided excellent support are the members of our research group, who discussed drafts of different chapters and were a great source of ideas.

The authors wish to thank explicitly Manuel Abegglen, Alexander Beck, Yvonne Boose, Robert David, Remo Dietlicher, Sylvaine Ferrachat, Blaz Gasparini, Anina Gilgen, Franziska Glassmeier, Olga Henneberg, Jan Henneberger, Katty Huang, Luisa Ickes, Zamin Kanji, Christina Klasa, Monika Kohn, Larissa Lacher, Claudia Marcolli, Amewu Mensah, Angela Meyer, Baban Nagare, David Neubauer, Mikhail Paramonov, Fabiola Ramelli, Carolin Rösch, Christina Schnadt, Sarah Schöpfer, Berko Sierau, Janina Stäudle, Kathrin Wehrli and Heini Wernli for very valuable discussions and suggestions that greatly improved the textbook.

Besides, we are indebted to Björn Baschek, Lea Beusch, Sebastian Bretl, Joel Corbin, Betty Croft, Daniel Cziczo, Corinna Hoose, Hanna Joos, Miriam Kübbeler, Glen Lesins, Rebekka Posselt, Jacopo Riboldi, Vivek Sant, Linda Schlemmer, Peter Spichtinger, Eric Sulmoni, André Welti and Marc Wüest. Finally, we are grateful for all the valuable feedback we obtained from students of the atmospheric physics lectures, which has been essential for continuous improvements of the book.

The authors want to thank Remo Dietlicher, Simon Förster, Anina Gilgen, Franziska Glassmeier, Pascal Graf, Miriam Kübbeler, Jeremy Michael, David Neubauer, Sarah Schöpfer and André Welti for their help with figures.

Photographs illustrating various aspects within our textbook were kindly provided by Kouji Adachi, Laurent Barbe, Robert David, Martin Ebert, Blaz Gasparini, Christian Grams, Zachary Hargrove, Jan Henneberger, Otte Homan, Luisa Ickes, Brian Johnson, Laurie Krall, Larissa Lacher, Sandra LaCorte, Kenneth Libbrecht, Julian Quinting, Milos Vujovic and Thomas Winesett.

We specially thank the very helpful staff at Cambridge University Press, namely Susan Francis, Cassi Roberts and Zoë Pruce in guiding the development of this book and also our copy-editor Susan Parkinson for valuable comments and suggestions.

Symbols and acronyms

Symbols		
Symbol	Value/unit	Description
A, B		Different substances in Raoult's law
A_i, B_i		Empirical constants for the saturation vapor pressure over ice
A_w, B_w		Empirical constants for the saturation vapor pressure over water
A	m^2	Area
a	m	Coefficient of the curvature term in the Köhler equation
a_w		Water activity
B		Buoyancy term
B_λ	$W\ m^{-2}$	Black body source function
b	m^3	Coefficient of the solution term in the Köhler equation
b		Coefficient in the Hatch–Choate equation
C	cm^3	CCN concentration at 1% supersaturation
C	F/m	Capacitance for ice crystals
C_c		Cunningham correction factor
C_D		Drag coefficient
CKE	J	Collision kinetic energy
C_R		Constant for radar reflectivity
c_l	$4219.9\ J\ kg^{-1}\ K^{-1}$	Specific heat capacity of liquid water
c_i	$J\ kg^{-1}\ K^{-1}$	Specific heat capacity of ice
c_p	$1005\ J\ kg^{-1}\ K^{-1}$	Specific heat capacity of dry air at constant pressure
c_{pv}	$1884.4\ J\ kg^{-1}\ K^{-1}$	Specific heat capacity of water vapor at constant pressure
c_v	$718\ J\ kg^{-1}\ K^{-1}$	Specific heat capacity of dry air at constant volume
c_{vv}	$1418.4\ J\ kg^{-1}\ K^{-1}$	Specific heat capacity of water vapor at constant volume
D_a, D_v	$m^2\ s^{-1}$	Diffusivities of aerosol particles or water vapor in air
$E, \tilde{E}, \hat{E}$		Collision, collection, coalescence or coagulation efficiencies
E_{coal}	J	Total energy of coalescence
e	Pa	Partial pressure of water vapor
e_{mix}, e'_{mix}	Pa	Mean water vapor pressure of isobarically mixed air before and after condensation
$e_{s,i}, e_{s,w}$	Pa	Saturation vapor pressures with respect to ice or water
e_{s0}	611.2 Pa	Saturation vapor pressure at $T_0 = 273.15$ K
$e*$		Equilibrium vapor pressure over a solution
F	J	Helmholtz free energy
$\vec{F}$	N	Force vector

F_B	$m\ s^{-2}$	Buoyancy force per unit mass
$\vec{F}_C$	$m\ s^{-2}$	Coriolis force per unit mass
F_d	$s\ m^{-2}$	Vapor diffusion term in droplet radius growth equation
F_d^i, F_d^l	$m\ s\ kg^{-1}$	Vapor diffusion terms in the mass growth equations for ice crystals and cloud droplets
F_D	$kg\ m\ s^{-2}$	Drag force
$\vec{F}_F$	$m\ s^{-2}$	Dissipation term for momentum, per unit mass, i.e. friction
F_g	$kg\ m\ s^{-2}$	Gravity force
F_k	$s\ m^{-2}$	Thermodynamic term in droplet radius growth equation
F_k^i, F_k^l	$m\ s\ kg^{-1}$	Thermodynamic terms in the mass growth equations for ice crystals and cloud droplets
F_{LW}, F_{LW}^{cs}	$W\ m^{-2}$	Net longwave radiative fluxes at the TOA in all-sky and clear-sky conditions
$F_{LW}^{\uparrow}$, $F_{LW}^{\downarrow}$	$W\ m^{-2}$	Upward- and downward-directed longwave radiative fluxes
F_{Sun}	3.85×10^{26} W	Radiation emitted by the Sun
F_{SW}, F_{SW}^{cs}	$W\ m^{-2}$	Net shortwave radiative fluxes at the TOA in all-sky and clear-sky conditions
$F_{SW}^{\downarrow}$	$W\ m^{-2}$	Downward-directed shortwave radiative fluxes
$\vec{F}_{PG}$	$m\ s^{-2}$	Pressure gradient force per unit mass
f	s^{-1}	Coriolis parameter, i.e. planetary vorticity
f		Compatibility parameter for heterogeneous nucleation
f_{act}, f_f		Activation and frozen fractions
$\bar{f}_v$		Mean ventilation coefficient
G	J	Gibbs free energy
$G_{s,hom}$, $G_{v,hom}$, $G_{v,het}$	J	Surface and volume terms of the Gibbs free energy for a pure liquid droplet and for a solution droplet
$G_{ex}(n)$	J	Excess Gibbs free energy due to cluster formation
$G(n)$	J	Total Gibbs free energy of the cluster
g	$9.81\ m\ s^{-2}$	Acceleration due to gravity
g, g_v, g_l	$J\ kg^{-1}$	Specific Gibbs free energy in general and in the vapor or liquid phases
H	J	Enthalpy
H	$W\ m^{-2}$	Heat flux into the ocean
H	$m^2\ s^{-2}$	Helicity
h	m	Height above Earth's surface, vertical distance
I_λ	$W\ m^{-2}$	Wavelength-dependent intensity of radiation
I, I_0	$W\ m^{-2}$	Intensity of radiation in general and at TOA
i		Van 't Hoff factor
J, J_i, J_w	$cm^{-3}\ s^{-1}$	General nucleation rate and nucleation rates for ice in vapor and water in vapor
K	$cm^{-3}\ s^{-1}$	Kinetic prefactor for the nucleation rate
K	$J\ m^{-1}\ s^{-1}\ K^{-1}$	Coefficient of thermal conductivity in air
K	$m^3\ s^{-1}$	Collision or collection kernel

$\|K\|^2$		Modulus squared of the complex index of refraction
k	1.38×10^{-23} J K^{-1}	Boltzmann constant
k		Slope of the $CCN-S$ relationship
$k_{\lambda,abs} = k_{abs}$	m^{-1}	Wavelength-dependent absorption coefficient for greenhous gases, aerosol particles and cloud hydrometeors
$k_{\lambda,ext} = k_{ext}$	m^{-1}	Wavelength-dependent extinction coefficient for greenhous gases, aerosol particles and cloud hydrometeors
$k_{\lambda,scat} = k_{scat}$	m^{-1}	Wavelength-dependent scattering coefficient for greenhous gases, aerosol particles and cloud hydrometeors
L	m	Characteristic length scale for geostrophic flow
$L_{2,1}$	J kg^{-1}	Latent heat for the phase change from phase 1 to phase 2
L_f	J kg^{-1}	Latent heat of fusion
$L_{f0} = L_f(T_0)$	0.333×10^6 J kg^{-1}	Latent heat of fusion at $T_0 = 273.15$ K
L_s		Latent heat of sublimation
$L_{s0} = L_s(T_0)$	2.834×10^6 J kg^{-1}	Latent heat of sublimation at $T_0 = 273.15$ K
L_v	J kg^{-1}	Latent heat of vaporization
$L_{v0} = L_v(T_0)$	2.501×10^6 J kg^{-1}	Latent heat of vaporization at $T_0 = 273.15$ K
M	m s^{-1}	Absolute momentum
MF	kg s^{-1}	Mass flux
M_d	28.96 g mol^{-1}	Molecular weight of dry air
M_l	kg m^{-3}	Cloud liquid water content in units of mass per unit volume
M_m	g mol^{-1}	Molecular weight of moist air
M_s	g mol^{-1}	Molecular weight of a solute
M_w	18 g mol^{-1}	Molecular weight of water
MF_d, MF_u	kg s^{-1}	Downward and upward mass fluxes
m, m_w, m_s	kg	Masses of air parcel, bulk water and solute
m_d, m_m, m_v	kg	Masses of dry air, moist air and water vapor
m_0	kg	Masses of one water molecule
m_i, m_l, m_R, $m(r)$	kg	Masses of an ice crystal, cloud droplet and collector drop and of hydrometeors in general
m_a, m_{tot}	kg	Total mass of aerosol particles and of the solution droplet
N	s^{-1}	Brunt–Väisälä frequency
N		Number of molecules
N_A	6.022×10^{23} mol^{-1}	Avogadro's constant
N_0	cm^{-4}	Intercept parameter for hydrometeor size distributions
N, N_a, N_{CCN}, N_c, N_d, N_i, N_r	cm^{-3}	Number concentrations in general and of aerosol particles, CCN, cloud droplets, drizzle drops, ice crystals and raindrops
N_j		Number of particles of type j

n		Number of moles or molecules
n_i, n_v	m^{-3}	Number concentrations of gas molecules and of water vapor molecules
$n_m^e(r)$	$g\ cm^{-3}$	Mass concentration of hydrometeors
$n_N(r)$	$cm^{-3}\ \mu m^{-1}$	Number concentration of aerosol particles per micrometer size
$n_N(r)$	$m^{-3}\ mm^{-1}$ or $m^{-3}\ m^{-1}$	Number concentration of hydrometeors per unit length
$n_N^e(r)$	cm^{-3}	Number concentration of aerosol particles on a logarithmic scale
n_s, n_w		Numbers of solute and water molecules
$n_S(r)$	$\mu m^2\ cm^{-3}\ \mu m^{-1}$	Surface concentration of aerosol particles per micrometer size
$n_S^e(r)$	$\mu m^2\ cm^{-3}$	Surface concentration of aerosol particles on a logarithmic scale
$n_V(r)$	$\mu m^3\ cm^{-3}\ \mu m^{-1}$	Volume concentration of aerosol particles per micrometer size
$n_V^e(r)$	$\mu m^3\ cm^{-3}$	Volume concentration of aerosol particles on a logarithmic scale
P		Probability
$\overline{P}_R$	W	Power received at a radar antenna
p	Pa	Atmospheric pressure
p_0	1000 hPa	Reference pressure
p_c	Pa	Pressure of the lifting condensation level
p_i, p_A, p_B	Pa	Partial pressure of particle type i and of substances A and B
p_n	Pa	Pressure within a cluster
p_s	Pa	Saturation vapor pressure
p_{tot}	Pa	Total vapor pressure of a solution
Q	J	Heat energy
Q		Generic quantity in thermodynamics
Q_1	m^{-1}	Thermodynamic variable in supersaturation equation
Q_2		Thermodynamic variable in supersaturation equation
$Q_{\lambda,ext}$		Extinction efficiency
q	$J\ kg^{-1}$	Specific heat energy
q_i	$kg\ kg^{-1}$	Cloud ice mass mixing ratio
q_l	$kg\ kg^{-1}$	Cloud liquid water mass mixing ratio
$q_s, q_{s,i}$	$kg\ kg^{-1}$	Saturation specific humidity with respect to water and ice
q_v	$kg\ kg^{-1}$	Specific humidity: mass of water vapor per unit mass of moist air
$q_{v,cl}$	$kg\ kg^{-1}$	Specific humidity in a cloudy air parcel
$q_{v,env}$	$kg\ kg^{-1}$	Specific humidity in the environment
$q_{v,mix}$	$kg\ kg^{-1}$	Mean specific humidity in well-mixed air
q_x	$kg\ kg^{-1}$	Condensate mixing ratio: mass of condensate per unit mass of dry air
R	m	Radius of a collector drop
R	$mm\ h^{-1}$	Precipitation or rain rate
R_d	$287\ J\ kg^{-1}\ K^{-1}$	Gas constant of dry air
R_i	m	Melted radius of a snowflake

R_m	$\mathrm{J\,kg^{-1}\,K^{-1}}$	Gas constant of moist air
R_{Sun}	6.98×10^8 m	Radius of the Sun
R_v	$461.5\ \mathrm{J\,kg^{-1}\,K^{-1}}$	Gas constant of water vapor
R^*	$8.314\ \mathrm{J\,mol^{-1}\,K^{-1}}$	Universal gas constant
R_{50}	m	Radius at which 50% of aerosol particles are activated
Re		Reynolds number
RH, RH_i	%	Relative humidities with respect to water and ice
Ro		Rossby number
r	m	Radius, distance in spherical coordinates
r_{act}	m	Critical radius for activation
r_c	m	Critical radius
r_d, r_h, r_i, r_r	m	Radii of a droplet, hydrometeor, ice particle and raindrop
$\overline{r}_d$	m	Mean volume radius
r_{eq}	m	Equivalent radius for a raindrop or for the drop that is formed when a snowflake melts
r_{dry}	m	Dry radius of an aerosol particle
r_{Earth}	6.371×10^9 m	Radius of the Earth
r_{max}	m	Maximum raindrop radius
$r_{Sun-Earth}$	1.5×10^{11} m	Sun–Earth distance
S	$\mathrm{J\,K^{-1}}$	Entropy
S	$\mathrm{m^2}$	Surface area
$S = S_w, S_i$		Ambient saturation ratios with respect to water and ice
S_a	$\mathrm{\mu m^2\,cm^{-3}}$	Total aerosol surface area concentration
S_{act}		Activation saturation ratio
S_c	J	Surface energy
S_{cry}		Crystallization saturation ratio
S_{del}		Deliquesence saturation ratio
S_{max}		Maximum supersaturation reached in a cloud
$S(r)$		Size-dependent saturation ratio of a solution droplet with radius r
S_0	$1360\ \mathrm{W\,m^{-2}}$	Solar constant
Sc		Schmidt number
s	$\mathrm{J\,kg^{-1}\,K^{-1}}$	Specific entropy
s	m	Path length
s, s_{cl}, s_{env}	$\mathrm{J\,kg^{-1}}$	Moist static energies in general and for cloudy or environmental air
s	%	Supersaturation
s_{act}	%	Supersaturation required for activation
T	K	Temperature of an air parcel
T	$\mathrm{m\,s^{-1}}$	Characteristic time scale for geostrophic flow
T_0	273.15 K	Melting point temperature (0 °C)
T_{atm}	K	Average temperature of the atmosphere
T_{cl}	K	Temperature of a cloudy air parcel
T_{env}	K	Temperature of the environment
T_c	K	Isentropic condensation temperature
T_d	K	Dew point temperature
T_e	K	Effective temperature
T_e, T_{es}	K	Equivalent and saturation equivalent temperature

T_{in}, T_{out}	K	Input and output temperature of the Carnot cycle
T_{mix}, T'_{mix}	K	Mean temperature in isobarically mixed air before and after condensation
T_p	K	Temperature of the air parcel
T_{p0}, T_{tropo}	K	Temperatures at 1000 hPa and at the tropopause
T_{rd}	K	Temperature at a droplet's surface
T_s	K	Annual global mean temperature at the Earth's surface
T_{Sun}	5769.56 K	Temperature of the Sun
T_v, $T_{v,env}$	K	Virtual temperature in general and of the environment
T_w	K	Wet-bulb temperature
t	s	Time
U	J	Internal energy
U	m s^{-1}	Characteristic velocity scale for geostrophic flow
u	J kg^{-1}	Specific internal energy
u, v	m s^{-1}	Horizontal velocity components in x- and y- directions
u_g, v_g	m s^{-1}	Geostrophic wind components in x- and y- directions
V, V_n	m^3	Volume in general and of the cluster
V_a	μm^3 cm^{-3}	Total aerosol volume concentration
V_{sweep}	m^3	Sweep-out volume
$\vec{v}$	m s^{-1}	Three-dimensional velocity vector
$\vec{v}_h$	m s^{-1}	Horizontal velocity vector
v, v_i, v_w	m^3 kg^{-1}	Specific volume in general and of ice and water
v_0	m^3	Volume of an individual molecule
v_D	m s^{-1}	Doppler velocity
v_h	m s^{-1}	Fall velocity of hydrometeors
v_r	m s^{-1}	Relative velocity of two falling hydrometeors
v_T, $v_{T,snow}$	m s^{-1}	Terminal velocity in general and for snow
W	J	Work
w	J kg^{-1}	Specific external work
w, w_{cl}	m s^{-1}	Vertical velocity in general and for a cloudy air parcel
w_j		mass fraction
w_l	kg kg^{-1}	Adiabatic cloud liquid water mixing ratio
w_s	kg kg^{-1}	Saturation water vapor mixing ratio
w_v	kg kg^{-1}	Water vapor mixing ratio: mass of water vapor per unit mass of dry air
x		Dimensionless size parameter for scattering
x_0	m	Impact parameter within which a collision is certain to occur
Z	mm^6 m^{-3}	Radar reflectivity factor
Z_e	mm^6 m^{-3}	Equivalent radar reflectivity factor
z	m	Height
α, α_i, α_l, α_v	m^3 kg^{-1}	Specific volumes of air, ice, liquid water and water vapor
α_m		Accommodation coefficient for ice crystal growth
α_c		Cloud albedo
α_p		Planetary albedo
β	m^{-1} s^{-1}	Meridional gradient of Coriolis parameter f
γ, Γ	K m^{-1}	Lapse rates of the ambient air and of an air parcel
γ		Radii ratio of collector drop to smaller droplet

Γ_d	9.8 K km^{-1}	Dry adiabatic lapse rate
Γ_s	K km^{-1}	Pseudoadiabatic lapse rate
ΔF = RF	W m^{-2}	Radiative forcing
ΔF_{LW}, ΔF_{SW}	W m^{-2}	Changes in F_{LW} and F_{SW} at the TOA
ΔH	W m^{-2}	Heat uptake by the ocean
ΔG	J	Change in Gibbs free energy
ΔG^*, ΔG^*_{het}	J	Gibbs free energy barrier in general and for heterogeneous nucleation
ΔG_s, ΔG_v	J	Surface and volume terms of the change in Gibbs free energy
$\Delta G_{s,hom}$, $\Delta G_{s,sol}$	J	Surface terms of the change in Gibbs free energy in pure water and in a solution
$\Delta G_{v,hom}$, $\Delta G_{v,sol}$	J	Volume terms of the change in Gibbs free energy in pure water and in a solution
$\Delta G_{w,v}$, $\Delta G_{i,w}$, $\Delta G_{i,v}$	J	Changes in Gibbs free energy between water and vapor, ice and water, and ice and vapor
ΔR	W m^{-2}	Net radiative imbalance at the TOA
ΔT_s	K	Change in global mean surface temperature
δ		infinitesimal change
ϵ, ϵ_λ		Emissivity in general and wavelength-dependent emissivity
ϵ		Entrainment
ϵ	0.622	Ratio $R_d/R_v = m_w/m_d$
ζ	s^{-1}	z-component of the relative vorticity
η	s^{-1}	x-component of the relative vorticity
η		Thermodynamic efficiency of the Carnot process
θ		Contact angle
θ	K	Potential temperature
θ_e, θ_{es}	K	Equivalent and saturation equivalent potential temperatures
θ_{mix}	K	Mean potential temperature in well-mixed air
θ_w	K	Wet-bulb potential temperature
κ	0.286	Ratio R_d/c_p
κ		Hygroscopicity parameter
Λ	m^{-1}	Entrainment rate
Λ	cm^{-1}	Slope of hydrometeor size distributions
Λ_B	s^{-1}	Scavenging coefficient
λ	m	Wavelength
λ	W m^{-2} K^{-1}	Climate sensitivity parameter
λ_0	W m^{-2} K^{-1}	Null climate sensitivity parameter
μ		Shape parameter of hydrometeor size distributions
μ	kg m^{-1} s^{-1}	Dynamic viscocity of air
μ	J s^{-1}	Chemical potential

μ, μ_S, μ_V, μ_m	m	Number mean radius and arithmetic means of the surface, volume and mass radii
$\mu_g, \mu_{g,S}, \mu_{g,V}, \mu_{g,m}$	m	Geometric means of the number, surface, volume and mass distributions of aerosol particles
$\widetilde{\mu}, \overline{\mu}$	m	Number mode radius and radius of average mass
ν	$m^2\ s^{-1}$	Kinematic viscocity
ξ	s^{-1}	*y*-component of the relative vorticity
ρ_a	$kg\ m^{-3}$	Density of an aerosol particle
ρ_d	$kg\ m^{-3}$	Density of dry air
ρ_{env}	$kg\ m^{-3}$	Density of ambient air
ρ_l, ρ_i, ρ_s	$kg\ m^{-3}$	Densities of liquid water, ice and a solution droplet
ρ, ρ_m	$kg\ m^{-3}$	Density in general and of moist air
$\rho_v, \rho_{v,r_d}, \rho_{vs}$	$kg\ m^{-3}$	Ambient water vapor density, water vapor density at a droplet's surface and saturation water vapor density
σ		Arithmetic standard deviation
σ	$5.67 \times 10^{-8}\ W\ m^{-2}\ K^{-4}$	Stefan–Boltzmann constant
σ	$N\ m^{-1}$	Surface tension
$\sigma_{i,a}, \sigma_{i,v}$	$N\ m^{-1}$	Surface tension of ice in air or vapor
$\sigma_{w,a} = \sigma_w$	$N\ m^{-1}$	Surface tension of water in air
$\sigma_{w,v}$	$0.0756\ N\ m^{-1}$	Surface tension of water in water vapor at 273.15 K
$\sigma_{i,w}$	$N\ m^{-1}$	Surface tension between ice and water
$\sigma_{INP,i}, \sigma_{INP,v}, \sigma_{INP,w}$	$N\ m^{-1}$	Surface tensions between an INP and ice, vapor or water
σ, σ_i	m	Standard deviations of the aerosol size distribution in mode *i*
σ_g		Geometric standard deviation of the aerosol size distribution
τ		Optical depth
τ_{AP}		Aerosol optical depth
ϕ	°, degrees	Latitude
$\varphi(V_n)$	J	Gibbs free energy of the interface between a cluster and the parent phase
Ω	$7.29 \times 10^{-5}\ s^{-1}$	Earth's rotation
ω	$Pa\ s^{-1}$	Vertical velocity in the *p*-system
$\vec{\omega}$	s^{-1}	Three-dimensional vorticity vector
$\vec{\omega}_h$	s^{-1}	Horizontal vorticity vector

Acronyms	
AEJ	African easterly jet
AEROCOM	Aerosol Intercomparison project
AERONET	Aerosol Robotic Network
AOD	Aerosol optical depth
AOGCM	Coupled atmosphere–ocean general circulation model
AR4	Fourth Assessment Report of the Intergovernmental Panel on Climate Change
AR5	Fifth Assessment Report of the Intergovernmental Panel on Climate Change
aci	Aerosol–cloud interactions
ari	Aerosol–radiation interactions
a.u.	arbitrary units
BC	Black carbon
BWER	Bounded weak echo region
CALIPSO	Cloud–Aerosol Lidar and Infrared Pathfinder Satellite Observation
CAPE	Convectively available potential energy (J kg^{-1})
CCB	Cold conveyor belt
CCL	Convective condensation level
CCN	Cloud condensation nuclei
CCNC	Cloud condensation nuclei counter
CERES	Capacité de Renseignement Electromagnétique Spatial
CIN	Convective inhibition
CKE	Collision kinetic energy
CNT	Classical nucleation theory
CRE	Cloud radiative effect
CRH	Crystallization relative humidity
CTP	Cloud top pressure
DCAPE	Downdraft convectively available potential energy (J kg^{-1})
DJF	December, January, February
DMS	Dimethyl sulfide
DRH	Deliquescence relative humidity
DU	Mineral dust particles
EC	Elemental carbon
ECA	Emission controlled area
ECHAM6 GCM	Global climate model from the Max Planck Institute for Meteorology in Hamburg, Germany
ECMWF	European Centre for Medium-Range Weather Forecast
ERA-interim	ECMWF Re-analysis interim
ERFaci	Effective radiative forcing due to aerosol–cloud interactions
ERFari	Effective radiative forcing due to aerosol–radiation interactions
ERFacitari	Effective radiative forcing due to aerosol–cloud and aerosol–radiation interactions
GCCN	Giant CCN
GCM	Global climate model
GeoMIP	Geoengineering Model Intercomparison Project
GHG	Greenhouse gases

GPCC	Global Precipitation Climatology Centre
HTI	Height–time indicator
INP	Ice nucleating particle
IPCC	Intergovernmental Panel on Climate Change
ISA	International standard atmosphere
ISCCP	International Satellite Cloud Climatology Project
ITCZ	Intertropical Convergence Zone
IWC	Ice water content
JJA	June, July, August
LAADS	Level-1 and Atmosphere Archive and Distribution System
LCRE	Longwave cloud radiative effect
LCL	Lifting condensation level
LFC	Level of free convection
LH	Latent heat flux
LNB	Level of neutral buoyancy
LW	Longwave
LWC	Liquid water content
MCC	Mesoscale convective complex
MCS	Mesoscale convective system
MEE	Mass extinction efficiency
MISR	Multi-angle imaging spectroradiometer
MODIS	Moderate resolution imaging spectroradiometer
NCFR	Narrow cold frontal rainband
NOAA	National Oceanic and Atmospheric Administration
OC	Organic carbon
OLR	Outgoing longwave radiation
PBL	Planetary boundary layer
POA	Primary organic aerosol
POM	Particulate organic matter
PPI	Plan position indicator
QLL	Quasi-liquid layer
Radar	Radio detection and ranging
RCP	Representative concentration pathway
RF	Radiative forcing
RFaci	Radiative forcing due to aerosol–cloud interactions
RFacitari	Radiative forcing due to aerosol–cloud and aerosol–radiation interactions
RFari	Radiative forcing due to aerosol–radiation interactions
RH	Relative humidity
SCE	Stochastic coalescence equation
SCRE	Shortwave cloud radiative effect
SH	Sensible heat flux
SOA	Secondary organic aerosols
SPCZ	South Pacific Convergence Zone
SRM	Solar radiation management
SST	Sea surface temperature
SS	Sea salt

SU	Sulfate
TC	Tropical cyclone
TDE	Thermodynamic equilibrium
TOA	Top of the atmosphere
WCB	Warm conveyor belt
WCFR	Wide cold frontal rainband
WFR	Warm frontal rainband
WMGHG	Well-mixed greenhouse gases
WMO	World Meteorological Organization

1 Clouds

Clouds are fascinating to watch for their myriad of shapes. They are also scientifically challenging because their formation requires both knowledge about the large-scale meteorological environment as well as knowledge about the microphysical processes involved in cloud droplet and ice crystal formation.

In this chapter we introduce clouds. In Section 1.1 we highlight their importance for Earth's energy budget and the hydrological cycle. In Section 1.2 we discuss the main cloud types, with their macroscopic properties, as defined by the World Meteorological Organization (WMO), and other, less common cloud types. After this macroscopic description of clouds, we turn to their microphysical properties in Section 1.3.

1.1 Definition and importance of clouds

A cloud is an aggregate of cloud droplets or ice crystals, or a combination of both, suspended in air. For a cloud to be visible, the cloud particles need to exist in a sufficiently large concentration. This definition has its origin in operational weather forecasting, where observers indicate the fraction of the sky that is covered with clouds. A more precise definition of cloud cover is used when the information is derived from satellite data, which nowadays provide a global picture of the total cloud cover. Satellites define clouds on the basis of their optical depth, which is the amount of radiation (in our case from the Sun) removed from a light beam by scattering and absorption (Chapter 12).

There are several global cloud climatologies; most of them derived from satellite data, so-called satellite retrievals (Stubenrauch *et al.*, 2009). In the global annual average, roughly 70% of Earth's surface is covered with clouds. The cloud cover is 5%–15% higher over oceans than over land (Table 1.1). The oldest satellite data are from the International Satellite Cloud Climatology Project (ISCCP) (Rossow and Schiffer, 1999), which has cloud information dating back to 1983. The ISCCP satellite picture (Figure 1.1) shows that clouds cover more than 90% of the sky in the storm tracks of the Southern Ocean and the semi-permanent Aleutian and Icelandic low pressure regions in the north Pacific and north Atlantic, respectively, as shown in Figure 1.2. High cloud amounts are also seen in the Intertropical Convergence Zone (ITCZ) between the equator and 10°–15° N and in the South Pacific Convergence Zone (SPCZ), which is a northwest to southeast oriented band starting at 120° E and the equator and extending to 120° W and 30° S, as a result of

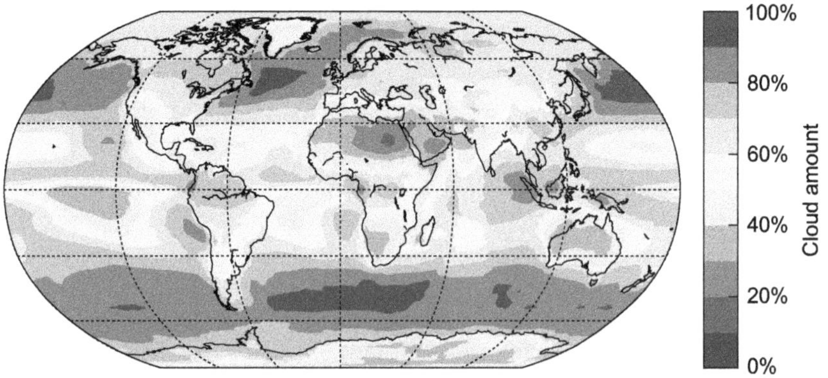

Fig. 1.1 Annual mean total cloud cover [%] averaged over 1983–2009. Data were obtained from the International Satellite Cloud Climatology Project (ISCCP) web site (http://isccp.giss.nasa.gov/) in December 2014 and are described in Rossow and Schiffer (1999). A black and white version of this figure will appear in some formats. For the color version, please refer to the plate section.

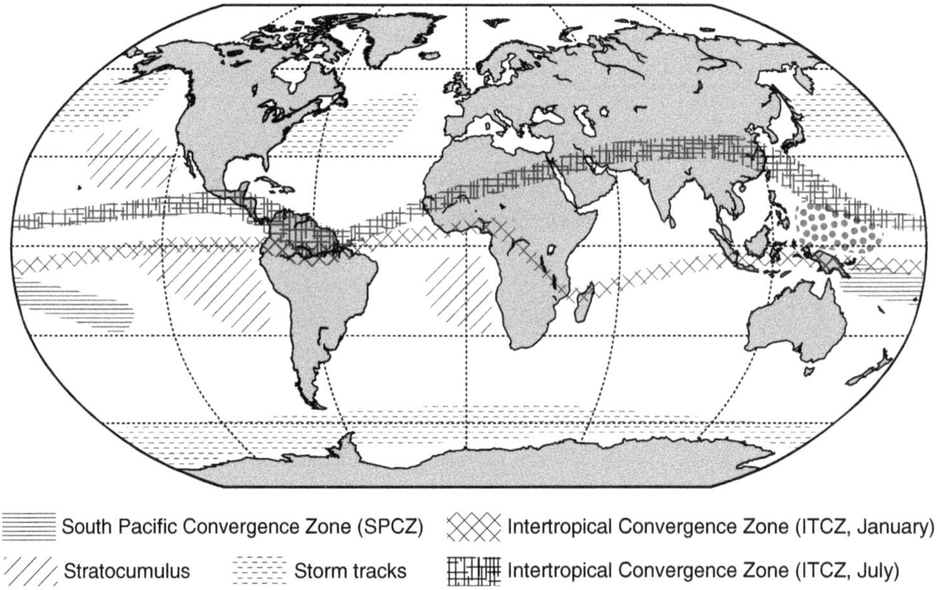

Fig. 1.2 Schematic of the regions where clouds occur most often.

convective activity. The location of the ITCZ has an annual cycle that follows the position of the Sun's zenith but is modulated by the distribution of land masses; see Figure 1.2.

The extensive cloud cover off the west coasts of North and South America and Africa is associated with stratiform clouds that form under subtropical high pressure systems over cold ocean currents. Apart from these regions, the subtropics are characterized by small cloud amounts, in particular over the main deserts, such as the Sahara and the Kalahari as well as the desert and arid regions of Australia and the Arabian Peninsula. Satellite retrievals still have problems in identifying clouds over ice-covered surfaces. Therefore cloud amounts over polar regions are rather uncertain. In addition, the area of lower cloud amount over the Indian Ocean seen in Figure 1.1 is an artifact due to the poorer satellite coverage in this region.

Clouds are an integral part of Earth's atmosphere and are a major factor in Earth's radiation budget. Their influence on the radiation budget, i.e. their radiative impact, differs for the solar (shortwave) and terrestrial (longwave) radiation and depends on the cloud type and altitude, as will be discussed in Chapter 11.

As part of the hydrological cycle, clouds deliver water from the atmosphere to Earth's surface as rain or snow (Section 9.6). Precipitation removes soluble gases and aerosol particles (Chapter 5) from the atmosphere and deposits them onto the surface. Moreover, clouds provide a medium for aqueous-phase chemical reactions, meaning reactions that take place inside cloud droplets. As an example, sulfate aerosols can be produced by oxidation of gaseous sulfur dioxide upon its uptake into cloud droplets. Furthermore, vertical motions associated with clouds, called updrafts and downdrafts, largely determine the vertical redistribution of trace species, temperature and moisture.

1.2 Macroscopic cloud properties and cloud types

Clouds can be grouped into three basic categories, following the French naturalist Jean Baptiste Lamarck (Lamarck, 1802) and English chemist Luke Howard (Howard, 1803), namely cumulus, stratus and cirrus. The Latin term "cumulus" means heap or pile and denotes dense isolated clouds, which appear as white tuberose flowers, composed of individual cloud elements. In a group of cumulus clouds the individual clouds are disconnected from each other. The tower-like structure of cumulus clouds indicates that they usually extend farther in the vertical than in the horizontal. In fact, cumulus clouds form due to convection, i.e. the rising of air masses in locally unstable air. Therefore cumulus clouds are often referred to as convective clouds, where their vertical extent is given by the depth of the instability. In contrast, stratus clouds represent clouds with dimensions that are much larger in the horizontal than in the vertical, as indicated by "stratus", which means flat. Consequently, they appear as uniform cloud layers when seen from Earth's surface. When stratus clouds cover the whole sky, the sky appears gray and may not have any structure. In contrast with cumulus clouds, stratus clouds are usually formed by large-scale vertical air motions in statically stable air (Rogers and Yau, 1989), i.e. there is little internal vertical motion. Cumulus and stratus can consist entirely of cloud droplets, of ice crystals or of both, depending on temperature and other parameters as will be discussed in Chapters 7 and 8.

Lastly, the word "cirrus" means hair or curl and is used to describe clouds that appear wispy and fibrous, composed of delicate filaments. Cirrus clouds consist purely of ice crystals, which give them their characteristic hair-like appearance. Cirrus clouds have horizontal dimensions that are much larger than their vertical dimension. Because of their low ice water content (Table 1.3), they appear almost transparent. Ice crystals can leave the cirrus clouds as precipitation, which causes the cloud edges to become optically thin and

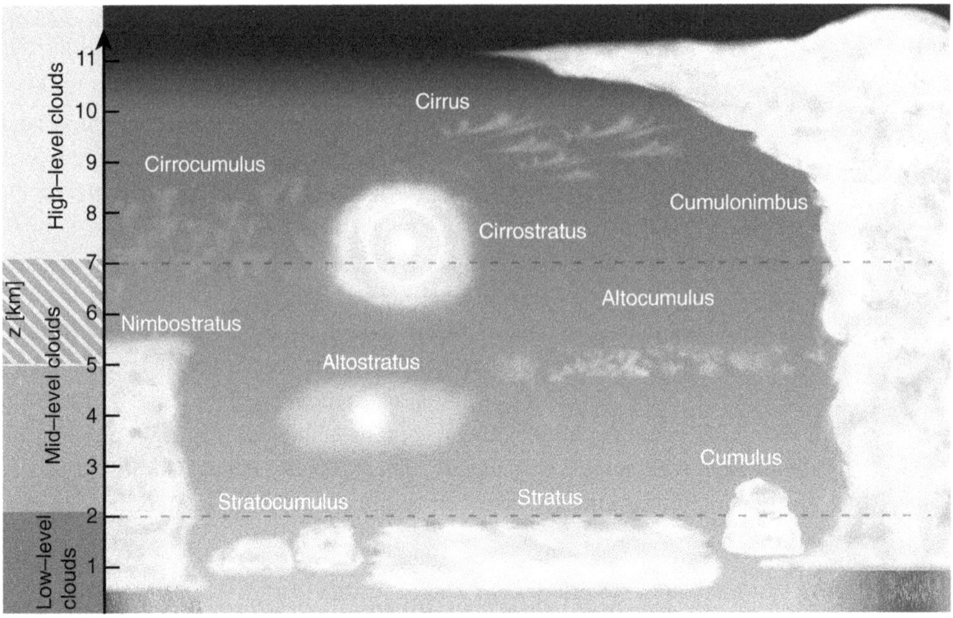

Fig. 1.3 Schematic figure of the ten cloud types according to WMO and as described in Table 1.1. A black and white version of this figure will appear in some formats. For the color version, please refer to the plate section.

to appear diffuse (Penner *et al.*, 1999). In contrast, cloud droplets are much smaller than ice crystals (Table 1.3) and do not leave the cloud as precipitation until they have grown to raindrop size. Therefore the edges of cumuliform clouds (see below) are quite sharp as long as they consist of cloud droplets.

The WMO further distinguishes clouds in these three categories and specifies them first in terms of the cloud type they belong to, also called "genera". This defines the main category to which a cloud belongs (WMO, 1975). On the basis of the cloud shape, i.e. the arrangement of the elements that compose it, a cloud type can be further subdivided into "species". Finally the concept of "varieties" allows one to characterize the spatial arrangement of the macroscopic cloud elements and the associated degree of transparency. Here we confine ourselves to a description of clouds only in terms of genus and species.

A cirrus cloud can be subdivided into the species "uncinus" or "floccus". Cirrus uncinus describes a cloud which has a comma-like shape (Figure 1.6c). Cirrus floccus, however, occurs as small tufts of cloud.

In total, ten exclusive cloud types exist according to the WMO. They are shown in Figure 1.3. It has proven useful to divide these cloud types into subgroups on the basis of the height at which they occur. The subgroups contain low-level, mid-level and high-level clouds, respectively. Here "level" refers to the altitude of the cloud base above mean sea level. Cloud base heights differ for clouds in polar regions, mid-latitudes and tropical regions and so this term should be considered somewhat flexible. Another important characteristic of clouds is their vertical extent, given by the distance between cloud base and cloud top.

In order to distinguish the various cloud types it helps to ask the following questions: How much of the sky is covered by this cloud? Is the Sun visible and, if so, is the Sun's disk sharply defined or diffuse? Is rain or snow falling from the cloud? If so, is the precipitation widespread or concentrated in narrow shafts? Are particular patterns such as small cloud elements, rolls or undulations visible or does the cloud base appear uniformly gray? Table 1.1 summarizes the ten main cloud types along with their abbreviations, their appearances, their base heights above Earth's surface as well as their global annual averages.

In general, the annual mean cloud coverage is higher over the ocean, 68%, as compared with that over land, 54%. The coverage of stratus, stratocumulus and cumulus clouds is 8%–10% higher over the oceans (Table 1.1) and their cloud bases are at lower altitudes because of the higher relative humidity in the oceanic boundary layer. Cirrus clouds have a larger coverage over land, however. This could be due to a real physical difference or to difficulties in observing cirrus clouds over the oceans (Eastman *et al.*, 2011). One physical explanation for a higher cirrus coverage over land areas is cirrus formation caused by gravity waves triggered when air flows over mountains.

1.2.1 Low-level layered clouds

Stratus (St) and stratocumulus (Sc) clouds (Figure 1.4d, e) comprise the low-level layered cloud types. These clouds are shallow stratiform clouds, usually less than 1 km in vertical extent (Rangno, 2002). In terms of appearance St cannot be distinguished from "high fog", a phenomenon that is common in Alpine valleys. While clouds form due to adiabatic expansion and cooling in rising air parcels or air masses (Section 2.2), fog forms as a result of isobaric cooling and it touches the ground.

Most commonly, fog forms by radiative cooling at night and it occurs most often in the early mornings during autumn and winter. This fog is called radiation fog. At night, the longwave radiation emitted from Earth's surface causes the air to cool. If it is sufficiently moist, the relative humidity can reach 100% with respect to water, allowing condensation to set in. Usually fog does not extend much in the vertical (a few hundred meters at most) so that heating by solar radiation during the day causes the fog droplets to evaporate and the fog to disappear. During winter, when the solar radiation reaching the fog layer is not strong enough to dissipate it, high fog may persist for days. Other types of fog are advection fog, which occurs when moist air passes over a cooler surface, and mixing fog, which occurs when two air masses with different temperatures and high relative humidity mix. An example of mixing fog is steam fog, discussed in Chapter 4.

Stratocumulus clouds show some degree of patchiness, which is why cumulus enters the name. In contrast with cumuliform clouds they are layered and have flattened cloud tops. While neighboring cumuliform clouds can be seen as clearly separate from each other, individual stratocumulus cloud elements are connected, forming a cloud layer. In contrast with stratus clouds, which are purely gray, stratocumulus clouds consist of a pattern of gray and white colors, where the rounded air masses of individual cloud elements can be recognized.

Table 1.1 The ten cloud types and their acronyms, according to the WMO definition, along with fog. Their typical altitude ranges in polar regions (pr), mid-latitudes (ml) and tropical regions (tr) and their characteristic features (WMO, 1975) are given. Global annual averages of the amounts of the different cloud types obtained from satellite observations (Raschke *et al.*, 2005) and from surface observations over land (1971–2009) and ocean (1954–2008), and their average base heights in km above the surface, are taken from Eastman and Warren (2013) and Khvorostyanov and Curry (2014).

Cloud type/genera	Description	Amount [%]		Base height [km]	
		Land	Ocean	Land	Ocean
Fog	a "cloud" that touches the ground	1	1	0	0
Low-level layered clouds: 0–2 km (pr, ml, tr)		17	35		
Stratus (St)	very low, gray, uniform layer; Sun outline very distinct when visible	2–5	2–13	0.5	0.4
Stratocumulus (Sc)	low, gray-white, patch or layer with elements, rolls or rounded masses	8–13	14–22	1	0.6
Low-level with vertical extent: 0–2 km (pr, ml, tr)		14	24		
Cumulus (Cu)	white, detached, dense elements with shape outlines and vertical growth	5–8	13–15	1.1	0.6
Cumulonimbus (Cb)	very deep, dense and precipitating, with flattened top	3–5	2–6	1	0.5
Nimbostratus (Ns)	gray, dark, diffuse, uniform cloud with steady precipitation	2–4	3–5	0.1–3	
Mid-level clouds: 2–4 km (pr), 2–7 km (ml), 2–8 km (tr)		21	24		
Altostratus (As)	uniform or striated gray/blue sheet; Sun can be seen as through translucent glass, i.e. there is no clear outline	4–12	6–7	3–5 (ml)	
Altocumulus (Ac)	gray or white broken sheets, elements, bands, rounded masses	5–17	3–18	2–6 (ml)	
High-level clouds: 3–8 km (pr), 5–13 km (ml), 6–18 km (tr)		6–22	6–14		
Cirrus (Ci)	detached white filaments or patches with fibrous appearance or silky sheen			7–10 (ml)	
Cirrostratus (Cs)	thin white translucent veil, either fibrous or smooth in appearance (halo)			6–8 (ml)	
Cirrocumulus (Cc)	thin white sheet or patch without shading, composed of very small ripples, grains			6–8 (ml)	
Total cloud cover		54	68		

Stratus and stratocumulus clouds tend to produce drizzle in clean air masses that are characterized by low concentrations of aerosol particles. These are primarily found over the oceans in the absence of anthropogenic pollution. Drizzle denotes light rain and has drop radii between 25 μm and 0.25 mm (Table 1.2). They are large enough to fall by the action of gravity but are still smaller than raindrops, with radii between 0.25 and 5 mm (Houze, 1993). Rain and drizzle formation will be discussed in Chapter 7. Sometimes drizzle inhibition over the oceans can be observed in so-called ship tracks, which owe their existence to exhaust from ships. Ship tracks are visible as bright lines (Figure 1.4f). They broaden with time, i.e. with increasing distance from the moving ship, due to the turbulent mixing of environmental air masses with the ship track. This causes the dilution of the ship

Fig. 1.4 (a) Cumulus humilis, Cu hum, (b) cumulonimbus, Cb, (c) nimbostratus, Ns, (d) stratus, St, (e) stratocumulus, Sc, as viewed from above, and (f) satellite image of ship tracks off Europe's Atlantic coast (MODIS Land Rapid Response Team). Photographs taken by Fabian Mahrt (a), (b), (d), (e) and Larissa Lacher (c). A black and white version of this figure will appear in some formats. For the color version, please refer to the plate section.

track until it either evaporates or cannot be distinguished from the background cloud any longer.

Ship exhausts create a much higher number concentration of aerosol particles than is normally found over the oceans. As will be discussed in Chapter 6, a higher number concentration of aerosol particles causes the cloud to consist of more cloud droplets; this normally implies that the droplets will be smaller as more of them have to compete for the water vapor available for condensation. In Chapter 7 we will see that it takes much longer for small droplets to form precipitation than for large droplets. The time it takes for precipitation formation in a cloud with a high cloud droplet number concentration, such as in a ship track, can be longer than the lifetime of the cloud, so that it dissipates before any precipitation has been formed.

Also, graupel particles may be formed in stratus and stratocumulus clouds (Table 1.3). Graupel refers to heavily rimed snow particles, also called snow pellets when their radius is less than 2.5 mm; above this radius heavily rimed snow particles are called hailstones (Table 1.2). Hailstones may be spheroidal, conical or irregular in shape.

1.2.2 Low-level clouds with vertical extent

Convective clouds (cumulus and cumulonimbus) and nimbostratus clouds are characterized by a low base height and a large vertical extent. Though nimbus is the Latin word for a cloud, the term is normally used to denote precipitating clouds. Nimbostratus is always found at mid levels but can extend down to low levels or up to high levels. Since its base height is usually around 2 km, we regard it as having a low-level cloud base. Nimbostratus is a formless cloud that is almost uniformly dark gray; it is thick enough to block the sunlight. Nimbostratus is typically associated with long-lasting stratiform precipitation (Figure 1.4c) and forms when air masses are lifted along a warm front in a low pressure system. Nimbostratus clouds will be discussed in detail in Chapter 9.

Convective clouds, however, develop in unstable air due to buoyancy. In the atmosphere, stability or instability is mainly determined by the vertical temperature gradient. In an unstable atmosphere this gradient is large enough that a rising air parcel (Chapter 2) stays warmer than the surrounding air, in spite of the cooling due to adiabatic expansion. If the air is sufficiently moist, buoyantly rising air parcels can cause the formation of convective clouds, as will be discussed in Chapter 3. Upon cloud formation, the air parcel gains additional buoyancy from the latent heat released during cloud droplet activation and condensational growth (Chapters 6 and 7).

An example of a cumulus cloud, a cumulus humilis (Cu hum), is shown in Figure 1.4a. The designation "Cu hum" is a combination of the genus and the species to which the cloud belongs, where "Cu" denotes that the cloud belongs to the "cumulus" genus and "hum" indicates the species humilis. Cumulus humilis is the smallest cloud of the cumulus genus, with a vertical extent of less than 1 km. Its vertical growth can be restricted by a temperature inversion, i.e. a layer of air in which the temperature increases with increasing altitude and which terminates the upward motion of an air parcel. However, a temperature inversion is not required to stop the growth of cumuli. An air parcel stops rising when its temperature is colder than that of the environment. This height can be predicted by

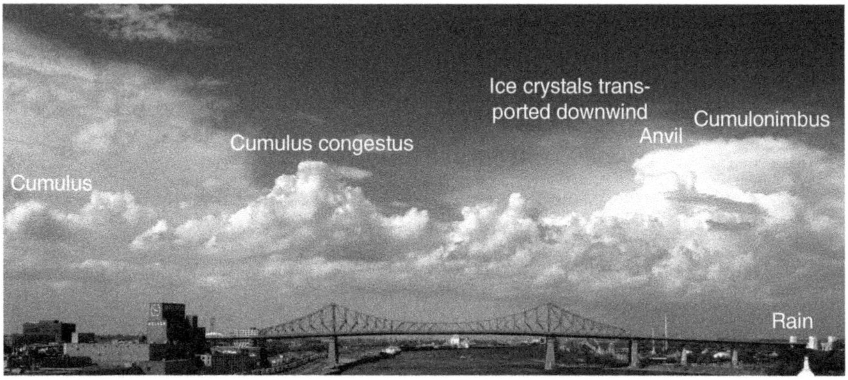

Fig. 1.5 Photograph of cumulus clouds in different stages of development, taken by Fabian Mahrt.

comparing how the environmental temperature changes with height (the ambient lapse rate) with the air parcel's lapse rate (Chapter 3).

Capping inversions or smaller ambient lapse rates cause the uniform cloud top heights of Cu hum; they are a feature of this cumulus species only. A capping inversion often coincides with the top of the planetary boundary layer, which is the lowest layer of the troposphere. Its vertical extent ranges from 50 m during wintertime in the Arctic to over a few hundred meters in mid-latitudes at night and up to 2 km in the tropics.

A cumulus cloud that has a slightly higher buoyancy than the surrounding cumuli is able to penetrate through the capping inversion. There it may develop into the slightly taller cumulus mediocris (Cu med) if the atmosphere above the inversion is unstable, allowing the cloud to extend further vertically. Taller cumuli also develop if the atmosphere is unstable throughout. Like Cu hum, Cu med clouds do not precipitate and frequently develop into cumulus congestus (Cu con, Figure 1.5). This growth occurs over the course of the day if the Sun continues to heat the surface, leading to thermally induced upward motions. Moreover, Cu con can form behind a cold front where the atmosphere becomes increasingly unstable and rising air parcels acquire more buoyancy. These clouds can grow up to heights of 6 km above their cloud bases. They appear as cauliflowers, where the cauliflower elements are indicative of the exchange between cloudy and environmental air in the form of entrainment and detrainment (Chapter 4). They may produce abundant precipitation, especially in the tropics (Johnson *et al.*, 1999). Their sharp outlines suggest that they still consist mainly of cloud droplets even at temperatures below 0 °C.

Cumulus congestus can further develop into cumulonimbus (Cb, Figures 1.4b and 1.5) if the atmosphere becomes even more unstable and the cloud extends high enough that ice crystals form. Part of the buoyancy in cumulonimbus originates from the freezing of cloud droplets into ice crystals and the associated release of latent heat, which warms the cloud relative to its environment and hence causes further upward motion. When Cbs reach the temperature inversion of the tropopause, further ascent is stopped by the stable stratification of the stratosphere. This causes the tops to flatten and the cloud to spread out horizontally leading to the formation of an anvil. Cumulonimbus clouds which have developed an anvil are characterized by a region of active convection on the upwind side and a stratiform anvil on the downwind side. The smooth fibrous anvils of Cbs are indicative of ice crystals falling with different speeds and sublimating below the cloud.

On a given day, it is common to observe cumuli at different stages of development (Figure 1.5) because of slight differences in buoyancy. After some time the seemingly random cumuli develop some organization, which enables the growth of taller cumuli. They in turn develop downdrafts which can either terminate their lifetime or lead to larger complexes such as multicells and supercells, as will be discussed in Chapter 10. Only Cbs can be accompanied by lightning, thunder or hail. As will be discussed in Chapter 10, the common prerequisite for lightning, thunder and hail is the coexistence of water, ice and strong updrafts.

1.2.3 Mid-level clouds

Mid-level clouds comprise altocumulus (Ac) and altostratus (As). They typically owe their existence to the slow upward lifting of air masses (Table 1.3) with updraft vertical velocities on the order of centimeters to tens of centimeters per second over a large area in the mid-troposphere (Rangno, 2002). Altocumulus are usually 200–700 m thick and altostratus clouds are usually 1–3 km thick (Khvorostyanov and Curry, 2014).

Like normal stratus clouds, As clouds are layered with a uniform appearance. Their cloud bases are usually at the mid level of the troposphere, while stratus clouds have base heights around 0.5 km. Even though As clouds are usually found at mid levels, they often extend higher. Moreover, As clouds have smaller optical depths than stratus clouds, i.e. they attenuate radiation less effectively and consequently appear less gray. They are low enough that the Sun's appearance is as if one were looking through translucent glass, meaning that the Sun does not have a clear outline (Figure 1.6a). However, the optical depth of As clouds is larger than that of cirrus clouds. If large-scale altostratus clouds follow cirrostratus clouds, this can often be taken as an indicator of an approaching warm front or occlusion, as will be discussed in Chapter 9.

Like other cumuliform clouds, Ac clouds are associated with local instabilities and convection. They consist of distinct cloud elements and can exist as either detached clouds (altocumulus castellanus) or as rolls in layers and patches (as shown in Figure 1.6b). Altocumulus castellanus clouds observed in the morning are a good indicator of afternoon convection. Altocumulus lenticularis provide a visualization of an oscillatory motion of the air in a statically stable atmosphere, as will be discussed in Section 3.2.

Sometimes precipitation falling from As and Ac clouds is visible in the form of fallstreaks, also called "virga". Virga is the term for precipitation that evaporates or sublimates before it reaches the ground (Section 7.3).

1.2.4 High-level clouds

The three cloud genera constituting in-situ-formed high-level clouds are cirrus (Ci), cirrostratus (Cs) and cirrocumulus (Cc), as shown in Figures 1.6c to 1.6e. Cirrocumulus is a finely granulated cloud, consisting of many small, similar looking, cloud elements. These small granular structures are often arranged in a co-joined regular pattern, where the

Fig. 1.6 (a) Altostratus, As, (b) altocumulus, Ac, (c) cirrus uncinus, Ci, (d) cirrostratus, Cs, with a 22° halo, and (e) cirrocumulus, Cc. Photographs taken by Robert David (a), Fabian Mahrt (b), (e) and Blaz Gasparini (c), (d). A black and white version of this figure will appear in some formats. For the color version, please refer to the plate section.

individual cloud elements are identifiable and intersected by small areas of blue sky. Cirrostratus appears as a whitish veil, having a waveless structure where no individual cloud elements can be identified. A distinct characteristic of Cs is the appearance of a halo (Figure 1.6d). Lastly, cirrus consists of delicate filaments. These three cloud genera and the associated species are solely comprised of ice crystals. Because of the decrease of water vapor with height in the troposphere, cirrus clouds have a much lower water content than mid-level or low-level clouds. Thus cirrus clouds are generally not optically dense enough to produce shading, except when the Sun is near the horizon (Rangno, 2015).

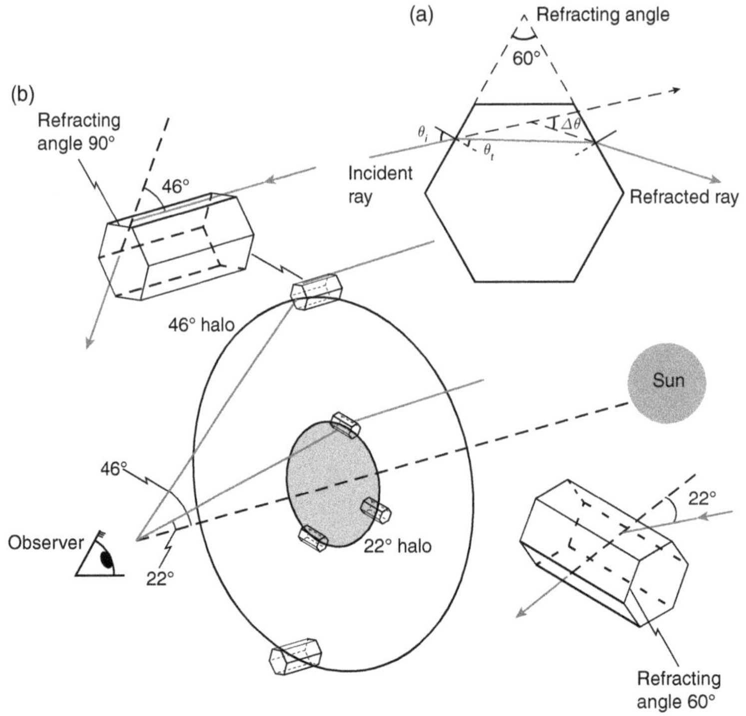

Fig. 1.7 Schematic of halo formation: (a) plan view of the refraction of light in a hexagonal ice crystal and (b) schematic of the formation of 22° and 46° halos.

Small ice crystals can be advected by the wind, causing cirrus clouds to become vertically sheared, as is the case for the cirrus uncinus shown in Figure 1.6c, with characteristic hooks at their upwind ends. However, not all high-level clouds are characterized by this fibrous pattern. Like cumulus clouds, the cirrocumuli shown in Figure 1.6e consist of individual cloud cells rather than of thin ice particle streamers. In fact this cloddy structure characterizes an early stage of cirrus development. When the cloud edges and the ice crystals are advected away from their initial cell, a fibrous appearance develops (Houze, 2014).

One characteristic of cirrus clouds, when viewed from the ground, is that they can produce halos (Figure 1.6d). A halo is an optical phenomenon consisting of a bright ring around the Sun. Figure 1.7a exemplarily shows the trace of a light ray through a hexagonal ice crystal with a refracting angle of 60° (perpendicular to the paper plane), where the incident ray deviates by $\Delta\theta$ from its original path. Light rays are concentrated in the vicinity of the angle of minimum deviation because here changes in the angle of the incident ray lead to hardly any change in the deviation angle. These concentrated rays appear as a bright ring around the Sun with an angular radius of 22° for visible radiation and prism angles of 60°. At angles smaller than the minimum deviation angle no light is refracted, so that the sky is darker inside the 22° halo (Jeske, 1988, Figure 1.7b). Not every cirrostratus produces a halo because it takes a sufficient number of hexagonal ice crystals to be aligned with their z-axes perpendicular to the Sun for the 22° halo, as shown in Figure 1.6d, to be seen. Occasionally a 46° halo also forms, caused by prism angles of 90°, which are found when the light rays enter a side (prism face) of the hexagonal ice crystal and leave it through the bottom (basal face) of the hexagonal crystal. The 90° angle between the prism and basal face causes the 46° halo to be much fainter than the 22° halo since most energy is lost in reflection.

In addition to in-situ-formed high-level clouds, Cbs leave stratiform anvil cirrus behind in their dissipating stage (Chapter 10). Deep convection transports moisture and condensate from the lower troposphere to the upper troposphere, causing anvil cirrus to have a higher ice water content than in-situ-formed cirrus. In addition, anvil cirrus often has a larger vertical extent, and since the ice crystals were formed in regions lower down in the cloud. During transport they have had time to grow to larger sizes than ice crystals in in-situ-formed cirrus (Heymsfield *et al.*, 2013). Because of their larger ice water content, they have a higher optical depth than in-situ-formed cirrus.

1.2.5 Other cloud types

There are some additional cloud types that are not included in the WMO cloud atlas. A cloud type that is restricted to isolated and sharp mountain peaks, such as the Matterhorn in Switzerland or Mount Everest in Nepal, is the banner cloud (Figure 1.8a). It is a cloud plume that extends downwind from such an isolated obstacle even on otherwise cloud-free days. Banner clouds form because an obstacle in a horizontal flow causes the air masses to split: part of the air is forced vertically upward over the obstacle while the remaining air diverges horizontally around it. The diverging air masses cause a lee vortex with a horizontal axis to form. This lee vortex is associated with pronounced upwelling close to the leeward face of the mountain (Voigt and Wirth, 2014). If this air is sufficiently moist, the contained water vapor will condense before the air reaches the top of the mountain and thus forms a banner cloud. This mechanism requires that the air forming the banner cloud on the leeward side of the mountain originates from lower levels than the air on the windward side. If this is not the case, a cloud would also form on the windward side (Schween *et al.*, 2007). After the cloud has been formed, it experiences the main wind flow around the mountain and hence is advected by it, leading to the formation of an elongated banner.

Another rarely observed cloud type is the so-called roll cloud that can be seen above the light horizon (Figure 1.8b). It is characterized by strong updrafts in front of it and downdrafts at its rear causing it to appear as if rolling. This cloud type occurs in connection with land–sea breezes (Section 9.6) and thunderstorm activity. Roll clouds are frequently observed in Australia, where they are associated with the land–sea breeze. The only place where they seem to occur regularly is over the southern part of the Gulf of Carpentaria in northern Australia in the southern hemispheric spring. Because such a cloud typically arrives in the morning at the coast of Queensland, it is also called "morning glory cloud". Roll clouds can also occur in connection with thunderstorms but are isolated from the thunderstorm cell from which they develop. They indicate the leading edge of thunderstorm outflow (Chapter 10). The roll cloud forms horizontally because of the circulating

Fig. 1.8 (a) Banner cloud, (b) roll cloud (the dark shape above a light horizon), (c) mammatus cloud, (d) asperatus cloud and (e) Kelvin–Helmholtz wave cloud. Photographs taken by Larissa Lacher (a), Sandra LaCorte (b), Robert David (c), Christian Grams (d) and Laurie Krall (e). A black and white version of this figure will appear in some formats. For the color version, please refer to the plate section.

(rolling) air masses. Along the direction in which the thunderstorm is moving a (cold) front is formed at the leading edge of the storm. It forces the warm air to be lifted and, if it is sufficiently moist, a roll cloud forms.

Mammatus clouds are a rare example of clouds in sinking air (Figure 1.8c). The term "mammatus" refers to the cellular pattern of pouches hanging underneath the cloud base, caused by negative buoyancy. Mammatus clouds typically develop below the anvil of a thunderstorm. The high concentration of heavy hydrometeors (Section 1.3.2) in anvils exerts a gravitational force, so that the air in the anvil starts sinking. The warming of the subsiding air and the associated decrease in relative humidity leads to evaporation or sublimation of the hydrometeors. This in turn causes cooling of the air owing to the latent heat consumption of evaporation or sublimation and counteracts the warming due to subsidence. If the amount of energy required for evaporation or sublimation is sufficient to cool the sinking air below the temperature of the surrounding air, the air will continue to sink. The subsiding air with the evaporating or sublimating hydrometeors eventually appears below the cloud base as rounded pouch-like structures.

A similar feature is seen in the so-called asperatus cloud (Figure 1.8d). The term "asperatus" means roughened or agitated and refers to their wavy cloud base. This unique feature distinguishes them from all other cloud types. Asperatus clouds are most closely related to undulatus clouds. Their wavy cloud base again indicates sinking cloudy air. They are particularly common in the Great Plains in North America during the morning or midday after thunderstorms. Even though asperatus clouds appear dark and storm-like, they are harmless leftovers of thunderstorms and will dissipate.

A Kelvin–Helmholtz instability arises from small perturbations at the interface of two fluids with different velocities, an example of which is waves on a lake or ocean. Such an instability can also occur within one medium. In the atmosphere the resulting waves form along the interface between two atmospheric layers with a sufficiently large difference in velocity. The instability is manifested in the form of Kelvin–Helmholtz billows (Figure 1.8e) if the air that originates from the more moist lower layer is lifted high enough for the air to reach saturation and condense.

1.3 Microphysical cloud properties

Besides the classification according to their form and altitude of occurrence, clouds can also be characterized in terms of their microphysical properties. Examples that can readily be observed and have been reported for different cloud types are the number concentrations, sizes and types of the hydrometeors within a cloud and its phase (liquid, ice or a mixture of both).

1.3.1 Cloud phase

Warm clouds consist purely of liquid water (cloud droplets, drizzle drops and raindrops). Their temperature can be above or below 0 °C as long as ice particles are absent. Warm clouds can exist down to temperatures close to −38 °C, below which water freezes homogeneously. We use the terms warm and liquid clouds synonymously. Cold clouds can consist purely of ice crystals (ice clouds) or they can contain a mixture of ice crystals and cloud droplets (mixed-phase clouds). Mixed-phase clouds can be found between 0 °C and −38 °C and pure ice clouds can occur at all temperatures below 0 °C.

Observations of the phase (liquid, mixed-phase, ice) of clouds with temperatures between 0 and −38 °C vary quite drastically. Clouds over the Arctic Ocean have been found to contain liquid water down to temperatures as low as −34 °C (Intrieri *et al.*, 2002). On the contrary, Sassen *et al.* (2003) observed an altocumulus cloud over Florida that was glaciated, meaning that it consisted of ice crystals even at temperatures between −5 and −8 °C. Cloud-phase observations from 11 years of ground-based lidar data over Leipzig, Germany, between 0 to −40 °C showed liquid clouds 42%, ice clouds 46% and mixed-phase clouds 12% of the time (Seifert *et al.*, 2010).

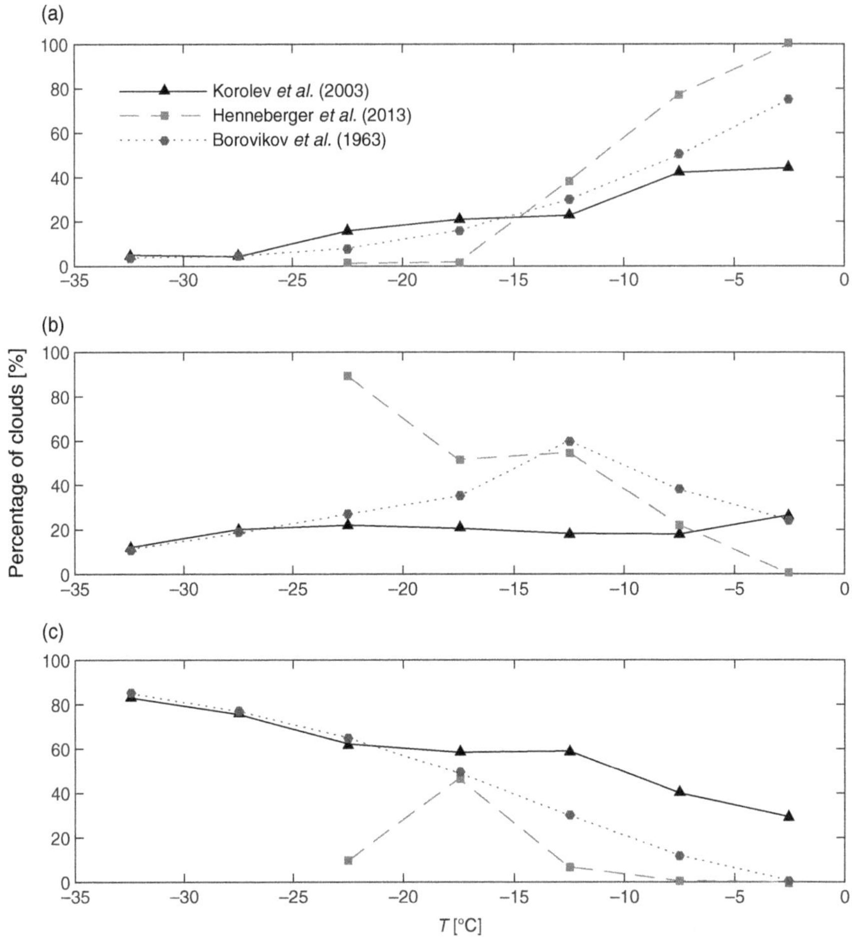

Fig. 1.9 Percentage of (a) liquid water, (b) mixed-phase and (c) ice clouds between −35 and 0 °C as composed from three different studies: frontal clouds over Canada (updated from Korolev *et al.*, 2003), various clouds over Israel (Borovikov *et al.*, 1963) and orographic clouds in the Swiss Alps (updated from Henneberger *et al.*, 2013). No clouds colder than −25 °C were sampled by Henneberger *et al.*

A composite of the percentage of occurrence of liquid water, ice and mixed-phase clouds between −35 and 0 °C obtained from three different in-situ studies is shown in Figure 1.9. A large variation exists between the different data sets. The occurrence of liquid clouds decreases with decreasing temperature, starting with between 40% and 100% of the time at temperatures above −5 °C and decreasing to below 30% at temperatures below −15 °C. A small fraction of liquid clouds is still found at temperatures below −30 °C.

Mixed-phase clouds have been observed over the entire temperature range. Their frequency of occurrence peaks between −20 and −25 °C with more than 80% in the orographic clouds studied in the Swiss Alps in the dataset extended from Henneberger *et al.* (2013). Their occurrence exceeds 50% between −10 and −15 °C in the clouds sampled over Israel and in the Swiss Alps but remains below 30% at all temperatures in the frontal clouds sampled over Canada by Korolev *et al.* (2003).

Table 1.2 Typical sizes and size ranges of hydrometeors in millimeters.

Size parameter	Typical value	Typical range
Cloud droplet radius	0.005	0.002–0.025
Drizzle drop radius	0.1	0.025–0.25
Raindrop radius	1	0.25–5
Ice crystal length	0.1	0.001–1.5
Snowflake maximum dimension	10	0.1–100
Graupel radius		< 2.5
Hailstone radius		> 2.5

The frequency of occurrence of ice clouds increases with decreasing temperature, starting from between 0 to 30% at temperatures above −5 °C to more than 80% at temperatures colder than −30 °C, in data sets obtained over Canada and Israel. Their frequency of occurrence drops to ∼10% below −20 °C in the orographic clouds studied in the Swiss Alps. These differences in occurrence can be explained by differences in their formation mechanisms. The sampled orographic clouds probably encountered rather high updraft velocities and hence high supersaturation, which permitted the simultaneous existence of cloud droplets and ice crystals (Sections 7.1 and 8.3).

1.3.2 Size distributions and water contents

Cloud particles are also called hydrometeors, but the term "hydrometeor" is more general as it also includes precipitation particles that leave the cloud. Therefore cloud particles are sometimes also referred to as cloud hydrometeors. Typical sizes and size ranges of the various hydrometeors are summarized in Table 1.2. Cloud droplets are the smallest cloud particles, with a typical radius of 5 μm and a range in radius between 2 and 25 μm. Drops between 25 and 250 μm in radius are called drizzle drops. Raindrops are of millimeter size.

In the ice phase, some hydrometeors such as ice crystals and snowflakes are better described in terms of a maximum dimension because their shapes can deviate significantly from a spheroid. Ice particles cannot be categorized by size as neatly as liquid hydrometeors. Ice crystal lengths range between 1 μm and 1.5 mm, which overlaps with the maximum dimension of snowflakes, between 0.1 mm and 10 cm. Only rimed hydrometeors can be separated by size, graupel denoting particles with radii less than 2.5 mm and hailstones anything larger.

The various cloud types consist of differing types of hydrometeors, which have differing size distributions. The number size distributions (also referred to as number distributions or size distributions for brevity) of cloud droplets and ice crystals are commonly described in terms of a three-parameter gamma distribution $n_N(r)$, given as

$$n_N(r) = N_0 r^{\mu} \exp(-\Lambda r). \tag{1.1}$$

Here r is the radius of the hydrometeor and N_0, Λ and μ are the intercept, slope and shape parameters, respectively. Frozen hydrometeors are described either in terms of their equivalent radius, i.e. the radius of the water sphere that results when the frozen hydrometeor is melted, or in terms of their maximum dimension.

By integrating over the size distribution, the number concentration N_x in m^{-3} and the mass mixing ratio q_x in kg per kg of air (from now on referred to as kg kg^{-1} for simplicity) of the different hydrometeors, cloud droplets, ice crystals, raindrops, snowflakes, graupel or hailstones, indicated by subscript x, can be obtained as

$$N_x = \int_0^\infty n_N(r)dr = \int_0^\infty N_0 r^\mu \exp(-\Lambda r)\, dr, \tag{1.2}$$

$$q_x = \int_0^\infty m(r) n_N(r)dr = \int_0^\infty m(r) N_0 r^\mu \exp(-\Lambda r)\, dr, \tag{1.3}$$

where $m(r)$ is the hydrometeor mass as a function of the hydrometeor radius r. Raindrops and snowflakes tend to follow an exponential distribution (Chapter 9), which is equivalent to a gamma distribution with shape parameter zero (i.e. $\mu = 0$).

Instead of mass mixing ratios, in-situ observations often report water content, for example the liquid water content (LWC) or ice water content (IWC). These refer to the amount of water or ice per cubic meter of air and are usually given in g m^{-3}.

The formation of precipitation depends on the water content of a cloud as well as on the cloud droplet or ice crystal number concentration. The LWC and IWC together with the cloud droplet and ice crystal number concentration determine the average sizes of the cloud droplets and ice crystals. In addition to being important for the initiation of precipitation-sized particles, the sizes of hydrometeors also determine the optical properties of the cloud, for instance the fraction of the sunlight that it reflects back to space (Chapter 12).

Typical values and ranges of these microphysical properties for the cloud types introduced above are summarized in Table 1.3. Generally, the smallest cloud droplets and the highest number concentrations N_c of cloud droplets are found in clouds over land. The value of N_c is determined by the number concentration of aerosol particles that serve as cloud condensation nuclei (CCN) (Chapter 6). Generally CCN concentrations are higher over land. In addition, N_c depends on the water vapor excess over the equilibrium value, the so-called supersaturation. Supersaturation is produced by adiabatic cooling in rising air parcels and thus is determined by the updraft velocity (Chapter 7). Updraft velocities are generally higher over land than over ocean because the land heats up more quickly than the ocean owing to its smaller heat capacity. This is turn produces more buoyancy and higher vertical velocities, which activates more and smaller aerosol particles (Chapter 6). Therefore cloud droplets of the same cloud genera are generally more numerous and smaller over land than over oceans.

An example of typical observed cloud droplet number concentrations in marine and continental cumuli is given in Figure 1.10 (Squires, 1958). It shows that N_c is much smaller in marine cumuli than in continental cumuli because there are fewer CCN in marine air. In addition, the typically lower updraft velocities in marine air cause lower supersaturations. Therefore only the larger aerosol particles, which are less numerous (Chapter 5), can become activated and form cloud droplets (Chapter 6). As a result, N_c rarely exceeds 200 cm^{-3} in marine cumuli, while N_c reaches up to 950 cm^{-3} in continental cumuli. Owing to the lower N_c values in marine clouds, cloud droplets experience less competition for the available water vapor. Therefore cloud droplets are capable of growing to larger sizes more

Table 1.3 Typical macroscopic and microphysical properties for various cloud types (compiled from the WMO cloud atlas (WMO, 1975), Pruppacher and Klett (1997), Miles *et al.* (2000), Lehmiller *et al.* (2001), Quante (2004), Deierling and Petersen (2008), Krämer *et al.* (2009), Cotton *et al.* (2011), Wood (2012), Khvorostyanov and Curry (2014) and Linda Schlemmer, personal communication); "oc" refers to clouds over oceans and "la" to clouds over land.

Parameter	Typical value	Typical range
Fog		
N_c [cm^{-3}]	150	1–600
Cloud droplet radius [μm]	7.5	1–15
LWC [g m^{-3}]	< 0.2	0.05–0.5
Vertical velocity [m s^{-1}]	0.01	
Vertical extent [km]	0.1	0.05–0.6
Lifetime [h]	4	2–6
St, Sc: rain (Sc), drizzle (Sc,St), snow possible		
N_c [cm^{-3}]	75 (oc), 290 (la)	10–300 (oc), 40–1000 (la)
Cloud droplet radius [μm]	7 (oc), 4 (la)	2–12.5
LWC [g m^{-3}]	0.2	0.1–0.6 (oc), 0.05–0.5 (la)
Vertical velocity [m s^{-1}]	0.01	
Vertical extent [km]	0.5	0.2–0.7 (oc), 0.2–0.9 (la)
Lifetime [h]	6	2–12
Cu hum, Cu med		
N_c [cm^{-3}]		100–1000
Cloud droplet radius [μm]	14 (oc), 6 (la)	2–25
LWC [g m^{-3}]	0.3	0.2–1
Vertical velocity [m s^{-1}]		1–3
Vertical extent [km]	1.5	
Lifetime [min]	20	10–40
Cu con: rain, snow and graupel possible		
LWC [g m^{-3}]		0.5–2.5
Vertical velocity [m s^{-1}]		3–10
Vertical extent [km]	5	
Lifetime [min]	30	20–45
Cb: rain (always), snow, graupel and hail possible		
LWC [g m^{-3}]		1.5–14
Vertical velocity [m s^{-1}]		5–50
Vertical extent [km]	9	6–12
Lifetime [min]	60	45–180
Ns: rain (always), snow and ice pellets possible		
LWC [g m^{-3}]		0.2–0.5
Vertical velocity [m s^{-1}]	1	1–2
Vertical extent [km]	4	2–10
Lifetime [h]		6–12

Table 1.3 (Continued)

Parameter	Typical value	Typical range
Ac, As: rain, snow, ice pellets		
N_c [cm^{-3}]	100	30–1000
Cloud droplet radius [μm]	4	2–10
LWC, IWC [g m^{-3}]	0.2	0.01–0.5
Vertical velocity [m s^{-1}]	<0.5 (As),<1 (Ac)	0.01–1
Vertical extent [km]	0.5 (Ac), 1 (As)	0.2–0.7 (Ac), 1–3 (As)
Lifetime [h]		6–12
Cc, Ci, Cs		
N_i [cm^{-3}]	0.2	10^{-4}–10
Ice crystal length [μm]	100	1–1500
IWC [g m^{-3}]	0.02	10^{-4}–0.5
Volume mean ice crystal radius [μm]	20	2–80
Vertical velocity [m s^{-1}]	0.1	0.01–1.5
Vertical extent [km]		0.2–0.4 (Cc), 0.1–3 (Cs)
Lifetime [h]		0.2–12

quickly in marine clouds than in continental clouds. In the marine cumuli shown in Figure 1.10b, cloud droplets grow up to 30 μm in radius at higher altitudes whereas they remain below 10 μm in radius in the continental cumuli at all altitudes (Figure 1.10c).

Similar differences, in terms of size distribution and number concentration, to those between marine and continental clouds are found between polluted and clean clouds. An example of an observed size distribution of cloud droplets (N_c) and drizzle drops (N_d) in marine clouds off the coast of Nova Scotia, Canada, is shown in Figure 1.11 (Peng *et al.*, 2002). These data are sorted according to their aerosol concentration (N_a), measured with an instrument which sampled aerosol radii larger than 0.07 μm. In this study, clouds with $N_a < 300$ cm^{-3} were considered clean and those with $N_a > 300$ cm^{-3} were considered polluted. The polluted clouds have average concentrations $N_a = 810$ cm^{-3}, $N_c = 260$ cm^{-3} and $N_d = 8 \times 10^{-4}$ cm^{-3}. In contrast, the clean clouds have average concentrations $N_a = 160$ cm^{-3}, $N_c = 70$ cm^{-3} and $N_d = 9.4 \times 10^{-3}$ cm^{-3}.

Figure 1.11 shows that the concentration of small cloud droplets (with $r < 5$ μm) is larger in the polluted clouds than in the clean clouds. At a radius of about 7.5 μm the curves intersect so that the concentration of larger drops is larger in the clean clouds. The gap in the data between radii 25 μm and 55 μm is caused by use of different instruments to obtain the size distributions and their respective measurement size ranges.

The differences between clean versus polluted or marine versus continental clouds show that "typical" cloud droplet size distributions for a given cloud type do not exist. In addition to differences in aerosol concentration and vertical velocity at the cloud base, clouds of the same cloud type can have different values of the temperature at the cloud base, which determines the growth rate of the cloud droplets (Chapter 7). They can also experience different levels of mixing with the ambient air (Chapter 4). Lastly, they could have been sampled at different stages of their life cycle and at different heights above cloud base. Therefore, large variations exist between the individual size distributions of the various cloud types.

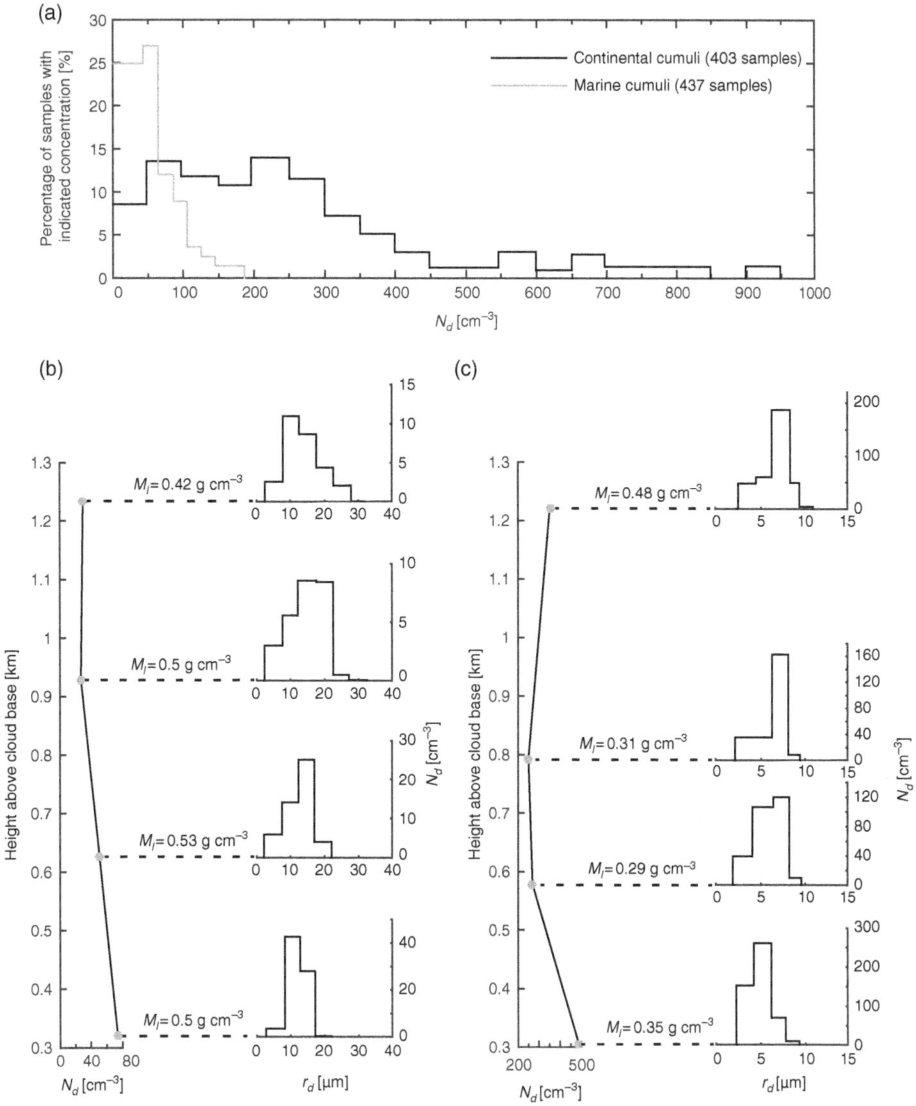

Fig. 1.10 Observed cloud droplet number concentrations in marine and continental cumuli. (a) Percentages of marine and continental cumuli with indicated droplet concentrations. (b), (c) Change in the droplet concentration and size distribution with height above cloud base in marine and continental cumuli; M_l denotes the liquid water content. Note the different x-axis scales in (b) and (c). Figure adapted from Squires (1958).

A few general differences between certain cloud types can be stated, with caution, keeping the above points in mind. As shown in Figure 1.10, cloud droplets are larger higher up in a cloud; thus one can expect that as a fair weather cumulus continues to develop into a cumulus congestus, the cloud droplets become larger and fewer in number, mainly owing to their growth by collision–coalescence between cloud droplets (Section 7.2). Also, a comparison of stratus and stratocumulus with Cu hum and Cu med shows that cumuli have higher updraft velocities than stratus clouds, causing N_c to typically be higher in cumulus

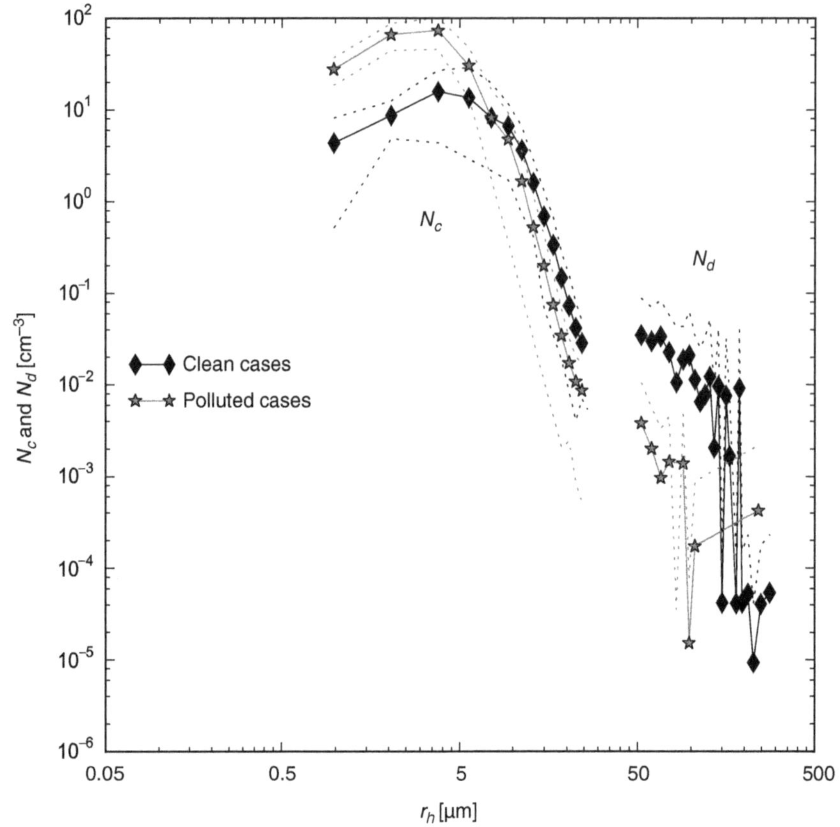

Fig. 1.11 Cloud droplet N_c and drizzle drop N_d size distributions for clean clouds (diamonds) and polluted clouds (asterisks). Each point refers to the averaged concentration over all clean or all polluted clouds. The solid and dotted lines give the averages and the standard deviations for the clean and polluted cases, respectively. The values of N_d minus one standard deviation are not shown because they extend below the lowest values shown. Figure adapted from Peng *et al.* (2002).

than in stratus and stratocumulus. Clouds which can produce precipitation (mainly Ns, Cb and Cu con) typically have broader drop size distributions, with drop radii exceeding 15 μm. In addition they often have an ice phase.

Because aerosol number concentrations decrease with altitude, the number concentration of CCN also decreases, which results in smaller cloud droplet number concentrations in mid-level clouds (Ac and As) than in the respective low-level clouds (Sc and St). As discussed above, the water vapor mixing ratio decreases exponentially with altitude in the troposphere, owing to the decrease in temperature with increasing altitude. If less water vapor is available, less can condense or deposit and form cloud droplets or ice crystals. Therefore, the liquid and ice water content also decreases with increasing height of the cloud base. The decrease of water vapor with increasing altitude is generally stronger than the decrease in the number of CCN with the result that mid-level clouds have smaller cloud droplets than low-level clouds.

Cirrus clouds are characterized by the lowest number concentrations and the largest average ice crystal size. However, there is a large spread in the microphysical properties of cirrus clouds (Tables 1.2 and 1.3). This is partly due to the different mechanisms by which cirrus clouds form and partly due to uncertainties in measurements. The variety of ice crystal shapes and sizes, as will be discussed in Chapter 8, additionally complicates the characterization of ice clouds.

1.4 Exercises

Single/multiple choice exercises

1. Clouds form an important part of the hydrological cycle and the Earth's radiation budget. Which of the following is/are true?
   (a) In the global annual average, around 70% of the Earth's surface is covered with clouds.
   (b) The North Atlantic storm tracks, the Intertropical Convergence Zone and the subtropics are associated with high cloud amounts.
   (c) Satellites can distinguish between clouds and sea ice.
   (d) All cloud types have a net cooling effect on the Earth's surface because they reflect solar radiation.
2. Clouds are classified into three basic categories, namely cumulus, stratus and cirrus clouds. Which of the following is/are true?
   (a) Stratus clouds form in statically stable air.
   (b) Cumulus clouds usually have a larger vertical than horizontal dimension.
   (c) Cumulus and stratus clouds consist purely of cloud droplets, whereas cirrus clouds consist of ice crystals.
   (d) Clouds are often classified according to their cloud top height.
3. Low-level clouds are characterized by a low cloud base height. Which of the following is/are true?
   (a) Stratus and high fog cannot be distinguished in terms of their appearance and formation mechanisms.
   (b) Nimbostratus forms when air is lifted along a warm front and is associated with convective precipitation.
   (c) Convective clouds form in unstable air masses.
   (d) Lightning and thunder can occur in cumulonimbus and nimbostratus.
4. High-level clouds have a base height of 5–13 km in mid-latitudes. Which of the following is/are true?
   (a) Both cirrus and low stratus clouds can produce halos.
   (b) All cirrostratus produce halos.
   (c) Cirrus clouds have a high optical depth.
   (d) Gravity waves can lead to the formation of cirrus clouds.
5. Assign the following attributes to the corresponding cloud type: (1) halo, (2) convective precipitation, (3) wavy cloud base, (4) touches the ground. Cloud types: (i) stratus, (ii)

cirrostratus, (iii) nimbostratus, (iv) cumulonimbus, (v) mammatus cloud, (vi) asperatus cloud, (vii) cumulus, (viii) fog.

(a) 1i, 2iv, 3vi, 4viii
(b) 1i, 2iii, 3v, 4viii
(c) 1ii, 2iv, 3vi, 4viii
(d) 1ii, 2iii, 3v, 4i
(e) 1ii, 2iv, 3vii, 4i

6. Clouds consist of cloud droplets and/or ice crystals. Which of the following is/are true?
   (a) Warm clouds consist purely of liquid water, and thus the temperature in these clouds is above 0 °C.
   (b) Mixed-phase clouds are the most frequent type.
   (c) The temperature in cold clouds cannot exceed 0 °C.
   (d) The number concentrations and sizes of cloud droplets and ice crystals determine optical cloud properties.
7. The microphysical properties differ for clouds over land and clouds over ocean. Which of the following is/are true?
   (a) Cloud droplets over land are usually smaller than over the ocean.
   (b) The cloud droplet number concentration in continental cumulus clouds can reach up to 950 cm^{-3}.
   (c) Cumulus clouds have an average lifetime of less than 1 hour.

Long exercises

1. Figure 1.12 shows examples of the different cloud types introduced. Which types of cloud are represented in the photographs? Describe their characteristics and classify them as either low-level, mid-level or high-level cloud.
2. Which cumulus cloud is associated with capping inversions? How does the capping inversion influence the appearance of that cloud type?
3. What is a halo and how does it form?
4. Assume a low level cloud with the following properties: vertical extent 1.5 km, horizontal extent 1 km, cloud droplet number concentration 90 cm^{-3} air and mean diameter 10 μm. Estimate the mass of the cloud by assuming that all the cloud droplets are spherical and have the same size. Take 1000 kg m^{-3} to be the density of water.
5. The average pressure at Earth's surface is 985 hPa. Knowing the radius of the Earth r_{Earth} = 6370 km and using the definition of pressure, estimate the total mass of the atmosphere.
6. Imagine a pure water cloud which consists of cloud droplets of uniform size with r_d = 5 μm. Assume a uniform spacing of the cloud droplets, with a number concentration of $N_c = 170\ cm^{-3}$.
   (a) Calculate the average distance apart $\bar{d}$ of the cloud droplets, using

$$\frac{\bar{d}}{\bar{r}_d} = \frac{0.893}{M_l^{1/3}}, \tag{1.4}$$

   where M_l denotes the liquid water content of the cloud.

Fig. 1.12 Photographs of different cloud types, taken by Fabian Mahrt.

(b) For the same M_l as calculated above, what would happen to the average distance between cloud droplets if their number concentration was tripled?

7. Consider a developing Cb cloud in mid-latitudes that grows vertically, experiencing an updraft velocity $w = 12.5 \text{ m s}^{-1}$. Assume a cloud base at 500 m.
   (a) How long does it take until the Cb cloud reaches the tropopause at 13 km?
   (b) How does this time compare with a typical time scale on which a thunderstorm forms in mid-latitude summer?
   (c) What happens once the rising air parcel (cloud top) reaches the tropopause?
8. A stratus cloud has a typical vertical velocity of $w = 0.01 \text{ m s}^{-1}$. Assume typical values for the vertical extent of a stratus cloud of 900 m over land and 500 m over the ocean.
   (a) How long does it take for an air parcel to be transported up the vertical extent of the cloud, i.e. from cloud base to cloud top of the stratus cloud, over land and ocean, respectively?
   (b) How do these times compare with the lifetimes of a stratus cloud in Table 1.3? How do you explain the differences?
   (c) Now repeat the calculations for typical values of a Cu con cloud in Table 1.3 and compare with those obtained for the stratus cloud. Interpret your results.

2 Thermodynamics

Thermodynamics describes the behavior of matter on a macroscopic scale, looking at it as a continuum rather than the sum of its atomic or molecular constituents. In this chapter we review the thermodynamics necessary for the understanding of cloud formation. We start by defining in Section 2.1 some fundamental terms, which will be used throughout the textbook. We discuss the physical laws that describe processes in dry air and introduce the concept of air parcels in Section 2.2. Since thermodynamic processes that involve adiabatic changes can best be visualized with the help of thermodynamic charts, the use of tephigrams is introduced in Section 2.3. We then move to moist air, i.e. a mixture of dry air and water vapor. Water is a unique substance because it exists in all three phases (ice, liquid water and water vapor) in the atmosphere. The transitions between its different phases are crucial for the understanding of liquid and ice cloud formation and microphysical processes inside clouds. They are the subject of Section 2.4. We end the chapter by extending the laws introduced for dry air to moist air, in order to analyze how water vapor influences certain processes; this enables us to predict the conditions under which a cloud is formed (Section 2.5).

2.1 Basic definitions

2.1.1 Thermodynamic states and variables of state

The aim of thermodynamics is to describe the spatially and temporally averaged properties of macroscopic systems of gases, liquids or solids. The concepts of thermodynamics are suitable for systems containing a number of molecules on the order of the Avogadro constant ($N_A = 6.022 \times 10^{23}$ mol^{-1}) and changing slowly in time. Considering a large, macroscopic, quantity Q of a substance also justifies the assumption of negligible edge effects, which is usually made in thermodynamics.

In reality the quantity Q consists of a large number of molecules N each of whose instantaneous state can be described by $6N$ coordinates (three for the spatial position x, y, z, and three for the momentum components of each molecule, mu, mv, mw). Such a state is called a "microstate" of the system. For most practical purposes, to describe a system in terms of its microstates is too complicated, i.e. to solve the equations of motion with $6N$ degrees of freedom. Furthermore, it would be meaningless because microstates cannot be directly observed. Instead, in thermodynamics we describe a system via macroscopic, observable, quantities like atmospheric pressure p, temperature T, volume V, number of moles n, entropy S and internal energy U. The main advantage of a thermodynamic, macroscopic, description of a system, as opposed to a microscopic, "atomistic", description, is that a macrostate given e.g. by p, T, V and n summarizes an infinite number of microstates leading to the same macrostate. For example, the microstate of a quantity of n moles of a gas contained in a cylinder with volume V at pressure p and temperature T changes very fast with the Brownian motion of the molecules, whereas its macrostate remains constant.

The macroscopic, observable, quantities describing a macrostate are termed variables of state or state variables. A set of state variables unambiguously describes a state of thermodynamic equilibrium of a system. Here, how the system reached that state is of no relevance for the value of the state variables. In contrast, work is an example of a quantity that depends on the process connecting two states of a system and therefore cannot serve as a state variable. The number of state variables required depends on the system under consideration. Because different relations exist between variables of state, they depend on each other. For example, the ideal gas law (eq. 2.4) relates the variables of state p, T, and ρ, leaving two of these variables to be varied independently.

2.1.2 Intensive and extensive variables

Extensive variables are variables of state whose value is proportional to the size of a system, whereas intensive variables describe an inherent state of matter that is irrespective of the size of the system. For example, the state variables internal energy U, volume V, number of moles n and entropy S scale with the amount of matter the system contains; hence they are extensive state variables. Examples of intensive variables are pressure p, temperature T and the chemical potential μ. We can assign to every extensive variable x_i a conjugate intensive variable a_i by taking the partial derivative for instance of the internal energy U with respect to x_i while holding all other extensive variables constant:

$$a_i = \left(\frac{\partial U}{\partial x_i}\right)_{x_{j \neq i}}. \tag{2.1}$$

Another example is a gas described by the state variables p, V, T and S. The first law of thermodynamics states that:

$$dU = dW + dQ = -pdV + TdS. \tag{2.2}$$

If we consider an infinitesimal change dV of the volume, while all other extensive variables are held constant, so that in this case $dS = 0$, we have $dU = -pdV$, i.e.

$$p = -\left(\frac{\partial U}{\partial V}\right)_S. \tag{2.3}$$

Hence p is the conjugate intensive variable of V. This implies that at higher pressures, the internal energy reacts more to a change in the gas volume, dV, since more work is associated with a given change in volume.

In addition to their conjugate intensive variables, we can assign an intensive variable to a given extensive variable by dividing it by the mass contained in the system, resulting in the corresponding specific quantity. Specific quantities are independent of the size of the system and are usually denoted with small letters, e.g. we write the specific internal energy

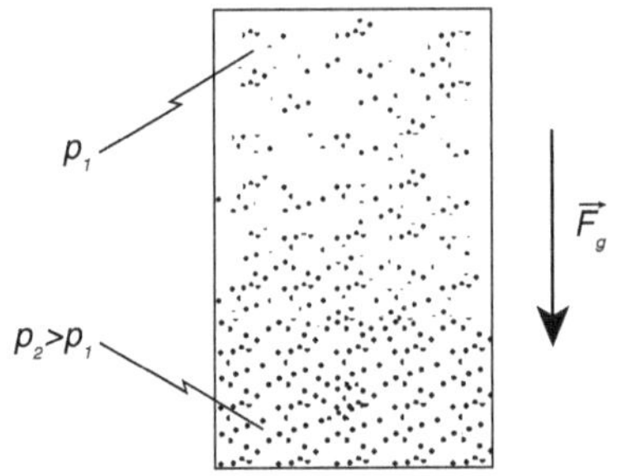

Fig. 2.1 Schematic of an isolated system in the presence of a downward oriented gravity force $\vec{F}_g$, where the pressure p increases downward.

as u, the specific entropy as s and the specific volume as v. Note that we adopt the usual notation from thermodynamics where dX denotes an infinitesimal change in the quantity X. Infinitesimal variations allow us to study the behavior of a system without moving it out of TDE.

2.1.3 Open, closed and isolated systems

The boundaries of an *open* system are defined to be permeable to both mass and energy. A *closed* system on the other hand can be thought of as having "rigid" walls that do not allow the exchange of mass with its surroundings. However, a closed system may exchange energy with its surroundings, which is the case when it is in contact with a heat reservoir. If a system is *isolated*, as shown in Figure 2.1, neither mass nor energy can be exchanged via the boundaries of the system. This does not exclude any interaction with the outside world. Long-range forces, like gravity $\vec{F}_g$, although having their origin outside an isolated system can nevertheless have an effect on it.

2.1.4 Thermodynamic equilibrium

The state of thermodynamic equilibrium (TDE) of an isolated system is characterized by the absence of any net macroscopic flow of matter or energy within it. This implies that all observable quantities, i.e. all state variables, are constant in time once the system has reached TDE. This does not apply to the microstate of the system, which will continuously change by microscopic fluctuations even in TDE. In thermodynamics, it is assumed that a system starting off in any non-equilibrium state spontaneously moves towards its state of TDE, which is reached after a sufficiently long (usually very long) time. Thermodynamic equilibrium is the only thermodynamically well-defined state of a system, i.e. the only state which is unambiguously determined by the values of the system's variables of state. Spatial invariance, i.e. homogeneity, of the intensive variables is not equivalent to the system being in TDE. If more than one phase of a substance coexists in TDE, some intensive variables, e.g. the density, may be inhomogeneous throughout the system. As an illustration, a closed system containing liquid water and water vapor in TDE shows a jump in density in going from the gas to the liquid phase of the system or vice versa.

If the system is subject to long-range forces like gravity, these may give rise to gradients in some internal variables. Consider an isolated system consisting of a gas in a container

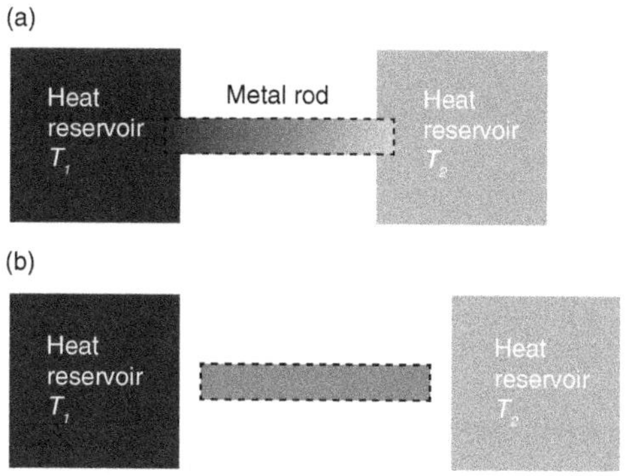

Fig. 2.2 Schematic of (a) the stationary state of a metal rod coupled to two heat reservoirs at either end with $T_1 > T_2$ and (b) the situation after the heat reservoirs have been decoupled.

subject to an external constant gravitational force (Figure 2.1). In TDE a pressure gradient exists within the gas along the direction of the gravitational force. Temperature is the only intensive variable which is always homogeneous throughout a system in TDE, because any temperature gradient is inevitably associated with a net flux of heat, which is in contradiction with TDE.

Note that a system in TDE is in a stationary state, whereas a system in a stationary state is not necessarily in TDE. As a simple example, consider a metal rod with its two ends kept at different temperatures (Figure 2.2a). As long as those temperatures do not change, the system is in a stationary state. However, the inhomogeneity of temperature along the metal rod and the consequent net heat flux through it from the end with higher temperature T_1 towards the end with lower temperature T_2 are incompatible with the definition of TDE. Only after the heat reservoirs have been removed and the temperature along the metal rod is the same everywhere is the rod in TDE (Figure 2.2b).

The concept of TDE can also be applied to the interaction of two or more systems with each other, called "mutual TDE" (Figure 2.3). The defining conditions of the absence of any net flux of matter or energy still holds, but is no longer sufficient. Consider two gas containers with one mutual wall, in which a shutter is implemented. If the shutter is closed, and both containers are at the same temperature, no flux of matter or energy exists. Nevertheless, the system comprising both containers is not in TDE, since the shutter acts as a mechanical constraint between the two containers. This constraint artificially prevents the systems from exchanging gas, if the two containers are at different pressures. In fact, TDE requires all kinds of equilibria between systems to hold simultaneously. In other words, the systems have to be simultaneously in mechanical, thermal, chemical and radiative equilibrium. In the example in Figure 2.3 the requirement of mechanical equilibrium is not met.

2.1.5 Reversible and irreversible processes

A reversible process connects two states of TDE with infinitesimal steps on a path where every intermediate state is also a state of TDE. Note that such a process is idealized, as it requires all changes to occur much more slowly than the time required for the state

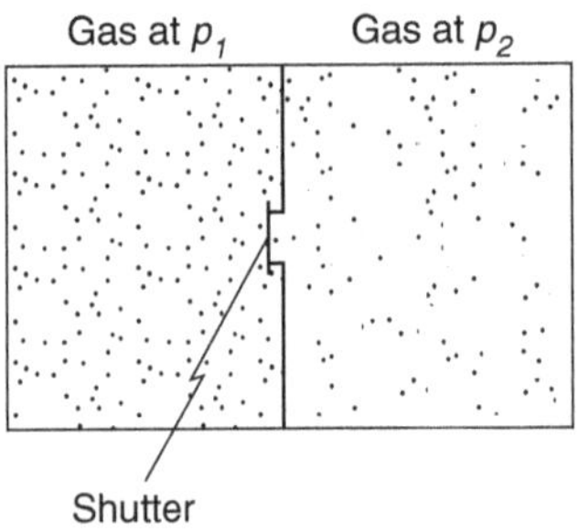

Fig. 2.3 Schematic of a system with two gas containers at different pressures p_1 and p_2 being separated by a shutter. Each container individually is in TDE, but with the shutter closed, no mutual TDE of the containers can be established.

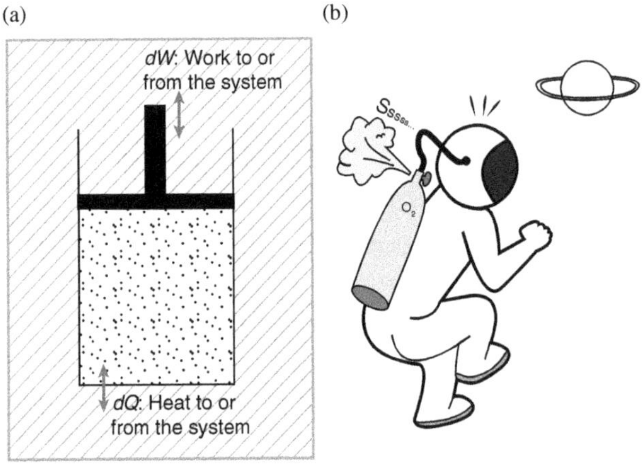

Fig. 2.4 Schematic of (a) a reversible isothermal process and (b) an irreversible isothermal process.

variables to equilibrate within the system – otherwise the process would immediately be irreversible. In turn, an irreversible process contains at least some intermediate states along its path that are not states of TDE. Usually, an irreversible process moves a system from a state that is out of TDE into a state that is in TDE.

An example of a reversible process is the isothermal expansion of a gas, contained in a cylinder with a movable piston, which is in contact with a heat reservoir as shown in Figure 2.4a. During the expansion the piston does work against the external force applied by the piston. This work dW is balanced by heat delivered from the heat reservoir dQ. Reversing the process is possible by simply moving the piston back and doing the necessary work, allowing the released heat to be transferred to the reservoir. However, if a gas is allowed to escape from a container, there is no simple way to reverse this process. Its irreversibility is intuitively plausible because it is highly unlikely that the random motions of the escaped gas molecules would bring all of them simultaneously back to their original position within the container. An example of an irreversible isothermal expansion is illustrated in Figure 2.4b.

Another example of an irreversible process is heat conduction. Consider the metal rod in Figure 2.2, which is initially in contact at each end with heat reservoirs having different temperatures. When the rod is isolated from the heat reservoirs, heat will flow in such a way that the temperature equilibrates along the rod, moving the rod from a state of non-equilibrium into a state of TDE. The opposite process will never occur, as heat never flows from colder to warmer temperatures. The Carnot engine is the only way to reversibly transfer heat between two reservoirs at different temperatures (Section 2.2.4).

A powerful, yet less intuitive, criterion for the reversibility of a process is its entropy balance as will be discussed in Section 2.2.6. In a reversible process, no net entropy is generated by the system and its surroundings. In the above example of reversible expansion the entropy of the expanding gas increases but at the same time, the entropy of the connected heat reservoir decreases by the same amount, leading to zero change in the total entropy. During the irreversible free expansion of a gas, however, there is no heat transfer from a reservoir to balance the increase of entropy in the expanding gas and so net entropy is generated by this irreversible process.

Most processes in the real world are irreversible (Figure 2.4b). The reversibility of the expansion of a gas in a container relies on the assumption that the piston moves without any friction, which is not realistic. Friction irreversibly converts some work into heat. As soon as such losses are encountered, the process under consideration cannot be considered reversible; the more important the losses are, the larger is the deviation of the process from reversibility. A process that comes very close to reversibility in the real world is the evaporation of a liquid, e.g. water, which coexists in TDE with its vapor phase (i.e. at its boiling point). The addition of some quantity of heat to the system will evaporate some liquid. As long as the addition of heat occurs slowly enough, the temperature of the system will not move from boiling point (assuming that the pressure of the gas phase stays constant and that its volume is allowed to increase) and the liquid stays in coexistence with its gas phase in TDE at every intermediate point along the process. In practice, it is more likely that phase transitions occur in a close to reversible manner than processes involving mechanical work. In the latter, mechanical action inevitably entails frictional losses.

2.2 Dry air

The goal of this section is to obtain the law that relates the changes in the intensive state variables pressure p and temperature T of an air parcel in dry air and thus explains the processes occurring in dry air. In volume mixing ratios, dry air is mainly composed of nitrogen, N_2 (78.08%) and oxygen, O_2 (20.95%). Argon (Ar) follows in third place with 0.93%. In addition to these gases, Earth's atmosphere contains a number of trace gases, the most important of which is carbon dioxide, CO_2 (0.04%).

Here we assume that the atmosphere can be treated as an ideal gas. Ideal gases are defined as gases in which the volume of the individual molecules can be neglected and where no interactions between the molecules exist besides collisions, i.e. there are no forces acting between the point-like molecules except when they collide.

2.2.1 Ideal gas law

The ideal gas law for dry air is given by

$$\frac{p}{\rho_d} = R_d T. \tag{2.4}$$

Here p is the atmospheric pressure in Pa, ρ_d the density of dry air in kg m^{-3}, R_d the specific gas constant of dry air (R_d = 287 J kg^{-1} K^{-1}) and T the temperature in K. The ideal gas law is also referred to as the equation of state for an ideal gas, which means that it describes the state of an ideal gas in TDE by defining a relationship between the state variables T, p and ρ_d.

The ideal gas law is an approximation to reality since, in a real gas at least, van der Waals forces (rather weak interactions between atoms or molecules that decay with the sixth power of the distance) act between the gas molecules. Nevertheless, the atmosphere can usually be regarded as an ideal gas as the deviation of $p/\rho_d R_d T$ from 1 is less than 0.2% (Iribane and Godson, 1981). This assumption is no longer valid above 100 km (50 Pa) owing to ionic interactions of the air molecules and the breakdown of the local TDE.

2.2.2 First law of thermodynamics

One of the most important laws is the first law. It states that energy is conserved and that heat is a form of energy. We start by defining the specific internal energy u. In atmospheric science we use specific quantities, which are quantities normalized to unit mass. As mentioned earlier, they are denoted by small letters. From here on the term "specific" is dropped for brevity. A change in internal energy can result from a change in the kinetic energy of the molecules in at least one of the available degrees of freedom or by a change in the forces of attraction between the molecules (Curry and Webster, 1999). Any of these changes will affect the temperature of the system. For ideal gases such as air, u is proportional to T because it consists only of the kinetic energy of the molecules since the assumption of an ideal gas neglects intermolecular forces.

As the first law of thermodynamics states that energy is conserved, any change in the internal energy du has to be balanced by changes in other quantities of energy. These could be a quantity of heat dq supplied to or subtracted from a unit mass of air or a quantity of work dw done by or exerted on the same unit mass of air, or both. The basic form of the first law of thermodynamics can thus be described as follows:

$$du = dq + dw. \tag{2.5}$$

In this form of the first law all net transfer rates to the system are positive and all net transfer rates away from the system are negative. Equation (2.5) is not very practical for applications in atmospheric sciences because changes in q, u and w cannot readily be observed. The goal is thus to rewrite the first law in such a way that temperature and pressure become the independent variables, as their changes can easily be measured in the atmosphere.

Let us again consider a gas in a cylinder with a movable piston (Fig 2.4a). The work dW done on the gas expanding against an external pressure p is given by:

$$dW = \vec{F} \cdot \vec{ds} = -Fds = pAds = -pdV \cdot \tag{2.6}$$

where $\vec{F}$ is the external pressure force, $\vec{ds}$ the displacement of the piston and A its area. The minus-sign arises because the external pressure force and the displacement are in opposite directions to each other. Normalizing eq. (2.6) by the mass leads to an expression for the work done on a unit mass of gas:

$$dw = -pd\alpha. \tag{2.7}$$

Here α is the volume per unit mass, which is also called the specific volume. It is the inverse of the density: $\alpha = \rho^{-1}$. Note that p denotes both the pressure within the system and the external pressure, since we are dealing with reversible processes where every state along the process is a state of TDE. A state with $p_{int} \neq p_{ext}$ would violate the definition of TDE. In real world processes, the equality $p_{int} = p_{ext} = p$ is rarely fulfilled.

If we add a quantity of heat dq to a unit mass of material then T increases by dT, in the absence of phase transitions, according to

$$dq = cdT, \tag{2.8}$$

where c is the specific heat capacity (J kg^{-1} K^{-1}). For gases, c depends upon whether work is done while heat is added. If no work is done on or by the system the volume remains constant, i.e. $d\alpha = 0$. Then the first law reduces to

$$du = dq. \tag{2.9}$$

We now rewrite eq. (2.8) in terms of derivatives, indicating that volume is kept constant by the use of the subscript α. We obtain:

$$c_v = \left(\frac{dq}{dT}\right)_\alpha = \left(\frac{du}{dT}\right)_\alpha . \tag{2.10}$$

This expression relates the specific heat at constant volume, c_v, to the change in internal energy with temperature. We can also define the specific heat at constant pressure, c_p, which is given as

$$c_p = \left(\frac{dq}{dT}\right)_p . \tag{2.11}$$

Note that in an isobaric process ($dp = 0$), the volume is no longer constant. Thus, if heat is supplied to an air mass then the air expands, and part of the supplied heat dq is transferred to the work dw required to expand a unit mass of air against the ambient pressure. Here work is done by the system on the environment. In an isochoric process ($d\alpha = 0$) the entire amount of supplied heat is transferred to internal energy; consequently the temperature change is larger in an isochoric process than in an isobaric process for the same dq. Therefore the heat capacity at constant pressure is larger than that at constant volume. At 273 K, c_p and c_v take the following values for dry air:

$$c_p = 1005 \text{ J kg}^{-1} \text{ K}^{-1}, \tag{2.12}$$

$$c_v = 718 \text{ J kg}^{-1} \text{ K}^{-1}. \tag{2.13}$$

In general, neither the pressure nor the volume are constant, and there are contributions to both the internal energy and to the work exerted by the system. However, the change in u is always given by

$$du = c_v dT. \tag{2.14}$$

We can now rewrite the first law of thermodynamics by replacing dw and du with eqs. (2.7) and (2.14) and obtain

$$dq = c_v dT + p d\alpha. \tag{2.15}$$

Adiabatic processes are those in which no heat or mass exchange with the environment occurs. They are good approximations for ascending and descending air masses or air parcels in the atmosphere. In order to obtain the equation describing adiabatic processes in dry air, we need to combine the first law of thermodynamics with the ideal gas law. Then the change in the specific volume $d\alpha$ in the first law of thermodynamics can be related to the change in pressure. For that, we take the total differential of the ideal gas law (eq. 2.4):

$$p d\alpha + \alpha dp = R_d dT, \tag{2.16}$$

and insert the expression obtained for $p d\alpha$ into eq. (2.15):

$$dq = (c_v + R_d) dT - \alpha dp. \tag{2.17}$$

Note that the difference between the specific heat of an ideal gas at constant pressure and its specific heat at constant volume is given by the specific gas constant. For dry air,

$$R_d = c_p - c_v. \tag{2.18}$$

This results from the definition of c_p (eq. 2.11) and differentiating eq. (2.17) with respect to temperature at constant pressure. Inserting eq. (2.18) into eq. (2.17) results in:

$$dq = c_p dT - \alpha dp. \tag{2.19}$$

This is the preferred form of the first law of thermodynamics in atmospheric science because the independent variables on the right-hand side (T, p) can easily be observed.

2.2.3 Special processes

From the first law of thermodynamics we can deduce some special processes. For an isobaric process ($dp = 0$, line AE in Figure 2.5), the first law (eq. 2.19) reduces to

$$dq = c_p dT. \tag{2.20}$$

For an isothermal process ($dT = 0$, line AD in Figure 2.5), it reduces to

$$dq = -\alpha dp. \tag{2.21}$$

For an isochoric process ($d\alpha = 0$, line AB in Figure 2.5), we obtain

$$dq = c_v dT = du. \tag{2.22}$$

For an adiabatic process ($dq = 0$, line AC in Figure 2.5), we obtain

$$c_p dT = \alpha dp. \tag{2.23}$$

The various types of process are shown schematically in Figure 2.5. In an isothermal process, the system can be thought of as connected to a heat reservoir keeping the temperature of the system constant; the work done by the gas upon expansion is taken entirely from the heat reservoir. In such an isothermal process an increase in volume leads to only a moderate decrease in pressure because the temperature remains constant (line AD). Since for an infinitesimal change dV the work done by the system is defined as pdV, the total work done is given by the gray shaded area.

In an adiabatic process the system is isolated from its surroundings, so that no transfer of heat can occur. As the gas expands adiabatically from point A to point C, it does work at the expense of its internal energy and thus its temperature decreases. Owing to the ideal

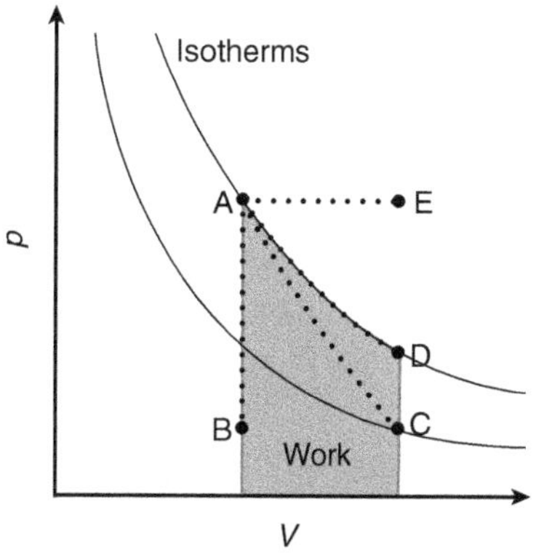

Fig. 2.5 Schematic of the various thermodynamic processes in a p–V diagram: isobaric process (AE), isothermal process (AD), isochoric process (AB), adiabatic process (AC). The gray area denotes the work done by the gas in the isothermal expansion from A to D.

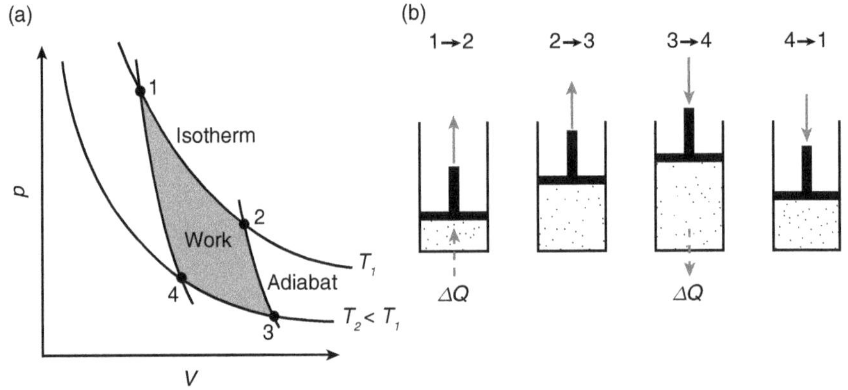

Fig. 2.6 (a) Schematic of the different thermodynamic processes in a Carnot cycle for an ideal gas in a p–V diagram: isothermal expansion ($1 \longrightarrow 2$), adiabatic expansion ($2 \longrightarrow 3$), isothermal compression ($3 \longrightarrow 4$), adiabatic compression ($4 \longrightarrow 1$). (b) A piston model of the Carnot process, showing the associated changes in ΔQ.

gas law, this leads to an additional decrease in pressure for the same volume change (line AC) in comparison with the isothermal case.

2.2.4 The Carnot process

We have seen earlier that thermodynamic processes can transform heat into mechanical work, and vice versa. For instance, during a reversible isothermal expansion of a gas from volume V_1 to volume $V_2 > V_1$, the first law of thermodynamics implies that an amount ΔQ of heat from the reservoir is transformed into mechanical work $-\Delta W$ exerted by the gas on the surroundings, given by:

$$\Delta Q = -\Delta W = \int_{V_1}^{V_2} pdV. \tag{2.24}$$

Invoking the ideal gas law in the form $pV = nR^*T$, upon integration we get

$$\Delta Q = nR^*T \ln\left(\frac{V_2}{V_1}\right). \tag{2.25}$$

Here n is the number of moles of gas and $R^* = 8.314\ \mathrm{J\,mol^{-1}\,K^{-1}}$ is the universal gas constant. In a cyclic process, a thermodynamic system (termed an engine in the following) performs a closed cycle in the phase space of its variables of state, i.e. it returns to the state where it started from. We obtain such a periodic cycle by combining isothermal expansion with its reverse, isothermal compression. However, this type of engine can only achieve a net transformation of heat to mechanical work if the isothermal compression is done at a lower temperature.

The Carnot cycle is the idealized concept of a heat engine as it has the maximum theoretically attainable efficiency (see below). The Carnot process consists of four subprocesses, taking place at two different temperatures, as shown in Figure 2.6:

$1 \rightarrow 2$ *Isothermal expansion* at the higher temperature T_1: $\Delta Q_1 = nR^*T_1 \ln(V_2/V_1) = -\Delta W_1$ of heat from reservoir 1 is transformed to mechanical work done by the system.

2 → 3 *Adiabatic expansion:* At point 2 in Figure 2.6a the system is decoupled from heat reservoir 1 and expands adiabatically until the lower temperature T_2 is reached at point 3. Since for an adiabatic process $\Delta Q = 0$, the work done by the system during the expansion reduces the internal energy U of the system by $\Delta U = U(T_2) - U(T_1)$.

3 → 4 *Isothermal compression* at the lower temperature, T_2: At point 3, the system is coupled to heat reservoir 2. Isothermal compression at the lower temperature T_2 proceeds until point 4; an amount $\Delta W_2 = nR^* T_2 \ln (V_3/V_4) = -\Delta Q_2$ of mechanical work done on the system is transformed to heat, which is absorbed by reservoir 2.

4 → 1 *Adiabatic compression:* At point 4, the system is decoupled from heat reservoir 2 and compressed adiabatically until the higher temperature T_1 is reached back at point 1. The work put into the system to achieve this compression increases U by $\Delta U = U(T_1) - U(T_2)$.

The total work done by the system during the cycle is given by the area enclosed by the cycle. The mechanical work done by the system and that put into the system during the adiabatic subprocesses cancel. The total heat absorbed by the system and transformed to mechanical work done by the system is therefore

$$\Delta Q = -\Delta W = \Delta Q_1 + \Delta Q_2 = nR^* T_1 \ln\left(\frac{V_2}{V_1}\right) - nR^* T_2 \ln\left(\frac{V_3}{V_4}\right). \tag{2.26}$$

Using the properties of adiabatic processes, one can show that $\ln (V_2/V_1) = \ln (V_3/V_4)$, i.e.

$$\Delta Q = -\Delta W = nR^*(T_1 - T_2) \ln\left(\frac{V_2}{V_1}\right). \tag{2.27}$$

Unlike the individual subprocesses, the cycle cannot convert the entire heat absorbed from reservoir 1 to mechanical work: a part ΔQ_2 of it is lost to reservoir 2. The efficiency η of the Carnot engine is the ratio of the "converted" heat ΔQ and the added heat ΔQ_1:

$$\eta = \frac{\Delta Q}{\Delta Q_1} = \frac{T_1 - T_2}{T_1}. \tag{2.28}$$

Whereas it is always possible to fully convert mechanical work to heat (e.g. by friction), eq. (2.28) shows that even the most idealized heat engine works at an efficiency $\eta < 1$. Real heat engines do not reach the efficiency given by eq. (2.28), as the Carnot process assumes all subprocesses to be fully reversible, which is not possible in the real world. From a different point of view, one can say that the Carnot process transports heat from the warmer to the colder reservoir but some of it is lost as mechanical work. The transport of heat by heat conduction, however, is irreversible and does not convert any heat to mechanical work. Whereas heat never flows from colder to warmer temperatures by itself, the inverted Carnot process can transport heat from the colder to the warmer reservoir upon the investment of mechanical work as is done in refrigerators.

Such periodic transformations as in the Carnot process are made use of e.g. in steam or combustion engines and, on a much larger scale, they can be considered as the source of the destructive power of tropical cyclones, as will be discussed in Section 10.5.

2.2.5 Air parcels

Adiabatic processes are of special relevance for the atmosphere, as the rising of any air parcel is an adiabatic process, at least initially. An air parcel is an imaginary volume of air to which the basic dynamic and thermodynamic properties of air are assigned. On the one hand, an air parcel must be large enough to contain a very large number of molecules. On the other hand, it must be small enough that the macroscopic properties assigned to it are approximately uniform within it and that its motion with respect to the surrounding atmosphere does not induce noticeable compensatory movements in the environment. One could visualize such an air parcel as having dimensions of some tens of centimeters in each direction. An air parcel is assumed to be thermally insulated from its environment, so that its temperature changes adiabatically as it rises or sinks. It must always be at the same pressure as the environmental air at the same altitude.

In order to establish the desired relation between T and p for an adiabatically changing air parcel and hence to be able to describe the changes of state of the air parcel, we start from eq. (2.23) and replace α using the ideal gas law. We then obtain

$$c_p dT = R_d T \frac{dp}{p}. \tag{2.29}$$

Now we rearrange this equation so that only the temperature appears on the left-hand side:

$$\frac{dT}{T} = \frac{R_d}{c_p}\frac{dp}{p}. \tag{2.30}$$

Upon integration of eq. (2.30) from a reference point (p_0, T_{p_0}) to (p, T) we arrive at

$$\ln\left(\frac{T}{T_{p_0}}\right) = \frac{R_d}{c_p}\ln\left(\frac{p}{p_0}\right) = \ln\left[\left(\frac{p}{p_0}\right)^{R_d/c_p}\right],$$

$$\frac{T}{T_{p_0}} = \left(\frac{p}{p_0}\right)^{\kappa}. \tag{2.31}$$

Here κ denotes the fraction R_d/c_p, which is given as 0.286 for dry air. Usually the reference pressure p_0 is defined by $p_0 = 1000$ hPa for adiabatic processes and T_{p_0} refers to the temperature that an air parcel would attain if it was brought dry adiabatically to the reference pressure p_0; "dry adiabatically" refers to an absence of phase changes. The temperature T_{p_0} is also called the potential temperature, θ. Note that eq. (2.31) holds only for dry adiabatic processes where no phase transitions are involved. From the definition of θ in eq. (2.31) it can be seen that θ is a conserved quantity for dry adiabatic transformations, i.e. it remains constant; it can thus be regarded as a measure of the energy content of an air parcel, i.e. as a characteristic property of an air parcel, as long as the parcel undergoes only dry adiabatic transformations. Equation (2.31) describes e.g. how a dry adiabatically ascending air parcel cools with decreasing pressure. It also describes the temperature profile in a well-mixed atmospheric boundary layer (Section 4.2.2).

2.2.6 Specific entropy

In thermodynamics, introducing a different thermodynamic quantity sometimes helps to show the relationships between variables more clearly or to simplify the description of certain processes. In order to point out an important property of dry adiabatic processes, we introduce the specific entropy s, i.e. the entropy per unit mass, and its relation to θ. An increase in s is related to the addition of heat dq to a unit mass of a gas at temperature T as follows:

$$ds = \frac{dq}{T}, \tag{2.32}$$

which is valid for reversible processes. In an adiabatic process, where $dq = 0$ and no heat exchange with the environment takes place, the change in (specific) entropy is also zero. In contrast, the (specific) entropy increases for irreversible processes, as stated in the second law of thermodynamics:

$$ds \geq 0. \tag{2.33}$$

Now replace dq in the definition of entropy using the first law of thermodynamics (eq. 2.19):

$$ds = \frac{1}{T}(c_p dT - \alpha dp). \tag{2.34}$$

Again, we replace α with the help of the ideal gas law:

$$ds = c_p \frac{dT}{T} - R_d \frac{dp}{p}. \tag{2.35}$$

Now insert κ into the second term on the right-hand side so that c_p can be placed in front of the parentheses:

$$ds = c_p \left(\frac{dT}{T} - \kappa \frac{dp}{p} \right). \tag{2.36}$$

The term in parentheses turns out to be the same as $d\theta/\theta$. This becomes clear if we start from the definition of the potential temperature (eq. 2.31),

$$\theta = T \left(\frac{p_0}{p} \right)^{\kappa}. \tag{2.37}$$

Taking the total derivative, we arrive at:

$$\begin{aligned}
d\theta &= \frac{\partial \theta}{\partial T} dT + \frac{\partial \theta}{\partial p} dp \\
&= dT \left(\frac{p_0}{p} \right)^{\kappa} - T\kappa \left(\frac{p_0^{\kappa}}{p^{\kappa+1}} \right) dp \\
&= dT \left(\frac{p_0}{p} \right)^{\kappa} - \kappa \left(\frac{p_0}{p} \right)^{\kappa} \frac{T}{p} dp \\
&= \left(\frac{p_0}{p} \right)^{\kappa} \left(dT - \kappa \frac{T}{p} dp \right)
\end{aligned}$$

$$= \frac{\theta}{T}\left(dT - \kappa\frac{T}{p}dp\right)$$
$$= \theta\left(\frac{dT}{T} - \kappa\frac{dp}{p}\right). \tag{2.38}$$

With this, eq. (2.36) reduces to

$$ds = c_p\frac{d\theta}{\theta}. \tag{2.39}$$

This equation can be integrated, yielding:

$$s = c_p \ln\theta + c \tag{2.40}$$

where c is a constant. This means that a dry adiabatic process (θ = constant) is also an isentropic process, i.e. a process in which the entropy remains constant. Note that as soon as phase transitions occur (wet adiabatic processes, Section 2.5), the simple relation (eq. 2.40) between s and θ does not hold any longer, since latent heat needs to be taken into account in the first law (eq. 2.19).

2.3 Thermodynamic charts

In atmospheric sciences, especially in weather analysis and forecasting, thermodynamic charts are used to visualize adiabatic processes occurring in the vertical direction, e.g. the ascents and descents of air parcels. They can be used to obtain easily different thermodynamic properties, e.g. θ and moisture quantities such as the specific humidity, from a given radiosonde ascent. In addition they can be used to study isobaric or isothermal processes. Various types of thermodynamic chart exist. The most common are the Stüve diagram, the Skew-T-Log-p diagram and the tephigram. Each has lines of constant p, T, θ, and saturation specific humidity and saturated adiabats. They only differ in the way in which these lines are plotted. The thermodynamic chart used in this book is the tephigram, as shown in Figure 2.7. It can be downloaded from www.cambridge.org/clouds.

The name tephigram evolved from the original name T–S-gram, where S refers to entropy. It emphasizes that entropy (or θ according to eq. 2.40) and temperature are the two basic variables of this diagram. In a tephigram the lines of constant temperature (isotherms) are perpendicular to the lines of constant θ (dry adiabats) or entropy (isentropes). As we will see later, this is of advantage when a moist adiabatic ascent is to be plotted. Knowing T and θ enables the calculation of all the other lines in this diagram.

An example of a tephigram, where the various types of line are highlighted, is shown in Figure 2.7. The lines of constant pressure are nearly parallel to the x-axis (highlighted medium-broken line). Lines of constant temperature rise from the lower left to the upper right (highlighted solid line with positive slope), perpendicular to the lines of constant θ, which rise from the lower right to the upper left (highlighted solid line with negative slope). Thus if an air parcel follows a dry adiabatic ascent, its potential temperature remains constant, i.e. it rises from the lower right to the upper left of the diagram. Once

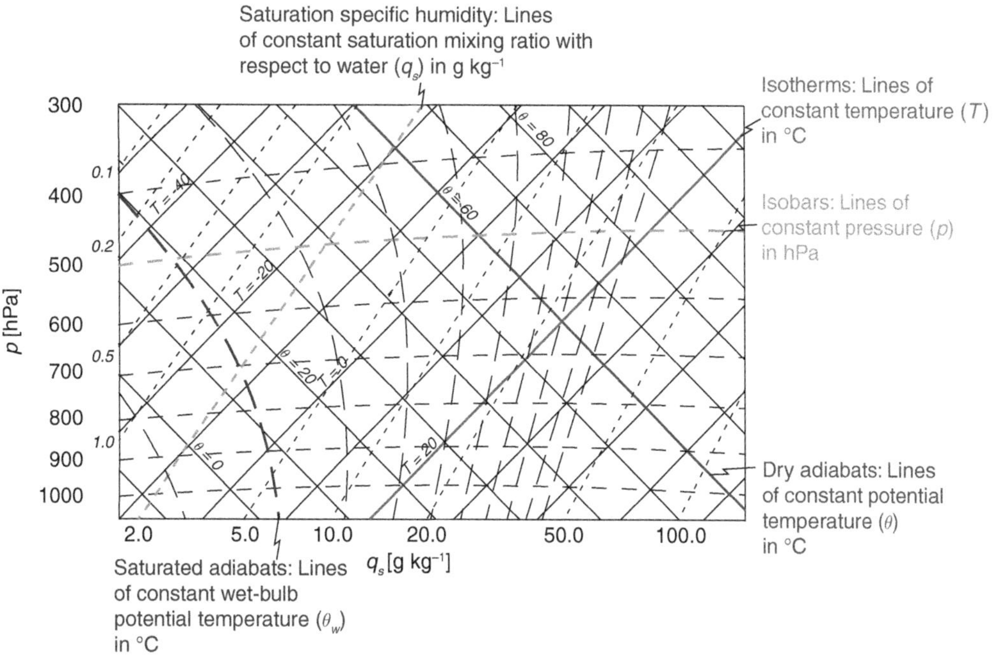

Fig. 2.7 Tephigram in which examples of the various types of lines are highlighted; p is the atmospheric pressure and q_s is the saturation specific humidity (Section 2.5.3). A black and white version of this figure will appear in some formats. For the color version, please refer to the plate section.

the air parcel is saturated with respect to water, it follows a saturated adiabat (highlighted large-broken curve line). Whether an air parcel is saturated can be determined from the saturation specific humidity (highlighted short-broken line).

2.4 Thermodynamics of phase transitions

As water can exist in all three phases in the atmosphere (ice, liquid water and water vapor), phase transitions, also called phase changes, between these states are crucial for the understanding of microphysical processes in clouds. The phase transitions occurring in clouds are condensation (vapor → liquid), evaporation (liquid → vapor), deposition (vapor → solid), sublimation (solid → vapor), freezing (liquid → solid) and melting (solid → liquid). In this chapter the role of the Gibbs free energy G, also referred to as the Gibbs potential, in phase transitions will be discussed. The Gibbs free energy is used to derive the Clausius–Clapeyron equation, which describes the coexistence of two phases in TDE (Section 2.4.2). Furthermore, G can be used to retrieve information about the direction of phase transitions under given conditions and about kinetic inhibitions of phase transitions. Therefore it is fundamental to the understanding of nucleation processes, which will be discussed in Chapters 6 and 8.

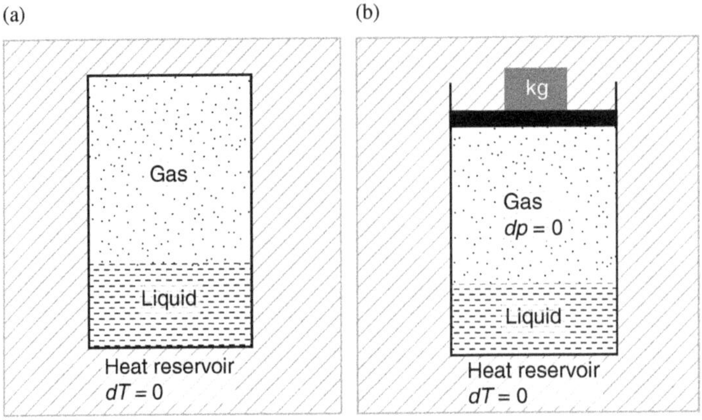

Fig. 2.8 Schematic of (a) an isothermal system and (b) an isothermal and isobaric system with a movable piston in which a gas and liquid coexist.

2.4.1 The role of the Gibbs free energy in phase transitions

In thermodynamics, various quantities are used to gather information about the state of a system. Which state variable is most suitable to describe the state of the system depends on the boundary conditions. The right choice of state variables often allows us to formulate a minimum principle for that quantity, which can be derived from the second law of thermodynamics and can be used to describe the TDE of the system under consideration.

Let us consider a gas coexisting with its liquid phase in a closed container which is in thermal contact with a large heat reservoir. In this case the system's temperature is constant, as shown in Figure 2.8a. For a system at constant T which is unable to do work on its surroundings or to "receive" work from its surroundings (i.e. a system at constant volume), a minimum principle can be formulated in terms of the Helmholtz free energy F, which is given by

$$F = U - TS, \tag{2.41}$$

where S denotes the entropy. The quantities on the right-hand side will adjust in such a way that F takes a minimum. This corresponds to the state of TDE. For our system the only free parameter is the ratio of the liquid and the gaseous amounts of the substance. An adjustment towards TDE means that liquid evaporates or gas condenses until the pressure in the container equals the saturation vapor pressure of the substance at a given temperature T. Bringing the system from a non-equilibrium state to TDE therefore means changing the pressure in the system.

It is important to note that the boundary condition $dT = 0$, as required by eq. (2.41) is not sufficient to describe atmospheric phase transitions. Hence, the minimum principle for F is not useful for atmospheric phase transitions. In the atmosphere the additional boundary condition of constant pressure has to be imposed. In our exemplary system, we can think of this additional boundary condition as corresponding to the top wall of the container being replaced by a movable wall, i.e. a piston, with a weight placed on top to ensure constant pressure in the system (Figure 2.8b). Instead of an increase in pressure, the evaporation of some quantity of liquid will increase the volume, i.e. the system does work on its surroundings.

For a system at constant T and p a different minimum principle for TDE can be formulated with the help of the Gibbs free energy, which is given by

$$G = U - TS + pV. \tag{2.42}$$

Since all quantities on the right-hand side are state variables, the Gibbs free energy is a state variable itself. The internal parameters, i.e. the ratio of the amounts of liquid and gas and the volume, in our example with the movable piston, will arrange themselves so that G is minimized.

Let us now illustrate the role of the Gibbs free energy for a closed system containing a substance in two different phases, namely the gas and the liquid phase, which can exchange matter and energy. Thus we need to consider the infinitesimal change of G in both phases. Furthermore, the internal energy of each of the two phases, which are not closed with respect to particle number N_j, has to be modified compared with eq. (2.2). This is done by adding a term to the right-hand side of eq. (2.2) that takes the change in particle number of the phase j into account:

$$dU_j = TdS_j - pdV_j + \mu_j dN_j. \tag{2.43}$$

Here μ_j denotes the *chemical potential* of phase j. Inserting eq. (2.43) into eq. (2.42), we arrive at:

$$\begin{aligned} dG &= \sum_{j=1}^{2}(dU_j - TdS_j - S_j dT + pdV_j + V_j dp) \\ &= \sum_{j=1}^{2}(-S_j dT + V_j dp + \mu_j dN_j) \\ &= \sum_{j=1}^{2}\mu_j dN_j, \end{aligned} \tag{2.44}$$

since T and p are constant for our system. As the combination of the two phases is closed with respect to the particle number, we have

$$dN_1 = -dN_2, \tag{2.45}$$

or

$$dG = dN_1(\mu_1 - \mu_2). \tag{2.46}$$

Because the system always tends towards a state with minimum G, it will increase phase 2 at the expense of phase 1 (i.e. $dN_1 < 0$) if $\mu_2 < \mu_1$. Therefore the minimum principle of G yields information about the direction in which a transition between two coexisting phases occurs under isothermal and isobaric conditions in order to reach TDE, if the difference in chemical potential between the two phases ($\Delta\mu = \mu_2 - \mu_1$) is known. For water vapor and liquid water, $\Delta\mu$ can be written as (Pruppacher and Klett, 1978)

$$\Delta\mu = -kT \ln S. \quad (2.47)$$

Here the reader should note that, in eq. (2.47), S is not the entropy but the saturation ratio $S = e/e_{s,w}$ of the vapor phase with respect to liquid water, i.e. the ratio of the pressure of the vapor phase and the equilibrium vapor pressure at temperature T. For a supersatured vapor phase, $S > 1$ and $\Delta\mu < 0$. That is, the chemical potential of the vapor phase is higher than that of the liquid phase and condensation will take place in order for the system to minimize G.

It should be stressed that the difference in chemical potential $\Delta\mu$ between two phases out of TDE has nothing to do with the latent heat involved in the phase transition. A latent heat flux is always required for a phase transition, even if the phase transition occurs in TDE, i.e. if $\Delta\mu = 0$.

Generally, the chemical potential μ_j can be viewed as the Gibbs free energy per molecule in phase j. In chemical reactions differences in chemical potential correspond to different compounds (products and reactants) rather than to different phases; the differences in chemical potential indicate in which direction the reaction tends to take place. Note that the sign of a difference in chemical potential gives no information about whether the reaction releases heat into the surroundings or takes heat from it. For instance, for a reaction which converts a large part of the reactants to products to take place, the chemical potential of the products must be lower than that of the reactants (Figure 2.9). It does not matter whether the reaction is exothermic (in which energy is released) or endothermic (in which energy is consumed).

The Gibbs free energy G has the important property that in an infinitesimal, reversible process it is insensitive to the work done by the system and to the heat exchanged between the system and a connected heat reservoir. Here the corresponding terms pdV and TdS in the calculation of dG (eq. 2.44) cancel out.

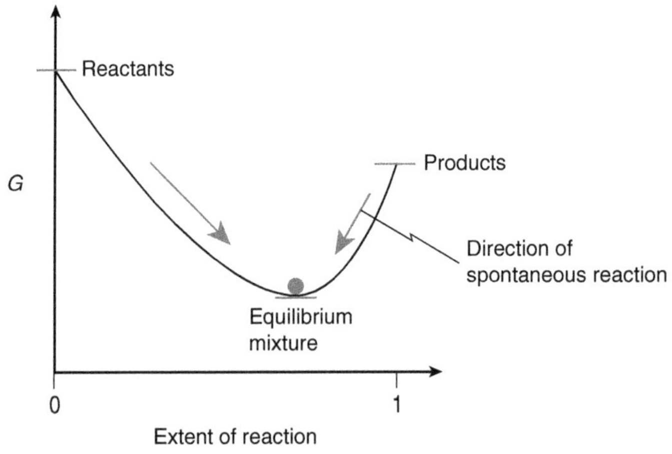

Fig. 2.9 Schematic of the Gibbs free energy G for a chemical reaction. The gray disk denotes the equilibrium point, corresponding to a minimum in the Gibbs free energy. At this point the formation rates of products from reactants (forward reaction) and of reactants from products (backward reaction) are equal.

Generally, the thermodynamic potential which measures the energy required or released during phase transitions in TDE is the enthalpy H:

$$H = U + pV. \tag{2.48}$$

Enthalpy is also a thermodynamic state variable, since it only depends on other state variables, namely U, p and V. Taking the total differential of eq. (2.48), we arrive at

$$dH = dU + pdV + Vdp. \tag{2.49}$$

Considering isobaric conditions ($dp = 0$), an infinitesimal change is given by:

$$dH = dU + pdV = TdS - pdV + pdV = TdS = dQ. \tag{2.50}$$

Here the non-specific form of eq. (2.32) was used for the last step.

Depending on the conditions (i.e. on T and p), the TDE of our two-phase system containing a gas and a liquid phase of a substance, is reached either at a state where only phase 1 or only phase 2 is present or at a state where both phases coexist. In the latter case, p and T are related by the Clausius–Clapeyron equation, as will be explained in the following.

2.4.2 Phase transitions in thermodynamic equilibrium: the Clausius–Clapeyron equation

If two phases coexist in TDE, their chemical potentials must be equal, i.e. $\mu_1 = \mu_2$, and therefore $dG = 0$ (eq. 2.46), during the transition of an infinitesimal amount of a substance from one phase to the other. The Gibbs free energy is constant during a phase transition as long as the system remains in TDE. Note, however, that heat is required for the phase to change from liquid to vapor because energy is required to break the intermolecular hydrogen bonds. This agrees with experiences in everyday life: stepping out of the water after a bath feels chilly, even on a hot summer day, because the water droplets on the skin withdraw energy from their environment during evaporation.

In order to derive the Clausius–Clapeyron equation, we use the fact that the specific Gibbs free energies of the individual phases are equal in TDE:

$$g_1 = g_2. \tag{2.51}$$

The dependence of g on p and T can be obtained by taking the total derivative of g (see eq. 2.44)

$$dg = du + pd\alpha + \alpha dp - Tds - sdT. \tag{2.52}$$

Since $du + pd\alpha = dq = Tds$, according to the first law of thermodynamics, eq. (2.52) reduces to

$$dg = \alpha dp - sdT. \tag{2.53}$$

For phase transitions in TDE, the change in the specific Gibbs free energy in the first phase equals that in the second phase, $dg_1 = dg_2$, so that

$$\alpha_1 dp - s_1 dT = \alpha_2 dp - s_2 dT. \tag{2.54}$$

Rearranging the terms and using

$$L_{2,1} = T \int_{h_1}^{h_2} \frac{dh}{T} = T \int_{q_1}^{q_2} \frac{dq}{T} = T(s_2 - s_1) \tag{2.55}$$

yields

$$\frac{dp}{dT} = \frac{s_2 - s_1}{\alpha_2 - \alpha_1} = \frac{L_{2,1}}{T(\alpha_2 - \alpha_1)}, \tag{2.56}$$

where $L_{2,1}$ denotes the latent heat [J kg^{-1}] involved in going from phase 1 to phase 2. Equation (2.56) is the general form of the Clausius–Clapeyron equation. It states the condition on T and p for the coexistence of the two phases in TDE. In particular, it describes the dependence of the saturation vapor pressure above a flat and pure liquid or solid water surface on temperature. The saturation vapor pressure is the vapor pressure at which the liquid or solid phase is in equilibrium with the vapor phase. Here, equilibrium means that, on average, the same number of molecules go from the liquid to the vapor phase as go from the vapor to the liquid phase per unit time interval.

The saturation vapor pressure for water is given by the curves shown in Figure 2.10. This figure is called a phase diagram because it shows in which phase water would exist in TDE as a function of pressure and temperature; in TDE is the key phrase here, as water may well be present in different phases in the real world. The lines in the phase diagram are phase transition curves. Crossing one of these lines is accompanied by a complete transformation from one phase to the other. At the conditions described by the phase transition curves, two phases can coexist in equilibrium.

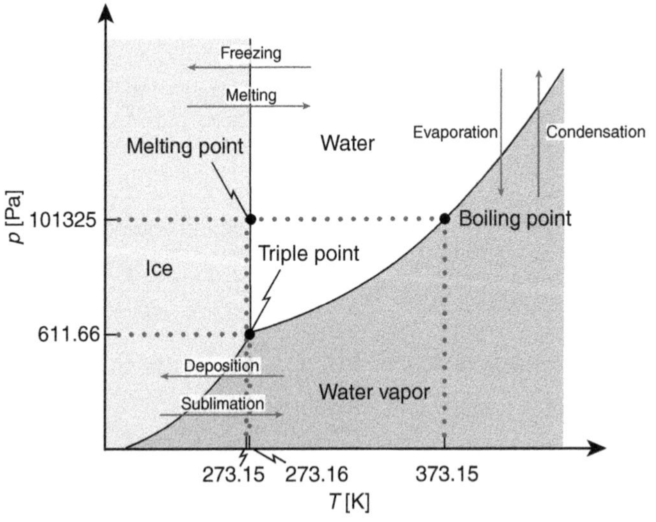

Fig. 2.10 Schematic phase diagram of water as a function of temperature and pressure. The temperatures and pressures of the melting, triple and boiling point are indicated by dotted lines.

Coexistence in TDE of all three phases of water is only possible for a single combination of pressure and temperature. This is defined as the triple point, with temperature 273.16 K = 0.01 °C. At the triple point the saturation vapor pressures over water and ice are equal ($e_{s,w} = e_{s,i}$ = 611.66 Pa).

The triple point temperature is 0.01 K higher than the melting point temperature, where liquid water and ice coexist under atmospheric pressure. Unlike at the triple point, the vapor phase cannot be in equilibrium with both the ice and the liquid water phase at the melting point. In this book, we are concerned with the phase transitions of water not as a pure substance but as a component in air at atmospheric pressure. Therefore, the melting point temperature of 273.15 K defines the coexistence of ice and liquid water in the remainder of this book.

One anomaly of water is the negative slope of the melting curve, indicating that water expands when it freezes, i.e. $\alpha_l - \alpha_i < 0$. Note that this negative slope is hardly visible in Figure 2.10 because only a small section of the whole phase diagram is shown here. Most other substances have a positive slope, meaning that their specific volume is reduced upon freezing. The negative slope of the melting curve of water is due to the fact that the density of ice is lower than the density of liquid water and explains why ice forms on top of lakes and not at the bottom. The negative slope of the melting curve further explains why ice can be melted by increasing the pressure. Against common belief this is not the explanation of why an ice skater glides on the ice, because the weight of a human being is not sufficient to melt ice. Also the quasi-liquid layer (QLL, Section 6.3) that exists on ice at temperatures warmer than −25 °C (Li and Somorjai, 2007) is too thin. Instead ice skaters glide because friction creates heat which causes the ice to melt (Rosenberg, 2005).

The general form of the Clausius–Clapeyron equation (2.56) was derived for a one-component system (water in two different phases). It also holds for partial pressures in multi-component systems, where p in the equation refers to the total pressure. In a one-component system the total pressure p equals the partial pressure p_i. In moist air e refers to the partial pressure of the water vapor and, in the case of a one-component system consisting only of water, $p = e$. The Clausius–Clapeyron equation is valid for every phase boundary in the phase diagram of water shown in Figure 2.10. For the equilibrium curve describing the coexistence between the liquid and solid phases, we obtain

$$\frac{dp}{dT} = \frac{L_f}{T(\alpha_l - \alpha_i)}, \tag{2.57}$$

where L_f denotes the latent heat of fusion. At temperatures below 0 °C, $\alpha_l < \alpha_i$ for water and hence $dp/dT < 0$, resulting in the negative slope of the melting curve evaluated at 0 °C:

$$\frac{dp}{dT} = \frac{3.33 \times 10^5\,\mathrm{J\,kg^{-1}}}{273.15\,\mathrm{K}(10^{-3}\,\mathrm{m^3\,kg^{-1}} - 1.091 \times 10^{-3}\,\mathrm{m^3\,kg^{-1}})} = -1.34 \times 10^7\,\mathrm{Pa\,K^{-1}}. \tag{2.58}$$

Equation (2.58) shows that large increases in pressure are needed for a small increase in the freezing temperature.

For the phase transition between water vapor and liquid water, we can simplify eq. (2.56) by neglecting the specific volume of liquid water. This is justified because the specific volume of water vapor at 0 °C and 100% relative humidity is

$$\alpha_v = \frac{R_v T}{e_{s,w}} = \frac{461.5\,\mathrm{J\,kg^{-1}\,K^{-1}} \times 273.15\,\mathrm{K}}{611.2\,\mathrm{Pa}} = 206.4\,\mathrm{m^3\,kg^{-1}},$$

which is five orders of magnitude larger than α_l, which is 10^{-3} m^3 kg^{-1}. Neglecting α_l in eq. (2.56) and using the saturation vapor pressure of water vapor with respect to liquid water $e_{s,w}$ yields the typical form of the Clausius–Clapeyron equation:

$$\frac{de_{s,w}}{dT} = \frac{L_v}{T\alpha_v} = \frac{L_v e_{s,w}}{R_v T^2}, \tag{2.59}$$

where L_v denotes the latent heat of vaporization. Here we have made use of the ideal gas law for water vapor under saturated conditions ($e = e_{s,w}$). In fact, the Clausius–Clapeyron equation (2.59) describes the local slope of the $e_{s,w}$ curve shown in Figure 2.10, i.e. the local change of the equilibrium saturation vapor pressure with respect to water for a change in temperature at temperatures > 273.15 K. To a first approximation, the Clausius–Clapeyron equation describing the equilibrium between the liquid and the vapor phase can be integrated by regarding L_v as constant, resulting in:

$$\ln \frac{e_{s,w}}{e_{s0}} = \frac{L_v}{R_v}\left(\frac{1}{T_0} - \frac{1}{T}\right), \tag{2.60}$$

where e_{s0} = 611.2 Pa at T_0 = 273.15 K. The latent heat of vaporization L_v near 0 °C is approximately 2.5×10^6 J kg^{-1} (Table 2.1).

2.4.3 Saturation vapor pressure below 273.15 K

The usual phase diagram, as shown in Figure 2.10, includes only the stable phases (solid, liquid, gaseous) of the respective substance, as only the stable phases are in TDE. Metastable phases exist only in systems out of TDE and therefore only on a limited time scale. Nevertheless, metastable phases are of high atmospheric relevance, occurring frequently, e.g. when water vapor becomes temporarily supersaturated in a cooling parcel of moist air or when cloud droplets get supercooled prior to freezing in a cloud that has cooled below 273.15 K.

Although supercooled water cannot be in TDE with the vapor phase (as it does not exist in TDE), a steady state in which supercooled water and water vapor coexist can still be established, meaning that there is no net flux between the two phases if no heat is added to or taken from the system. In that respect, this steady state is similar to the equilibrium described by the Clausius–Clapeyron equation for the coexistence of water and water vapor in TDE above 273.15 K. However, the Clausius–Clapeyron equation was derived assuming TDE, which justified the required identity of the specific Gibbs free energies of the two phases (eq. 2.51). Indeed, the identity $g_1 = g_2$ is still valid for the two phases of supercooled water and water vapor, even though the former is metastable, and the Clausius–Clapeyron equation can be used to describe the equilibrium between liquid water and water vapor even at temperatures below 273.15 K. Note that the finite difference in Gibbs free energies that exists between the metastable supercooled water and the stable ice phase does not affect the coexistence of supercooled water and the vapor phase.

Table 2.1 Saturation vapor pressures with respect to water, $e_{s,w}$, and to ice, $e_{s,i}$, in Pa according to eqs. (2.63) and (2.64), latent heat of vaporization L_v and sublimation L_s in 10^6 J kg^{-1} according to eqs. (2.72)–(2.74), and specific heat capacities at constant pressure of water vapor, c_{pv} (Lemmon, 2015) and of ice, c_i, in 10^3 J kg^{-1} K^{-1} based on the parameterization of Murphy and Koop (2005) as function of temperature T in °C.

T	$e_{s,w}$	$e_{s,i}$	L_v	L_s	c_{pv}	c_i
−70	0.48	0.26		2.834		1.587
−65	0.96	0.54		2.835		1.623
−60	1.87	1.08		2.836		1.658
−55	3.49	2.09		2.837		1.695
−50	6.32	3.94		2.838		1.731
−45	11.09	7.21		2.838		1.768
−40	18.91	12.84		2.839		1.806
−35	31.41	22.35	2.591	2.839		1.843
−30	50.94	38.01	2.576	2.839		1.881
−25	80.78	63.28	2.562	2.838		1.916
−20	125.50	103.25	2.549	2.838		1.958
−15	191.31	165.29	2.537	2.837		1.997
−10	286.45	259.59	2.525	2.837		2.036
−5	421.76	401.76	2.513	2.835		2.075
0	611.21	611.15	2.500	2.834		2.114
0.01	611.66	611.66	2.501	2.834	1.884	2.114
5	872.60		2.488		1.889	
10	1228.26		2.476		1.895	
15	1705.89		2.464		1.900	
20	2339.40		2.452		1.906	
25	3169.94		2.441		1.912	
30	4246.81		2.429		1.918	
35	5628.62		2.417		1.924	
40	7384.31		2.405		1.931	

At temperatures below 273.15 K, we have therefore two saturation vapor pressure curves, one for supercooled water, given by eq. (2.59) and one for ice given by:

$$\frac{de_{s,i}}{dT} = \frac{L_s}{T\alpha_v} = \frac{L_s e_{s,i}}{R_v T^2}. \tag{2.61}$$

For the approximation of a constant latent heat of sublimation, eq. (2.61) gives an analytical solution to the Clausius–Clapeyron equation in a similar manner to eq. (2.60) for liquid water:

$$\ln \frac{e_{s,i}}{e_{s0}} = \frac{L_s}{R_v}\left(\frac{1}{T_0} - \frac{1}{T}\right). \tag{2.62}$$

As stated above, the temperature dependence of the latent heat L_s makes an analytical solution of eq. (2.61) impossible. Hence, different empirical formulas for the saturation vapor pressure as a function of temperature have been developed. Currently, the most accurate empirical formulas for the saturation vapor pressures over water and ice are those provided by Murphy and Koop (2005):

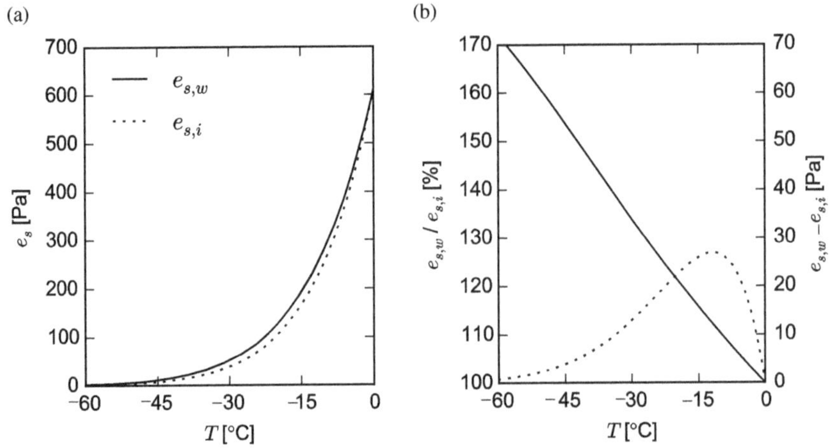

Fig. 2.11 (a) Saturation vapor pressure [Pa] over water and ice at temperatures below 0 °C, according to eqs. (2.63) and (2.64). (b) Difference in saturation vapor pressures [Pa] (dotted line, right-hand axis) and the ratio $e_{s,w}/e_{s,i}$ [%] (solid line, left-hand axis).

$$e_{s,w}(T) = \exp\left\{54.842763 - \frac{6763.22}{T} - 4.210 \ln T + 0.000367T + \tanh[0.0415(T - 218.8)]\left(53.878 - \frac{1331.22}{T} - 9.44523 \ln T + 0.014025T\right)\right\}, \tag{2.63}$$

$$e_{s,i}(T) = \exp\left(9.550426 - \frac{5723.265}{T} + 3.53068 \ln T - 0.00728332T\right). \tag{2.64}$$

The results for $e_{s,w}$ are valid for temperatures between 123 and 332 K, those for $e_{s,i}$ between 110 and 273.16 K. They are shown in Figure 2.11 for temperatures below 273.15 K. A simpler approximation to $e_{s,w}$ is given in Rogers and Yau (1989):

$$e_{s,w}(T) = A_w \exp\left(-\frac{B_w}{T}\right). \tag{2.65}$$

Here $A_w = 2.53 \times 10^{11}$ Pa and $B_w = 5420$ K include the constants from eq. (2.60). A similar approximation to $e_{s,i}$ is found by fitting an exponential to $e_{s,i}(T)$ from eq. (2.64):

$$e_{s,i}(T) = A_i \exp\left(-\frac{B_i}{T}\right). \tag{2.66}$$

Here $A_i = 3.64 \times 10^{12}$ Pa and $B_i = 6148$ K. This equation is valid for the temperature range between 213 and 273.16 K.

Note that L_v depends weakly on temperature, changing by about 6% from −30 °C to 30 °C (Table 2.1). We can deduce the temperature dependence of L_v by differentiating

$$L = L_{2,1} = u_2 - u_1 + p(\alpha_2 - \alpha_1) \tag{2.67}$$

with respect to temperature. Note that $L_{2,1}$ corresponds to a difference in the enthalpy $H = U + pV$. For now we neglect α_l compared with α_v and replace $\alpha_v e_{s,w}$ with $R_v T$, according to the ideal gas law for water vapor (eq. 2.77). We then obtain

$$\frac{dL_v}{dT} = \frac{du_v}{dT} - \frac{du_l}{dT} + R_v = c_{vv} - c_l + R_v = c_{pv} - c_l. \tag{2.68}$$

Here we used eqs. (2.10) and (2.18) for water vapor instead of for dry air, c_{vv} refers to the specific heat capacity of water vapor at constant volume, c_l is the specific heat capacity of liquid water and c_{pv} is the specific heat capacity of water vapor at constant pressure. Note that the specific heat capacities also vary slightly with temperature but at this point we regard them as constant. The typical values used for the specific heats at 273.16 K are c_{vv} = 1418.4 J kg^{-1} K^{-1}, c_{pv} = 1884.4 J kg^{-1} K^{-1} and c_l = 4219.9 J kg^{-1} K^{-1} (Lemmon, 2015). If the temperature dependence of the specific heats is neglected, eq. (2.68) can be integrated to yield an approximation for $L_v(T)$:

$$L_v = (c_{pv} - c_l)(T - T_0) + L_{v0}. \tag{2.69}$$

The constant of integration L_{v0} can be taken as $L_v(T_0) = 2.501 \times 10^6$ J kg^{-1}. Accordingly, for ice,

$$L_s = (c_{pv} - c_i)(T - T_0) + L_{s0}, \tag{2.70}$$

with $L_{s0} = L_s(T_0) = 2.834 \times 10^6$ J kg^{-1}. These expressions for L can now be introduced into the Clausius–Clapeyron equations (2.60) and (2.62), which can then be integrated more precisely. For the saturation vapor pressure with respect to water, this yields the following result:

$$\ln \frac{e_{s,w}}{e_{s0}} = \frac{L_{v0} - (c_{pv} - c)T_0}{R_v}\left(\frac{1}{T_0} - \frac{1}{T}\right) + \frac{c_{pv} - c}{R_v} \ln \frac{T}{T_0}. \tag{2.71}$$

Although the temperature dependence of L_v might be neglected within some applications, it is often more convenient to use parameterizations, such as the one by Murphy and Koop (2005):

$$L_v = 56579 - 42.212T + \exp(0.1149(281.6 - T))\frac{1}{M_w}, \tag{2.72}$$

with L_v in J kg^{-1} K^{-1} for T in K and M_w in kg mol^{-1}. Equation (2.72) is valid between 30 and 273.16 K, i.e. for supercooled water. For higher temperatures an approximate dependence is given by Hess (1959):

$$L_v = (2.5 \times 10^6) - (2.37 \times 10^3)(T - 273.15). \tag{2.73}$$

Similarly an equation for the temperature dependency of the latent heat of sublimation is given by:

$$L_s = 46782.5 + 35.8925T - 0.07414T^2 + 541.5 \exp\left\{-\left(\frac{T}{123.75}\right)\right\}^2 \frac{1}{M_w}. \tag{2.74}$$

Equation (2.74) is valid between 236 and 273.16 K.

2.5 Moist air

2.5.1 Water in the atmosphere

Because air contains between 0% and 4% water vapor by mass, it is necessary to take the water vapor content into account when for example calculating the density of air or obtaining cloud base height. The highest amount of water vapor is found near the surface over the tropical oceans. It decreases to nearly 0% with increasing altitude. The average water vapor concentration declines rapidly with height because (i) its major sources and sinks are near Earth's surface and (ii) because the saturation vapor pressure decays exponentially with decreasing temperature. The lifetime of water vapor in the troposphere is on the order of ten days.

Water vapor is also the most important natural greenhouse gas. Because Earth has an atmosphere, its global average surface temperature is 33 K higher than it would be without an atmosphere. Water vapor is responsible for 60% or 20 K of this natural greenhouse effect. It has a large latent heat of condensation and therefore dominates the energy fluxes in the Earth–atmosphere system (Chapter 11).

The presence of water vapor on the one hand influences basic properties of air such as its density. On the other hand, phase transitions between the different phases of water alter atmospheric processes, e.g. the adiabatic ascent of air parcels. In contrast with dry adiabatic ascent, where the potential temperature θ remains constant, θ increases in a moist adiabatic ascent once the air is saturated with water vapor and the water vapor starts to condense. Therefore the temperature T decreases more slowly in a moist adiabatic ascent, owing to the release of latent heat of condensation, as compared with a dry adiabatic ascent. This can be seen from the difference in the slopes of the dry adiabats and the saturated adiabats in Figure 2.7. At very cold temperatures the saturated adiabats approach the dry adiabats because of the small moisture content in cold air. The calculation of the saturated adiabats will be outlined in Section 3.3.

The moisture content of an air mass can be described by different quantities. Moist air is a two-component (air and water vapor) system. For gas mixtures, the most practical quantity to describe the relative amount of each component in the mixture is the partial pressure of the respective component, which we consider in the following.

2.5.2 Partial pressure

The partial pressure of a gas component is defined as the pressure that component i would exert on the walls of a volume V if the volume V was occupied by only component i, with all other components removed. The ideal gas law directly links the partial pressure p_i with the number concentration n_i of gas molecules of component i in the mixture:

$$n_i = \frac{p_i}{kT} = \frac{p_i N_A}{R^* T}, \qquad (2.75)$$

where $k = 1.38 \times 10^{-23}$ J K^{-1} is the Boltzmann constant, $N_A = 6.022 \times 10^{23}$ mol^{-1} is the Avogadro's number and $R^* = N_A k = 8.314$ J mol^{-1} K^{-1} is the universal gas constant.

The sum of the partial pressures of all components constituting the system is equal to the pressure of the gas mixture:

$$\sum_i p_i = p. \tag{2.76}$$

This is known as Dalton's law.

In moist air the partial pressure of water vapor, e, is often used to characterize the water vapor content in the atmosphere. Alternatively the mass density of water vapor, ρ_v, also referred to as the absolute humidity, can be used. As water vapor approximately behaves as an ideal gas, and the ideal gas law applies to the partial pressures of the individual components in a gas mixture, ρ_v can be linked to e via the ideal gas law:

$$e = R_v \rho_v T, \tag{2.77}$$

where $R_v = R^*/M_w = 461.5$ J kg^{-1} K^{-1} is the specific gas constant of water vapor and $M_w = 18$ g mol^{-1} is the molecular weight of water.

2.5.3 Water vapor mixing ratio and specific humidity

Often it is convenient to express the moisture content of the atmosphere in terms of the water vapor mixing ratio w_v, which is the mass of water vapor m_v per unit mass of *dry* air m_d. It can be obtained from

$$w_v = \frac{m_v}{m_d} = \frac{\rho_v V}{\rho_d V} = \frac{\rho_v}{\rho_d} = \frac{e/R_v T}{(p-e)/R_d T} = \frac{\epsilon e}{p-e} \approx \epsilon \frac{e}{p}, \tag{2.78}$$

where

$$\epsilon = \frac{R_d}{R_v} = \frac{M_w}{M_d} = 0.622 \tag{2.79}$$

and $M_d = 28.96$ g mol^{-1} is the molecular weight of dry air. The approximation in eq. (2.78) is justified because the water vapor pressure is much smaller than the atmospheric pressure: $e \ll p$.

The water vapor mixing ratio defined above is similar to the specific humidity q_v, which is the mass of water vapor m_v per unit mass of *moist* air m_m:

$$q_v = \frac{m_v}{m_m} = \frac{\rho_v}{\rho_m} = \frac{\rho_v}{\rho_d + \rho_v} = \frac{e/R_v T}{(p-e)/R_d T + e/R_v T} = \frac{\epsilon e}{p - e + \epsilon e} \approx \epsilon \frac{e}{p}, \tag{2.80}$$

where again we have made use of $e \ll p$.

Instead of using moisture quantities to describe the amount of moisture in the atmosphere, there are also various temperature measures that can be used for this purpose.

2.5.4 Virtual temperature

The virtual temperature T_v is introduced to avoid having a gas constant in the ideal gas law that depends on the moisture content of the air. The mass density of moist air is lower than that of dry air because the molecular weight of water vapor is smaller than that of dry air. As an example, if we have a mixture of 4% (per volume) water vapor and 96% dry air, the molecular weight of the moist air M_m would be $M_m = 0.96 M_d + 0.04 M_w = 28.526$

g mol^{-1}. In order to obtain the density of moist air we divide the sum of the molecular weights of the components by the volume that one mole of an ideal gas would occupy. It is given as $v = 22.4$ l mol^{-1} at 273.15 K and 1013.25 hPa and is independent of the composition of the gas. Hence we obtain

$$\rho_d(q_v = 0) = \frac{M_d}{v} = \frac{28.96 \text{ g mol}^{-1}}{22.4 \text{ l mol}^{-1}} = 1.29 \text{ kg m}^{-3}, \tag{2.81}$$

$$\rho_m = \frac{M_m}{v} = \frac{28.53 \text{ g mol}^{-1}}{22.4 \text{ l mol}^{-1}} = 1.27 \text{ kg m}^{-3}. \tag{2.82}$$

Note that in this example ρ_d was used under the assumption that the air is completely dry ($q_v = 0$ kg kg^{-1}). If air does not contain any condensate but does contain water vapor then it has the density of moist air given by ρ_m. For simplicity, from here on the density of moist air will just be denoted ρ.

Application of the ideal gas law to moist air involves a gas constant which depends on the absolute humidity in the air. In order to be able to use the ideal gas law for moist air with the gas constant R_d for dry air, a correction to the temperature is applied, leading to the definition of the virtual temperature T_v:

$$p = R_d \rho_m T_v. \tag{2.83}$$

Here T_v is the temperature that dry air would have if it had the same density as the moist air: if we compare eq. (2.4) with eq. (2.83), we see that the product $\rho_d T$ has to equal $\rho_m T_v$ at the same pressure. Because moist air has a lower density than dry air (see the example above), $\rho_d T = \rho_m T_v$ implies that the temperature of moist air must be higher. Thus T_v is higher than the normal temperature and increases with increasing water vapor mixing ratio.

In order to derive T_v, we start with the gas constant of moist air R_m, which can be obtained from the mass-weighted gas constants of dry air and water vapor:

$$R_m = \frac{m_d R_d + m_v R_v}{m} = (1 - q_v)R_d + q_v R_v, \tag{2.84}$$

where $m = m_d + m_v$ is the total mass, i.e. the mass of moist air. Applying the relationship $R_d = \epsilon R_v$ and rearranging terms yields

$$R_m = \left[1 + \left(\frac{1}{\epsilon} - 1\right) q_v\right] R_d = (1 + 0.608 q_v) R_d. \tag{2.85}$$

Instead of using R_m the term in parenthesis on the right-hand side is now combined with the temperature, leading to the definition of the virtual temperature T_v:

$$T_v = T(1 + 0.608 q_v). \tag{2.86}$$

2.5.5 Relative humidity

Knowing the saturation vapor pressure allows us to define the relative humidity RH, which is a measure of how much water vapor a quantity of air contains in comparison with its "capacity" for water vapor at that temperature. The relative humidity RH is normally expressed in % and is defined as follows:

$$\mathrm{RH} = \frac{e}{e_{s,w}} \times 100. \tag{2.87}$$

The relative humidity is defined only with respect to a reference phase. For example, air with a partial water vapor pressure of 125.5 Pa is saturated with respect to liquid supercooled water at −20 °C (i.e. RH = 100%) but supersaturated by 21.5% with respect to ice (i.e. $\mathrm{RH}_i = 121.5\%$). At temperatures below 0 °C, the relative humidity with respect to ice is defined analogously:

$$\mathrm{RH}_i = \frac{e}{e_{s,i}} \times 100. \tag{2.88}$$

The parameter RH_i is important in ice crystal formation (Chapter 8). For air that is saturated with respect to liquid water at temperatures below 0 °C, the ratio of $e_{s,w}$ and $e_{s,i}$ is the relative humidity with respect to ice. It can be obtained by combining eqs. (2.60) and (2.62):

$$\frac{e_{s,w}}{e_{s,i}} = \exp\left[\frac{L_f}{R_v T_0}\left(\frac{T_0}{T} - 1\right)\right]. \tag{2.89}$$

Here L_f is the latent heat of fusion, which is the difference between L_s and L_v. At 0 °C, L_f amounts to $L_{f0} = 0.33 \times 10^6$ J kg^{-1}. From $T < T_0$ it follows that the ratio $e_{s,w}/e_{s,i}$ always exceeds 1 and that $e_{s,w}/e_{s,i}$ steadily increases as T decreases, as seen in Figure 2.11b, where $e_{s,w}/e_{s,i}$ is plotted as a function of temperature. Any atmosphere saturated with respect to liquid water is therefore supersaturated with respect to ice. The maximum difference between $e_{s,w}$ and $e_{s,i}$ occurs at −12 °C (Figure 2.11b) and this has important implications for precipitation formation via the ice phase, as will be discussed in Section 8.3.

If a cloud consists of supercooled cloud droplets at temperatures below 0 °C then the air surrounding the droplets is in equilibrium with the droplets, i.e. its relative humidity with respect to liquid water is approximately 100%. It is thus supersaturated with respect to ice, as shown in Figure 2.11. As a rule of thumb, per 10 K supercooling the supersaturation with respect to ice increases by 12% when the relative humidity is at 100% with respect to liquid water. If some ice crystals form in this supercooled water cloud, they experience an environment that is supersaturated with respect to ice. This supersaturation causes a flux of water vapor onto the ice crystals leading to a rapid growth of these crystals. Over time, the growth of the ice crystals will decrease the vapor pressure below saturation with respect to liquid water, causing the cloud droplets to evaporate. This process is referred to as the Wegener–Bergeron–Findeisen process (Wegener, 1911). It is illustrated in Figure 2.12 and will be discussed in Section 8.3.

2.5.6 Dew point temperature

As the saturation water vapor pressure $e_{s,w}$ decreases exponentially with decreasing temperature, sufficient cooling of any air parcel with constant specific humidity q_v (or mixing ratio w_v) brings it to condensation, i.e. to saturation with respect to liquid water. The temperature at which saturation is reached if the air cools isobarically and q_v remains

Fig. 2.12 Photograph illustrating the Wegener–Bergeron–Findeisen process, taken from www.snowcrystals.com and used by permission and courtesy of Kenneth Libbrecht.

constant is called the dew point temperature, T_d. Mathematically T_d can be obtained from inverting the following equation:

$$q_v = q_s(p, T_d), \tag{2.90}$$

where q_s is the saturation specific humidity. This isobaric cooling of moist air is representative of fog formation. The dew point temperature can be measured with a dew point mirror (Figure 2.13). It is one of the oldest and most precise measures of the moisture content of air. Within the dew point mirror, the moist air travels over a small cooled mirror. A light beam is directed onto the mirror, and the intensity of the reflected light beam is measured. When condensation sets in, it is recognized by the changes in reflection from the mirror. Leaving aside direct measurements, T_d can be calculated iteratively or numerically from p and q_v (eq. 2.90). A good analytical approximation is obtained by inverting eq. (2.65) to obtain T_d:

$$T_d = T_d(q_v, p) = -\frac{B_w}{\ln(e_{s,w}/A_w)} = \frac{B_w}{\ln(A_w\epsilon/q_v p)}. \tag{2.91}$$

Here ϵ is given by eq. (2.79).

The analog of the dew point temperature for saturation with respect to ice is called the frost point temperature. Note that the dew point can be measured with a dew point mirror only at temperatures above 0 °C. When moist air cools down at the surface of a cooled mirror below 0 °C, saturation with respect to ice is reached before saturation with respect to liquid water, because $e_{s,i} < e_{s,w}$. Therefore deposition of ice on the mirror starts and this prevents the air from reaching saturation with respect to (supercooled) water. Thus the dew point mirror detects the frost point temperature at temperatures below 0 °C. The larger the difference between the ambient temperature and the dew point temperature, the drier the air mass. The dew point temperature is also a good indicator for human discomfort in warm humid weather. A dew point temperature above 20 °C is felt as uncomfortable, and above 24 °C as sticky or muggy.

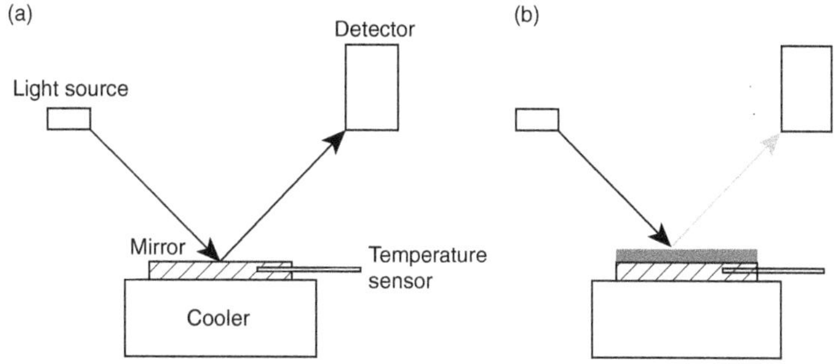

Fig. 2.13 Schematic of a dew point mirror (a) before and (b) after water condenses on the mirror.

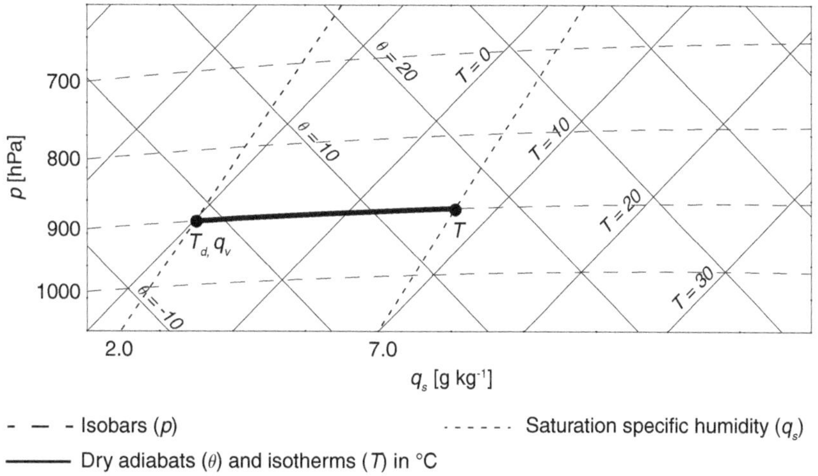

Fig. 2.14 Tephigram showing the dew point temperature T_d for an air parcel at 900 hPa (bold solid line) with temperature 7 °C and specific humidity $q_v = 2$ g kg^{-1}.

If the specific humidity q_v and the pressure of an air parcel are known, its dew point temperature can easily be determined graphically on a tephigram. The dotted lines in the tephigram (highlighted small-broken line in Figure 2.7) that rise more steeply from lower left to upper right than the isotherms are lines of constant saturation specific humidity q_s. They are not parallel to the isotherms because the saturation specific humidity depends not only on the saturation vapor pressure but also on atmospheric pressure (eq. 2.80).

Let us consider an air parcel that starts at $T = 7$ °C with a specific humidity $q_v = 2$ g kg^{-1} at 900 hPa (Figure 2.14). In order to cool the air parcel isobarically until saturation with respect to liquid water is reached, we follow the bold solid line to the left until we reach the temperature, for which q_v = 2 g kg^{-1} corresponds to the saturation specific humidity q_s. This is the dew point temperature T_d, which in this case is approximately −10 °C.

2.5.7 Wet-bulb temperature, wet-bulb potential temperature and lifting condensation level

Another method to obtain the relative humidity or the moisture content of the air is by measuring the wet-bulb temperature T_w; T_w is the temperature at which a parcel of air cooled isobarically reaches saturation. The specific humidity q_v increases owing to evaporation from a water reservoir (wet bulb) until it is in equilibrium with the wet bulb and the air in the environment. The latent heat of evaporation required for the evaporation of water is supplied by the air parcel and cools it. In this process internal energy is converted to latent heat, whereas in an adiabatic ascent internal energy is converted to the work of expansion.

The wet-bulb temperature T_w is measured with a psychrometer, which consists of two thermometers one of which is surrounded by a wet cloth and is ventilated. A normal temperature measurement from the thermometer is necessary in order to obtain RH. The ventilation of the wet thermometer ensures that sufficient water evaporates from the wet bulb to establish an equilibrium between the water vapor surrounding the thermometer and the water in the wet cloth. As evaporation consumes energy, the air in the vicinity of the wet bulb cools until it is saturated. At this time, the temperature of the wet bulb reaches a steady value, the wet-bulb temperature T_w. Psychrometers are used in weather stations for humidity measurements, as follows.

The decrease in temperature at the wet bulb is a function of the ratio of the latent heat flux away from the bulb and the sensible heat flux towards the bulb. Whereas the latent heat flux depends on the difference of the saturation specific humidity at the wet bulb and the specific humidity in the vicinity of the wet bulb, the sensible heat depends on the temperature difference between the wet bulb and the ambient air. The increase in both the sensible and latent heat fluxes with increasing ventilation speed is the same. Thus the ratio stays constant and consequently the wet-bulb temperature is independent of the ventilation speed. Therefore T_w can also be obtained from a tephigram (Figure 2.15).

On the tephigram we obtain T_w by first determining the lifting condensation level (LCL) and then following the saturated adiabat back down to the starting pressure level (long-broken line in Figure 2.15). The LCL is the level to which an air parcel needs to be cooled dry adiabatically with constant specific humidity for the relative humidity to reach 100% and for condensation to occur. In other words, the air parcel is cooled dry adiabatically until the saturation specific humidity q_s approaches the specific humidity q_v of the air parcel. Hence the LCL is purely determined by the parcel properties at the point where the parcel starts rising. To determine it on a tephigram, we follow the dry adiabat that goes through the starting temperature of our air parcel (which equals the ambient temperature) $T = 5$ °C ($\theta = 13$ °C) upwards until it intersects with the line of constant specific humidity q_v of 2 g kg^{-1}. Note that while the short-broken lines are lines of constant saturation specific humidity q_s, for an ascending air parcel it is q_v that remains constant and is a conserved quantity and not q_s, which is an intrinsic property of water vapor depending only on temperature. The LCL marks the cloud base. In our example in Figure 2.15, the LCL is reached at approximately 710 hPa.

The wet-bulb temperature T_w is approximately 0 °C with a specific humidity q_v of 4 g kg^{-1}. It lies in between the values for T_d and the ambient temperature T at that pressure. This relationship is generally valid. Assume that the subsaturated air approaching the wet bulb has specific humidity q_v and that the air leaving the wet bulb has a specific humidity q'_v. If the air that approaches the wet bulb is not saturated then $q'_v > q_v$ because of evaporation. In this case the wet-bulb temperature will be higher than the dew point temperature ($T_w > T_d$). However, T_w is lower than the ambient temperature ($T_w < T$) because water had to be evaporated, yielding the following relationship between the dew point, wet-bulb and ambient temperatures:

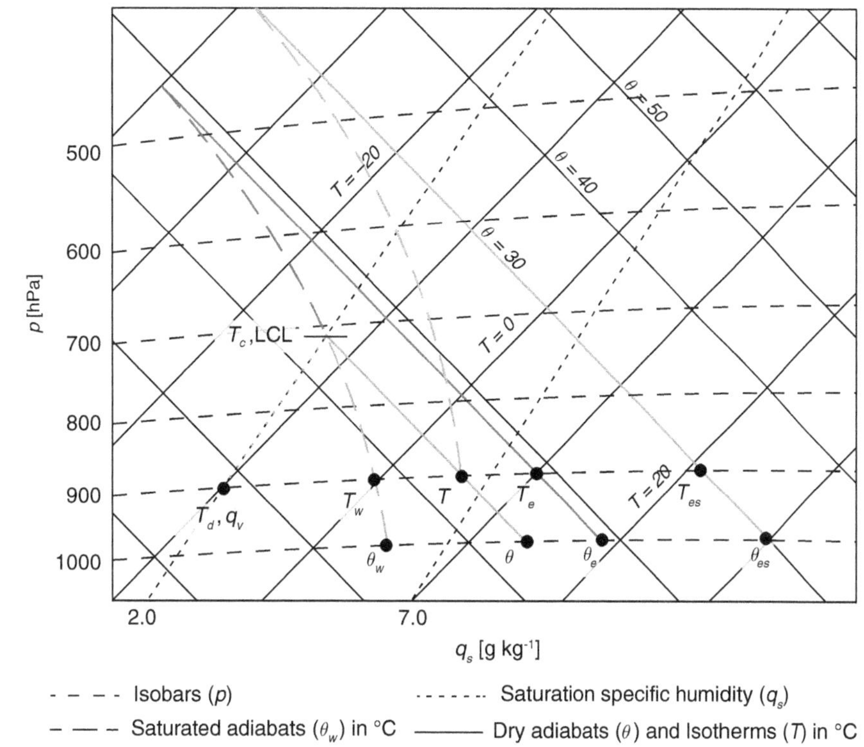

Fig. 2.15 This tephigram depicts the following temperatures for an air parcel at 900 hPa with $T = 5\,°C$ and $q_v = 2\,g\,kg^{-1}$, and the pathways by which they can be obtained: T_d is the dew point temperature, T_w is the wet-bulb temperature, T_c is the isentropic condensation temperature, i.e. the temperature at the lifting condensation level (LCL), T_e is the equivalent temperature and T_{es} the saturated equivalent temperature (Section 2.5.8), θ the potential temperature, θ_w the wet-bulb potential temperature, θ_e the equivalent potential temperature and θ_{es} the saturated equivalent potential temperature.

$$T_d \leq T_w \leq T, \tag{2.92}$$

where the equal sign applies only under saturated conditions.

In addition, a wet-bulb potential temperature θ_w exists, which can be obtained on a tephigram by following the saturated adiabat from T_w to 1000 hPa. In fact the saturated adiabats in tephigrams are lines of constant θ_w. In the example in Figure 2.15, θ_w is approximately 4 °C; θ_w is the conserved quantity when layers of moist air are lifted (Chapter 3).

2.5.8 Isentropic condensation temperature

The most common way in which a cloud is formed in the atmosphere is by the cooling of an air parcel due to lifting and the associated adiabatic expansion. The temperature at which the adiabatic cooling of an air parcel leads to saturation is called the isentropic condensation temperature T_c. Thus T_c is the temperature at the LCL. Because no mass is exchanged between the air parcel and the environment in an adiabatic process, the specific humidity of the air parcel remains constant. According to eq. (2.31) T_c is given as follows:

$$T_c = \theta \left(\frac{p_c}{p_0} \right)^{\kappa}, \tag{2.93}$$

where p_c is the pressure at which saturation is reached, i.e. at the LCL and p_0 is the reference pressure. In the example in Figure 2.15 with $p_c = 710$ hPa, T_c is -13°C. Above this point water vapor will condense and latent heat will be released. Any further ascent of the air parcel will then follow a saturated adiabat (the long-broken line in Figure 2.7).

2.5.9 Equivalent potential temperature and equivalent temperature

Unlike in a dry adiabatic process, θ is no longer conserved if phase transitions occur. In the latter case the ascent of an air parcel follows a saturated adiabat θ_w instead of θ (Figure 2.7). Because of the latent heat release, the decrease in temperature with decreasing pressure is less along saturated adiabats than along dry adiabats (Chapter 3).

In the presence of phase transitions the *equivalent potential temperature* θ_e defines a second conserved quantity in addition to θ_w; θ_e is the temperature that an air parcel would attain if it was brought to 1000 hPa with all water vapor contained in the parcel condensing. Owing to the latent heat release, however, θ_e is higher than θ. Unlike the dry adiabatic process, in which θ can be found by bringing the air parcel to 1000 hPa, there is no atmospheric process in which the air parcel reaches 1000 hPa while condensing out all the water vapor. Thus, θ_e is a hypothetical moisture quantity. The value of θ_e can be obtained with the help of a thermodynamic chart: (i) first lift the air parcel dry adiabatically to the LCL, (ii) further lift it along the saturated adiabat until the point where all water vapor is condensed out, i.e. where θ_w can be approximated by θ, and (iii) bring the parcel back down to 1000 hPa along the corresponding dry adiabat. In Figure 2.15, we lift the air parcel dry adiabatically from $T = 5$ °C and $q_v = 2$ g kg^{-1} at 900 hPa to the LCL at 710 hPa. Thereafter we follow the air parcel along the long-broken line up to approximately 450 hPa where the saturated adiabat meets the dry adiabat and then follow it along the dry adiabat (solid line) down to 1000 hPa. The value of θ_e is approximately 19 °C. Whereas θ_e remains constant in this process, θ increases from 14 °C to 19 °C.

To derive θ_e mathematically, we start from eq. (2.32), $ds = dq/T$. In a process involving condensation, the expression (2.19) for dq has to be modified to

$$dq = c_p dT + L_v dq_s - \alpha\, dp, \tag{2.94}$$

where q_s is the saturation specific humidity. Equation (2.94) now accounts for the latent heat release. Since in an adiabatic process $dq = 0$, we have

$$-L_v dq_s = c_p dT - \alpha\, dp. \tag{2.95}$$

It was shown in Section 2.2.6 that $c_p dT - \alpha dp = c_p d\theta/\theta$, so that

$$-\frac{L_v dq_s}{T} = \frac{c_p d\theta}{\theta}. \tag{2.96}$$

Note that some authors prefer to express θ_e in terms of the saturation water vapor mixing ratio, because w_s is the mass of water vapor at saturation per unit mass of dry air. Thus the denominator (unit mass of dry air) remains constant during phase changes, whereas the unit mass of moist air that is used in the denominator of q_s changes during phase transitions involving the vapor phase. Because q_s and w_s are approximately the same, we stick to q_s to be consistent with the moisture quantity on the tephigram. Following North and Erukhimova (2009), the left-hand side can be approximated using:

$$\frac{dq_s}{T} \approx d\left(\frac{q_s}{T}\right), \tag{2.97}$$

so that we can integrate eq. (2.96) to obtain

$$-\frac{L_v q_s}{T} = c_p \ln \theta + c. \tag{2.98}$$

If we identify the constant with $-c_p \ln \theta_e$, take the exponential of this equation and re-arrange the terms, we obtain

$$\theta_e = \theta \exp\left(\frac{L_v q_s}{c_p T}\right) \tag{2.99}$$

If the air parcel is initially not saturated then T and q_s have to be taken at the LCL. Since q_s at LCL is the same as q_v at the starting level of the air parcel, θ_e can be calculated from

$$\theta_e = \theta \exp\left(\frac{L_v q_v}{c_p T_c}\right), \tag{2.100}$$

where T_c is the isentropic condensation temperature (eq. 2.93).

Analogously, we can define an *equivalent temperature* T_e:

$$T_e = T \exp\left(\frac{L_v q_v}{c_p T_c}\right). \tag{2.101}$$

In Figure 2.15, the value of T_e at 900 hPa is approximately 10 °C.

2.5.10 Saturation equivalent potential temperature and saturation equivalent temperature

Another characteristic of moist air that is useful when analyzing the stability of the atmosphere (Chapter 3), is the *saturation equivalent potential temperature* θ_{es}, from which the *saturation equivalent temperature* T_{es} can be obtained. They are defined as

$$\theta_{es} = \theta \exp\left(\frac{L_v q_s}{c_p T}\right), \tag{2.102}$$

$$T_{es} = T \exp\left(\frac{L_v q_s}{c_p T}\right). \tag{2.103}$$

The temperatures θ_{es} and T_{es} are obtained by following the saturated adiabat that goes through the starting temperature T at 900 hPa (long-broken line in Figure 2.15) upward until all moisture is condensed and precipitated out and then following the dry adiabat (solid line) downward to the starting level to 1000 hPa. In Figure 2.15 the specific humidity at the starting temperature is close to 5 g kg^{-1}. This yields a T_{es} value of about 21 °C and a θ_{es} value of 30 °C. Note that θ_{es} is not conserved during a partly unsaturated process; during a fully saturated process, in which the air parcel is already saturated at the starting point (T_s, p), it is conserved and is equal to θ_e.

Fig. 2.16 Photograph of a Foehn wall cloud, by Laurent Barbe.

2.5.11 Wet adiabatic processes

In a wet adiabatic process θ_e is conserved instead of θ. We distinguish between reversible and irreversible wet adiabatic processes. In the former the assumption is made that all condensed liquid water remains in the parcel. The process is reversible because, upon reversal of its direction, it will descend along the same line of constant θ_w, while the formerly released latent heat is consumed to evaporate the condensed liquid. In an irreversible wet adiabatic or pseudoadiabatic process the assumption is made that all condensed liquid immediately leaves the parcel as precipitation. In this case θ_w is not conserved.

In order to calculate the change in temperature in a wet adiabatic process, we start from the first law of thermodynamics:

$$dq = c_p dT - \alpha dp + L_v dq_s, \tag{2.104}$$

which corresponds to eq. (2.19) but takes into account the latent heat released or consumed by an amount of water vapor dq_s that needs to condense or evaporate, respectively, due to the changing saturation mixing ratio. Since the process is adiabatic, $dq = 0$, and we obtain

$$dT = \frac{1}{\rho c_p} dp - \frac{L_v}{c_p} dq_s. \tag{2.105}$$

We will revisit this equation in Chapter 3 when we calculate the saturated adiabatic lapse rate, i.e. the rate at which an air parcel cools with increasing altitude in a wet adiabatic ascent. The cloud water content formed during this process is called the adiabatic liquid water mixing ratio, w_l, where

$$dw_l = -dq_s. \tag{2.106}$$

To be precise the increase in w_l is given by the decrease in w_s rather than in q_s but as mentioned in Section 2.5.9 we neglect this difference in order to use the same moisture quantity as that used in a tephigram. The adiabatic liquid water mixing ratio w_l is the maximum cloud water that can be contained in the air parcel in a wet adiabatic ascent; it can only be contained in the air parcel in the absence of precipitation formation and the absence of dilution by dry environmental air, called entrainment (Chapter 4).

In atmospheric sciences irreversible wet adiabatic or pseudoadiabatic processes are more common than reversible wet adiabatic processes. Processes of this kind can be responsible for warm fall winds, such as the Foehn wind in the Alps, the Chinook wind at the eastern slopes of the Rocky Mountains and the Santa Ana desert wind in California. After the descent on the downslope (lee) side the air is warmer than it was before the ascent on the upslope (windward or luv) side because the air has been cooled moist adiabatically above the LCL at a rate of less than 1 K per 100 m on the upslope side, but is warmed by 1 K per 100 m dry adiabatically on the downslope side. Nowadays, however, it is recognized that this thermodynamic "Foehn theory" cannot explain all Foehn events. In addition, dynamical aspects such as the cold air in the luv that needs to be eroded have to be taken into account (Drobinski *et al.*, 2007).

A spectacular visible example of Foehn winds is a so-called Foehn wall, as shown in Figure 2.16. Here the clouds are restricted to the luv of the mountain. If the clouds precipitate then the air that is blown over the mountain crest contains less moisture. Consequently, when this air sinks and is warmed adiabatically, the remaining cloud droplets are evaporated at a higher altitude (the LCL is higher in the lee, after the cloud has precipitated than in the luv). This process can lead to a distinct boundary between the clouds and the cloud-free air, in which case the clouds form a Foehn wall.

2.6 Exercises

Single/multiple choice exercises

1. Imagine an air parcel which undergoes isothermal expansion from V to $V+dV$. Which of the following is/are true?
   (a) The gas does work as it expands.
   (b) The work done on the system is given by $dw = -PdV$.
   (c) The work done on the system is given by $dw = PdV$.
   (d) The internal energy of the system decreases, as a result of the work it has done on the surrounding.
2. Non-adiabatic processes are not important in the atmosphere.
   (a) True because changes involving latent heat release can be neglected.
   (b) False because radiative heating is the driver for cloud formation.
   (c) False because e.g. radiative heating and cooling, turbulence and dissipation can be large.
3. Given the universal gas constant $R^* = 8.314$ J mol^{-1} K^{-1} and the gas constant of dry air $R_d = 287$ J kg^{-1} K^{-1}, what is the molecular weight of dry air?
   (a) 28.96 g mol^{-1}
   (b) 38.96 g mol^{-1}
   (c) 38.96 kg mol^{-1}

4. What is the approximate lifetime of water vapor in the atmosphere?
   (a) 1 day
   (b) 1 week
   (c) 1 month
   (d) 1 year
5. In the following lists of variables which list(s) only include extensive variables?
   (a) T, p, μ
   (b) V, U, T
   (c) V, U, n
   (d) V, n, S
6. To determine whether a system is in thermodynamic equilibrium, we can use the Gibbs free energy. Which of the following is/are true?
   (a) In thermodynamic equilibrium, all variables are homogeneously distributed.
   (b) Temperature can be inhomogeneously distributed in thermodynamic equilibrium.
   (c) A gradient in pressure can exist in thermodynamic equilibrium, e.g. when the system is under gravitational force and the pressure gradient counteracts these forces to maintain thermodynamic equilibrium.
7. The ideal gas law …
   (a) …is only valid for non-compressible gases when the molecules are defined as points that interact with each other only with inelastic impact.
   (b) …describes the relationship between the variables p, ρ and T.
   (c) …describes only the state of dry air.
8. The first law of thermodynamics …
   (a) …is based on the consumption of energy.
   (b) …balances the internal energy of a system and the work done on the system for isothermal processes.
   (c) …can be used to describe adiabatic processes in the atmosphere.
9. The entropy describes in how many specific ways a thermodynamic system can be arranged. Which of the following is/are true?
   (a) During adiabatic processes, entropy is conserved and no temperature changes will occur.
   (b) Entropy does not increase during a reversible thermodynamic process, as implied by the second law of thermodynamics.
   (c) Entropy increases during melting processes because heat is lost while the temperature remains constant.
   (d) An increase in entropy means that a system changes into a state with higher disorder.
10. Which of the following is/are true?
    (a) The Clausius–Clapeyron equation describes under which conditions (temperature, pressure) two phases can coexist in equilibrium.
    (b) If two phases are in equilibrium, both their chemical potentials and their specific Gibbs free energies are equal.
    (c) The triple point of water is equal to the freezing point in the atmosphere.

(d) At −20 °C, cloud droplets or ice crystals can exist, but only the ice crystals can be in thermodynamic equilibrium.

11. Attribute the following statements to the temperatures listed below in the correct order: (i) is helpful to predict fog formation; (ii) is helpful to predict whether raindrops are converted to snow due to evaporative cooling; (iii) is larger than or equal to the ambient temperature.
    (a) T_v, T_d, T_w
    (b) T_v, T_w, T_d
    (c) T_d, T_w, T_v
    (d) T_w, T_v, T_d
    (e) T_d, T_v, T_w

Long exercises

1. Consider a gas mixture that comprises 78% N_2, 2% H_2O and 20% O_2. The total pressure of the mixture is 1060 hPa. Further assume a constant latent heat of vaporization $L_v = 2.5 \times 10^6$ J kg^{-1}.
    (a) Calculate the dew point temperature of the gas mixture.
    (b) What volume percentage does the water vapor occupy after the gas mixture is cooled isobarically to 10 °C?
    (c) What happens to the excess water vapor?
2. In Figure 2.17, a tephigram with a profile of the environmental temperature T_{env} and the dew point temperature T_d is shown.
    (a) Assume that an air parcel at 1000 hPa is in equilibrium with the environment. If the parcel is lifted dry adiabatically, at which pressure level will it reach its LCL?
    (b) Determine the wet-bulb temperature T_w of the parcel.
3. An air parcel at temperature T_0 = 20 °C and with relative humidity 40% starts rising at sea level. Use a tephigram to determine the temperature and the pressure at which condensation sets in. A tephigram can be downloaded from here: www.cambridge.org/clouds.
4. The saturation vapor pressures with respect to ice and water differ for temperatures below 273.16 K. Use a suitable programming language to plot $e_{s,w}$ and $e_{s,i}$ as a function of T and to obtain the temperature at which the difference between the vapor pressures is largest.
5. A psychrometer gives the following values: a temperature $T = 2$ °C and a wet-bulb temperature $T_w = -1$ °C. The measured pressure is p = 1000 hPa.
    (a) Use a tephigram to obtain the relative humidity. Assume that $p - e \approx p$. A tephigram can be downloaded from here: www.cambridge.org/clouds.
    (b) In this situation, during a clear night with light breeze, some pools on the road freeze despite the positive ambient temperature. Why?
6. At the bottom of a valley at sunset, a meteorological station measures the following data: T = 12 °C, p = 1000 hPa, q_v = 6 g kg^{-1}.
    (a) Using the expression for e_s calculate the relative humidity.

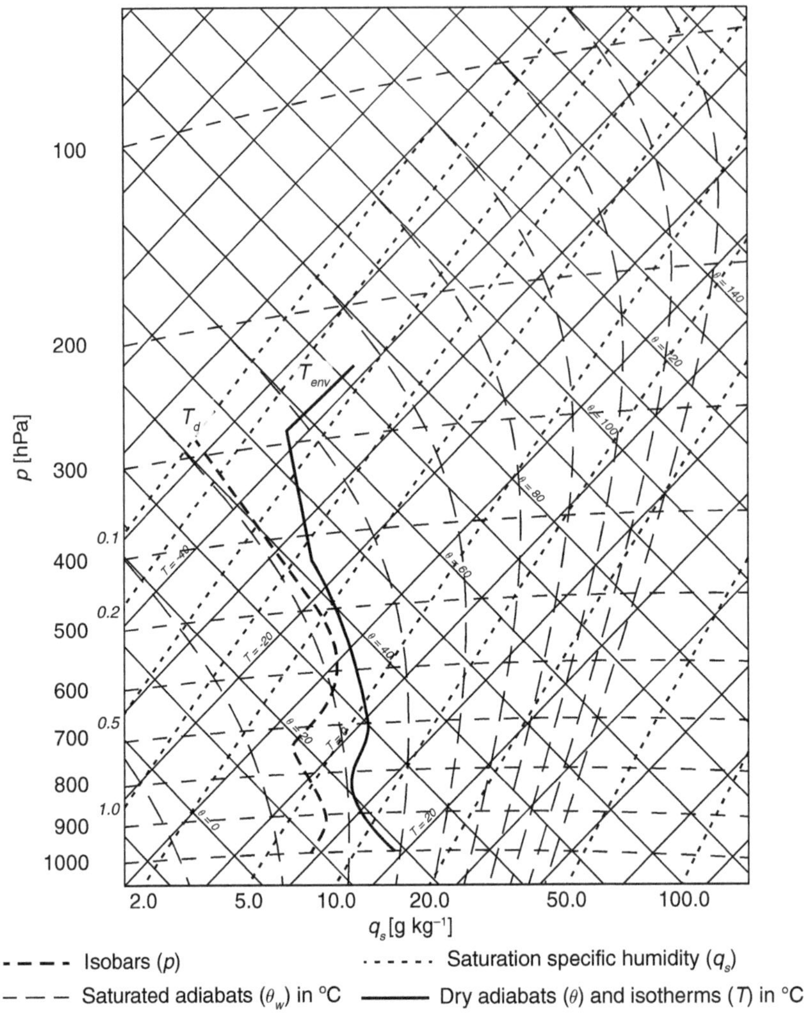

Fig. 2.17 A tephigram containing an environmental temperature profile and a profile for the dew point temperature T_d. The smaller numbers along the vertical axis refer to the q_s lines.

(b) Calculate the dew point temperature T_d using the analytical approximation given in eq. (2.90).

(c) During the night, the soil starts cooling and the temperature reduces to 5 °C. The atmospheric pressure remains constant. Should fog be expected?

7. Rain falling from a cloud almost always encounters an unsaturated layer between the cloud base and the ground. Therefore, a fraction of the liquid water in the raindrops evaporates.

(a) How can the vertical profile of the wet-bulb temperature T_w be used to forecast the precipitation type (rain, snow, freezing rain) at the surface? Sometimes it is observed that precipitation starting as rain can turn into wet snow during intense showers. Why is this the case?

(b) Below a cloud there is a subsaturated layer with T = 2 °C and RH = 70% and heavy precipitation is forecast. Supposing that the precipitation reaches the ground, should the local meteorological office emit an alert for snow? (*Hint: Compute the dew point temperature at* 1000 hPa *and, using a tephigram, determine the wet-bulb temperature.*) A tephigram can downloaded from here: www.cambridge.org/clouds.

8. In tropical marine conditions the air can become very hot and moist and conditions can be very uncomfortable for humans. Regions which suffer considerably from this problem are the coastal areas of the Red Sea and Persian Sea. The highest dew point recorded in the region was T_d = 35 °C at Dharhan, Saudi Arabia, at 3 pm on July 8, 2003. The temperature at that time was T = 42.2 °C. (These data are taken from the "Weather Climate Extremes" Army Corps of Engineers TEC-0099 report.)
   (a) How large is the relative humidity under these extreme conditions?
   (b) Compute T_v, assuming a surface pressure of 990 hPa. Does it differ greatly from T? Explain why or why not.
   (c) The human body retains its internal temperature by the evaporation of sweat. It carries away the energy produced by metabolic process, which typically has a value around F = 100 W for a resting body. The internal temperature of humans is on average T_{int} = 37 °C and does not adapt to climate, but the skin has to be slightly cooler (T_{skin} = 35.5 °C) to allow efficient conduction of the metabolic heat towards the exterior. Given below are simplified equations for the sensible and the total heat fluxes, SH and F, respectively:

$$\mathrm{SH} = \alpha(T_{skin} - T),$$
$$F = \mathrm{SH} + \mathrm{LH} = \alpha(T_{skin} - T_w).$$

   Calculate how much water has to evaporate (per hour) from a human body to keep the body temperature constant, assuming that T = 40 °C and T_w = 35 °C. Is your result realistic?
   (d) Under the extreme conditions detailed above, the wet-bulb temperature was estimated to be close to 38 °C. What would happen according to the simplified model in part (c)?

9. Consider a volume of air, having a mass of 350 kg, which has a uniform temperature of −13 °C, a relative humidity with respect to water of 100% and a pressure of 700 hPa. Initially only cloud droplets are present in the cloud. Suddenly ice crystals are formed. What mass of water vapor could theoretically be deposited onto these ice crystals via the Wegener–Bergeron–Findeisen process?

3 Atmospheric dynamics

As mentioned in Chapter 1, convective clouds such as cumulus and cumulonimbus develop in unstable air due to the buoyancy force. Buoyancy is an upward force experienced by objects that are less dense than their environment. In the context of cloud formation it mainly refers to air parcels (Chapter 2) that are warmer than the ambient air. While buoyancy is by definition positive, in the atmospheric sciences negative buoyancy, in the downward direction, experienced by objects that are denser than their environment is equally important. To better emphasize this difference, the buoyancy in the upward direction is sometimes called positive buoyancy.

In this chapter we start by describing in Section 3.1 the basic equations of the reference state, including the hydrostatic and hypsometric equations as well as the Navier–Stokes equation. We then introduce the concept of buoyancy. We focus on the vertical direction because in the atmosphere that is the direction in which the main component of the buoyancy force acts. Thereafter we explain the meaning of stability and instability in the atmosphere, starting with dry air in Section 3.2 before considering condensation in Section 3.3. Not only can rising air parcels experience the stability of the atmosphere; whole layers can be lifted, as discussed in Section 3.4. In the atmosphere horizontal restoring forces are also at play; these are subject of Section 3.5. We end this chapter by discussing the combination of vertical and horizontal restoring forces that gives rise to the slantwise displacement of air parcels (Section 3.6).

3.1 Basic equations and buoyancy force

In the atmospheric sciences we use forces per unit mass, which is consistent with the use of specific quantities, i.e. quantities normalized to unit mass, as introduced in Chapter 2.

3.1.1 Navier–Stokes equation

The Navier–Stokes equation describes the equation of motion of a fluid. In its three-dimensional form, it is given per unit mass as

$$\frac{d\vec{v}}{dt} = -\frac{1}{\rho}\nabla p - f\vec{k} \times \vec{v} - g\vec{k} + \vec{F}_F, \tag{3.1}$$

where $\vec{i}, \vec{j}, \vec{k}$ denote the unit vectors in the x-, y- and z-directions. The x-direction is defined as from west to east, the y-direction from south to north and the z-axis points upwards. The three dimensional velocity vector is $\vec{v} = u\vec{i} + v\vec{j} + w\vec{k}$ and t is time. In eq. (3.1), d/dt denotes the total derivative:

$$\frac{d}{dt} = \frac{\partial}{\partial t} + \vec{v} \cdot \nabla,$$

where the first term denotes the local time derivative and the second term denotes the three-dimensional advection. The advection term expresses the fact that a variable of the fluid element, in this case the velocity, changes as a result of the displacement of the fluid element. The total derivative is the same as the Lagrangian time derivative. It represents the total change in a quantity as seen by an observer moving with the particular fluid element. The total derivative is also referred to as the inertial term or inertial force per unit mass.

The first term on the right-hand side of eq. (3.1) is the pressure gradient force $\vec{F}_{PG}$, which represents the force acting on a fluid element along the pressure gradient. The second term is the Coriolis force per unit mass, also referred to as the Coriolis term $\vec{F}_C$; $f = 2\,\Omega \sin\varphi$ is the Coriolis parameter, where φ is the latitude of the location at which the Coriolis force is calculated and $\Omega = 7.29 \times 10^{-5}\ \mathrm{s}^{-1}$ is the angular velocity of the Earth's rotation. The third term on the right-hand side of eq. (3.1) is the acceleration due to gravity, which acts vertically; $\vec{F}_F$ denotes the drag due to small-scale turbulent fluctuations, also referred to as friction in the case of the atmosphere.

3.1.2 Coriolis and centrifugal force

The Coriolis force $\vec{F}_C$ on an object results from the rotation of the Earth and per unit mass is given as

$$\vec{F}_C = -2\vec{\Omega} \times \vec{v}. \tag{3.2}$$

It is perpendicular both to the axis of rotation and to the relative velocity (i.e. the object's velocity without the contribution from the Earth's rotation). Because it is perpendicular to the relative velocity, the Coriolis force does no work and hence is an "apparent" force, seen only in a rotating frame. As the axis of the Earth's rotation goes from south to north, the three components of $\vec{\Omega}$ are given as:

$$\Omega_x = 0, \qquad \Omega_y = \Omega\cos\phi, \qquad \Omega_z = \Omega\sin\phi. \tag{3.3}$$

Thus the three components of the Coriolis force are:

$$\vec{F}_C = -2\begin{pmatrix} \Omega\cos\phi w - \Omega\sin\phi v \\ \Omega\sin\phi u \\ -\Omega\cos\phi u \end{pmatrix}. \tag{3.4}$$

Because the large-scale vertical air motion is smaller than the horizontal air motion, the contribution of the vertical air motion to the x-component of the Coriolis force can be neglected. The contribution of the Coriolis force to the z-component of the Navier–Stokes equation (eq. 3.1) can also be neglected because the acceleration due to gravity and the vertical pressure gradient force are orders of magnitude larger.

The two horizontal components of the Coriolis force are then given as

$$F_{C,x} = 2\Omega \sin\varphi v = fv,$$
$$F_{C,y} = -2\Omega \sin\varphi u = -fu. \tag{3.5}$$

The Coriolis force causes a moving air parcel in the northern hemisphere to be deflected to the right and in the southern hemisphere to the left.

In a rotating frame the centripetal force is a reaction to the fictitious centrifugal force generated by the rotation. In the case of a ball swung in a loop on a string, the centripetal force is provided by the tension in the string and balances the centrifugal force exactly, so that the ball stays in orbit and does not fly off. In the case of a satellite orbiting the Earth, the centripetal force is exerted by gravity and balances the radially outward directed pull generated by the centrifugal force.

In the Earth's rotating frame, the centrifugal force on an object is not balanced by a reactionary centripetal force, i.e. there is no string and gravity acts in a different direction (apart from at the equator). The centrifugal force is small compared with other forces exerted on an object in Earth's atmosphere (e.g. gravity or the Coriolis force) and therefore is often neglected and does not appear in the Navier–Stokes equation (eq. 3.1).

3.1.3 Hydrostatic equation

The z-component of the Navier–Stokes equation (eq. 3.1) can be simplified in order to derive the hydrostatic equation. Neglecting the Coriolis force, it reduces to

$$\frac{dw}{dt} = -\frac{1}{\rho}\frac{\partial p}{\partial z} - g + F_{F,z}. \tag{3.6}$$

To a first approximation the atmosphere is in a hydrostatic balance, which results from the balance of the vertical pressure gradient force and the acceleration due to gravity:

$$\frac{\partial p}{\partial z} = -g\rho. \tag{3.7}$$

Equation (3.7) is called the hydrostatic equation. The difference between eqs. (3.6) and (3.7) comprises the vertical acceleration and friction, which are negligible on the large scale but can be important locally.

3.1.4 Hypsometric equation

It is common in atmospheric science to express the air density in terms of p and the virtual temperature T_v (eq. 2.86) using the ideal gas law. Then we obtain

$$\frac{\partial p}{\partial z} = -g\frac{p}{R_d T_v}. \tag{3.8}$$

In order to replace the partial derivative with the total derivative we need to assume that T_v is not a function of p. If we restrict the integration of eq. (3.8) to a sufficiently small z-interval, we can replace T_v by its average value $\overline{T_v}$ over that z-interval. Therefore, rearranging terms we get

$$\frac{dp}{p} = -\frac{g}{R_d\overline{T_v}}\,dz. \tag{3.9}$$

Integration of this equation between z_0 and z on the right-hand side and the corresponding pressure interval between p_0 and p on the left-hand side yields the hypsometric equation,

$$p = p_0 \exp\left[-\frac{g}{R_d\overline{T_v}}(z - z_0)\right]. \tag{3.10}$$

Using the hypsometric equation, the pressure p at an altitude z can be calculated if the pressure p_0 at a starting altitude z_0 is known.

3.1.5 Buoyancy force

An air parcel will undergo vertical displacements from the ambient air if a buoyancy force F_B acts on it. The buoyancy force in the atmosphere is caused by the difference in density between an air parcel and the surrounding atmosphere. It results from a perturbation of the reference state of the environmental air, which we will assume to be in hydrostatic equilibrium. Furthermore, we will assume that the air parcel is always at the same pressure as the environmental air at any given altitude.

The starting point is the vertical component of the Navier–Stokes equation (eq. 3.1), but this time we use it in its perturbation form, which means that we are interested in deviations from the mean quantities. In the case of the vertical velocity, we denote the mean vertical velocity by $\overline{w}$ and the deviation from the mean, i.e. the perturbation, by $\tilde{w}$:

$$w = \overline{w} + \tilde{w}. \tag{3.11}$$

Neglecting friction, the vertical component (eq. 3.6) of the Navier–Stokes equation is

$$\frac{d}{dt}(\overline{w} + \tilde{w}) = -\frac{1}{\overline{\rho} + \tilde{\rho}}\frac{\partial \overline{p} + \tilde{p}}{\partial z} - g. \tag{3.12}$$

The large-scale environment, i.e. the mean state of the ambient air, is assumed to be in hydrostatic balance. Using the hydrostatic balance equation (3.7) and noting that $d\overline{w}/dt = 0$ yields

$$\begin{aligned}
\frac{d\tilde{w}}{dt} &= -\frac{1}{\overline{\rho} + \tilde{\rho}}\frac{\partial(\overline{p} + \tilde{p})}{\partial z} + \frac{1}{\overline{\rho}}\frac{\partial \overline{p}}{\partial z} \\
&= \frac{\tilde{\rho}}{\overline{\rho} + \tilde{\rho}}\frac{\partial \overline{p}}{\overline{\rho}\partial z} - \frac{1}{\overline{\rho} + \tilde{\rho}}\frac{\partial \tilde{p}}{\partial z} \\
&= -\frac{\tilde{\rho}}{\rho}g - \frac{1}{\rho}\frac{\partial \tilde{p}}{\partial z} \\
&= -\left(\frac{\rho - \overline{\rho}}{\rho}\right)g - \frac{1}{\rho}\frac{\partial \tilde{p}}{\partial z}.
\end{aligned} \tag{3.13}$$

We now associate the quantities of the ambient atmosphere (index "*env*") with the mean state (e.g. $\overline{\rho}$) and the parcel's properties with the actual state (e.g. ρ). In eq. (3.13), the first term on the right-hand side is the acceleration that arises due to buoyancy. This acceleration per unit mass is referred to as the buoyancy force F_B and is given as

$$\frac{d\tilde{w}}{dt} = \frac{d^2\tilde{z}}{dt^2} = F_B = g\left(\frac{\rho_{env} - \rho}{\rho}\right). \tag{3.14}$$

Differences in density between the air parcel ρ and the ambient air ρ_{env} give rise to buoyancy. These density fluctuations are due to perturbation in temperature, pressure, water vapor and condensed water. The buoyant force is positive, i.e. it is directed upwards, when the parcel is less dense than the ambient air and is negative when the parcel is denser than the ambient air. For cloud formation the temperature perturbations are dominant. Neglecting the other perturbations, eq. (3.14) reduces to

$$F_B = g\left(\frac{T - T_{env}}{T_{env}}\right), \tag{3.15}$$

where T_{env} is the ambient temperature. This simplified buoyancy equation will be used in this chapter. Its limitations will be discussed in Section 4.2.4. In order to use this formula for moist air, T needs to be replaced by the virtual temperature T_v because moist air has a lower density than dry air.

3.2 Stability in dry air

The question of stability or instability refers to the reaction of a system to perturbations. A system is in a situation of stable equilibrium if after a perturbation (such as a displacement, an impulse, etc.), a restoring force brings the system back to the initial state. The onset of a perturbation in a state of unstable equilibrium, however, leads to further deviation from the initial conditions. A ball on a sharp summit of a mountain is an example of the latter state, as a small displacement will cause it to fall down the mountain.

In the atmosphere, vertical stability reduces to the question whether an air parcel is colder or warmer than the surrounding air. According to eq. (3.15), this leads to a respectively negative or positive buoyancy force and the air parcel respectively sinks or rises.

In subsaturated air the atmosphere is referred to as "dry". In a dry atmosphere, the temperature of the parcel at the perturbed position is determined by eq. (2.31). The vertical stability therefore depends on the vertical temperature distribution in the atmosphere. The change in the environmental temperature with height is called the ambient lapse rate γ.

3.2.1 Dry adiabatic lapse rate

For the dry adiabatic ascent of a parcel we can define a dry adiabatic lapse rate Γ_d, which describes the rate at which the temperature of the parcel decreases with increasing altitude;

Γ denotes a process lapse rate that describes the motion of an air parcel, while the ambient lapse rate γ represents the change of the ambient temperature with height.

The dry adiabatic lapse rate Γ_d can be derived from the first law of thermodynamics (eq. 2.19), assuming that the air parcel undergoes adiabatic transformations, i.e. $dq = 0$, and that the atmosphere is in hydrostatic equilibrium:

$$dq = 0 = -\alpha dp + c_p dT. \tag{3.16}$$

Rearranging terms and making use of the hydrostatic equation (3.7) yields

$$c_p dT = \alpha dp = \frac{1}{\rho} dp = -g dz. \tag{3.17}$$

Dividing by $c_p dz$ then gives Γ_d:

$$\Gamma_d \equiv -\left(\frac{dT}{dz}\right)_{dry\ parcel} = \frac{g}{c_p} \sim 9.8\,\mathrm{K\,km^{-1}}. \tag{3.18}$$

In the atmosphere, an ambient lapse rate as large as the dry adiabatic lapse rate is only found in areas where the moisture content is very small. Examples of this are the boundary layer in deserts or the upper troposphere. The condensation of water vapor reduces the ambient lapse rate owing to the release of latent heat. This will be discussed in Section 3.3.

3.2.2 Lapse rate and stability

The ambient lapse rate can be related to the vertical gradient of the potential temperature θ of the ambient air. In order to see that, we start from the partial derivative of θ with altitude using the definition of θ (Section 2.2.5 and eq. 2.37):

$$\frac{1}{\theta}\frac{\partial \theta}{\partial z} = \frac{1}{\theta}\frac{\partial}{\partial z}\left[T\left(\frac{p_0}{p}\right)^{\kappa}\right] = \frac{1}{T}\frac{\partial T}{\partial z} - \frac{\kappa}{p}\frac{\partial p}{\partial z}. \tag{3.19}$$

Now replace $\partial p/\partial z$ by the hydrostatic equation and insert the definitions of the ambient lapse rate and κ to obtain

$$\frac{1}{\theta}\frac{\partial \theta}{\partial z} = -\frac{\gamma}{T} + \frac{R_d}{c_p\, p} g\rho. \tag{3.20}$$

The last term can be simplified by making use of the definition of the parcel's lapse rate and the ideal gas law. This yields the desired relationship between the vertical gradient of the potential temperature and atmospheric stability:

$$\frac{1}{\theta}\frac{\partial \theta}{\partial z} = -\frac{\gamma}{T} + \frac{\Gamma_d}{T} = \frac{1}{T}(\Gamma_d - \gamma). \tag{3.21}$$

Equation. (3.21) states that an increase in the potential temperature with altitude ($\partial\theta/\partial z > 0$) corresponds to an ambient lapse rate γ that is smaller than the dry adiabatic lapse rate Γ_d of the air parcel. This is a situation of stable equilibrium, where buoyancy F_B is the restoring force. Under these circumstances a small displacement in the positive z-direction cools the air parcel more than the surrounding environment, leading to a

Table 3.1 Static stability criteria in non-condensing air.

$\gamma < \Gamma_d \leftrightarrow \partial\theta/\partial z > 0$	Stable atmosphere
$\gamma = \Gamma_d \leftrightarrow \partial\theta/\partial z = 0$	Neutral atmosphere
$\gamma > \Gamma_d \leftrightarrow \partial\theta/\partial z < 0$	Unstable atmosphere

negative F_B that pushes the air parcel down to its original position. The same effect, causing a positive F_B, is produced by displacements in the negative z-direction.

In a statically unstable atmosphere, lifted air parcels continue to rise. Here the parcel's lapse rate Γ_d is smaller than that of the ambient air, γ, and $\partial\theta/\partial z < 0$. An unstable atmosphere is extremely short-lived because first turbulence, and if this is not sufficient also convection, (Section 4.2) mix the air vertically until it is neutral again.

If the two lapse rates are equal, i.e. the potential temperature does not change with altitude ($\partial\theta/\partial z = 0$), the air parcel remains in hydrostatic equilibrium after a small displacement, and no force acts upon it. This situation is called a statically neutral stratification. The stability criteria are summarized in Table 3.1.

3.2.3 Brunt–Väisälä frequency

The air parcel's equation of motion can be derived if F_B is related to the static stability, i.e. to the change in θ with height. Assume that the parcel is subject to a small displacement around the starting point z_0, so that we can use a Taylor series expansion to obtain a linear term for the change in the parcel's temperature T with altitude z,

$$T(z) \approx T(z_0) + \left.\frac{dT}{dz}\right|_{z=z_0} (z - z_0), \tag{3.22}$$

and likewise for the ambient temperature,

$$T_{env}(z) \approx T_{env}(z_0) + \left.\frac{dT_{env}}{dz}\right|_{z=z_0} (z - z_0). \tag{3.23}$$

Assume that the temperature of the air parcel and the ambient environment are the same in equilibrium at z_0 ($T(z_0) = T_{env}(z_0)$), so that

$$T(z) - T_{env}(z) \approx \left(\left.\frac{dT}{dz}\right|_{z=z_0} - \left.\frac{dT_{env}}{dz}\right|_{z=z_0} \right) (z - z_0). \tag{3.24}$$

Inserting this into the numerator of eq. (3.15) yields:

$$F_B = \frac{d^2\tilde{z}}{dt^2} \approx \frac{g}{T_{env}} \left(\left.\frac{dT}{dz}\right|_{z=z_0} - \left.\frac{dT_{env}}{dz}\right|_{z=z_0} \right) (z - z_0). \tag{3.25}$$

We further approximate T_{env} by T, in the denominator, insert the expressions for the lapse rates and note that $z - z_0 = \tilde{z}$:

$$\frac{d^2\tilde{z}}{dt^2} = \frac{g}{T}(-\Gamma_d + \gamma)\tilde{z}. \tag{3.26}$$

Last, we replace the difference in lapse rates (eq. 3.21) by the vertical gradient in θ:

$$\frac{d^2\tilde{z}}{dt^2} = -\frac{g}{\theta}\left(\frac{\partial\theta}{\partial z}\right)\tilde{z}, \tag{3.27}$$

and introduce the Brunt–Väisälä frequency N in s^{-1}, which is given as

$$N = \sqrt{\frac{g}{\theta}\frac{\partial\theta}{\partial z}}. \tag{3.28}$$

Then eq. (3.27) can be associated with the equation of motion of a harmonic oscillator (see for instance Seaborn, 2002) with frequency N in a stable atmosphere:

$$\frac{d^2\tilde{z}}{dt^2} = -N^2\tilde{z}. \tag{3.29}$$

The solution is of the form $\tilde{z}(t) = A\exp(\pm iNt)$. If $N^2 > 0$, i.e. in a statically stable atmosphere, the air parcel performs an oscillatory motion, with the buoyancy force acting as a restoring force. If $N^2 = 0$, the atmosphere is in neutral equilibrium with no forces acting on the parcel in a perturbed position. For $N^2 < 0$, i.e. in unstable conditions, there is no oscillatory motion, but $\tilde{z}(t)$ increases exponentially with time until the air parcel reaches stable conditions again; in this case the association of N^2 with a frequency is misleading.

A typical value of N in a statically stable atmosphere is 0.012 s^{-1}, so that the period of oscillation is $\tau = 2\pi/N = 9$ min. The oscillatory motion of the air in a statically stable atmosphere can be observed if altocumulus lenticularis clouds form behind a mountain ridge, as shown in Figure 3.1. In this case the oscillatory motion is triggered as the air

Fig. 3.1 Altocumulus (Ac) lenticularis cloud. Photograph taken by Jeremy Michael. A black and white version of this figure will appear in some formats. For the color version, please refer to the plate section.

mass crosses a mountain ridge and is moist enough to reach the LCL. Note that in the case where the oscillatory motion reaches the LCL its equation of motion is no longer governed by $\Gamma_d - \gamma$ or $(1/\theta)\partial\theta/\partial z$ but by $\Gamma_s - \gamma$ or $(1/\theta_e)\partial\theta_e/\partial z$ (see Section 3.3). This means that no single Brunt–Väisälä frequency can be associated with this motion: it differs between the clear sky and the cloud.

3.3 Stability in condensing air

We have seen that, in the absence of condensation, the atmosphere is stable if the ambient lapse rate is smaller than the process lapse rate for dry adiabatic ascent. We can formulate similar conditions for the stability of the atmosphere with respect to the ascent of a saturated air parcel, i.e. a parcel that is above its LCL. For this we need to compare the ambient lapse rate with the process lapse rate Γ_s of the saturated adiabatic ascent, i.e. the rate of temperature change with height of an air parcel undergoing a wet adiabatic process (Section 2.5). We start from the first law of thermodynamics, accounting for latent heat release (eq. 2.94), and assuming hydrostatic balance:

$$dq = c_p dT - \alpha dp + L_v\, dq_s = c_p dT + g dz + L_v\, dq_s. \tag{3.30}$$

Now divide both sides by $c_p dz$ and rearrange terms:

$$\frac{dT}{dz} = -\frac{L_v}{c_p}\frac{dq_s}{dz} - \frac{g}{c_p}. \tag{3.31}$$

Replacing g/c_p with Γ_d and write the change in q_s as a function of temperature instead of altitude:

$$\frac{dT}{dz} = -\frac{L_v}{c_p}\frac{dq_s}{dT}\frac{dT}{dz} - \Gamma_d. \tag{3.32}$$

Rearranging terms then yields the saturated adiabatic lapse rate Γ_s:

$$\Gamma_s \equiv -\left(\frac{dT}{dz}\right)_{saturated\ parcel} = \Gamma_d\left(1 + \frac{L_v}{c_p}\frac{dq_s}{dT}\right)^{-1}. \tag{3.33}$$

In analogy to the non-condensing case, the atmosphere will counteract the ascent of a saturated air parcel if $\gamma < \Gamma_s$ and promote it if $\gamma > \Gamma_s$. Note that, in contrast with Γ_d, Γ_s is not constant but depends on T and on p because q_s depends on p (eq. 2.80). As dq_s/dT is always positive, because it is proportional to $de_{w,s}/dT$, which in turn is always positive according to eq. (2.59), Γ_s is always smaller than Γ_d. This is intuitive because the release of latent heat continuously heats the rising air parcel, so that the net cooling rate is lower than in the case of a dry adiabatic ascent.

A typical value of Γ_s between the surface and the tropopause is 6.5 K km^{-1}, which is the value corresponding to the international standard atmosphere (ISA) model. This model provides values of pressure, temperature and density that roughly correspond to the average

Table 3.2 Pressure, density and temperature as a function of height, as provided by ISA.

z [km]	p [hPa]	ρ [kg m^{-3}]	T [K]
0	1013.25	1.2250	288.150
1	898.76	1.1117	281.651
2	795.01	1.0066	275.154
3	701.21	9.0925×10^{-1}	268.659
4	616.60	8.1935×10^{-1}	262.166
5	540.48	7.3643×10^{-1}	255.676
6	472.18	6.6011×10^{-1}	249.187
7	411.05	5.9002×10^{-1}	242.700
8	356.51	5.2579×10^{-1}	236.215
9	308.00	4.6706×10^{-1}	229.733
10	264.99	4.1351×10^{-1}	223.252
11	226.99	3.6480×10^{-1}	216.774
12	193.99	3.1194×10^{-1}	216.650
13	165.79	2.6660×10^{-1}	216.650
14	142.70	2.2786×10^{-1}	216.650
15	121.11	1.9576×10^{-1}	216.650
16	103.52	1.6647×10^{-1}	216.650
17	88.497	1.4320×10^{-1}	216.650
18	75.652	1.2165×10^{-1}	216.650
19	64.674	1.0400×10^{-1}	216.650
20	55.293	8.8910×10^{-2}	216.650
21	47.289	7.5715×10^{-2}	217.580

values found in the atmosphere and can be taken as being representative of conditions at 40° N (Table 3.2). The saturated adiabatic lapse rate Γ_s can decrease to 4 K km^{-1} near the ground in warm humid air in the tropics. Near the tropopause and above, Γ_s approaches Γ_d because at low temperatures the moisture capacity of air is small.

Having derived Γ_s, we can now combine the result with the non-condensing case and obtain more general stability criteria, as shown in Figure 3.2 and summarized in Table 3.3. Instead of a dry stable atmosphere, we now have an absolutely stable atmosphere in the case where the ambient lapse rate is smaller than that of the saturated adiabat, which automatically also means that it is smaller than that of the dry adiabat. If, however, the ambient lapse rate exceeds that of the dry adiabat, we have absolutely unstable conditions; "absolutely" in this case means that the situation is unstable no matter whether the air parcel ascends dry or saturated adiabatically.

We now consider the case where the ambient lapse rate lies between that for the dry adiabat and the saturated adiabat. In this case the stability depends on whether the air parcel is saturated and ascends saturated adiabatically or whether it is unsaturated and ascends dry adiabatically. If the air parcel ascends dry adiabatically then the atmosphere is statically stable, but it would be unstable for a saturated ascent. This situation is called "conditionally unstable" to indicate that the stability of the atmosphere depends on the

Table 3.3 General static stability criteria in moist air.

$\gamma < \Gamma_s$	Absolutely stable
$\gamma = \Gamma_s$	Saturated neutral
$\Gamma_s < \gamma < \Gamma_d$	Conditionally unstable
$\gamma = \Gamma_d$	Dry neutral
$\gamma > \Gamma_d$	Absolutely unstable

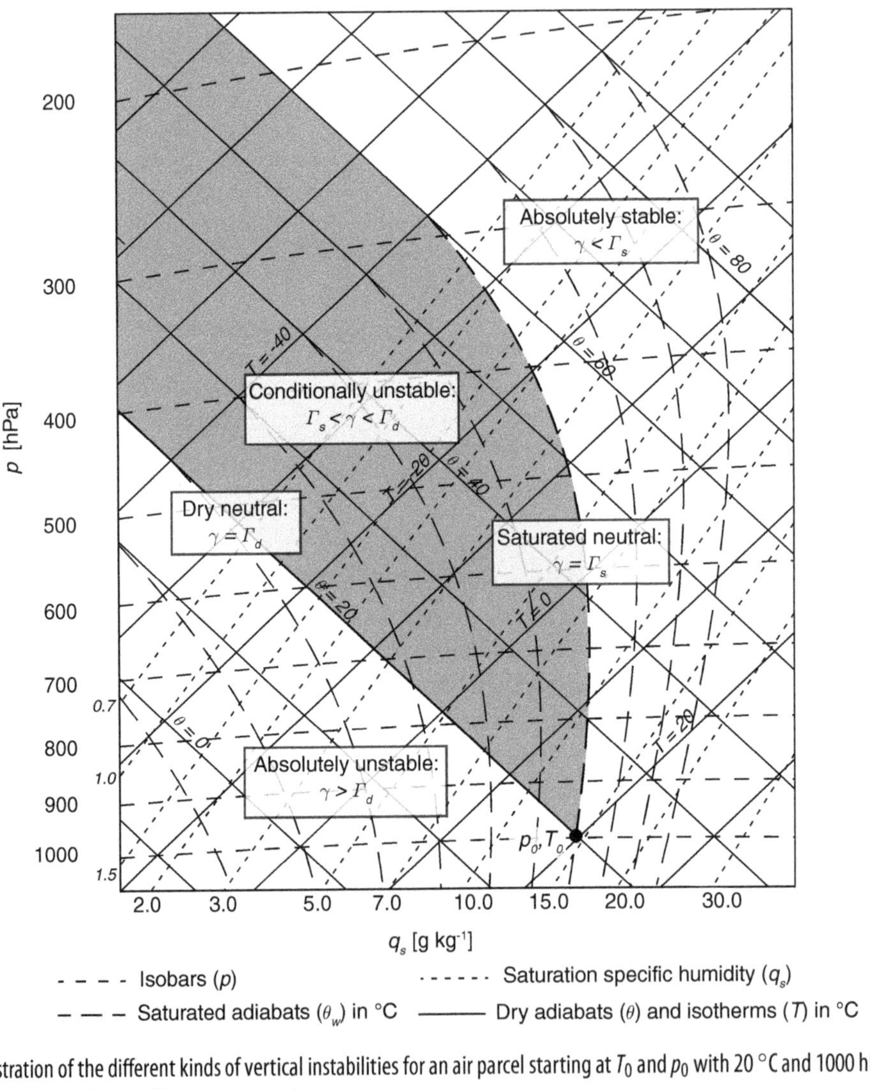

Fig. 3.2 Illustration of the different kinds of vertical instabilities for an air parcel starting at T_0 and p_0 with 20 °C and 1000 hPa. The conditionally unstable area is shown in gray.

condition of the air parcel. If the atmospheric lapse rate equals that of the dry adiabat, we have the dry neutral case where no forces support or inhibit the dry adiabatic ascent of an air parcel. If the atmospheric lapse rate equals the saturated adiabat, we have the saturated neutral case.

3.4 Instability of layers

A different kind of vertical instability is potential instability, which is also referred to as convective instability. The stability of a layer of air can change if the entire layer is lifted. The lifting of an entire layer is likely to happen when an air mass is forced to rise owing to orography or when warm air is lifted above cold air at warm or cold fronts.

We start with the lifting of layers in dry air. The weight of an air column per unit area is $\int \rho dz$. Using the hydrostatic equation this equals $\int (1/g)/dp$. Because g is constant, we can integrate this expression to $(p_1 - p_2)/g$. In a vertical displacement, $\Delta p = p_1 - p_2$ is constant owing to the conservation of mass. If we consider a vertical air column embedded in the lifted layer and small enough that the density ρ can be considered constant over the column, we can apply the hydrostatic equation again:

$$\Delta p = -g\rho\Delta z, \tag{3.34}$$

where $\Delta z = z_1 - z_2$. At the lower end, let the potential temperature be θ_1, and at the higher end θ_2. After lifting, the potential temperatures at the lower and higher ends still have the values θ_1 and θ_2 since the lifting occurs dry adiabatically in a dry atmosphere. Therefore $\Delta\theta = \theta_1 - \theta_2$ also remains constant. Since ρ decreases with altitude, the lifting of the layer results in a stretching of the column, i.e. $\Delta z' > \Delta z$, according to eq. (3.34). As $\Delta\theta$ can be written as $\Delta\theta = (\partial\theta/\partial z)/\Delta z$, this means that after the lifting $|\partial\theta/\partial z|$ is smaller than before the lifting (Figure 3.3).

This means that an initially unstable layer (where $\partial\theta/\partial z < 0$) becomes less unstable and an initially stable layer (where $\partial\theta/\partial z > 0$) becomes less stable, i.e. the lifting shifts the lapse rate towards that of the dry adiabat. Note that although the degree of stability or instability changes upon lifting, an initially stable layer cannot become unstable upon being lifted in a dry atmosphere, and likewise for an unstable layer. This is illustrated in Figure 3.3. Only the stability of neutrally stable air is unaffected by vertical displacement.

The situation becomes more complicated in moist air, where condensation can occur during the lifting process. Whether there is potential instability depends on the distribution of humidity in the atmosphere. Thus, the lifting of a layer of moist air can transform an initially stable air mass into an unstable or conditionally unstable state. The presence of a potential instability can then lead to vigorous convection and thunderstorms. Potential instability is defined in terms of the wet-bulb potential temperature θ_w defined in Section 2.5.7. A given value of θ_w can unambiguously be attributed to an individual saturated adiabat. The higher the specific humidity of an air parcel at a given temperature, the sooner will the parcel reach its LCL upon ascending and therefore the higher the wet-bulb potential

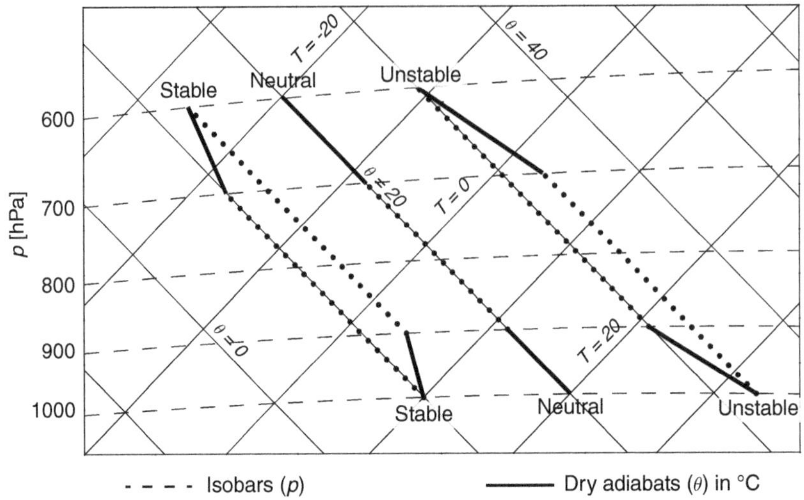

Fig. 3.3 Schematic of the lifting of a layer of dry air from between 1000 and 900 hPa to between 700 and 600 hPa for three cases: from left to right, initially stable, neutral layer and unstable layer. The bold solid lines represent the lapse rate of the layer before and after the lifting. The dotted lines show the ascent of the top and the bottom of the layer.

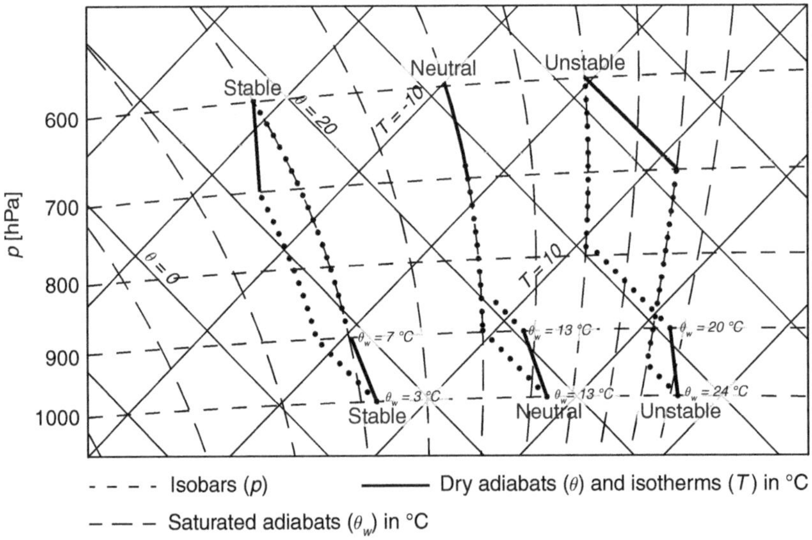

Fig. 3.4 Schematic of the lifting of a layer of moist air from between 1000 and 900 hPa to between 700 and 600 hPa for three cases, that where the layer is potentially stable, potentially neutral or potentially unstable. The bold solid lines represent the lapse rate in the layer before and after lifting. The dotted lines show the ascent of the top and the bottom of the layer.

temperature. The vertical gradient of θ_w is a measure of the vertical distribution of moisture in the layer.

The examples shown in Figure 3.4 are chosen in such a way that the layer between 1000 and 900 hPa is potentially stable (left ascent), potentially neutral (middle ascent) and

Table 3.4 Stability criteria for the lifting of layers of moist air.

$\partial\theta_w/\partial z > 0$	Potentially stable layer
$\partial\theta_w/\partial z = 0$	Potentially neutral layer
$\partial\theta_w/\partial z < 0$	Potentially unstable layer

potentially unstable (right ascent). In the potentially stable case the water vapor mixing ratio is higher at the top than at the bottom of the layer, so the lifting condensation level is reached immediately at the top of the layer but only after some dry adiabatic ascent at the bottom of the layer. Condensation occurs at 900 hPa throughout the layer. At the top of the layer $\theta_w^{top} = 7$ °C and at its base $\theta_w^{base} = 3$ °C. Here θ_w increases with altitude, which is indicative of a potentially stable layer.

In the potentially neutral case (the middle ascent in Figure 3.4), the moisture distribution is such that condensation is reached at 900 hPa at the base of the layer and at 840 hPa at the layer top. Here $\theta_w = 13$ °C is constant throughout the layer.

In the potentially unstable case (the right ascent in Figure 3.4), the layer is very dry at the top and very moist at the bottom. In this case the top of the layer cools dry adiabatically until approximately 790 hPa, which corresponds to $\theta_w^{top} = 20$ °C, whereas the base reaches its lifting condensation level at 960 hPa, which corresponds to $\theta_w^{base} = 24$ °C. Thus θ_w decreases with altitude, which is indicative of a potentially unstable layer. These criteria for potential instability are summarized in Table 3.4.

3.5 Horizontal restoring forces

In this section we consider the horizontal displacement of air. Instability with respect to horizontal displacements is referred to as inertial instability.

3.5.1 Geostrophic wind

The major horizontal forces acting in the atmosphere are the Coriolis force $\vec{F}_C$ and the horizontal pressure gradient force $\vec{F}_{PG}$. The Navier–Stokes equation (eq. 3.1) cannot be solved analytically. However, using approximation methods such as scale analysis, it is possible to gain insights on the major features of atmospheric motions. Conducting a scale analysis means finding the approximate relative magnitude of the various terms of an equation, given the particular environment in which the equation is applied. A very common and useful estimate of eq. (3.1) is the geostrophic approximation, extracted from a scale analysis of the Navier–Stokes equation for large-scale atmospheric flow. As an example, the scaling for the zonal wind component u of the Navier–Stokes equation can be obtained from

$$\frac{\partial u}{\partial t} + u\frac{\partial u}{\partial x} = -\frac{1}{\rho}\frac{\partial p}{\partial x} + fv + F_{F,x}. \tag{3.35}$$

Table 3.5 Scale analysis for geostrophic flow; L denotes the characteristic length scale, T the characteristic time scale and U the characteristic velocity scale.

$\Delta x = L$	~ 1000 km
$u, v, \Delta u, \Delta v = U$	~ 10 m s^{-1}
ρ	~ 1 kg m^{-3}
Δp	~ 10 hPa
f	$\sim 10^{-4}$ s^{-1}
$\Delta t = T = L/U$	$\sim 10^5$ s

Neglecting friction, typical values of the terms in the Navier–Stokes equation (i.e. spatial scale, velocity, density, horizontal pressure gradient, Coriolis parameter f) on the synoptic scale are given in Table 3.5. We assume a characteristic length scale $L = 1000$ km and that the characteristic horizontal velocity scale $U = 10$ m s^{-1} can be used for u, v and the gradients of u and v. That leads to a characteristic time scale $T = L/U \sim 10^5$ s or approximately one day. With these numbers, the following orders of magnitude can be obtained for the individual terms in eq. (3.35):

$$F_{PG,x} \approx \frac{\Delta p}{\rho L} = 10^{-3}\,\mathrm{m\,s^{-2}}, \tag{3.36}$$

$$F_{C,x} \approx fU = 10^{-3}\,\mathrm{m\,s^{-2}}, \tag{3.37}$$

$$\frac{\partial u}{\partial t} \approx \frac{U}{T} = 10^{-4}\,\mathrm{m\,s^{-2}}, \tag{3.38}$$

$$u\frac{\partial u}{\partial x} \approx \frac{U^2}{L} = 10^{-4}\,\mathrm{m\,s^{-2}}. \tag{3.39}$$

Thus the inertial term (the sum of the advection term and the time derivative) is seen to be an order of magnitude smaller than $F_{C,x}$ and $F_{PG,x}$. Neglecting the inertial term leads to simplified horizontal momentum equations:

$$fu_g = -\frac{1}{\rho}\frac{\partial p}{\partial y}, \tag{3.40}$$

$$fv_g = \frac{1}{\rho}\frac{\partial p}{\partial x}, \tag{3.41}$$

which are used to define the so-called geostrophic wind components:

$$u_g = -\frac{1}{f\rho}\frac{\partial p}{\partial y}, \qquad v_g = \frac{1}{f\rho}\frac{\partial p}{\partial x}. \tag{3.42}$$

The geostrophic wind blows parallel to the isobars and thus requires straight isobars as shown in Figure 3.5. An air parcel which is initially at rest does not experience any Coriolis force (point A in Figure 3.5). Hence it starts to move down the horizontal pressure gradient. With increasing speed, the Coriolis force also increases until the parcel's trajectory is

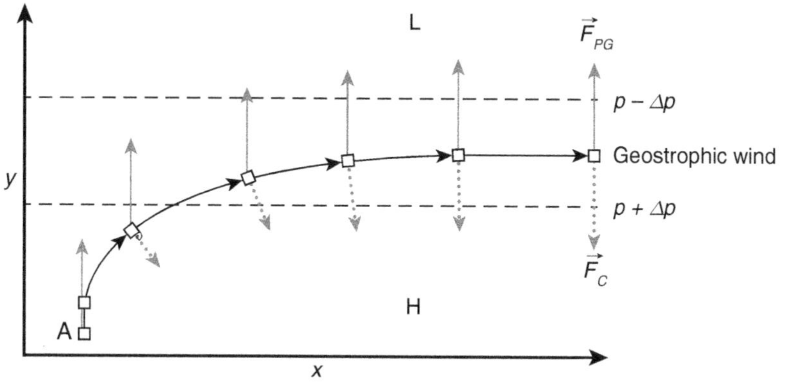

Fig. 3.5 Schematic of an air parcel approaching geostrophic wind balance in the northern hemisphere: An air parcel initially at rest at point A starts to move along the negative pressure gradient until $\vec{F}_{PG}$ (solid gray arrows) and $\vec{F}_C$ (dotted gray arrows) are of equal magnitude but opposite sign. The broken lines denote isobars, H is a high pressure region and L a low pressure region.

deflected in such a way that the direction of the Coriolis force is of opposite sign but equal in magnitude to the pressure gradient force. This implies that the trajectory is perpendicular to the pressure gradient (eq. 3.42) and that the wind blows parallel to the isobars. Because f is zero at the equator, the geostrophic wind is not defined at the equator. The above geostrophic balance is not arbitrary but results from the speed at which Earth rotates. If Earth were to rotate much more slowly, friction rather than the Coriolis force would limit the wind speed. If it rotated faster then the geostrophic balance would qualitatively be the same, but because f would be larger in that case the balance would be achieved at smaller velocities.

It should be noted, however, that the wind is not geostrophic for flows in which the centrifugal force or friction are important. This is especially true for winds near Earth's surface. There, a frictional drag is exerted on the air by the surface, slowing the air down. This slowing down of the air reduces the Coriolis force, making a balance between Coriolis and pressure gradient forces impossible. The difference between geostrophic flow and the actual flow is the ageostrophic flow. Whether the geostrophic wind is applicable can be estimated by means of the so-called Rossby number Ro. It is the ratio of the inertial term and the Coriolis term in the Navier–Stokes equation (eq. 3.1). A scale estimate of the inertial term is obtained by taking the larger of the two terms: $U/T, U^2/L$. Taking U^2/L to be the larger term, the Rossby number is given as

$$Ro = \frac{U^2}{fLU} = \frac{U}{fL}. \tag{3.43}$$

The geostrophic flow is applicable for $Ro < 1$, i.e. at large scales where the Coriolis force dominates the inertial force. If this condition is not met then the appropriate wind balance is the gradient wind. It results from a balance between the Coriolis force fv, the pressure gradient force and the centrifugal force v^2/r and is given in cylindrical coordinates as

$$\frac{1}{\rho}\frac{\partial p}{\partial r} = \underbrace{\frac{v^2}{r}}_{\text{centrifugal}} + \underbrace{fv}_{\text{Coriolis}} \tag{3.44}$$

3.5.2 Thermal wind

In a barotropic atmosphere the horizontal pressure gradient and the geostrophic wind are constant with altitude, because the density depends only on pressure. In a real atmosphere, the density depends on both pressure and temperature so that the warmer the temperature, the larger the distance between two pressure levels (eq. 3.10). This causes the horizontal pressure gradient to vary with altitude, which implies that the geostrophic wind also varies with altitude. The cause for this variation is the non-uniform temperature distribution in the horizontal or, put differently, that isotherms are not parallel to isobars. Such an atmosphere is termed baroclinic. The variation in the geostrophic wind with altitude in a baroclinic atmosphere is called the geostrophic wind shear or the thermal wind, because it has its origin in the horizontal temperature gradient. The thermal wind components are (Holton, 2004)

$$p\frac{\partial u_g}{\partial p} = \frac{R_d}{f}\left(\frac{\partial T}{\partial y}\right)_p,$$
$$p\frac{\partial v_g}{\partial p} = -\frac{R_d}{f}\left(\frac{\partial T}{\partial x}\right)_p, \tag{3.45}$$

where R_d is the gas constant for dry air. Applying the hydrostatic equation to replace partial derivatives with respect to pressure with those with respect to z and making use of the ideal gas law yields

$$f\frac{\partial u_g}{\partial z} = -\frac{g}{T}\left(\frac{\partial T}{\partial y}\right)_p,$$
$$f\frac{\partial v_g}{\partial z} = \frac{g}{T}\left(\frac{\partial T}{\partial x}\right)_p. \tag{3.46}$$

Because the partial derivatives of temperature with respect to x and y are to be taken at constant pressure, T can be replaced with θ so that the thermal wind equations become

$$f\frac{\partial u_g}{\partial z} = -\frac{g}{\theta}\left(\frac{\partial \theta}{\partial y}\right)_p,$$
$$f\frac{\partial v_g}{\partial z} = \frac{g}{\theta}\left(\frac{\partial \theta}{\partial x}\right)_p. \tag{3.47}$$

As eq. (3.47) shows, the vertical gradient in the geostrophic wind is related to the horizontal gradient of θ. That is, if θ is not uniform in the horizontal, the geostrophic wind will change with altitude. A typical illustration of the thermal wind in mid-latitudes in the northern hemisphere is shown in Figure 3.6. The subtropics are characterized by high pressure systems, e.g. the Azores high, and warm temperatures. In polar latitudes we have

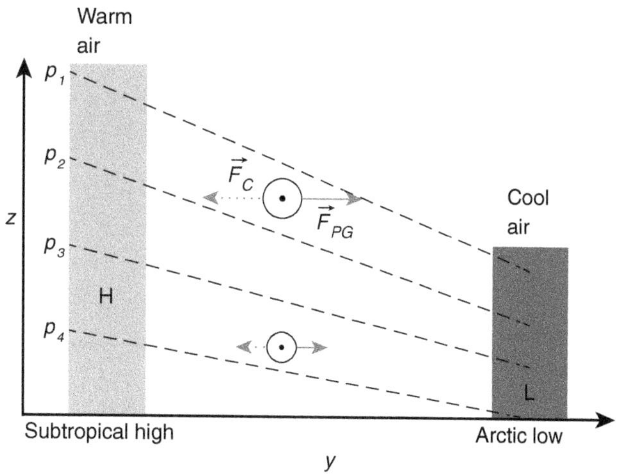

Fig. 3.6 Schematic of the thermal wind for a typical position in the mid-latitudes of the northern hemisphere. In this example the wind blows in the direction perpendicular to the page and towards the observer. Isobars are shown as broken lines, the solid gray arrows indicate the pressure gradient force $\vec{F}_{PG}$ and the dotted gray arrows the Coriolis force $\vec{F}_C$. The light gray region denotes warm air and high pressure and the dark gray region cool air and low pressure.

the Aleutian and Icelandic lows as part of the Arctic front, with cold temperatures. Thus there is a horizontal temperature gradient and a horizontal pressure gradient. Because of the hypsometric equation (3.10) the vertical distance between two isobars is larger in the warm subtropical air than in the cold polar air. With increasing altitude the horizontal pressure gradient will be larger, increasing the horizontal wind speed and hence the Coriolis force. Thus the geostrophic wind increases with increasing altitude and explains the high wind speeds of the polar and subtropical jet streams.

3.5.3 Inertial instability

While knowledge of the vertical distribution of the potential temperature allows us to characterize the static stability of the atmosphere, knowledge of the horizontal distribution of the absolute momentum M allows us to characterize the inertial stability of the atmosphere (Holton, 2004). For a flow in the x-direction, M is defined as the momentum due to Earth's rotation, fy, minus the geostrophic wind, u_g, i.e. the linear momentum, in a closed system (Section 2.1.3) with no friction:

$$M = fy - u_g. \tag{3.48}$$

Under these conditions M is conserved, i.e. $dM/dt = 0$. The conservation of M for horizontal displacements is the analog of the conservation of θ ($d\theta/dt = 0$) for vertical displacements in adiabatic conditions (eq. 2.23). The expression for M in the y-direction can be derived from the x-component of the Navier–Stokes equation (3.1) in the absence of friction and assuming a constant Coriolis parameter f:

$$\frac{du}{dt} = -\frac{1}{\rho}\frac{\partial p}{\partial x} + fv. \tag{3.49}$$

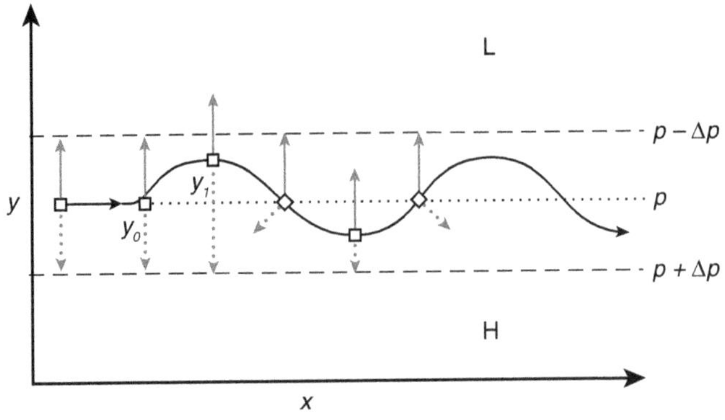

Fig. 3.7 Schematic of an inertial oscillation of an air parcel (solid curve) in an x–y plane. The solid gray arrows indicate the pressure gradient force $\vec{F}_{PG}$ and the dotted gray arrows the Coriolis force $\vec{F}_C$; H is a high pressure region, L is a low pressure region. The broken lines denote isobars. The initial position, from which the air parcel in a geostrophic flow is displaced, is denoted by y_0.

If we consider a two-dimensional geostrophic flow in the x-direction (Figure 3.7) then the pressure gradient $\partial p/\partial x$ is zero, so that

$$\frac{du_g}{dt} - fv = \frac{d}{dt}(u_g - fy) = -\frac{dM}{dt} = 0. \tag{3.50}$$

Suppose that the parcel is displaced in the y-direction without perturbing the pressure field (Figure 3.7). The y-component of the Navier–Stokes equation is given as

$$\frac{dv}{dt} = -\frac{1}{\rho}\frac{\partial p}{\partial y} - fu = f(u_g - u), \tag{3.51}$$

where we have made use of the definition of the geostrophic wind (eq. 3.42). This equation states that the air parcel experiences an acceleration in the y-direction if there is a positive imbalance between the geostrophic wind u_g and the actual zonal wind u.

For the situation depicted in Figure 3.7, $\partial p/\partial y$ is negative and thus u_g is positive. Assume an air parcel moving in the positive x-direction is accelerated towards the low pressure region at point y_0, i.e. in the positive y-direction. The component of the Coriolis force perpendicular to the direction of acceleration of the air parcel becomes larger, and hence u increases until it reaches its maximum at point y_1 in Figure 3.7. Here the Coriolis force component in the x-direction is zero but the component in the negative y-direction is non-zero causing the air parcel to move towards the high pressure region and so to experience a deceleration. Because the Coriolis force lags behind the acceleration from the pressure gradient force, the result is an oscillatory motion.

The horizontal shear of M is the absolute vorticity, which is the sum of the relative vorticity $\partial v/\partial x - \partial u/\partial y$ and the Coriolis parameter f. In a two-dimensional geostrophic flow in the x-direction, we obtain the horizontal shear by finding the derivative with respect to y:

$$\frac{\partial M}{\partial y} = \underbrace{f - \frac{\partial u_g}{\partial y}}_{\text{abs.vorticity}}. \tag{3.52}$$

Table 3.6 Criteria for inertial instability for a two-dimensional geostrophic flow in the x–y plane, where the oscillation is along the x-direction.

$\partial M/\partial y > 0$	Inertially stable
$\partial M/\partial y = 0$	Inertially neutral
$\partial M/\partial y < 0$	Inertially unstable

While the Brunt–Väisälä frequency determines the frequency of oscillation in a statically stable atmosphere, in an inertially stable atmosphere the frequency of oscillation is the product of the Coriolis parameter and the horizontal shear of the absolute momentum. In order to obtain the analogous equation to that for static oscillation (eq. 3.29), we assume that the air parcel is displaced from y_0 (Figure 3.7) by $\tilde{y}$, where the new geostrophic wind can be approximated using a truncated Taylor series,

$$u_g(y_0 + \tilde{y}) = u_g(y_0) + \left.\frac{\partial u_g}{\partial y}\right|_{y=y_0} (y - y_0); \tag{3.53}$$

likewise, for the new zonal velocity we have

$$u(y_0 + \tilde{y}) = u(y_0) + \left.\frac{\partial u}{\partial y}\right|_{y=y_0} (y - y_0). \tag{3.54}$$

Now assume that the zonal wind equals the geostrophic wind at y_0, i.e. $u(y_0) = u_g(y_0)$. At this point the zonal wind does not experience a pressure gradient force in the x-direction and we can obtain the change in u with y from eq. (3.49),

$$\frac{du}{dt} = fv = f\frac{dy}{dt}, \tag{3.55}$$

which, upon integration and taking the derivative with respect to y, yields: $\partial u/\partial y = f$. Inserting this together with $u(y_0) = u_g(y_0)$ into eq. (3.54) yields

$$u(y_0 + \tilde{y}) = u_g(y_0) + f(y - y_0). \tag{3.56}$$

Inserting eqs. (3.53) and (3.56) into eq. (3.51), associating v with $d\tilde{y}/dt$ and noting that $y - y_0 = \tilde{y}$ yields the equation for the inertial oscillation, we have

$$\frac{d^2\tilde{y}}{dt^2} + f\left(f - \frac{\partial u_g}{\partial y}\right)\tilde{y} = \frac{d^2\tilde{y}}{dt^2} + f\frac{\partial M}{\partial y}\tilde{y} = 0. \tag{3.57}$$

The solution is of the form: $\tilde{y}(t) = A\exp(\pm i\sqrt{f(\partial M/\partial y)}t)$. If M decreases with increasing y in the northern hemisphere, where $f > 0$, the situation is inertially unstable. If M increases with increasing y, as it does in Figure 3.7, we obtain stable sinusoidal oscillations (undamped harmonic oscillations). This means that an air parcel accelerated in the y-direction will experience a larger Coriolis force owing to its higher velocity, and it will be pulled back towards higher pressure. If the air parcel has too much momentum and overshoots towards the higher pressure, it experiences a negative acceleration and will again move towards the lower pressure. The stability criteria are summarized in Table 3.6.

3.6 Slantwise displacement

Considering inertial stability in addition to hydrostatic (often abbreviated to "static") stability allows us to consider instabilities that occur only if an air parcel or air mass is displaced both in the horizontal and vertical directions. They are referred to as slantwise instabilities. One prominent example of slantwise instability is symmetric instability, which is thought to be responsible for the mesoscale structure of rainbands in tropical cyclones or in mesoscale convective systems, which will be discussed in Chapter 10.

In a statically and inertially stable atmosphere, slantwise instability and displacement of an air parcel can nevertheless occur. This requires that the mean horizontal wind experiences a vertical shear, as shown in Figure 3.8. Because buoyancy is the vertical force for a slantwise displacement and because it leads to convection, slantwise displacement occurs in the form of slantwise convection. In addition to the vertical buoyancy force, horizontal forces that will initiate a horizontal motion are needed for slantwise convection to occur. Besides the horizontal Coriolis and pressure gradient forces discussed above, the centrifugal force can also contribute to a horizontal motion. The centrifugal force is important when the isobars are curved, as is the case around a low pressure system, a hurricane or a tornado.

Let us assume an atmosphere with a horizontal temperature gradient, in which the isentropic surfaces (constant-θ surfaces) are tilted (Figure 3.8). As shown in Figure 3.9, this is the normal situation in the troposphere. Such tilted isentropes arise because of a horizontal temperature gradient, indicated by the warm and cold air in Figure 3.8. In the absence of a horizontal temperature gradient in the y-direction, the isentropes would be horizontal. In

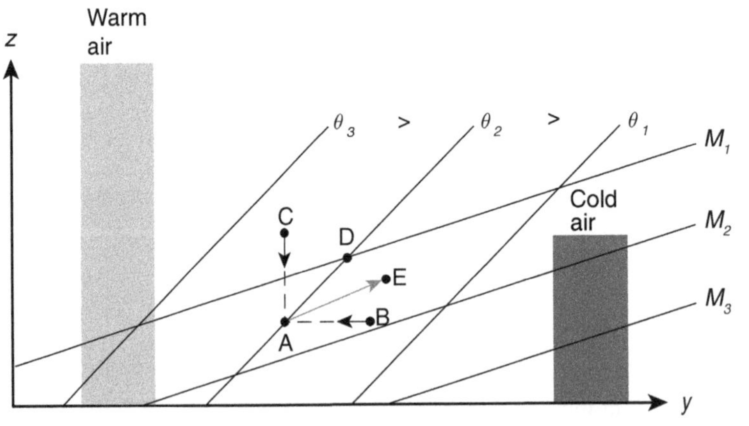

Fig. 3.8 Schematic showing how the symmetric instability of an air parcel leads it from an equilibrium condition at point A to point E, in a cross section in the northern hemisphere. Isolines are shown of absolute momentum M, increasing from M_1 to M_3, and of potential temperature θ, increasing from θ_1 to θ_3. A purely horizontal (A to B) or purely vertical (A to C) displacement would result in a counteracting movement back to point A because of inertial or static stability, respectively.

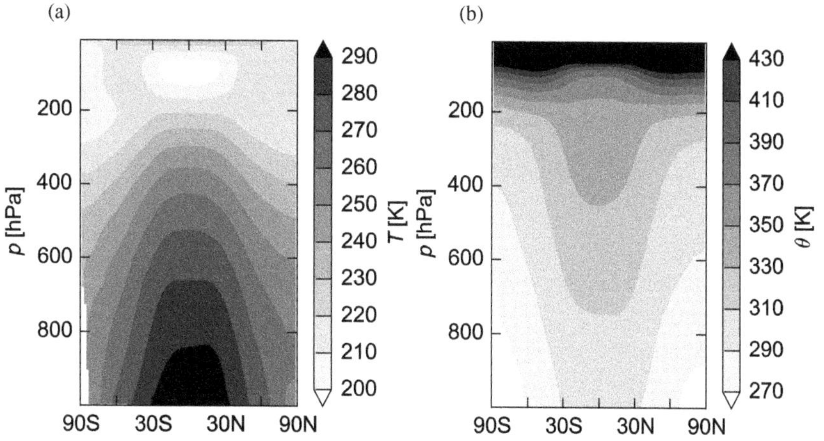

Fig. 3.9 Annual and zonal mean latitude versus pressure distribution of (a) temperature and (b) potential temperature.

such an atmosphere, a horizontal displacement in the y-direction of an air parcel would be neutral because θ remains constant during the displacement. Vertical isentropes indicate a dry neutral atmosphere.

Suppose that a parcel of air at point A in Figure 3.8 is in equilibrium with the environment (it has the same T, p, u, v). Next, suppose that the air parcel is slantwise displaced to E. By buoyancy alone the air parcel would rise vertically; in order for it to be slantwise displaced from A to E, an imbalance between the horizontal pressure gradient and the inertial forces is required such that the air parcel is also displaced in the meridional direction (in the positive y-direction for the example in Figure 3.8).

In addition to the gradient in potential temperature, we need to consider the gradient in the absolute momentum surface since this determines whether a horizontal displacement will be further displaced or experiences a restoring force. For a barotropic flow, the potential temperature surfaces are oriented horizontally and the absolute momentum surfaces are oriented vertically, in a cross section of latitude versus height. In such a flow the atmosphere is neutral both to static and to inertial instability. For a baroclinic flow in which the mean westerly flow increases with height (Figure 3.6), both the θ and M surfaces slope upward towards the pole as shown for both types of surfaces in Figure 3.8 and for the θ surfaces additionally in Figure 3.9.

If the buoyancy force is zero, i.e. if the parcel displacement occurs along an isentropic surface (e.g. along the θ_2 isentrope from points A to D in Figure 3.8) then the parcel has a lower absolute momentum than the environment, which implies that it is symmetrically unstable; see Table 3.7. The broken lines with the backward pointing arrows from points B and C indicate that a pure horizontal or pure vertical displacement results in a counteracting movement back to point A because of inertial or static stability, respectively. The condition for symmetric instability that the absolute momentum M decreases with increasing y on an isentropic surface is equivalent to the condition that θ needs to decrease with altitude on an M surface. This is the case if the M surfaces are tilted more from the vertical than the isentropes. The typical situation in the

Table 3.7 Stability criteria for symmetric instability

$\partial\theta/\partial z\|_M > 0 \leftrightarrow \partial M/\partial y\|_\theta > 0$	Symmetrically stable atmosphere
$\partial\theta/\partial z\|_M = 0 \leftrightarrow \partial M/\partial y\|_\theta = 0$	Symmetrically neutral atmosphere
$\partial\theta/\partial z\|_M < 0 \leftrightarrow \partial M/\partial y\|_\theta < 0$	Symmetrically unstable atmosphere

atmosphere is the opposite since the M surfaces are tilted less from the vertical than the isentropes.

Symmetric instability and slantwise convection can be associated with extratropical fronts. Slantwise convection can result in single or multiple bands that are isolated or embedded in larger structures. The bands vary in length between 100 km to more than 500 km and in width between 5 and 40 km. The concept of slantwise displacement will be revisited in Chapters 9 and 10.

3.7 Exercises

Single/multiple choice exercises

1. The Coriolis parameter f ...
   (a) ...equals $2\Omega \sin\varphi$ and describes the Coriolis force exactly in the Coriolis term $f\vec{k} \times \vec{v}$.
   (b) ...depends on the latitude and longitude of a moving air parcel on which it acts.
   (c) ...is largest (in terms of absolute value) at the poles.
   (d) ...is approximately 10^{-4} m s^{-1} at mid-latitudes.
2. Which of the following is/are true?
   (a) If we call the atmosphere dry, we mean that it contains no water vapor.
   (b) On the large scale, the hydrostatic equation is often not a good approximation.
   (c) The buoyancy force is positive when an air parcel is warmer than the ambient air.
   (d) The condensation of water vapor reduces the ambient lapse rate owing to the release of kinetic energy.
3. Lapse rates can be used to measure atmospheric stability. Which of the following is/are true?
   (a) The difference between the dry and saturated adiabatic lapse rates is largest at the tropopause.
   (b) The saturated adiabatic lapse rate Γ_s has the constant value 6.5 K km^{-1}.
   (c) If the ambient lapse rate lies between the lapse rates for the dry and saturated adiabats, the stability of the atmosphere with respect to an ascending air parcel depends on the moisture content and temperature of the air parcel.
   (d) An atmosphere which is statically and inertially stable is also symmetrically stable.

4. Consider an air parcel that is displaced upwards. If the lapse rates of the parcel and the environment are equal and non-zero, …
   (a) … in the absence of friction the parcel continues rising without acceleration.
   (b) … no net force acts on the air parcel and it is in hydrostatic equilibrium at the new position.
   (c) … the air parcel does not cool or warm during the displacement.
   (d) … the atmosphere is stable with respect to the air parcel's ascent and the air parcel performs an oscillatory motion around its initial position.
5. Which of the following is/are true?
   (a) According to the international standard atmosphere, the average lapse rate between the surface and the tropopause is approximately 6.5 K km^{-1} at 40° N.
   (b) The centrifugal force is included in the Navier–Stokes equation.
   (c) The hypsometric equation only holds over small z-intervals.
   (d) The troposphere is often absolutely unstable.
6. Scale analysis is a method of simplifying equations by analyzing the relative magnitudes of the various terms of an equation. This is done by assigning typical values to atmospheric variables (for example the velocity) of a specific flow system. Which of the following is/are true?
   (a) The vertical air motion is larger.
   (b) The horizontal air motion is larger.
   (c) The vertical and horizontal air motions are on the same order of magnitude.
7. The thermal wind relation arises from …
   (a) … the variation in the horizontal pressure gradient with altitude.
   (b) … the baroclinicity of the atmosphere.
   (c) … the increasing distance between isobars from the subtropics to the poles.
8. What is important for slantwise convection to occur?
   (a) A horizontal temperature gradient must exist, which causes the isentropes to be tilted.
   (b) Conditions of barotropic flow must be present, where the mean flow is westerly and increasing with height.
   (c) Both the potential temperature and absolute momentum surfaces need to have the same slope.
   (d) The absolute momentum surfaces must have a smaller slope, $\Delta z/\Delta y$, than the isentropes.

Long exercises

1. (a) Use the tephigram in Figure 3.10 to determine the static stability between the seven points (six layers) indicated for the case of (i) non-condensing air and (ii) moist air.
   (b) Indicate the path of an air parcel which is forced to rise from the surface (where $p = 980$ hPa) to 400 hPa.
   (c) At what height do you expect the parcel to experience the largest (positive) buoyant force? At what height is the buoyant force of the parcel zero? Indicate the levels in

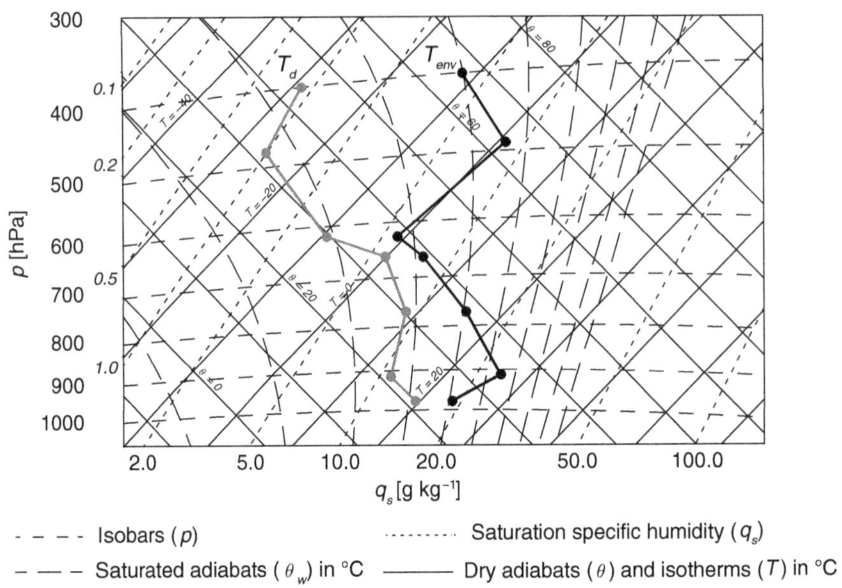

Fig. 3.10 Tephigram showing the vertical profiles of T_d and T_{env}.

the tephigram and give a qualitative explanation of what this means for the ascent of the parcel.

(d) Assume that the Sun heats up the surface and increases T_{env} to 40 °C at $p = 980$ hPa. If this air parcel rises, where will its LCL be? Will the parcel rise by itself after reaching the LCL? Explain your answer.

2. "When going higher up in the troposphere it gets colder and colder, therefore the air density increases. In fact cold air is denser than warm air."

   (a) Comment on this statement.

   (b) Starting from the hydrostatic approximation and the ideal gas law, derive the pressure profile $p(z)$ for an atmosphere with a constant linear lapse rate such that the temperature profile is given by $T = T_0 - \gamma z$, where γ is a constant. By neglecting the difference between T and T_v, show that

$$p(z) = p_0 \left(\frac{T_0 - \gamma z}{T_0} \right)^{g/R_d \gamma} .$$

   (c) Again using the hydrostatic assumption, derive the vertical density profile for such an ideal setting.

   (d) Such an atmosphere has a finite height. Compute it in the case of $T_0 = 295$ K and $\gamma = 6.5 \text{ K km}^{-1}$.

   (e) How large should the temperature lapse rate be to achieve a balance between temperature and pressure effects on density such that a constant density exists throughout the vertical extent of the atmosphere? What would happen if the lapse rate were even larger than this limit value?

3. The presence of narrow regions with strong winds in the upper troposphere, known as jet streams, is essentially due to horizontal temperature gradients at the surface. This can be shown using a simple idealized framework, as follows.

   Consider a region of the troposphere similar to a narrow band in the Atlantic Ocean: $L_x = 3000$ km, $L_y = 500$ km, $p_0 = 1000$ hPa (at sea level) and $p_{top} = 200$ hPa. Set the origin of a reference system (x, y, z) in the lower left-hand corner at the front of this "box". Consider a constant surface pressure field. The temperature field in the domain can be described as $T(x, y, z) = T_0 - ay - \gamma z$, with $T_0 = 295$ K, $a = \partial T/\partial y \approx 10\,\text{K}/500\,\text{km} = 0.02$ K km^{-1} and $\gamma = 6.5$ K km^{-1}. Assume that the hydrostatic assumption is valid and that the atmosphere is dry.

   (a) The geopotential height describes the height at which a certain value of pressure is reached. Find the expression for this quantity $z = z(p)$.
   (b) Compute the height of the 500 and 200 hPa pressure surfaces above the surface points $\vec{x_1} = (x, y) = (L_x/2, 0)$ and $\vec{x_2} = (x, y) = (L_x/2, L_y)$.
   (c) Compute the meridional slope dz/dy of the 500 and 200 hPa geopotential surfaces.

4. The squared Brunt–Väisälä frequency can be described by the following equation:

$$N^2 = \frac{g}{T}\left(\frac{\partial T}{\partial z} + \frac{g}{c_p}\right).$$

   Compute N^2 at a height of 1830 m for an atmosphere with a surface temperature of $T_s = 15$ °C, assuming the following:

   (a) An isothermal atmosphere with $T(z) =$ constant.
   (b) An atmosphere with $T(z) = T_s + \gamma z$ and a lapse rate γ equal to
      - -7 K km^{-1},
      - -13 K km^{-1} or
      - $-g/c_p$.

   Which atmosphere is stably stratified and which is unstable or neutral?

5. Titan, Saturn's largest moon, has a nitrogen atmosphere. It is believed to have a methane cycle similar to Earth's hydrological cycle, i.e. on Titan methane condenses, forms clouds and rains to the surface. Consider the condensation of methane in an adiabatically rising "air" parcel starting with a surface temperature of 94 K. The mass of the "air" parcel consist of 5% methane and 95% nitrogen, which corresponds to volume percentages of methane and nitrogen of 8% and 92%, respectively.

   On Titan, the surface pressure is 1450 hPa. The saturation vapor pressure of methane, e_{s,CH_4}, at 94 K is 180 hPa and the latent heat of vaporization, L_{v,CH_4}, is 450 kJ kg^{-1}. Take the specific heat of methane, c_{p,CH_4}, to be 2×10^3 J kg^{-1}K^{-1} and the specific heat of nitrogen, c_{p,N_2}, to be 1.041×10^3 J kg^{-1}K^{-1}. The acceleration on Titan, g_{Titan}, is 1.35 m s^{-2}.

   (a) Starting from the Clausius–Clapeyron equation;

$$\frac{de_{s,CH_4}}{dT} = \frac{L_{v,CH_4}}{T\Delta\alpha},$$

   find an expression for e_{s,CH_4} as a function of temperature ($\Delta\alpha$ is the specific volume change). You can neglect the temperature dependence of L_{v,CH_4}. Note that the gas constants for "dry" and "humid" air differ from our atmosphere: the specific gas constant of a gas i is given by R^*/M_i, where M_i is the molar mass of the gas i.

(b) At which temperature does saturation set in at the surface?

(c) Using the dry adiabatic lapse rate (considering only nitrogen in the calculation of c_p), estimate at which altitude you reach the temperature calculated in part (b). Explain why methane will not condense at that altitude.

(d) At which altitude do you expect methane to condense? (*Hint: Plot* e_{CH_4} *and* e_{s,CH_4} *as functions of* z*. If you use the hypsometric equation, note that you can replace the virtual temperature by the normal temperature by calculating* R_m*.*)

4 Mixing and convection

In this chapter we discuss the mixing of air masses (Section 4.1) and convection (Section 4.2). Mixing can either occur isobarically, i.e. at constant pressure, or adiabatically, when air parcels repeatedly move up- and downward. The isobaric mixing of different air masses can lead to fog formation when the air is sufficiently moist. In contrast, adiabatic mixing occurs when air parcels or air masses are mixed with the potential temperature constant. Adiabatic mixing can be observed when thermals rise from a hot surface in summer or when smoke from chimneys disperses.

Convection in the atmosphere refers to vertical motions of air parcels. Atmospheric convection may be divided into (i) (mechanically) forced convection and (ii) free convection, which is also referred to as thermal, gravitational or buoyant convection. An example of forced convection is the lifting of air along a cold or warm front or because of the orography. Here we are mainly interested in buoyant convection. This is best described in terms of the elementary parcel theory, which we will use to introduce the convectively available potential energy (CAPE) and the level of free convection (LFC). Because elementary parcel theory is rather simplified, its modifications, including entrainment, aerodynamic drag, compensating downdrafts and the weight of hydrometeors, are discussed at the end of this chapter.

4.1 Mixing

4.1.1 Isobaric mixing

Isobaric mixing provides one possibility for fog formation, creating the so-called "mixing fog". Mixing fog is most commonly observed if warm and cold air is mixed, when both air masses are close to saturation with respect to liquid water. An example of mixing fog is "steam fog" (Figure 4.1), which can often be observed in autumn when cool air moves over warm water. When the cool air mixes with warm moist air over the water, the warm moist air cools. If its RH reaches 100% then condensation sets in and fog forms. Other examples of isobaric mixing occur when you see your breath on a cold winter day or the steam above a hot cup of tea.

Let us consider two air masses with mass, temperature and specific humidity given by m_1, T_1, q_{v1} and m_2, T_2, q_{v2}. If these air masses are well mixed at constant pressure, we obtain a mixture that is the mass-weighted mean of their individual properties. The

Fig. 4.1 Photograph of steam fog taken by Julian Quinting.

properties of the mixed air mass in which we are interested are the intensive variables temperature and specific humidity, which do not depend on size or volume or mass. The mean specific humidity $q_{v,mix}$ is:

$$q_{v,mix} = \frac{m_1}{m_1 + m_2} q_{v1} + \frac{m_2}{m_1 + m_2} q_{v2}. \tag{4.1}$$

Likewise, we can obtain the mean water vapor mixing ratio $w_{v,mix}$. Using the approximation to the vapor pressure given by $e \approx q_v p/\epsilon$ (eq. 2.78), the mean vapor pressure e_{mix} can be calculated accordingly (Figure 4.2).

The relation between the heat dQ supplied to a given mass of humid air and its rise in temperature dT is given by:

$$dQ = mc_p dT + mq_v c_{pv} dT, \tag{4.2}$$

where c_p is the specific heat capacity of dry air at constant pressure and c_{pv} is the specific heat capacity of water vapor at constant pressure. If overall no heat is gained or lost, the amount of heat lost by the warmer air mass equals the heat gained by the colder air mass:

$$m_1(c_p + q_{v1}c_{pv})(T_1 - T_{mix}) = m_2(c_p + q_{v2}c_{pv})(T_{mix} - T_2). \tag{4.3}$$

Here T_{mix} is the resulting temperature of the combined air mass after mixing. Because the volumes of the specific parts m_1 and m_2 change, the specific heat capacity at constant pressure needs to be used in eq. (4.3). As $q_v c_{pv} \ll c_p$, eq. (4.3) can be approximated as

$$m_1(T_1 - T_{mix}) = m_2(T_{mix} - T_2) \tag{4.4}$$

or

$$T_{mix} = \frac{m_2 T_2 + m_1 T_1}{m_1 + m_2}. \tag{4.5}$$

This equation implies that, in a plot of e as a function of T, the final temperature of the mixed air mass lies on a straight line between T_1 and T_2, with its exact position given by the ratio m_1/m_2, as shown in Figure 4.2.

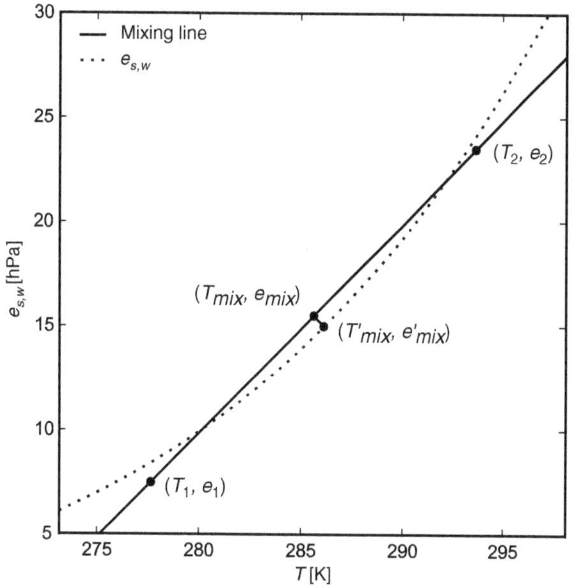

Fig. 4.2 Example of isobaric mixing for two air parcels with equal masses and with $T_1 = 4.5\,°C$; $e_1 = 7.5$ hPa and $T_2 = 20.5\,°C$; $e_2 = 23.5$ hPa. The dotted line, indicating the saturation vapor pressure $e_{s,w}(T)$, is obtained from the Clausius–Clapeyron equation (2.59).

The total volume V occupied by the air masses remains the same after mixing. From $V = V_1 + V_2$ it follows that $V = R_m/p\,(m_1T_1 + m_2T_2)$. The volume after mixing V_{mix} is

$$V_{mix} = \frac{R_m}{p} mT_{mix} = \frac{R_m}{p}(m_1 + m_2)T_{mix}. \tag{4.6}$$

It thus follows that $V = V_{mix}$ if T_{mix} is given by eq. (4.5).

Condensation occurs if the resulting value of e_{mix} is larger than the saturation vapor pressure at T_{mix}, as is the case in the example in Figure 4.2. The specific humidity will then decrease due to condensation until all the water vapor in excess of 100% relative humidity is condensed. The corresponding release of latent heat warms the air parcel; thus, in cases where the isobaric mixing of two air masses leads to condensation, the resulting temperature and specific humidity cannot simply be calculated according to eqs. (4.1) and (4.5), as the process of condensation and therefore the release of latent heat has to be taken into account. The changes in temperature and specific humidity due to condensation can be obtained from the first law of thermodynamics accounting for latent heat release (eq. 2.94), noting that $dp = 0$ in an isobaric process:

$$dq = c_p dT + L_v dq_s. \tag{4.7}$$

Making use of $q_v \approx \epsilon e/p$, the change in q_v is given as $dq_v = \epsilon de/p$ since p is constant. Thus we obtain the slope of the isobaric condensation line connecting the points (T_{mix}, e_{mix}) and (T'_{mix}, e'_{mix}) in Figure 4.2:

$$\frac{de}{dT} = \frac{p}{\epsilon}\frac{dq_v}{dT} = -\frac{pc_p}{\epsilon L_v}. \tag{4.8}$$

The intersection of this line with the saturation vapor pressure curve determines the resulting temperature T'_{mix} and vapor pressure e'_{mix} after the condensation process has terminated. The line connecting the points (T_{mix}, e_{mix}) and (T'_{mix}, e'_{mix}) is not necessarily perpendicular to the mixing line, but it is important to note that the temperature increases due to condensation.

In the example of Figure 4.2, the mixing of two air parcels with equal masses yields T_{mix} = 12.5 °C and e_{mix} = 15.5 hPa. Due to condensation, e_{mix} is lowered until saturation and T increases, leading to the final values T'_{mix} = 13 °C and e'_{mix} = 15 hPa at an ambient pressure of 780 hPa.

4.1.2 Adiabatic mixing

In contrast to fog, clouds form because air parcels or air masses cool adiabatically. In addition to the adiabatic cooling and expansion that occur when an entire air mass is lifted, small-scale turbulent adiabatic mixing can take place. In this case air parcels from different pressure levels mix after being brought together adiabatically. Temperature is not a good variable to describe adiabatic mixing because it changes under vertical motion. Instead the potential temperature θ is used because it is a conserved variable under dry adiabatic transformations.

As we now mix air parcels vertically, we need to take the differences in air density with altitude into account, and therefore to integrate vertically:

$$\theta_{mix} = \frac{1}{m}\int_{z_1}^{z_2} \rho\theta dz. \tag{4.9}$$

Here $m = \int_{z_1}^{z_2} \rho dz$ refers to the mass of the column in kg m^{-2}. Assuming that the air mass is in hydrostatic equilibrium, we can use the hydrostatic approximation (eq. 3.7) to replace dz with $-dp$, where we account for the minus sign by switching the limits of integration from z_1 and z_2 to p_2 and p_1:

$$\theta_{mix} = \frac{1}{\Delta p}\int_{p_2}^{p_1} \theta dp. \tag{4.10}$$

Here Δp refers to the difference in pressure between p_2 and p_1. As long as no condensation occurs, the specific humidity of the well-mixed boundary layer, $q_{v,mix}$, can be obtained accordingly because it is conserved under dry adiabatic transformations:

$$q_{v,mix} = \frac{1}{\Delta p}\int_{p_2}^{p_1} q_v dp. \tag{4.11}$$

The result of adiabatic mixing is a well-mixed planetary boundary layer, which originates from the continuous adiabatic mixing of countless up- and downward moving air parcels. As a consequence of this continuous mixing, the potential temperature θ_{mix} is constant with height in the entire well-mixed boundary layer.

4.2 Convection

Convection involves the conversion of potential energy to kinetic energy. In the context of convection, potential energy is used in a more general way, so that a system is said to have potential energy when it is not in its energetically lowest state. Convection involves a transition to an energetically lower state of the atmosphere in which ascending warmer and less dense air parcels are replaced by descending colder and denser air parcels. This is analogous to the behavior of an ascending air bubble in a glass of water, which minimizes the potential energy of the total system (the glass of water plus the air bubble).

Convection, in general, is the atmosphere's most efficient way to transport energy, momentum, moisture, trace gases and aerosol particles vertically. The rise of air on the windward side of a mountain is an example of forced convection, while the movement of air parcels induced thermally due to density differences in the ambient fluid is an example of free convection. In the atmosphere the latter dominates.

Convective clouds comprise all the cumuliform clouds (Cu, Ac, Cc and Cb, see Table 1.1). They form directly due to free (thermal) convection when the Sun's heating of the Earth's surface, or cooling aloft, creates an unstable air layer. Convective clouds can have large vertical extents (ranging from less than 1 km for altocumulus to 1.5 km for cumulus humilis, to 5 km for cumulus congestus to 12 km for cumulonimbus clouds; see Table 1.3).

4.2.1 Level of free convection

Free convection starts when the level of free convection (LFC) is reached. The LFC is the level beyond which an air parcel rises freely due to its positive buoyancy, i.e. above which its density is less than that of the environmental air. The LFC can be located in cloud-free air, or at the cloud base or somewhere inside the cloud.

If an air parcel that is lifted dry adiabatically is sufficiently moist, the lifting condensation level (LCL, see Section 2.5.7) will be reached at some point and further lifting will lead to cloud formation. If the LCL does not coincide with the LFC, mechanical lifting of the air parcel or air mass above the LCL, for example due to the orography or along a frontal surface, is needed for a cloud to form.

In the example in Figure 4.3, the air parcel would not rise by itself beyond the LCL, because the ambient temperature has a strong inversion and thus the air parcel is colder than the environment. Consequently no cloud can be formed and the LFC cannot be reached, unless the air parcel is mechanically forced further upwards until it reaches the LFC.

4.2.2 Convective condensation level

Instead of reaching the LFC by forced mechanical lifting as discussed above, an air parcel can also reach the CFC if it has sufficient positive buoyancy. The LFC reached in this latter way is also called the convective condensation level (CCL). It is defined as the point above which the atmosphere is conditionally or absolutely unstable, so that the air parcel

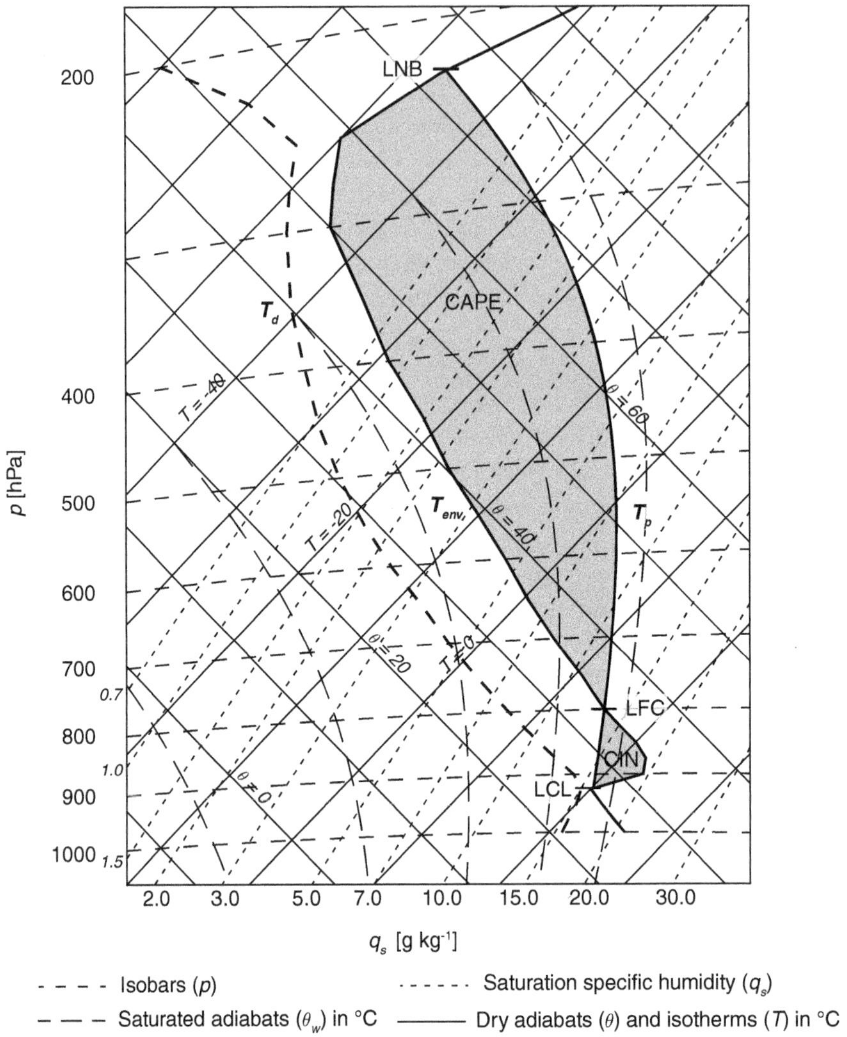

Fig. 4.3 Tephigram including examples of the LCL, LFC, LNB, CAPE and CIN for a given atmospheric temperature profile T_{env} (bold solid line) and dew point profile T_d (bold broken line). The air parcel's ascent beyond the LCL is shown as the solid line labeled T_p.

continues to rise along a saturated adiabat and a convective cloud forms. In other words, the CCL marks the base of a convective cloud.

The CCL can be obtained from a tephigram. Let us start again from the definition of the LCL. Figures 4.4a, b, c show an environmental temperature profile T_{env} (the bold broken-and-dotted line) that is absolutely stable and observable on any sunny summer morning. Under these conditions an air parcel would require mechanical lifting to reach the LCL. With $T_{env} = T_1 = 10$ °C and $q_{v,1} = 6$ g kg^{-1} at 1000 hPa, an air parcel lifted from the ambient temperature profile will reach its LCL at about 940 hPa (LCL$_1$ in Figure 4.4a). At LCL$_1$ the air parcel is much colder than the environmental temperature, so it will sink back and condensation will not occur. Additional mechanical lifting, e.g. forcing by orography, would be needed before it could cross the environmental temperature and rise due to its own positive buoyancy. Thus, reaching the LCL does not imply that convection occurs.

Now assume that the Sun heats the Earth's surface. This will cause the air above the surface to become absolutely unstable and will trigger vertical mixing within this column of air; as follows. First, heat is transferred by conduction from the surface to the air layer directly in contact with it. This results in a strong lapse rate between the surface air layer and the air above it. If the lapse rate becomes superadiabatic, i.e. it exceeds the value of the dry adiabatic lapse rate Γ_d, any small disturbance will lead to the ascent of air parcels. Together with the corresponding compensating downward motions (as will be discussed in more detail in Section 4.2.4.2) this results in mixing and the overturning of the heated layer. This process continues until the temperature profile is no longer unstable, i.e. it approaches the dry adiabatic lapse rate and θ becomes constant. At the same time the specific humidity q_v is vertically mixed until it approaches a constant value throughout the mixed layer.

In the example in Figure 4.4, the strong heating raises the surface temperature from $T_1 = 10$ °C to $T_2 = 24.5$ °C. This causes an air parcel to rise and heat to be convected upward, causing the mixed layer to expand vertically. The potential temperature θ then rises from 10 °C to 24.5 °C between the surface and the altitude at which the mixed layer has the same temperature as the environment, which in this case is 800 hPa. Below 800 hPa the new environmental temperature profile follows the dry adiabat to T_2 (Figure 4.4b). Also, q_v is mixed adiabatically and is constant at $q_{v,2} = 4$ g kg^{-1} from the surface up to the top of the well-mixed layer at 800 hPa. Thus in Figure 4.4b the T_d profile follows the solid line with $q_{v,2} = 4$ g kg^{-1} from the surface to 800 hPa, where T_d increases sharply to -2.4 °C (broken-and-dotted line); LCL_2 is reached at 700 hPa. At this point, no cloud has yet formed because the temperature at LCL_2 is still colder than the environmental temperature.

For an air parcel to reach the CCL and to form a cloud, a little extra heating from T_2 to $T_3 = 27$ °C is needed (Figure 4.4c). Above the CCL, which is found at 700 hPa in our example, further ascent will occur along a saturated adiabat; T_3 thus represents the critical temperature that the surface must reach in order for free convection to set in. Hence reaching T_3 is a prerequisite for the formation of cumuliform clouds. The temperature T_3 is also referred to as the convective temperature. The heat added to the mixed layer is proportional to the area spanned by T_1, T_3 and the temperature at the CCL (gray area in Figure 4.4c) because a tephigram is a "real" thermodynamic chart, where equal areas correspond to equal amounts of energy.

Note that in contrast with the LCL, which depends only on the properties of the air at the surface and does not entail any vertical mixing, the CCL describes a more realistic onset of convection, taking solar heating and vertical mixing into account. On a warm summer's day, as considered here, the CCL is higher than the initial LCL indicated in Figure 4.4a, because the surface has been heated by the Sun. As the air below the CCL is well mixed (with constant $q_{v,3}$ and $\theta = 27$ °C), the new LCL, i.e. LCL_3, coincides with the CCL.

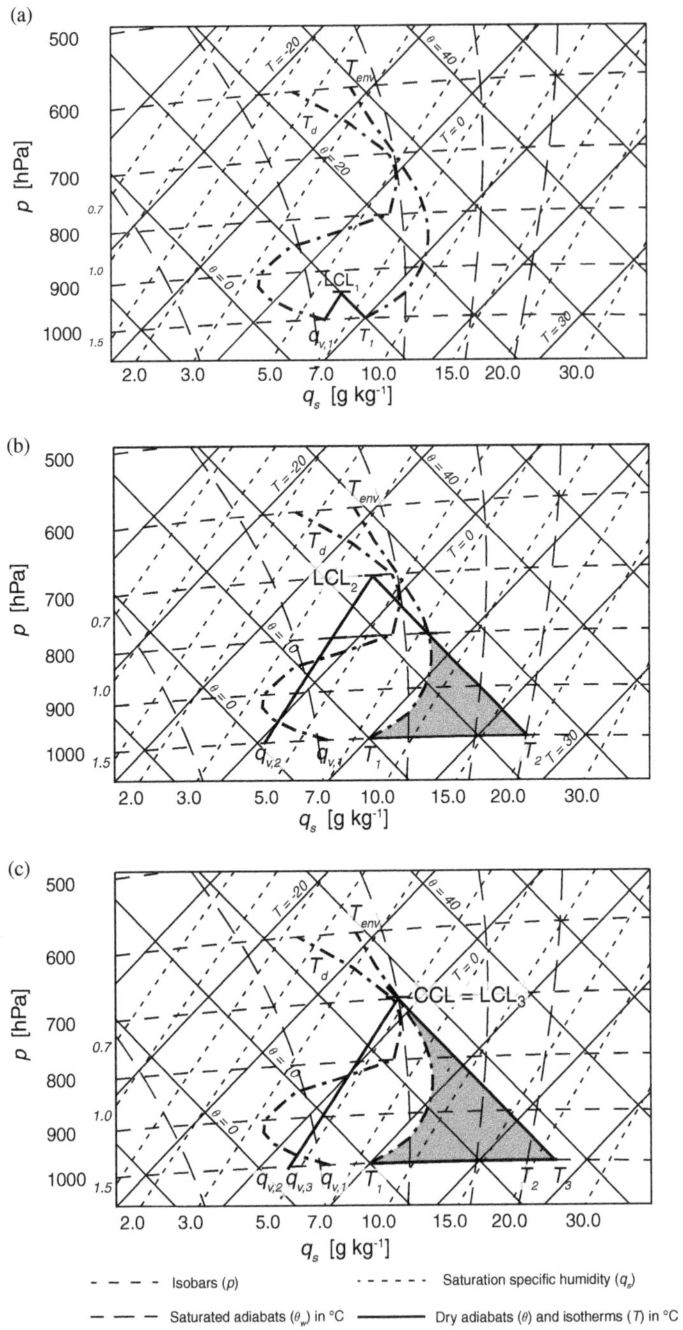

Fig. 4.4 Tephigram illustrating an air parcel reaching the CCL, owing to heating of the surface; it moves from T_1 via T_2 to T_3. The temperatures T_{env} and T_d before mixing are shown as bold broken-and-dotted lines. (a) The LCL for the air parcel which starts at the surface with T_1 and $q_{v,1}$ is reached at ~ 940 hPa as shown in solid lines (LCL_1). The gray areas in (b) between T_1, T_2 and T_{800} and in (c) between T_1, T_3 and the CCL are measures of the added heat. (c) Final temperature and moisture profiles (bold solid lines) from T_3 and $q_{v,3}$ to the CCL=LCL_3.

This mechanism of triggering convection by warming of the surface until it reaches a critical temperature is common in the tropics and mid-latitudes during summer. In reality, however, things are more complicated because heating often causes evaporation from Earth's surface, leading to a higher q_v and therefore a lower CCL.

4.2.3 Elementary parcel theory

In the last section the triggering of convection by solar heating of the surface was explained qualitatively. In the present section the intensity of convection for given atmospheric conditions will be calculated.

The easiest way to describe convection is to use elementary parcel theory. However, this theory has some limitations, which will be discussed in Section 4.2.4. In order to use elementary parcel theory, assume that an air parcel is a buoyant element of air with an unspecified size and shape. Furthermore, the buoyant element maintains its identity in thermodynamic processes and it does not interact with the environmental air, i.e. there is no friction and no exchange of mass and energy between an ascending (positively buoyant) air parcel and the surrounding air. The buoyant element has uniform properties throughout and its pressure instantly adjusts to the pressure of the surrounding air.

In order to estimate the strength of the potential convective activity, weather forecasters often use the convectively available potential energy (CAPE), measured in J kg^{-1}. It can be regarded as an estimate of the maximum kinetic energy that an air parcel may gain due to latent heat release during condensation on its convective ascent.

The convectively available potential energy can be obtained from a tephigram or can be calculated mathematically. On a tephigram, it is proportional to the area encompassed by the LFC, the level of neutral buoyancy (LNB) also called the equilibrium level, the environmental temperature and the saturated adiabat (Figure 4.3). An air parcel that rises beyond the LFC remains positively buoyant as long as its temperature is warmer than that of the environment. The altitude where both temperatures are again the same is the LNB. At this altitude the buoyancy force is zero, which does not imply that the air parcel's ascent abruptly terminates as will be discussed in Section 10.1.1.

The CAPE value obtained from a tephigram is often referred to as the pseudoadiabatic CAPE, because it assumes that the condensed water immediately falls out as precipitation (Section 2.5.11). For brevity, the pseudoadiabatic CAPE will be referred to simply as CAPE from here on.

The area that is encompassed by the profiles of the lifted air parcel and the environment, between the parcel's starting point and the LFC, is the work that needs to be done on the system to lift the air parcel to the LFC. It is called the convective inhibition (CIN) and is shown in Figure 4.3.

Mathematically, the CAPE value can be obtained by starting from the definition of the buoyancy force (eq. 3.15) and replacing the temperature by the virtual temperature T_v (eq. 2.86) in order to take moisture fluctuations into account. Thus an air parcel with the same temperature but higher moisture content than the environment will rise because of the lower density of water vapor compared with that of air. In order to calculate the CAPE value, we define $w = dz/dt$ as the vertical velocity and relate its change with time, i.e. the acceleration, to the buoyancy force per unit mass F_B:

$$\frac{dw}{dt} = gB = F_B, \qquad B = \frac{T_v - T_{v,env}}{T_{v,env}}, \tag{4.12}$$

where B is the buoyancy term. Multiplication of eq. (4.12) with dz yields

$$\frac{dw}{dt}dz = wdw = \frac{1}{2}d(w^2) = g\left(\frac{T_v - T_{v,env}}{T_{v,env}}\right)dz. \tag{4.13}$$

Integration between the LFC at z_0 and LNB at z and rearranging terms yields

$$w^2 = w_0^2 + 2g\int_{z_0}^{z}\left(\frac{T_v - T_{v,env}}{T_{v,env}}\right)dz = w_0^2 + 2 \times \text{CAPE}, \tag{4.14}$$

where w is the velocity at height z and w_0 is the velocity at z_0. The square root of the right-hand side of eq. (4.14) provides the upper limit for the vertical velocity of an ascending air parcel that would be reached in the absence of friction and dilution:

$$w_{max} = \sqrt{w_0^2 + 2 \times \text{CAPE}}. \tag{4.15}$$

Assuming that w_0 is negligible, w_{max} equals the square root of twice the CAPE value. From eq. (4.14) we see that then the CAPE value can be written as follows:

$$\text{CAPE} = g\int_{z_0}^{z}\frac{T_v - T_{v,env}}{T_{v,env}}dz. \tag{4.16}$$

Using the hydrostatic approximation (eq. 3.7) to replace dz by dp:

$$\text{CAPE} = -\int_{p(z_0)}^{p(z)}\frac{1}{\rho}\frac{T_v - T_{v,env}}{T_{v,env}}dp. \tag{4.17}$$

Then the ideal gas law (eq. 2.4) can be used to replace ρ:

$$\text{CAPE} = -\int_{p(z_0)}^{p(z)}\frac{R_d}{p}(T_v - T_{v,env})\,dp \tag{4.18}$$

or

$$\text{CAPE} = -\int_{\ln p(z_0)}^{\ln p(z)} R_d(T_v - T_{v,env})\,d\ln p. \tag{4.19}$$

As discussed above, CAPE is proportional to the area on a tephigram bounded by the process curve of the air parcel's temperature T_p and the ambient temperature profile T_{env}, from pressure $p(z_0)$ at the LFC to pressure $p(z)$ at the LNB (Figure 4.3). Because the LFC can occur in cloud-free air, the process curve of the air parcel in case of an absolutely unstable situation could first follow the dry adiabat up to the CCL and then follow the saturated adiabat above.

The CAPE value does not exactly equal this area because for the calculation T_v is needed, but only T is given in a tephigram. The proportionality of the CAPE value to the area in

Table 4.1 Typical values of the pseudoadiabatic CAPE [J kg^{-1}] in relation to static stability, type of convection, w_{max} from eq. (4.15) with $w_0 = 0$, the effective vertical velocity $w_{eff} = \sqrt{\text{CAPE}}$ and typical observed updraft velocity w_{obs}, all given in m s^{-1}.

CAPE	Stability	Type of convection	w_{max}	w_{eff}	w_{obs}
0	stable	no convection	0	0	0
0–1000	marginally unstable	hurricanes and weak to moderate convection	0–45	32	10
1000–2500	moderately unstable	ordinary thunderstorms	45–71	32–50	30
2500–3500	very unstable	tornadic storms, squall lines	71–84	50–59	40
>3500	extremely unstable	severe tornadic storms	> 84	> 59	> 40

a tephigram arises because the product of entropy times temperature is an energy and the tephigram is an $S-T$ diagram. In practice the integral in eq. (4.16) is approximated by a sum because a radiosonde provides observational data only at discrete levels:

$$\text{CAPE} = \sum_{i=1}^{n} R_d(\overline{T}_{v,i} - \overline{T}_{v,env,i})\frac{\Delta p_i}{\overline{p}_i}, \tag{4.20}$$

where the values $\overline{T}_{v,i}$, $\overline{T}_{v,env,i}$ and $\overline{p}_i$ represent the average values over the pressure interval Δp_i.

Typical values of CAPE in relation to the average static stability between the LFC and the LNB, the maximum vertical velocity and the associated type of convection are summarized in Table 4.1 (Jorgensen and Lemone, 1989).

Predicting the vertical velocity from eq. (4.15) and Table 4.1 results in an overestimation because the finite size of the rising air parcel, which exerts an aerodynamic drag, the mixing of the air parcel with environmental air, the compensating downward motions of the surrounding air and the weight of the condensed water, some of which is carried along with the parcel, have all been neglected. These effects can be taken into account by modifying the elementary parcel theory following Rogers and Yau (1989), as will be discussed next.

4.2.4 Modification of the elementary parcel theory

4.2.4.1 Weight of the condensed water

So far we have simplified the buoyancy term by assuming that once the relative humidity exceeds 100% and cloud droplets have formed, they were converted immediately to precipitation and left the air parcel. Now we will allow the condensed water to remain in the air parcel. If water is present in the condensed phase, i.e. in the form of cloud droplets or ice crystals, instead of in the vapor phase, its virtual temperature will be reduced with increasing amount of condensate present, owing to the weight of the latter. The virtual temperature in the presence of condensate, $T_{v,cl}$, can be defined as

$$T_{v,cl} \approx T(1 + 0.608q_v - q_l - q_i), \tag{4.21}$$

where q_l, q_i are the cloud liquid-water and cloud ice-mass mixing ratios (eq. 1.3). The buoyancy term for a thermal, i.e. a column of warm air rising through otherwise cloud-free air from a given reference virtual temperature $T_{v,env,0}$, taking the condensate into account, is given as

$$B = \frac{T_{v,cl} - T_{v,env,0}}{T_{v,env,0}}. \tag{4.22}$$

If the thermal rises inside a cloud, the virtual temperature of the environment needs to be replaced by the temperature inside the cloud, $T_{v,env,cl}$:

$$B = \frac{T_{v,cl} - T_{v,env,cl}}{T_{v,env,cl}}. \tag{4.23}$$

Taking the weight of the condensed water, i.e. the adiabatic liquid water content (eq. 2.106), into account means that we now consider a reversible saturated ascent instead of a pseudoadiabatic ascent. An analysis of 732 soundings over Florida in 2007 showed that the reversible saturated CAPE value, accounting for the condensate only in the form of liquid water, was on average 1000 J kg^{-1} smaller than the pseudoadiabatic CAPE values owing to the weight of the condensed water (Cotton *et al.*, 2011).

4.2.4.2 Compensating downward motion

Because of the finite vertical extent of a thermal, the upward displacement of air within it must be compensated by a downward motion in the ambient air. The descending air parcel can be characterized by a negative downward oriented CAPE, the so-called DCAPE. This quantity equals the absolute value of the negative CAPE for a saturated downdraft. It is equal to the amount of kinetic energy associated with a saturated downdraft (Markowski and Richardson, 2010). It can be obtained on a tephigram when following the moist adiabat down to the surface starting from the minimum value of θ_e found in the lowest 400 hPa of the sounding and calculating the positive and negative area. The higher DCAPE, the higher are the vertical velocities in the downdrafts. Just as the CAPE value is used to estimate updraft velocities, DCAPE can be used to estimate downdraft velocities. However, while it is clear that CAPE needs to be integrated from the LFC to the LNB, the integration limits for DCAPE are less obvious. Also, part of the descent of the air parcel may be unsaturated or subject to entrainment. Thus, predictions of downdraft velocities based on DCAPE are even more uncertain than updraft velocities estimated from CAPE.

Let us concentrate on the compensating descending air mentioned above. If the descending air is cloud free it will be warmed dry adiabatically. This will influence the temperature and, hence, the buoyancy of the rising thermal. Let A_{env} denote the horizontal area over which air is descending and A the area of the thermals. The mass flux MF of

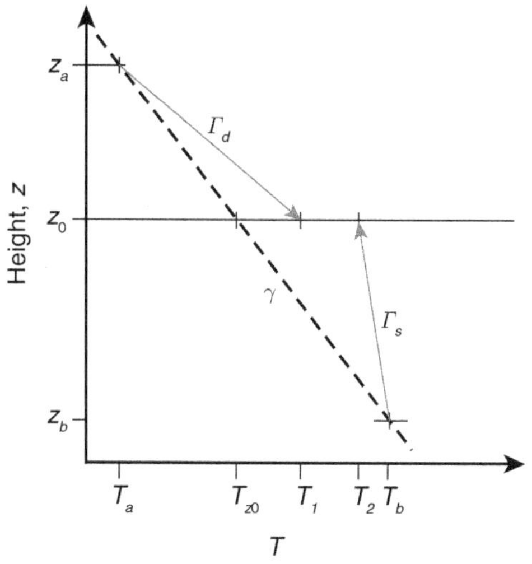

Fig. 4.5 Schematic of height versus temperature for a rising cloudy air parcel and the compensating dry adiabatic downward motion of the environment; T_{z0} refers to the initial temperature at z_0, γ to the ambient lapse rate, Γ_d to the dry adiabatic lapse rate and Γ_s to the saturated adiabatic lapse rate.

upward-moving air through z_0 is $\text{MF}_u = \rho w A$ [kg s^{-1}] and the downward mass flux is $\text{MF}_d = -\rho_{env} w_{env} A_{env}$ [kg s^{-1}]. If the area of consideration is large enough, the upward and downward mass fluxes will equal each other, so that

$$\frac{A}{A_{env}} = -\frac{\rho_{env} w_{env}}{\rho w} \sim -\frac{w_{env}}{w}, \tag{4.24}$$

where we have assumed that the densities ρ_{env} and ρ are equal. For a simple approximation, let us assume further that the ascending air is cloudy, i.e. it follows the saturated lapse rate Γ_s, while the descending air follows the dry adiabatic lapse rate Γ_d, as illustrated in Figure 4.5. Thus, in a short time dt, air will arrive at a level z_0 from a lower level $z_b = z_0 - wdt$. This air was originally at temperature $T_b = T_{z0} + \gamma wdt$ and will now have temperature $T_2 = T_{z0} + (\gamma - \Gamma_s)wdt$, where γ is the ambient lapse rate and T_{z0} is the initial temperature at z_0, i.e. before the upward and compensatory downward motions started.

Air arriving from the higher level $z_a = z_0 + w_{env}dt$, where the temperature is $T_a = T_{z0} - \gamma w_{env}dt$ now has temperature $T_1 = T_{z0} + (\Gamma_d - \gamma)w_{env}dt$. The situation is unstable when T_1 is smaller than T_2 (Figure 4.5). If the compensating downward motion were not taken into account, the temperature difference between the air parcel and the environment at z_0 would be $T_2 - T_{z0}$, which is larger than the difference $T_2 - T_1$ obtained in the presence of the downward motion. Thus the positive buoyancy of a cloudy thermal is reduced in the presence of compensating downward motions in the surrounding cloud-free air, and the criteria for instability have to be adjusted. Instability occurs if $T_2 > T_1$:

$$(\gamma - \Gamma_s)w > (\Gamma_d - \gamma)w_{env},$$

thus if

$$(\gamma - \Gamma_s)A_{env} > (\Gamma_d - \gamma)A. \tag{4.25}$$

In the limit in which A goes to zero, this is equivalent to $\gamma > \Gamma_s$, the previously defined statically unstable situation (Table 3.3). If the ambient air is dry and neutrally stratified, i.e. $\Gamma_d = \gamma$, the downward motion has no effect, because the temperature at z_0 within A_{env} does not change with time ($T = T_1 = T_{z0}$). This is reflected in eq. (4.25): if $\Gamma_d = \gamma$ then $(\gamma - \Gamma_s)A_{env} > 0$ for instability, which corresponds to the normal conditionally unstable case as defined in Table 3.3. Instead of $\gamma = \Gamma_s$, the condition for the neutral case for the rising cloudy air parcel in the presence of downward motion in the surrounding cloud-free air has to be written as

$$\frac{\gamma - \Gamma_s}{\Gamma_d - \gamma} = \frac{A}{A_{env}}. \tag{4.26}$$

If $A/A_{env} > 0$, that is if the thermals are not negligible in size, this equation can only be satisfied if $\Gamma_d > \gamma > \Gamma_s$. In other words, when the compensating downward motions are taken into account, the environmental lapse rate must be steeper for an instability to occur. On a tephigram part of the originally conditionally unstable area (Figure 4.6) now belongs to the absolutely stable region.

4.2.4.3 Dilution by entrainment

When a buoyant element ascends, some mixing is expected to take place through the boundaries of the air parcel, as shown schematically in Figure 4.7. Since the ambient air is generally cooler ($T_{env} < T_{cl}$) and drier ($q_{v,env} < q_{v,cl}$) than the buoyant element, such mixing will reduce the temperature and hence the positive buoyancy of the thermal and so lower its specific humidity q_v. This incorporation of environmental air and the associated changes in the air parcel's properties are referred to as entrainment. The reverse process, where the air parcel loses mass to the environment, is called detrainment.

To account for entrainment, we need to consider the energy exchange between the rising air parcel and its environment. As discussed in Chapter 2 this is best done by choosing the right thermodynamic variables to describe the state of our system and the changes in its properties. Here we introduce the moist static energy s to describe the properties of the air parcel, the environment and the exchange of energy between them. The quantity s can be shown to be given by:

$$s = c_p T + gz + L_v q_v. \tag{4.27}$$

We will use s_{cl} and s_{env} to denote the moist static energies of the cloud air parcel and of the environment, respectively. During a finite time Δt in the air parcel's ascent, a mass $(\Delta m)_\epsilon$ will be entrained from the environment, while the air parcel loses a mass $(\Delta m)_\delta$ owing to detrainment. Focusing on the properties of the air parcel inside a cloud, we are primarily interested in the entrainment process. The rate of change in the air parcel's moist static energy due to entrainment is given by

$$\frac{ds_{cl}}{dt} = \frac{1}{m}\left(\frac{dm}{dt}\right)_\epsilon (s_{env} - s_{cl}), \tag{4.28}$$

where m denotes the mass of cloudy air in the parcel; this consists of dry air, water vapor and condensed water. Now substituting s_{env} and s_{cl} by their definitions from eq. (4.27), we arrive at

$$\frac{d}{dt}(c_p T_{cl} + gz + L_v q_{v,cl}) = \frac{1}{m}\left(\frac{dm}{dt}\right)_\epsilon \left[c_p(T_{env} - T_{cl}) + L_v(q_{v,env} - q_{v,cl})\right]. \tag{4.29}$$

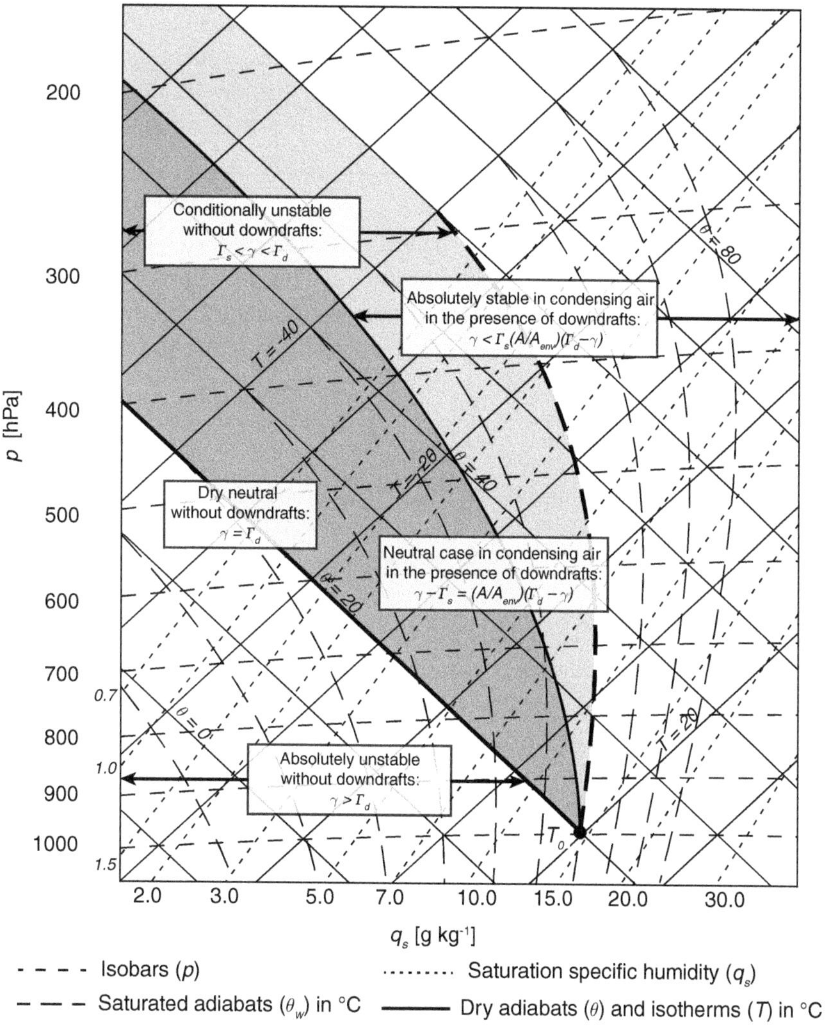

Fig. 4.6 The stability criteria in condensing air in the presence of downdrafts, for an air parcel starting at 20 °C and 1000 hPa. The conditionally unstable area in the case without downdrafts includes the total gray area and that in the presence of downdrafts is shown in the darker gray.

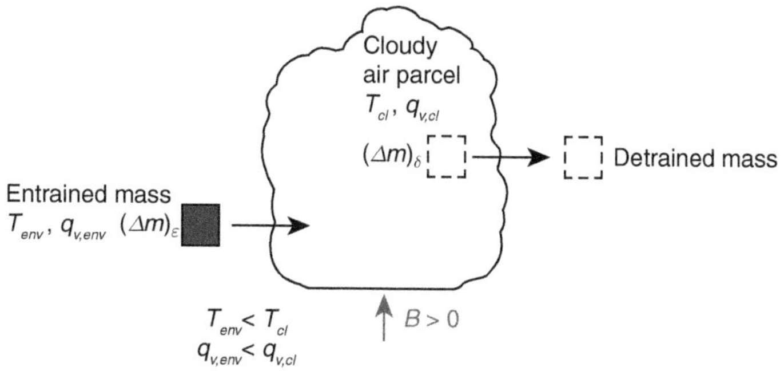

Fig. 4.7 Schematic of the entrainment process of a rising air parcel with positive buoyancy B.

Rearranging terms and noting that $w = dz/dt$ and that the specific humidity inside the cloudy parcel $q_{v,cl}$ equals the saturation specific humidity q_s, we obtain:

$$\frac{dT_{cl}}{dt} = \underbrace{-\frac{g}{c_p}w}_{\text{(i)}} \underbrace{-\frac{L_v}{c_p}\frac{dq_s}{dt}}_{\text{(ii)}} + \underbrace{\frac{1}{m}\left(\frac{dm}{dt}\right)_\epsilon \left[(T_{env} - T_{cl}) + \frac{L_v}{c_p}(q_{v,env} - q_s)\right]}_{\text{(iii)}}, \tag{4.30}$$

where the individual terms on the right-hand side of eq. (4.30) describe (i) the dry adiabatic cooling rate, (ii) the latent heating rate and (iii) the effects of entrainment. For an ascending air parcel we can obtain the change in temperature with height, instead of with time, on dividing eq. (4.30) by the positive vertical velocity dz/dt. The result is

$$\frac{dT_{cl}}{dz} = -\frac{g}{c_p} - \frac{L_v}{c_p}\frac{dq_s}{dz} + \Lambda\left[(T_{env} - T_{cl}) + \frac{L_v}{c_p}(q_{v,env} - q_s)\right], \tag{4.31}$$

where the entrainment rate Λ [m^{-1}] is given by

$$\Lambda = \frac{1}{m}\left(\frac{dm}{dz}\right)_\epsilon. \tag{4.32}$$

The case of no entrainment ($dm = 0$) yields the original saturated adiabatic lapse rate (eq. 3.31). In the presence of convection, the term in square brackets in eq. (4.31) is always negative, because the temperature and specific humidity are higher inside the air parcel than in the environment. This means that the temperature of an ascending air parcel decreases at a faster rate when entrainment takes place, i.e. its buoyancy is impaired by entrainment (Figure 4.8). Consequently, the CAPE value and the vertical velocity are significantly overestimated when entrainment is neglected.

The change in the vertical component of the momentum equation with height in the presence of entrainment, when pressure perturbations are ignored, can be shown to be given as

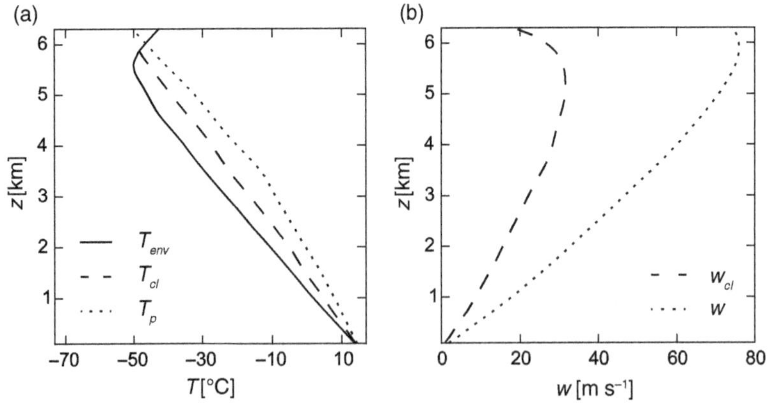

Fig. 4.8 Temperature and vertical velocity profiles for an air parcel rising beyond the level of free convection LFC at cloud base. (a) Temperature as a function of height for the environment, T_{env}, and for the air parcel without entrainment, T_p, and with entrainment, T_{cl}, using eq. (4.31). (b) Vertical velocity as a function of height without entrainment, w, and with entrainment, w_{cl}, using eqs. (4.33) and (4.15).

$$w\frac{dw}{dz} = B - \Lambda w^2; \tag{4.33}$$

here Λ can be approximated as $0.2/b$ (Houze, 2014), where b is the radius of the rising air parcel, which increases linearly with height as: $b = \alpha_e z$.

In order to solve eq. (4.31), the change in q_s with height must also be known. It can be shown to be given by:

$$\frac{dq_s}{dz} = -\frac{C}{w} + \Lambda(q_{v,env} - q_s) \tag{4.34}$$

where C is the condensation rate dq_x/dt [kg kg^{-1} s^{-1}], which is the mass of condensate that is produced per unit mass of air and per unit time.

An example of an air parcel rising above the level of free convection with and without entrainment is shown in Figure 4.8. The temperature profile including entrainment T_{cl} according to eq. (4.31) falls in between the environmental temperature profile T_{env} and the temperature profile for a pseudoadiabatic air parcel, T_p, without entrainment. The integrated pseudoadiabatic CAPE value without entrainment amounts to 2650 J kg^{-1} and leads to a maximum vertical velocity 76 m s^{-1}. If entrainment is taken into account as in eq. (4.33), the peak vertical velocity is reached lower in the cloud and is only 31 m s^{-1}; this is less than half the vertical velocity without entrainment, owing to the reduction in the integrated pseudoadiabatic CAPE value to 1275 J kg^{-1}.

4.2.4.4 Aerodynamic drag

A last point to be considered is that a rising air parcel has a finite size. An easy visualization of a rising air parcel is to assume that it has the shape of a thermal or a bubble. Regardless of its shape, however, it needs to push air out of its way when it rises. Owing to friction this creates a drag force counteracting the positive buoyancy force, which leads to a further reduction in the CAPE value and thus the vertical velocity.

4.2.4.5 Intermittence and inhomogeneity of entrainment

In all these modifications to the elementary parcel theory, we have considered the entrainment process to take place instantaneously and to affect the whole air parcel. Both assumptions are unrealistic, as observations have shown that entrainment takes place intermittently in time and inhomogeneously in space (Houze, 2014). Still, as shown in Table 4.1, it is better to consider entrainment even as an instantaneous and homogeneous process than to neglect it completely.

In summary, all the modifications to elementary parcel theory discussed above in Sections 4.2.4.1–4.2.4.4 reduce the positive buoyancy and thus the vertical velocity of a thermal. A simple approach to these modifications is to calculate w from $\sqrt{\text{CAPE}}$ instead of from $\sqrt{2 \times \text{CAPE}}$ (see w_{eff} in Table 4.1), which in effect assumes a 50% efficiency of conversion of CAPE into vertical kinetic energy (Zipser, 2003). These values of w are still higher than the typical vertical velocities observed in various types of convective clouds, because not only the total amount of CAPE but also its vertical distribution matter: for the same value of CAPE, convection is more severe in cases when CAPE is distributed over a shallower layer than if it is distributed over a deep layer, because the buoyancy maximum and the vertical acceleration are larger in the former case (Blanchard, 1998).

4.3 Exercises

Single/multiple choice exercises

1. For a given ambient temperature T_{env} and dew point temperature T_d ($< T$), which of the following statements about fog and clouds is/are true?
   (a) Saturation is reached at the same dew point temperature in fog and clouds.
   (b) The saturation mixing ratio is the same in fog and clouds.
   (c) The cloud base is the same in fog and clouds.
   (d) The thicknesses of fog and clouds are the same.
2. Elementary parcel theory is a simple way to describe convection. Which of the following are considered in this approach?
   (a) The aerodynamic drag.
   (b) The mass and energy exchange between the air parcel and the surrounding air.
   (c) The compensating downward motions.
   (d) The moisture content of the air parcel.
   (e) The weight of the condensed water.
3. Which of the following is/are true?
   (a) The absolute value of CAPE is always larger than that of CIN.
   (b) When an air parcel reaches the LFC, it rises until it reaches the LNB.
   (c) A tephigram is useful to describe isobaric mixing.
   (d) Reaching the LCL implies that convection occurs.
4. Which property(ies) is/are not constant with altitude in a vertically well-mixed cloud-free boundary layer?
   (a) The water vapor mixing ratio w_v.
   (b) The specific humidity q_v.
   (c) The temperature T.
   (d) The potential temperature θ.
5. Ascending air must be compensated by downward motion in the ambient air in such a way that ...
   (a) ... the velocity of the sinking air equals that of the ascending thermal.
   (b) ... the mass flux of the compensating air equals that of the ascending thermal.
   (c) ... the sinking air is warming at a rate equal to the cooling rate of the ascending thermal.
   (d) ... the specific humidity of the compensating air equals that of the ascending thermal.
6. Which of the following is/are true?
   (a) Above the CCL the atmosphere is absolutely unstable.

(b) The cloud top is found at the LNB, where the ascent of an air parcel is terminated; the cloud base is located at the LFC, where the air parcel is positively buoyant.
(c) The LCL can be found on a tephigram at the point where the line of constant saturation specific humidity intersects the line of constant temperature.
(d) On a warm day in spring over the ocean, we will usually find the CCL at lower levels than those expected from studying a tephigram.

7. Entrainment refers to the mixing between a buoyant element (e.g. a cloud) and the ambient air. Which of the following is/are true?
(a) Entrainment increases the buoyancy and the water vapor mixing ratio in a cloud.
(b) The temperature profile including entrainment lies between the moist adiabatic and the environmental lapse rate.
(c) Entrainment tends to increase the saturated adiabatic lapse rate.
(d) The calculated CAPE value and the vertical velocity are increased if entrainment is taken into account.

8. Complete the following sentence in as many ways as possible. Convection ...
(a) ... is the only way to transport atmospheric heat and moisture.
(b) ... includes transport and possibly phase changes.
(c) ... destabilizes the atmosphere, causing radiative cooling to occur.
(d) ... is underestimated by the elementary parcel theory.

Long exercises

1. Name the most important differences between cloud and fog formation by briefly explaining the terminology of mixing processes.

2. Consider two water reservoirs A and B which are separated by a wall. Reservoirs A and B together have volume 90 dm^3; the volume of A is twice as large as the volume of B. The water in reservoir A has temperature 12 °C and the water in reservoir B has temperature 20 °C. Calculate the mixing temperature T_{mix} after the removal of the wall. Assume that the term $w_v c_{pv}$ and the volume of the wall can be neglected.

3. Assume a cold winter day with an air temperature T_{env} = 0 °C, relative humidity RH = 90% and air pressure p = 1013 hPa. The air you exhale is at body temperature (T_{breath} = 37 °C) and is almost saturated with respect to water, having relative humidity RH$_{breath}$ = 95%.
(a) Neglecting condensation, calculate the temperature T, the vapor pressure e and the saturation vapor pressure e_s if equal masses of the exhaled air and ambient air mix. Would you expect the air to condense?
(b) Solve graphically for how T and e inside the cloud (the mixed air) change when the latent heat released by condensing droplets is taken into account.

4. Inspect the tephigram in Figure 4.9, showing a balloon sounding.
(a) Locate the LCL for an air parcel that starts rising at 1000 hPa. Will the air parcel continue ascending if it reaches its LCL?
(b) How much does the surface temperature have to increase in order that the LCL coincides with the CCL? You can neglect the vertical mixing of q_v.

Table 4.2 Pressure p, environmental temperature T_{env} and specific humidity ($q_{v,env}$) from a balloon sounding.

p (hPa)	T_{env} (°C)	$q_{v,env}$ (g kg^{-1})
1000	25.5	15
970	20	15
800	15	9
770	13	8.5
560	−6	2.0
360	−26	0.5
260	−37	0.5

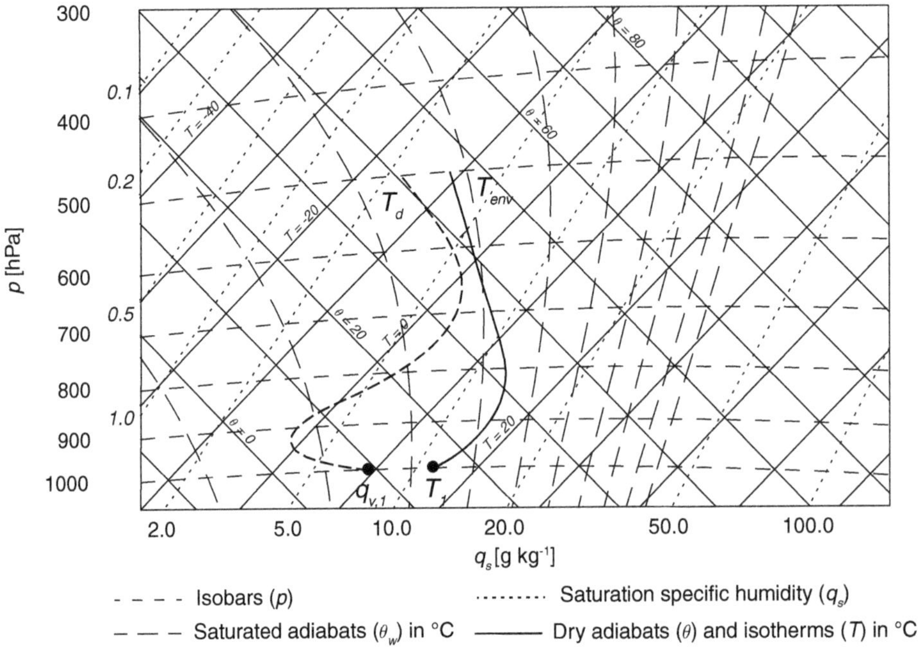

Fig. 4.9 Tephigram of a balloon sounding.

5. The compensating downward motion of the ambient air influences the buoyancy and the vertical velocity of an air parcel and therefore also its CAPE value. As a result, the criteria for instability, which is essential for the further lifting of the air parcel, need to be adjusted.
   (a) Draw the temperature profile from the sounding data given in Table 4.2 in a tephigram. Sketch and calculate the CIN and CAPE values.
   (b) Interpret your results. What type of convection do you expect if CIN could be overcome?

   A tephigram can be downloaded from here: www.cambridge.org/clouds.

5 Atmospheric aerosol particles

Thus far we have discussed the macrophysical requirements for cloud formation. Since each cloud droplet forms on an aerosol particle, these particles will be discussed in this chapter. An aerosol is defined as a dispersed system of solid or liquid particles suspended in a carrier gas, in our case air. Although technically the term "aerosol" includes both the particles and the carrier gas, in atmospheric science it is common to use the term "aerosols" just for the solid or liquid particles and to neglect the carrier gas.

Aerosol particles can be classified in terms of their chemical composition and according to their physical characteristics such as size, shape, density and mass, as will be discussed in Section 5.1. Because aerosol particles vary in size from a few nanometers to several micrometers, it has been found useful to describe their number, surface and mass concentrations in terms of size ranges, called modes, using log-normal size distributions (Section 5.2). Aerosol particles can be directly emitted into the atmosphere or form by gas-to-particle conversion. These mechanisms and the emissions of the most important aerosol types are the subject of Section 5.3. Aerosol particles are removed from the atmosphere by dry and wet scavenging (Section 5.4). The difference between their formation and removal rates determines their abundance (burden) in the atmosphere as well as their atmospheric lifetime. Their lifetime is strongly dependent on their size but also on the altitude in the atmosphere (Section 5.5). The chapter concludes with a summary of the most important aerosol processes in Section 5.6.

5.1 Chemical and physical characteristics of aerosol particles

Depending on the type of particle, aerosols have differing properties that determine their role in atmospheric processes. A number of fundamental properties can be used to characterize aerosol particles; these include chemical and physical properties.

5.1.1 Chemical characteristics

The chemical characterization of an aerosol particle at the level of chemical compounds would usually be too complicated. Whereas a sulfuric acid particle nucleated from its gaseous precursors can be described by a single chemical formula, a mineral dust particle or a combustion-generated carbonaceous particle consists of a vast multitude of different chemical compounds. Hence, it is more common to group aerosols into aerosol types (or species) relating the chemical characterization to the aerosol source. For example the term carbonaceous aerosol particles refers to combustion-generated particles that include many different chemical compounds. Depending on the application, chemical characterization can be refined to resolve differences within a specific aerosol species, such as between the fractions of black (BC) and organic carbon (OC) in carbonaceous aerosols. Organic carbon refers to any carbonaceous compound in which the carbon is chemically bound to hydrogen and other elements (e.g. O, S, N are typical elements for organic compounds) and is characterized by volatilization temperatures below 613 K (in an inert atmosphere). In contrast, BC refers to carbonaceous compounds which are produced during (incomplete) combustion processes (of fossil fuels and biomass) and are termed refractory owing to a volatilization temperature near 4000 K. Some authors loosely use BC and soot interchangeably, but this practice is discouraged.

Moreover, aerosol particles can exist in different mixing states. If different aerosol species get mixed in the atmosphere (e.g. desert dust and marine salt particles) but the individual particles retain their identity, the result is called an externally mixed aerosol. Each particle then only consists of one specific species (Figure 5.1a). In an internally mixed aerosol, every particle contains different species (Figure 5.1b) but the internal mixture can vary from one particle to the next. For instance a black carbon particle internally mixed with organic carbon can be next to a mineral dust particle coated with sulfate. Internal mixing may be the result of atmospheric processes such as the condensation of gaseous compounds on the surface of a particle or the evaporation of a cloud droplet which have collected particles of different species, leaving behind one particle combining the species previously present in the droplet.

5.1.2 Physical characteristics

The state of matter, i.e. whether aerosol particles are liquid or solid, determines the ability of the particles to act as cloud condensation nuclei (CCN) or as ice nucleating particles (INPs). Furthermore, this greatly controls their role in chemical reactions. As we will see in Chapters 7 and 8, aqueous particles can be activated to become cloud droplets in warm clouds, whereas solid or crystalline aerosol particles promote ice crystal formation in ice and mixed-phase clouds. An example of an aerosol type which usually is liquid under atmospheric conditions is provided by sulfuric acid particles. In contrast, mineral dust particles are always solid or at least have a solid core if coated with soluble material.

In terms of physical characteristics, shape, surface area, volume, density, mass or size are often used to characterize particles. Liquid aerosol particles are usually approximately spherical, while solid aerosol particles often have complex shapes. Black carbon particles, for instance, are often composed of long chain-like agglomerates.

In general, size is the most important property of aerosol particles, since a lot of other properties are related to it. However, the size of an aerosol particle is not unambiguously determined but depends on the measurement method. Calculation of the particle radius from the measured quantity usually requires an assumption on particle properties such as

Table 5.1 Aerosol modes with their corresponding radius ranges.

Aerosol mode	Range in radius
Nucleation mode	1.5 – 5 nm
Aitken mode	> 5 nm – 0.05 μm
Accumulation mode	> 0 .05 – 0.5 μm
Coarse mode	> 0.5 – 5 μm
Giant particles	> 5 μm

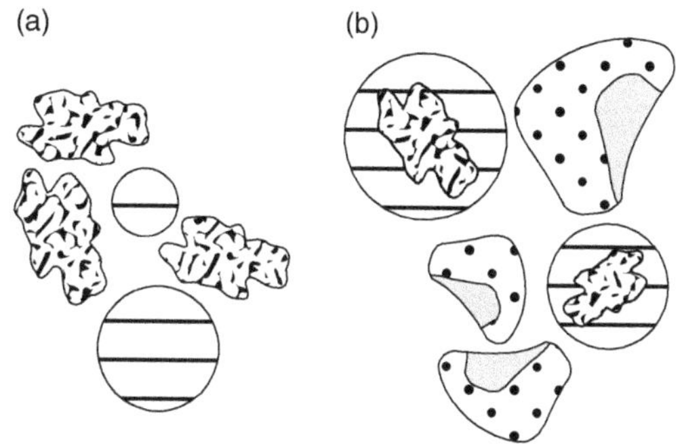

Fig. 5.1 Illustration of (a) externally mixed aerosol particles and (b) internally mixed aerosol particles.

shape, mass density and refractive index. Therefore, standard values for these properties are assumed in order to obtain an equivalent particle radius for the specific measurement method (Baron and Willeke, 2001). For example, the so-called equivalent aerodynamic particle radius obtained from a measurement of the reaction of a particle to inertial forces is defined as the radius of a sphere with unit density ($\rho = 1$ g cm^{-3}) having the same inertia as the measured particle.

Throughout this book, when using the term "size" we refer to the geometrical (physical) radius of the particle under consideration, if not stated otherwise. For simplicity we make the assumption that all particles are perfectly spherical when introducing the concept of size distributions.

Aerosol particles range in radii from nanometers to hundreds of micrometers. They are larger than typical atmospheric ions, which are usually below 0.5 nm in radius, but can be as large as cloud droplets ($r \sim 10$ μm, see Table 5.1). Cloud droplets and ice crystals generally occur in clearly defined entities (clouds and regions of precipitation). In contrast, in the troposphere and stratosphere a background aerosol concentration exists everywhere.

Different sources produce aerosol particles of different sizes. Because of the size dependence of many atmospheric processes, various maxima and minima can be found in aerosol size distributions; these are referred to as modes of the distribution. Therefore it has been found useful to define certain aerosol modes on the basis of the size ranges given in Table 5.1.

The lower boundary of the nucleation mode, 1.5 nm radius, has on the one hand been found to be a useful separator between large molecules and small aerosol particles and, on the other hand, corresponds to an experimental limit of traditional aerosol particle counters.

Particles below 0.5 μm in radius are sometimes summarized as fine mode particles and particles larger than 0.5 μm as coarse mode particles. Coarse mode particles are mostly directly emitted into the atmosphere, while particles in the nucleation mode form by nucleation from supersaturated vapors. These small particles grow further by the condensation of vapor and by coagulation with other particles, forming Aitken and accumulation mode particles.

Examples of the aerosol types in the various modes, as determined from electron microscope images, are shown in Figure 5.2. One can see that anthropogenic aerosol particles such as soot occur mainly in the submicron fine mode range and natural aerosol particles such as mineral dust occur mainly in the supermicron coarse mode range.

5.2 Aerosol size distributions

5.2.1 Discrete size distributions

Sometimes we are interested not only in the total number concentration of particles in an aerosol but also in how much the individual particle sizes contribute to the total concentration. Therefore, as mentioned above, we classify the particles according to size. In aerosol measurements this means that, from the entire aerosol entering the instrument, we select only particles with radii in a narrow size range $[r_i, r_{i+1}]$ and measure the concentration of these selected particles. This is repeated for the next size range until the whole range of interest is covered. The size distribution is then obtained by plotting the particle number concentration per size bin ΔN_i against the average radius r_i of the respective size bin. In practice the size bins often increase with increasing radius and vary between different instruments. In order to obtain an instrument-independent size distribution, normalization of the measured number concentration in each size bin by its width is required.

5.2.2 Size distribution function

Although normalization to the bin width allows the use of histograms for displaying aerosol size distributions in a comparable manner, discrete size distributions are often impractical because a large number of data need to be tabulated to describe a single size distribution. Statistics can help to reduce the measured data to quantities such as mean, median or mode radius (the radius where the size distribution peaks), standard deviation and total number concentration. However, information about the shape of the size distribution is lost if one cannot assume the sizes of the measured aerosol particles to be distributed according to a known distribution function.

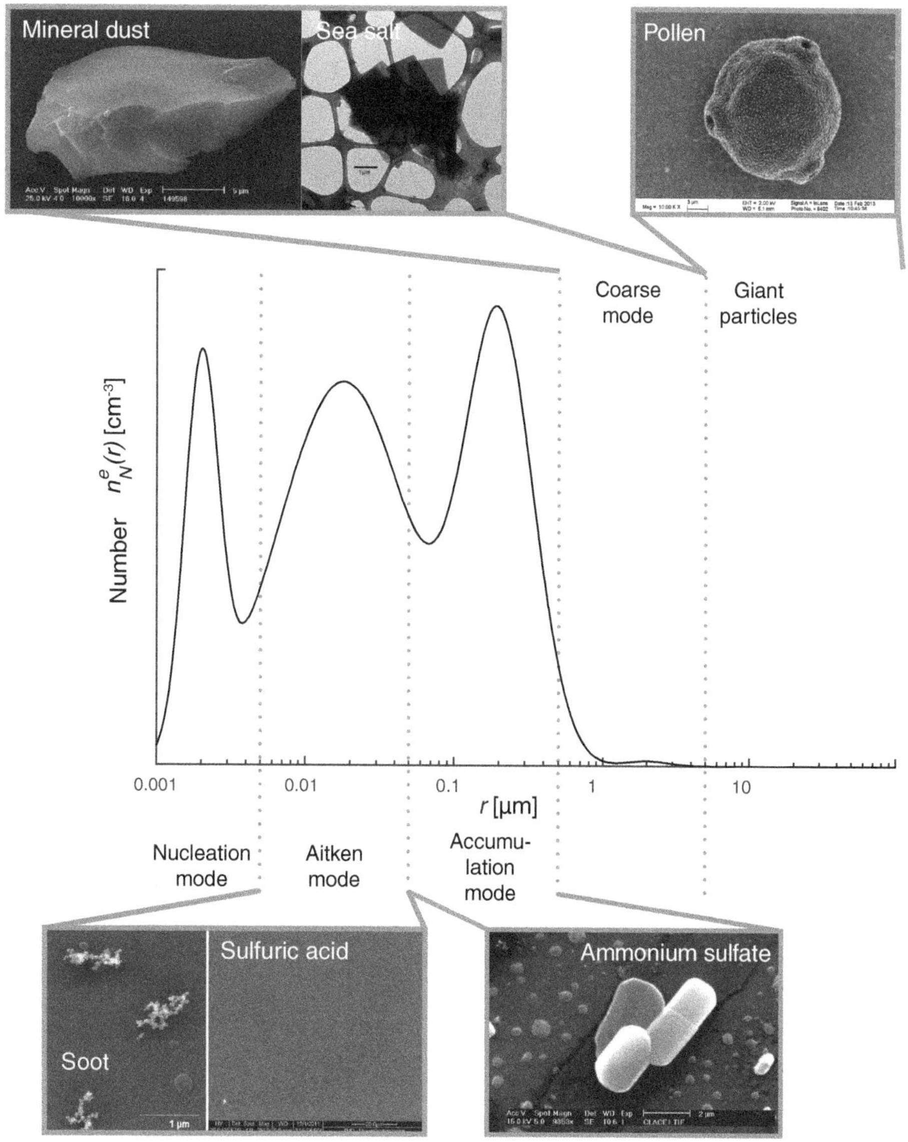

Fig. 5.2 Examples of different aerosol types from electron microscope images as a function of the aerosol mode. An artificial number distribution is shown to illustrate the different modes listed in Table 5.1. Photographs taken by Kouji Adachi (sea salt), Martin Ebert (soot, sulfuric acid, ammonium sulfate, mineral dust) and the Otte Homan (pollen).

Decreasing the width of the size bins Δr until Δr approaches dr allows us to describe the particle number distribution as a continuous function of the radius r, rather than using the discrete method introduced above. We can thus define the particle number distribution

$$n_N(r) = \frac{dN(r)}{dr}. \tag{5.1}$$

Here $n_N(r)$ denotes the number of particles per unit volume of air within a size range of unit length around radius r, having units of μm^{-1} cm^{-3} if one micrometer is chosen to be the unit length of particle radius. Hence the number density function is the derivative of the cumulative number distribution function $N(r)$.

The total number of particles N_a is then obtained by integrating the number density function $n_N(r)$ over the whole size range:

$$N_a = \int_0^\infty n_N(r)dr. \tag{5.2}$$

The arithmetic mean μ and the arithmetic standard deviation σ of $n_N(r)$ are given by

$$\mu = \frac{1}{N_a}\int_0^\infty r n_N(r)dr, \tag{5.3}$$

$$\sigma^2 = \frac{1}{N_a}\int_0^\infty (r-\mu)^2 n_N(r)dr. \tag{5.4}$$

A typical aerosol size distribution spans several orders of magnitude in particle radius, and is strongly skewed towards smaller particles. Therefore, the arithmetic mean and standard deviation are usually of little practical value in summarizing a size distribution; a few very large particles can dominate μ, whereas the many very small particles do not contribute significantly. Therefore, μ is often significantly larger than characteristic quantities, such as median or mode radius, which give intuitive information about the "typical" radius of the aerosol under consideration. Knowledge of σ is useful only in combination with knowledge about μ because σ is not invariant upon scaling of the size axis. For this reason, instead of μ and σ the geometric mean μ_g and geometric standard deviation σ_g are generally used to summarize a particle size distribution. For a given set of N_a particles, they are defined as follows:

$$\mu_g = \sqrt[n]{r_1 r_2 \cdots r_n}, \tag{5.5}$$

$$\sigma_g = \exp\left[\sqrt{\frac{\sum_{i=1}^{n}(\ln r_i - \ln \mu_g)^2}{n}}\right], \tag{5.6}$$

where the index i runs over all particles. The geometric standard deviation σ_g is invariant under scaling of the particle radii (i.e. multiplying the radius of every particle in the distribution by a constant factor). Therefore, σ_g gives size-independent information about the width of the distribution and allows us to compare the widths of particle populations with different typical radii. The values of σ_g range from around 1.1 for an aerosol population with a very narrow size distribution (a quasi-monodisperse aerosol, obtained by, for example, size selection) to more than 2 for a very broad size distribution, such as that for a polydisperse aerosol.

A further consequence of the large size range and the asymmetry of particle size distributions is that the distribution function $n_N(r)$ as defined in eq. (5.1) presents difficulties in visualizing the overall features of the size distribution, as shown in Figure 5.3a. For most aerosol populations, a sharp peak at a small radius is followed by a long, hardly visible, tail. As an example, consider a distribution $n_N(r)$ with a peak at 20 nm. The number concentration of particles falling into a size bin of 10 nm around 200 nm is negligible compared with the number concentration of particles in a 10 nm size bin around 20 nm. Nevertheless, the presence of a few large particles can be highly important for atmospheric processes such as cloud droplet formation or ice nucleation. A better representation of large particles in the tail of the distribution can be achieved if the distribution function is normalized to unit size bins on a logarithmic length scale instead of a linear length scale, as shown in Figure 5.3b.

This leads to the commonly used distribution function

$$n_N^e(r) = \frac{dN}{d\ln r} = \frac{dN}{dr}\frac{dr}{d\ln r} = rn_N(r), \tag{5.7}$$

with units cm^{-3}, i.e. $n_N^e(r)$ gives the number concentration of particles around radius r in a hypothetical size bin of unit length on a logarithmic scale. In a log-normal distribution, the logarithm of r is normally distributed.

Speaking mathematically, the functional form of a log-normal distribution can be seen only in $n_N(r)$, where the particle radius is log-normally distributed, not in $n_N^e(r)$. However, in atmospheric science both $n_N(r)$ and $n_N^e(r)$ are referred to as log-normal distributions. Size distributions referring to unit length on a logarithmic scale correspond to exponentially increasing size bins on a linear length scale. For example, using base 10 for logarithms, a size bin of unit length on the logarithmic scale ranges over 90 nm, from 10 nm to 100 nm, but also over 9 μm from 1 μm to 10 μm for large particles. An advantage of the use of $n_N^e(r)$ is that its values do not depend on the choice of a unit for the length scale; in contrast, the value of $n_N(r)$ changes if 1 nm is chosen as the width of the unit size bin instead of 1 μm. An advantage of $n_N(r)$ is that the area below the curve is proportional to the total particle number concentration, which is not the case for $n_N^e(r)$.

5.2.3 Logarithmic normal distributions

In general, an aerosol can be described reasonably well either by a single log-normal number size distribution (in the following referred to as a number distribution) or by a linear superposition of such distributions (Whitby and Sverdrup, 1980), where each is given by

$$n_N^e(r) = \frac{dN}{d\ln r} = \frac{N_a}{\sqrt{2\pi}\tilde{\sigma}} \exp\left[\frac{-(\ln r - \tilde{\mu})^2}{2\tilde{\sigma}^2}\right]. \tag{5.8}$$

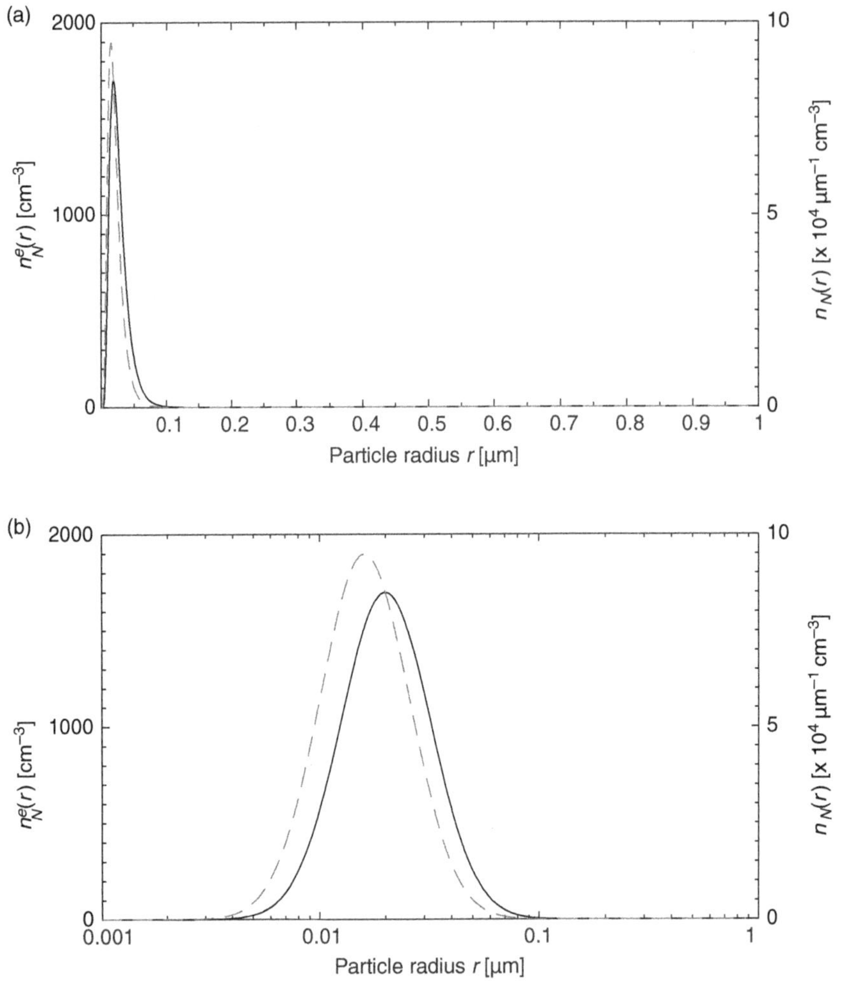

Fig. 5.3 Example of a log-normal number distribution, as given in eq. (5.1) normalized to a linear (broken line, right-hand axis) or a logarithmic (solid line, left-hand axis) size bin, respectively, shown for (a) a linear and (b) a logarithmic x-axis. The parameters of the distribution are $N_a = 2000\ \text{cm}^{-3}$, $\mu_g = 0.02\ \mu\text{m}$ and $\sigma_g = 1.6$.

Here $\tilde{\mu}$ is the arithmetic mean of the logarithms of the particle radii. It is the logarithm of the geometric mean radius, $\tilde{\mu} = \ln \mu_g$ (eq. 5.5). The standard deviation of $\ln r$ is given by

$$\tilde{\sigma} = \sqrt{\frac{1}{n} \sum_i (\ln r_i - \ln \mu_g)^2}.$$

Comparison with eq. (5.6) shows that $\tilde{\sigma} = \ln \sigma_g$. Hence, the number distribution of a typical aerosol can be written as

$$n_N^e(r) = \frac{dN}{d\ln r} = \frac{N_a}{\sqrt{2\pi}\ln\sigma_g}\exp\left[\frac{-(\ln r - \ln\mu_g)^2}{2\ln\sigma_g^2}\right]. \tag{5.9}$$

Figure 5.3 also shows the same log-normal number distribution plotted with a logarithmic underlying unit size bin. Note that the two curves do not peak at the same radius although they are merely different representations of the same number distribution with the same geometric mean and standard deviation. The log-normal curve $dN/d\ln r$ is below dN/dr for the entire displayed size range, since in this example the unit logarithmic size bin is smaller than the linear size bin (1 μm) for the entire size range.

There are some points that are worthwhile to note about log-normal particle number distributions, as follows.

- The median radius of a log-normal number distribution is equal to the geometric mean radius μ_g.
- In a log-normal number distribution, σ_g is the ratio of the radius below which 84.1% of all particles lie and the geometric mean radius. Thus, 68% of all particles lie in the range μ_g/σ_g to $\mu_g\sigma_g$ and 95% of all particles lie in the range μ_g/σ_g^2 to $\mu_g\sigma_g^2$.
- An aerosol with a narrow size distribution can also be described by a normal distribution, since the skewness of a log-normal distribution decreases with decreasing σ_g, i.e. the log-normal distribution approaches a normal distribution.
- The distribution $n_N^e(r)$ remains the same up to a constant factor upon a change in the base of the logarithm.

In an ambient aerosol, each of the five modes introduced in Section 5.1.2 is represented by an individual log-normal size distribution, and the source strength as well as the history of the aerosol in the atmosphere determine to what extent each mode contributes to the total size distribution.

5.2.4 Surface and volume distributions

So far we have characterized aerosol populations only in terms of their number distribution. However, other aerosol characteristics such as surface and volume are important as well, depending on the atmospheric process of interest. For instance, number distributions are of interest when the ability of the aerosol particles to act as CCN or INPs is studied. Surface distributions are the most relevant when one is looking at the radiative properties of aerosol particles since the surface area of an aerosol particle determines the amount of solar radiation that it scatters (Section 12.1). The surface distribution is also important in terms of chemistry, as the uptake of molecules on the surface of aerosol particles strongly depends on the available surface area. Volume or mass distributions are used to obtain information about the aerosol burden in the atmosphere, where "aerosol burden" refers to the vertically integrated aerosol mass, and about the amount of solar radiation absorbed by aerosol particles. Recalling that the surface of a sphere is given by $S = 4\pi r^2$ and the volume by $V = (4/3)\pi r^3$, we can compute surface and volume distribution functions based on eq. (5.9) from:

$$n_S^e(r) = \frac{dS}{d\ln r} = 4\pi r^2 n_N^e(r) = 4\pi r^2 \frac{N_a}{\sqrt{2\pi}\ln\sigma_g}\exp\left[\frac{-(\ln r - \ln\mu_g)^2}{2\ln\sigma_g^2}\right], \quad (5.10)$$

$$n_V^e(r) = \frac{dV}{d\ln r} = \frac{4}{3}\pi r^3 n_N^e(r) = \frac{4}{3}\pi r^3 \frac{N_a}{\sqrt{2\pi}\ln\sigma_g}\exp\left[\frac{-(\ln r - \ln\mu_g)^2}{2\ln\sigma_g^2}\right], \quad (5.11)$$

where $n_S^e(r)$ is given in $\mu m^2\ cm^{-3}$ and $n_V^e(r)$ in $\mu m^3\ cm^{-3}$. Figure 5.4 shows the number, surface and volume distributions of an aerosol population, using eqs. (5.9)–(5.11). The solid curve in Figure 5.4a denotes the number distribution of all aerosol particles in this population, which consists of a superposition of three individual log-normally distributed modes. The number size distribution shows the strongest mode around 0.02 μm, with smaller modes at 0.2 μm and 3.5 μm. They are referred to as Aitken, accumulation and coarse modes, respectively (Table 5.1). In general, atmospheric aerosol distributions can be described as the sum of three (Whitby and Sverdrup, 1980) or four log-normal distributions, depending on the source of the particles, as we will see later.

It is worth noting that the number, surface and volume distributions depicted in Figure 5.4 differ in terms of the modes that dominate each distribution. While the number distribution shows the highest abundance for Aitken mode particles, the surface and volume distributions are dominated by coarse mode particles, which are hardly visible in the number size distribution.

The surface and volume (or mass) distributions defined in eqs. (5.10) and (5.11) do not have the form of log-normal distributions. However, they can be transformed into log-normal distributions, with surface, volume and mass geometric mean radii $\mu_{g,S}$, $\mu_{g,V}$ and $\mu_{g,m}$, respectively:

$$\ln\mu_{g,S} \equiv \frac{\sum s_i \ln r_i}{S_a} = \ln\mu_g + 2\ln^2\sigma_g, \quad (5.12)$$

$$\ln\mu_{g,V} \equiv \frac{\sum v_i \ln r_i}{V_a} = \ln\mu_g + 3\ln^2\sigma_g, \quad (5.13)$$

$$\ln\mu_{g,m} \equiv \frac{\sum m_i \ln r_i}{m_a} = \ln\mu_g + 3\ln^2\sigma_g. \quad (5.14)$$

Here, r_i, s_i, v_i and m_i denote radius, surface, volume and mass of the ith particle, where i runs over all particles in the aerosol population, and S_a, V_a and m_a denote the total particle surface, volume and mass, respectively. In contrast with the number mean radius (geometric or arithmetic), the surface mean radius is the mean over all particle radii, each weighted according to the particle surface at that particle radius. For a log-normally distributed particle population, the surface geometric mean radius, or median surface radius, $\mu_{g,S}$ denotes the radius below which half of the particle surface is found. Analogous definitions hold for the median values of volume $\mu_{g,V}$ and mass $\mu_{g,m}$. The quantities $\mu_{g,S}$

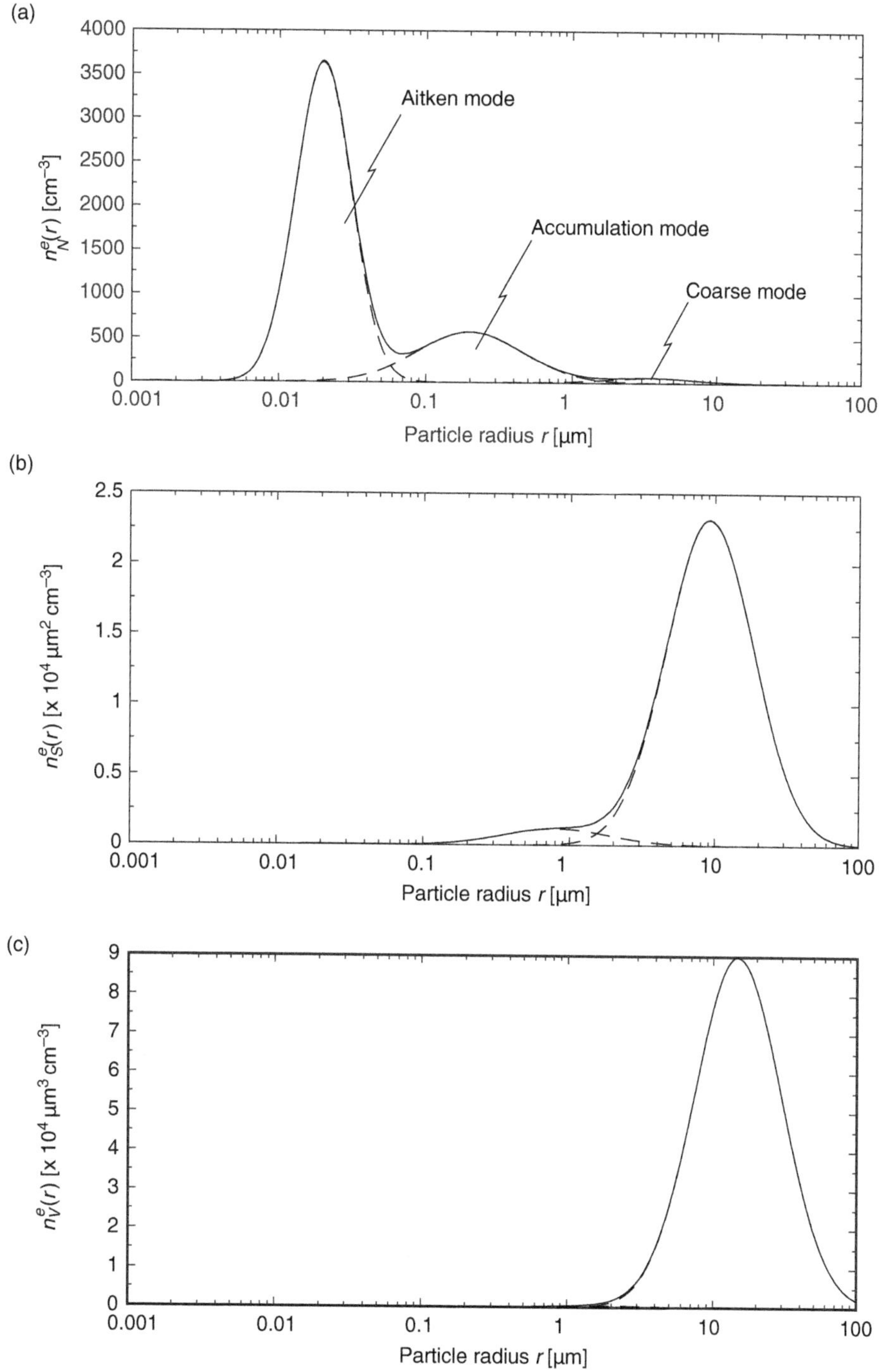

Fig. 5.4 (a) Number, (b) surface and (c) volume distributions, as obtained from eqs. (5.9)–(5.11). The solid line denotes the sum of three individual log-normal distributions given by $N_{a,1} = 4000\ \text{cm}^{-3}$, $\sigma_{g,1} = 1.55$, $\mu_{g,1} = 0.02\ \mu\text{m}$; $N_{a,2} = 1200\ \text{cm}^{-3}$, $\sigma_{g,2} = 2.3$, $\mu_{g,2} = 0.2\ \mu\text{m}$; $N_{a,3} = 100\ \text{cm}^{-3}$, $\sigma_{g,3} = 2$ and $\mu_{g,3} = 3.5\ \mu\text{m}$.

Table 5.2 Values of the factor b needed for the Hatch–Choate conversion eq. (5.15), of number geometric mean radius to other average radii.

μ_X	b
Mass median radius or mass geometric mean radius $\mu_{g,m}$	3
Mass mean radius μ_m	3.5
Surface median radius or surface geometric mean radius $\mu_{g,S}$	2
Surface mean radius μ_S	2.5
Volume median radius or volume geometric mean radius $\mu_{g,V}$	3
Volume mean radius μ_V	3.5
Number mode radius $\tilde{\mu}$	−1
Number mean radius μ	0.5
Radius of average mass $r_{\bar{m}}$	1.5

and $\mu_{g,V}$ are larger than μ_g, since the weight of the large particles is larger in the surface and volume distributions than in the number size distribution. It can be shown that the geometric standard deviation of the surface and volume distribution is identical to that for the number distribution, σ_g. This property of the log-normal distribution allows us to calculate all the characteristic average radii μ_X from the number geometric mean radius and the universal geometric standard deviation according to the Hatch–Choate conversion equation:

$$\mu_X = \mu_g \exp(b \ln^2 \sigma_g). \tag{5.15}$$

The values of the factor b for some average radii μ_X are given in Table 5.2 taken from Hinds (1999).

5.2.5 Observed aerosol size distributions

The observed number, surface and volume distributions for a biomass-burning aerosol and a marine aerosol are given for comparison in Figure 5.5. The biomass-burning aerosol distributions were obtained from more than 8000 size distributions in the Amazon basin at different seasons (Rissler *et al.*, 2006). The highest concentrations are found in the dry season with more than 10 000 particles cm^{-3} in the accumulation mode when biomass-burning emissions are highest. They are lowest during the wet season, with less than 2000 particles cm^{-3}, when only little biomass burning takes place and aerosol particles are strongly removed by wet deposition (Section 5.4.2). In all seasons the distributions are composed of three log-normal distributions with a contribution from the nucleation, Aitken, and accumulation modes. The geometric mean radius for the number distribution as well as the number concentration for the nucleation mode remain almost the same for the different seasons. The Aitken and accumulation mode radii increase by around 50% from the wet season to the dry season. The number concentrations in the Aitken and accumulation modes increase by factors of 13 and 7, respectively. The distinction

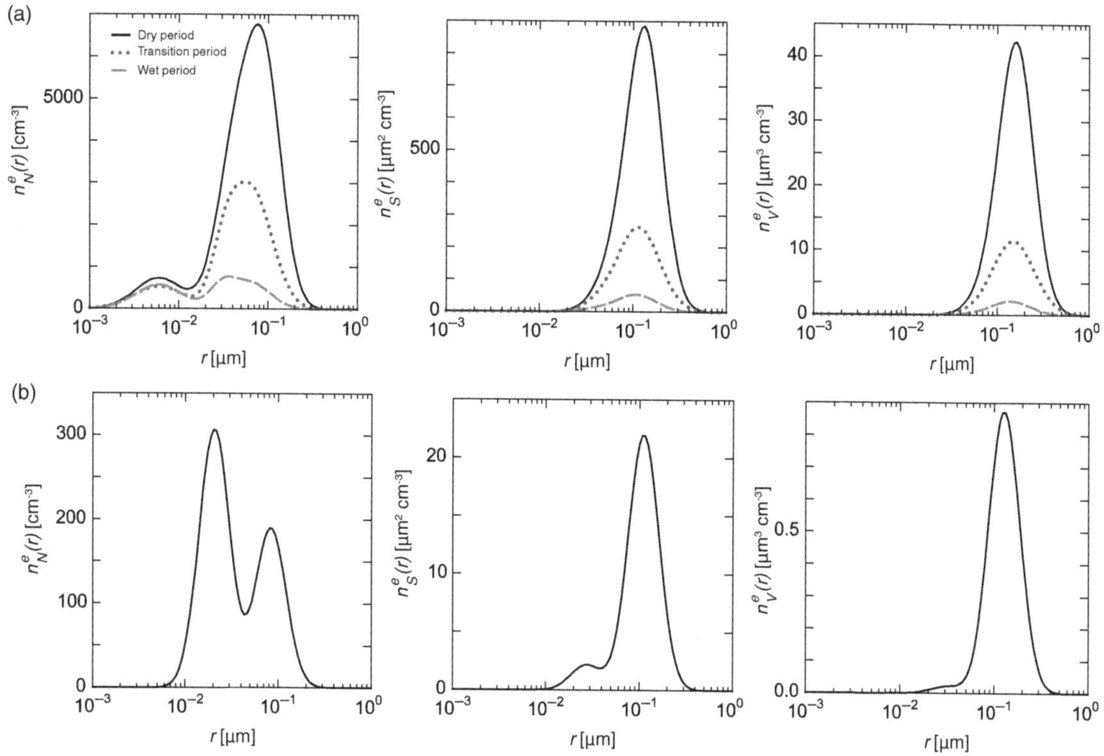

Fig. 5.5 Number distribution, $n^e_N(r) = dN/d\ln r$, surface distribution, $n^e_S(r) = dS/d\ln r$ and volume distribution, $n^e_V(r) = dV/d\ln r$, as functions of the aerosol radius for (a) a biomass burning aerosol (Rissler *et al.*, 2006) and (b) a submicron aerosol distribution found over the ocean (Heintzenberg *et al.*, 2000). Note the differing vertical scales.

between the Aitken and accumulation mode is not clearly visible as their mode radii are only about a factor 2 apart. In contrast, the geometric nucleation mode radius, 12 nm, is a factor 5–8 smaller than that for the Aitken mode. The contribution of the nucleation mode vanishes in the surface and volume distributions. Vice versa, the particles contributing most to the volume distribution do not have a visible effect on the number distribution.

A submicron near-global aerosol distribution over the ocean was compiled from various ship observations (Heintzenberg *et al.*, 2000); see Figure 5.5b. It is bimodal, with the larger number concentration in the Aitken mode and the secondary peak in the accumulation mode. The large Aitken mode in the number distribution reveals that the aerosol sample from over the ocean has contributions from anthropogenic sources, because it includes data taken in coastal regions. The number concentration of roughly 750 cm^{-3} for the Aitken mode is, however, considerably smaller than the number concentration of the biomass-burning aerosol. The Aitken mode of the aerosol over the ocean is still visible in the surface distribution but becomes negligible in the volume distribution.

Vertically averaged aerosol volume distributions can be inferred from remote sensing measurements, as shown in Figure 5.6 for various parts of the world. The measurements were obtained from the Aerosol Robotic Network (AERONET) (Dubovik *et al.*, 2002).

These vertical averages are dominated by the large aerosol concentrations in the boundary layer and are thus biased towards low altitudes.

The urban volume aerosol size distributions obtained near Washington, DC, Paris and Mexico City, and the biomass-burning size distributions obtained for the Amazonian forest

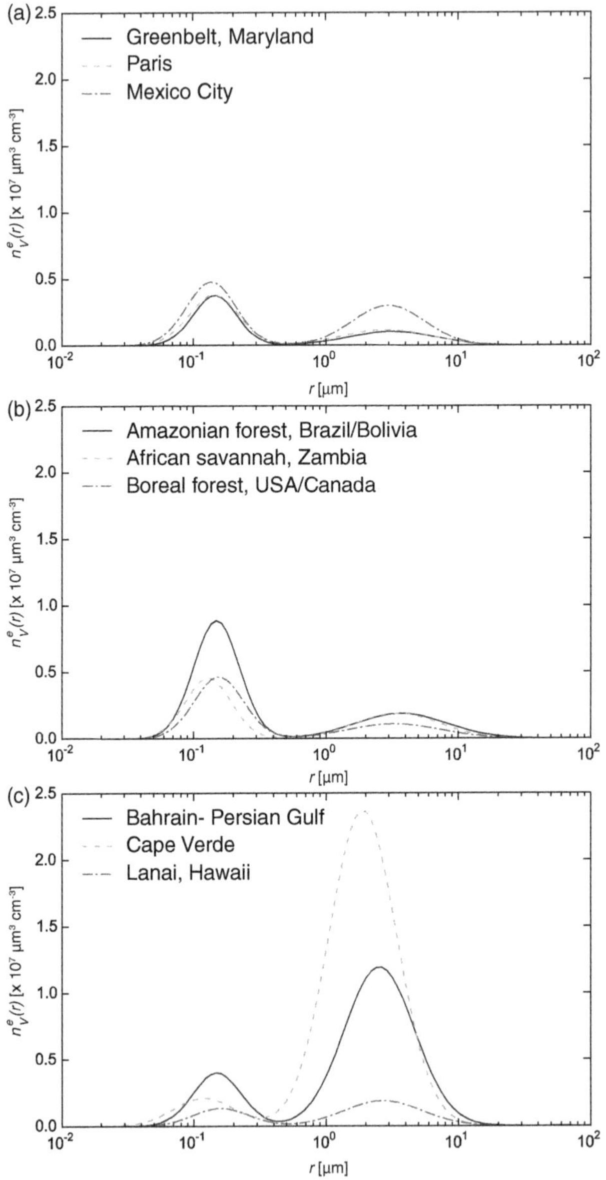

Fig. 5.6 Examples of vertically integrated aerosol volume distributions $dV/d \ln r$ ($\mu m^3\ cm^{-3}$) as a function of aerosol radius at various locations, as obtained from the Aerosol Robotic Network (AERONET): (a) urban or industrial aerosol volume distributions, (b) biomass-burning aerosol volume distributions, (c) desert dust and marine aerosol volume distributions. The values refer to vertical averages over the entire atmosphere. Figure adapted from Dubovik *et al.* (2002). Reprinted with permission of the American Meteorological Society.

in Brazil and Bolivia, for the African savannah in Zambia and for the boreal forest of the United States and Canada are dominated by a peak in the accumulation mode and a secondary maximum in the coarse mode. Biomass burning in the Amazonian forest in Brazil and Bolivia causes a higher concentration of accumulation mode aerosols than is found in the other biomass-burning regions in the African savannah and the boreal forests in Canada and the United States and as compared with the urban aerosol. The volume distribution of the marine aerosol at Hawaii has equal contributions from the accumulation and coarse modes. Desert dust, however, is dominated by a peak in the coarse mode. It has a higher peak over Cape Verde, which is influenced by Saharan dust, than in the Persian Gulf.

5.3 Aerosol sources

Aerosol particles can also be classified according to their geographical origin. Typical distinctions would be desert, marine, polar, remote continental (referring to the continental aerosol aside from big cities), urban and free tropospheric aerosols. All these aerosol types have their sources near the surface, even the free tropospheric aerosols, for which biomass burning is a strong contributor. Thus the aerosol mass and number concentrations typically decay with altitude.

5.3.1 Formation mechanisms of aerosol particles

Aerosol particles have a variety of different formation mechanisms. Accordingly, they are divided into primary and secondary particles. Primary particles are emitted directly as aerosol particles (either liquid or solid) into the atmosphere. They are to be distinguished from secondary aerosols, which form within the atmosphere from precursor substances (gases). Primary aerosol particles originate from bulk-to-particle conversion, such as wind-blown dust from arid regions. Such emissions of mineral dust from arid regions depend on wind speed, soil moisture and the bare soil fraction. Besides the emissions of pollen and fungal spores by plants, combustion processes are also examples of primary aerosol sources.

Sometimes liquid-to-particle conversion is distinguished from bulk-to-particle conversion. Sea salt aerosols that originate from drops ejected into the air when air bubbles in breaking waves burst at the ocean surface are an example of liquid-to-particle conversion. When a sea water droplet evaporates in the atmosphere, it leaves a sea salt particle behind.

Any type of pure secondary particle is the result of the nucleation of a new phase (liquid or solid) from a supersaturated gas phase. These processes are also referred to as gas-to-particle conversion. The most common form of particle nucleation is binary nucleation, where the aerosol forms from two gas-phase precursors. Homogeneous homo-molecular nucleation from the gas phase, which is the self-nucleation of a single species, does not take place in the atmosphere as that would require that the vapor pressure of this species alone is supersaturated. For binary nucleation to take place, each species itself can be subsaturated if the mixture is supersaturated. A prominent example is the reaction of n

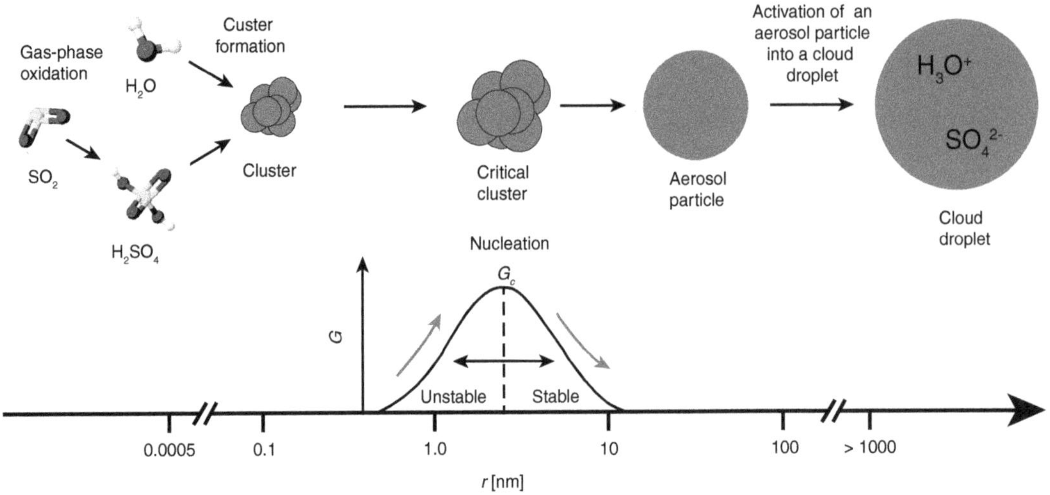

Fig. 5.7 Schematic representation of the nucleation and subsequent growth process for the atmospheric binary homogeneous nucleation of H_2SO_4 and H_2O. Once stable clusters are formed, other substances can take part in the growth process. Particles may grow to sizes large enough to activate into cloud droplets.

moles of water vapor, H_2O (g), with m moles of gaseous sulfuric acid, H_2SO_4 (g), where n and m are arbitrary numbers. This can result in the nucleation of liquid sulfuric acid aerosol particles:

$$n\,H_2O(g) + m\,H_2SO_4(g) \rightarrow (H_2O)_n(H_2SO_4)_m(aq). \tag{5.16}$$

The chain from cluster formation to aerosol nucleation, with subsequent growth to larger sizes and finally activation into a cloud droplet, is shown schematically in Figure 5.7. The first step is that, in the gas phase, molecules of H_2O and H_2SO_4 form a cluster. Once the cluster exceeds a critical size, nucleation occurs (Section 6.1) and a stable sulfuric acid aerosol particle is formed. The critical size of a cluster is the size above which it is more likely to grow further than to shrink and evaporate. Once the initial (liquid) particle is formed, other species such as low volatility organics can condense onto the liquid particle and participate in its growth. The restriction to low volatility ensures that the organics remain in the condensed phase and do not evaporate again. However, the concentrations of H_2SO_4(g) and H_2O(g) are too low to explain the nucleation rates in the planetary boundary layer (Sipilä *et al.*, 2010) by the simple binary nucleation considered above. Ions (Kirkby *et al.*, 2011), low-volatility organic carbon aerosols (Ehn *et al.*, 2014) and other ternary species can help to stabilize the initial cluster and decrease evaporation.

Nucleation takes place when the background aerosol particle concentration is low. Then the existing aerosol surface is too small for the condensable vapors to condense on and hence nucleation occurs. A typical example of an aerosol nucleation event as observed in a forest in Finland is shown in Figure 5.8 (Boy and Kulmala, 2002). Here nucleation sets in at around 9 h, as can be seen in the strong increase in particle number concentration of the smallest aerosol particles. Until 13 h, new particles are formed and grow rapidly in size.

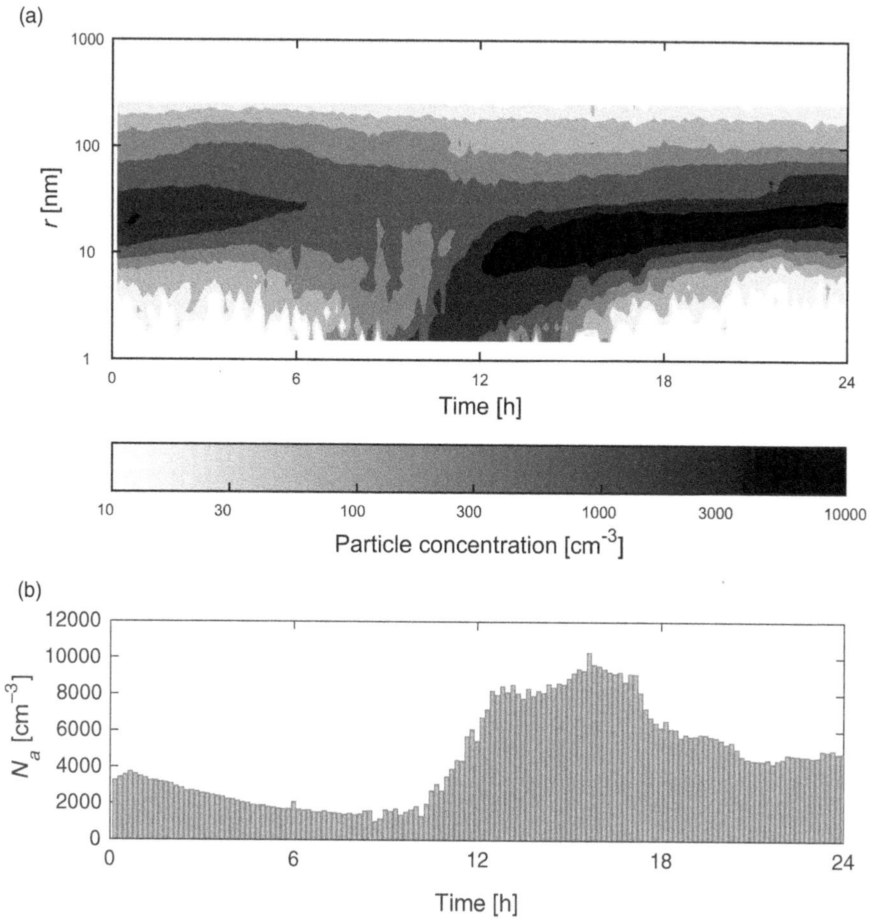

Fig. 5.8 Example of a typical gas-to-particle nucleation event as a function of time as observed in Hyytiälä, Finland, on May 19, 1999: (a) particle number size distribution, (b) total number concentration. Figure adapted from Boy and Kulmala (2002).

Thereafter the particle concentration remains approximately constant (until 17 h) but the particles still continue to grow, albeit at a lower rate.

The condensation of a supersaturated gas phase onto a primary particle converts the primary particle into a secondary particle. Examples of such secondary particles are internally mixed particles that contain condensed material, such as the secondary particle that results when sulfuric acid condenses on a mineral dust particle.

5.3.2 Aerosol emissions

In order to estimate the transport of aerosol particles and their influence on air quality and climate, we need to know the global distribution of aerosol emissions. The global source strengths of aerosol particles, their burdens and lifetimes are summarized in Table 5.3. The

Table 5.3 Global source strengths, burdens and lifetimes of the major aerosol species for the year 2000. The emissions for sea salt, mineral dust, dimethyl sulphate (DMS), SO_2, BC and particulate organic matter (POM) originate from Dentener *et al.* (2006) and the NO_x emissions from Stevenson *et al.* (2006). The lifetimes and burdens are taken from an aerosol model intercomparison (Textor *et al.*, 2006). The diversity of the model results is shown as standard deviations normalized by the all-models average. The burden and lifetime are given once per aerosol species only (for POM and SO_4 in the entry of anthropogenic emissions of POM or SO_2, respectively). The fluxes of DMS and SO_2 have been converted into Tg SO_4 yr^{-1} and those for NO_x into Tg NO_3 yr^{-1}.

Aerosol type	Flux [Tg yr^{-1}]	Burden [Tg]	Lifetime [d]
Primary natural emissions			
Dust (desert)	1678	19.2±40%	4.1±43%
Sea salt	7925	7.52±54%	0.5±58%
Secondary natural emissions			
DMS	54.6		
SO_2 from explosive volcanoes	6		
SO_2 from continuous degassing volcanoes	37.8		
NO_x from soils & lightning	52.8		
Biogenic SOA	19.1		
Total natural emissions	9773.3	26.72	
Primary anthropogenic emissions (fossil fuel, biofuel and wild fires)			
POM	47	1.7±27%	6.5±27%
BC	7.7	0.24±42%	7.1±33%
Secondary anthropogenic emission (fossil fuel, biofuel and wild fires)			
SO_2	337.8	1.99±25%	4.1±18%
NO_x	167.2		
Total anthropogenic emissions	559.7	3.93	
Sum of natural and anothropogenic emissions	10 333	30.65	

lifetime of an aerosol compound is calculated as its global mean burden divided by its emission rate. The emission data in Table 5.3 were compiled from different emission inventories as well as from model simulations of the aerosol intercomparison project (AEROCOM) (Dentener *et al.*, 2006). Emission inventories collect emission data for aerosol compounds from various sectors, such as biomass-burning and industrial activities, and for various countries. The AEROCOM emission data set distinguishes natural and anthropogenic emissions. The emissions of the five major aerosol species, i.e. sea salt, mineral dust, sulfate, black carbon (BC) and particulate organic matter (POM), obtained from Dentener *et al.* (2006) are representative for the year 2000. Since sulfate is a secondary aerosol, its emissions are given in terms of its precursor gases, dimethylsulfide (DMS) and sulfur dioxide (SO_2). Nitrate is a secondary aerosol that forms from the gas-to-particle conversion of NO_x. The values of emissions of biological particles, such as bacteria or pollen, are not included in the AEROCOM data base because they are associated with large uncertainties.

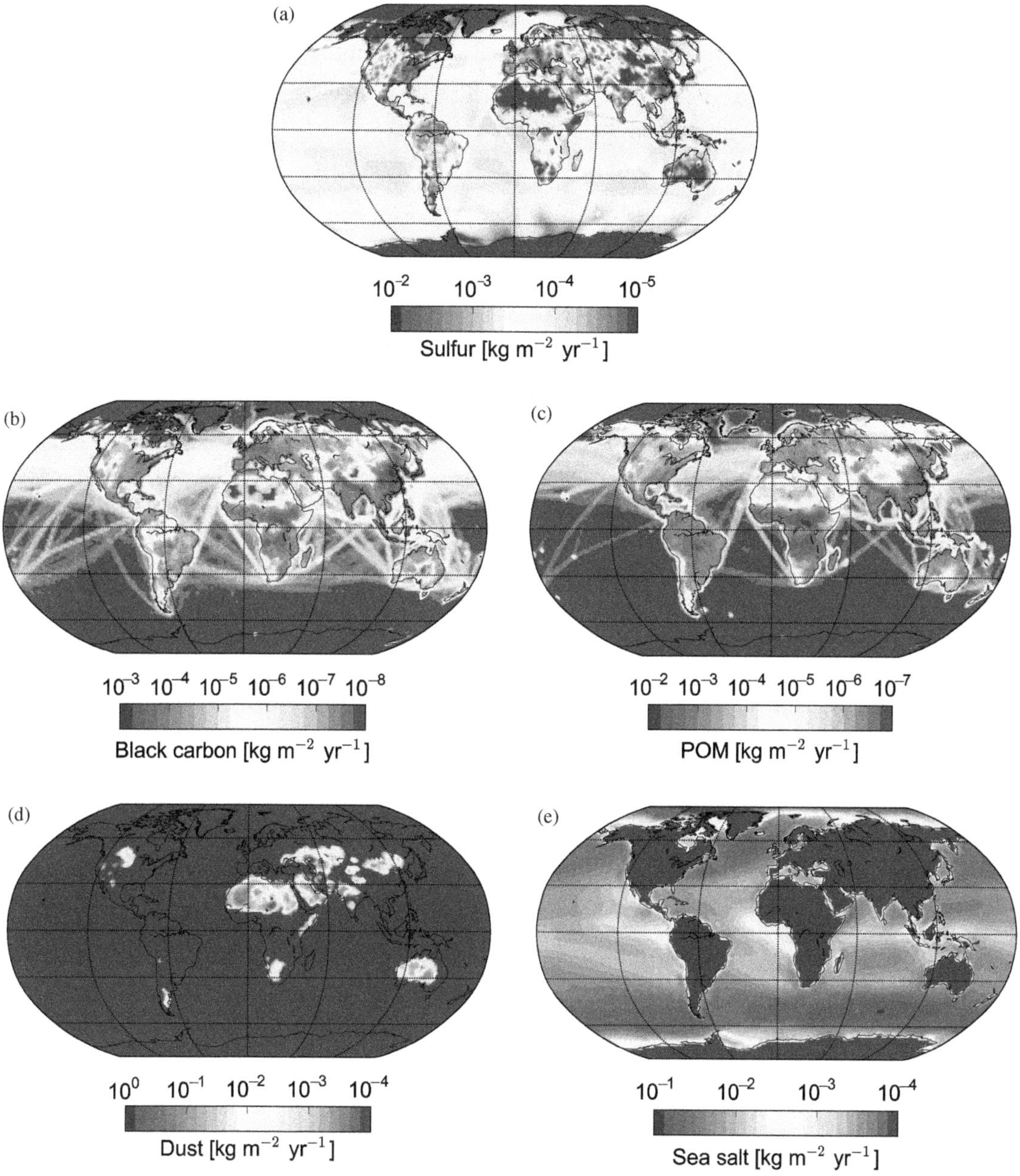

Fig. 5.9 Annual mean emissions of (a) sulfur, (b) black carbon particles, (c) particulate organic matter (POM), (d) mineral dust particles and (e) sea salt particles obtained from the ECHAM6-HAM global climate model (Neubauer *et al.*, 2014). Note that the scale is different for each aerosol type. A black and white version of this figure will appear in some formats. For the color version, please refer to the plate section.

Geographical distributions of the annual mean emissions from different aerosol sources (Neubauer *et al.*, 2014) are shown in Figure 5.9. Mineral dust and sea salt are the two most abundant aerosol types. They originate from bulk-to-particle and liquid-to-particle conversion, respectively, and are therefore primary particles. Sea salt has the largest flux

(Table 5.3) with almost 8000 Tg yr^{-1}. The main sources of sea salt aerosols are the oceans in the west wind zone, i.e. the Roaring Forties and the storm tracks over the northern hemispheric Atlantic and Pacific Ocean. Here emissions amount up to 100 g m^{-2} yr^{-1}. Mineral dust has the second largest emission flux. It is mainly a natural aerosol, which is predominantly produced through wind erosion in sparsely vegetated and arid regions. Globally the most important dust sources are deserts, especially the Sahara and Gobi deserts, where emissions can reach up to 1000 g m^{-2} yr^{-1}. Because the regions that emit mineral dust are much smaller than the source regions of sea salt, the global emissions of mineral dust amount to only about 1700 Tg yr^{-1}.

The other aerosol types displayed in Figure 5.9 are either assumed to be entirely of anthropogenic origin (BC) or have sources that are dominated by anthropogenic emissions (POM and sulfate). The anthropogenic sulfate emissions have been further subdivided into individual sectors (Dentener *et al.*, 2006). They are due to emissions from power plants (145.2 Tg S yr^{-1}), industry (117.6 Tg S yr^{-1}), domestic biofuel (28.8 Tg S yr^{-1}), shipping (23.4 Tg S yr^{-1}), wild fires (12.3 Tg S yr^{-1}), roads (5.7 Tg S yr^{-1}) and railroads and agricultural activity (4.8 Tg S yr^{-1}). Sulfate also has natural sources, e.g. some phytoplankton species emit dimethyl sulfide (DMS) when they die. In the atmosphere DMS is oxidized mainly to SO_2 and further to SO_4^{2-} (sulfate). These DMS emissions amount to 54.6 Tg S yr^{-1} (Table 5.3). There is also a small terrestrial source of DMS from vegetation, which is roughly 10% of the oceanic DMS source. Even though explosive volcanic eruptions emit a lot of SO_2 at once, major explosive eruptions are not that common. Therefore the volcanic SO_2 emissions are dominated by those from continuously outgassing volcanoes (Table 5.3) with 37.8 Tg S yr^{-1} as compared with only 6 Tg S yr^{-1} from explosive volcanoes.

The map of sulfur emissions clearly shows the major shipping lanes (Figure 5.9a). This results from the fact that ship fuel is regulated only inside emission controlled areas (ECAs); outside the ECAs ships can burn sulfur-rich fuel. As compared with the SO_2 emissions from ships, aircraft emissions of SO_2 are negligible (Skeie *et al.*, 2009). Relative sulfur emission maxima in the ocean are related to DMS emissions, whereas the highest emissions over land are associated mainly with fossil fuel combustion and occur predominantly in southeast Asia, Europe and North America.

Emissions of carbonaceous aerosol particles are subdivided into POM and BC. Black carbon originates from primary emissions, whereas POM can be subdivided into primary organic aerosol (POA) and secondary organic aerosol (SOA). Sometimes the category OC is used instead of POM; note that the POM mass can be calculated to be 1.4 times the OC mass.

Being negligible in terms of SO_2 emissions, biomass burning is an important source of BC and POM (Figures 5.9b, c). This is due to the often incomplete combustion of biomass; complete combustion would produce only CO_2 and H_2O. The relative amounts of BC and POM emitted during combustion depend on the availability of oxygen. Less oxygen being available for combustion leads to higher amounts of BC and vice versa. Because oxygen is always available in the atmosphere but not in sufficient amounts at the emission source, BC is always emitted together with POM so that these two emission maps look rather similar.

The POM emissions are dominated by anthropogenic emissions from wild fires (34.7 Tg yr^{-1}), biofuel (9.1 Tg yr^{-1}) and fossil fuel combustion (3.2 Tg yr^{-1}) that amount to 47 Tg yr^{-1} of POM in total. The anthropogenic POM emissions always occur in conjunction with the anthropogenic emissions of BC, which amount to 7.7 Tg C yr^{-1}. Some POM has natural sources. These are secondary organic aerosols, whose precursor gases are of biogenic origin, being emitted from certain plants. As compared with the dominant natural emissions of sea salt and mineral dust, the natural SOA emissions of 19.1 Tg yr^{-1} are negligible.

Precursor emissions for nitrate aerosols occur in the form of NO_x. The NO_x emissions are dominated by anthropogenic sources. Natural sources include emissions from soils due to microbial production and NO_x from lightning. They amount to 52.8 Tg N yr^{-1} whereas the anthropogenic sources of fossil fuel combustion and biomass burning amount to 167.2 Tg N yr^{-1}. As most climate models do not yet include nitrate aerosols, the nitrate burden and its lifetime are not shown in Table 5.3.

5.4 Aerosol sinks

The removal of aerosol particles from the atmosphere is called deposition or scavenging. Broadly, we can distinguish between two categories of scavenging: dry scavenging and wet scavenging. Further subclassification of the scavenging mechanisms depends on the physical processes that are responsible for the removal of the aerosol particle.

Dry scavenging (also referred to as dry deposition) constitutes the removal of aerosol particles from the atmosphere by sedimentation due to gravity and the subsequent impaction on Earth's surface. In contrast, the removal of aerosol particles by wet scavenging involves the presence of hydrometeors. On the one hand aerosol particles can be removed upon collision with cloud droplets, raindrops and snowflakes, a scavenging mechanism referred to as impaction scavenging. On the other hand, aerosol particles can become incorporated into hydrometeors, by acting as CCN (Chapter 6) or INPs (Chapter 8). The importance of dry versus wet scavenging has been estimated from global climate models (GCMs) coupled to an aerosol model. In a study with 12 different GCMs, dry scavenging was found to be the dominant sink for the coarse-mode natural aerosol particles, i.e. sea salt and dust, in the global annual mean. It removes 70%–80% of the mass (Textor *et al.*, 2007). For the smaller aerosol particles such as sulfate, black and organic carbon, wet scavenging is the most important sink; it removes 80%–90% of the aerosol mass.

5.4.1 Dry scavenging

Dry scavenging describes the transition in which particles still in the atmosphere, but very close to the surface, are deposited onto a surface. It has been found useful to describe dry deposition in terms of a dry deposition velocity (Figure 5.10). This is an effective velocity, which describes how quickly an aerosol particle makes contact with the surface. It depends on the aerosol's size, density, shape and other features. The dry deposition velocity is

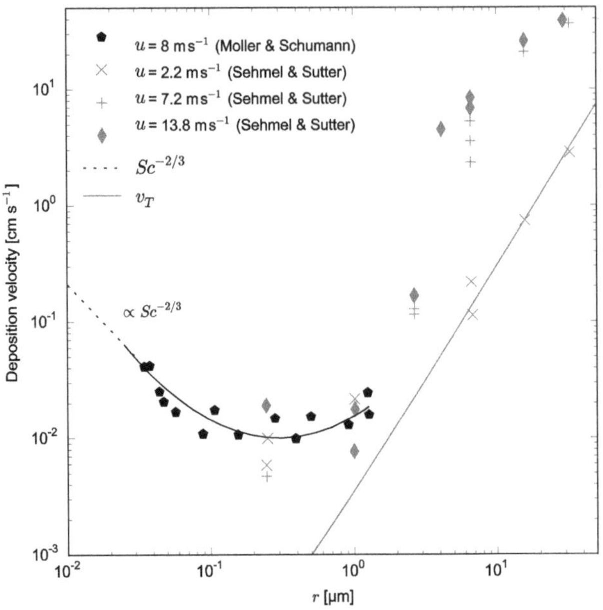

Fig. 5.10 Dry deposition velocity of aerosol particles as a function of their size, for particles depositing on a water surface at different wind speeds u, with data from Moller and Schumann (1970) and Sehmel and Sutter (1974). Also shown are estimates for gravitational settling, in terms of the terminal velocity v_T, and for the diffusional regime as a function of the Schmidt number $Sc^{-2/3}$. Figure adapted from Slinn *et al.* (1978) with permission from Elsevier.

smallest for accumulation-size aerosol particles and increases towards smaller sizes due to Brownian motion (left side of Figure 5.10). Brownian motion refers to microscopic diffusion on a single-particle level. It results from the continuous bombardment of an aerosol particle by the molecules of the surrounding medium: each time the particle is hit by a gas molecule, momentum is transferred to it, causing it to change its direction of movement. This leads to a "random walk" of the particles, called Brownian motion. Note however, that the particle diffusion is much slower than the diffusion of the gas molecules, making its diffusion coefficient orders of magnitude smaller. Brownian motion speeds up the dry deposition of the smallest particles because the random walk for such small particles has an amplitude large enough that particles in the vicinity of Earth's surface collide with it and are thus removed at a significant rate. This leads to a larger net averaged downward velocity (i.e. dry sedimentation velocity) than that for larger particles.

The dry deposition velocity in the diffusional regime dominated by Brownian motion is proportional to the Schmidt number Sc to the power 2/3, as indicated in Figure 5.10; the Schmidt number is the ratio of the viscous diffusion rate and the molecular diffusion rate. It can be described in terms of the kinematic viscosity in air, ν, and the mass diffusivity D_a of the aerosol particle in air:

$$Sc = \frac{\nu}{D_a} = \frac{\mu}{\rho D_a}. \tag{5.17}$$

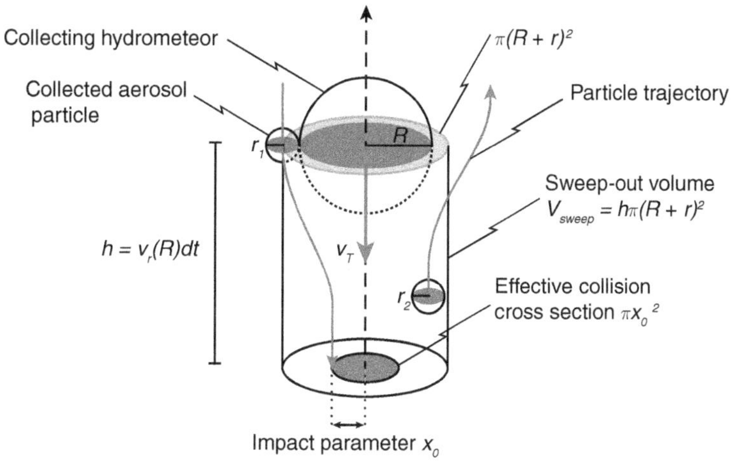

Fig. 5.11 Schematic outlining the collision process of a collector drop with radius R and two smaller particles with radius r. The sweep-out volume is given by $V_{sweep} = h\pi(R + r)^2$; x_0 denotes the impact parameter and πx_0^2 defines the effective collision cross section.

Note that D_a depends on the aerosol radius. Instead of the kinematic viscosity, the dynamic viscosity can be used. They are related by the density of air: $\mu = \rho\nu$.

The dry deposition velocity increases for particles with radii larger than 0.5 μm, when gravitational settling starts to become the dominant force (right-hand side of Figure 5.10). Dry deposition is the dominant sink for coarse mode aerosol particles. For smaller particles, the deposition velocity is small enough that the particles are likely to undergo one of the processes leading to wet deposition.

5.4.2 Wet scavenging

Wet scavenging (also referred to as wet deposition) is the dominant sink for Aitken- and accumulation-mode particles. It is divided into nucleation scavenging, which only takes place in-cloud, and impaction scavenging, which takes place in- and below-cloud.

5.4.2.1 Impaction scavenging

Impaction scavenging is usually described in terms of the collection efficiency $\tilde{E}$ between a collecting hydrometeor (also referred to as the collector drop) and the collected aerosol particle. It is the product of the collision efficiency E times the coagulation efficiency $\hat{E}$, where coagulation means that the two colliding particles remain together after collision. The coagulation efficiency $\hat{E}$ is also referred to as the sticking efficiency.

A schematic of the collision process is shown in Figure 5.11: a collecting hydrometeor with radius R falls down and sweeps out a volume of air V_{sweep}, which is the product of the distance fallen times the cross-sectional area of a disk whose radius is the sum of the radii of the falling hydrometeor and the collected particle. The collision process holds generally for all kinds of hydrometeors including cloud droplets, raindrops, ice crystals, snowflakes, graupel and hail.

In the absence of hydrodynamic interactions of the surrounding air with the hydrometeor and the particles, the hydrometeor would collide with all smaller particles in its path. However, taking hydrodynamics into account, smaller particles are driven away from the collector drop. The geometrical collision cross section can be obtained by taking the size of the collected particle into account. It defines the cross section within which the center of a (small) particle with radius r must lie such that it will collide with the (larger) collector drop with radius R. Allowing for contact at the edges of the two particles, it is given as $\pi(R+r)^2$. The sweep-out volume, V_{sweep}, is then given by the geometrical collision cross section times the height h: $V_{sweep} = h\pi(R+r)^2$. Here $h = v_r dt$ is the distance through which the collector drop falls during dt and v_r is the difference in velocity between the collecting hydrometeor and the collected aerosol particle.

As mentioned above, in a medium such as air, the falling collector drop displaces the surrounding air. Since small particles tend to follow the streamlines of the air, the (effective) collision process depends on two things, (i) the inertia of the small particles and (ii) the Reynolds number (eq. 7.36) of the collector drop. Only particles with enough inertia, such as particle r_1 in Figure 5.11, cross the streamlines of the air flow and thus collide with the collector drop if they are close enough to its center. Smaller particles and those further away from its center, such as that indicated by r_2, follow the movement of the medium around the collector drop and do not collide with it. A large Reynolds number causes turbulence around the collector drop, which increases the probability of particles colliding with it.

On the whole these effects lead to the net result that not every particle within V_{sweep} of a falling hydrometeor collides with it. This fact is taken into account by the collision efficiency E, which gives the fraction of particles with radius r that collide with the falling collector drop with radius R compared with the total number of particles in the sweep-out volume:

$$E(R,r) = \frac{\pi x_0^2}{\pi(R+r)^2}. \quad (5.18)$$

Here πx_0^2 defines the effective collision cross section, where x_0 is the impact parameter within which a collision is certain to occur whereas outside x_0 particles will be deflected out of the path of the collector drop (see the particle r_2 in Figure 5.11). Owing to the random positions of the droplets, E can also be interpreted as the probability that the collector drop will collide with a smaller particle located at random in the sweep-out volume.

The processes that can cause collision of an aerosol particle with a falling raindrop or snowflake are shown in Figure 5.12. Collisions can be caused by inertial impaction (Figure 5.12a), Brownian motion (Figure 5.12b) or phoretic forces i.e. those which cause the motion of aerosol particles via a temperature, water vapor or electrical gradient (Figure 5.12c). Scavenging by inertia and Brownian motion were discussed above. In

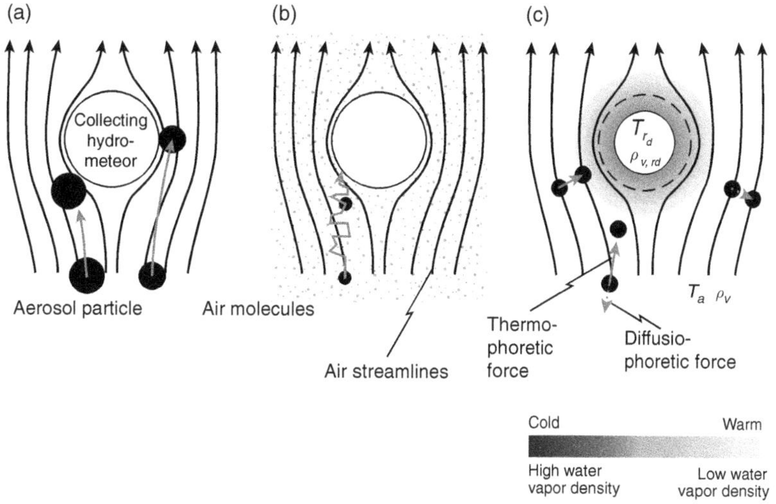

Fig. 5.12 Schematic of the different forces and processes that can contribute to a collision between an aerosol particle and a hydrometeor: (a) inertial impaction (left) and interception (right), (b) Brownian motion and (c) thermophoresis (left), diffusiophoresis (right) and combined phoresis (below). The gray arrows show the direction of the processes. As the different processes are important for aerosol particles of different sizes, they are shown in different sizes. The ambient air temperature is T_a, T_{r_d} is the temperature at the hydrometeor's surface, ρ_v is the ambient water vapor density and ρ_{v,r_d} is the water vapor density at the hydrometeor's surface.

addition to inertial impaction, aerosol particles can also be intercepted by the collector drop (Figure 5.12a, right). Scavenging by interception occurs when the particle is small enough to follow the streamlines around the collector drop but gets so close to it that it is "intercepted".

Scavenging due to phoretic forces describes a collision of an aerosol particle with a hydrometeor and can be initiated by thermophoresis, diffusiophoresis, electrophoresis or photophoresis. Photophoresis is negligible in the troposphere (Wurm and Krauss, 2008). The role of electrophoresis is still highly uncertain. Therefore we limit our discussion to thermo- and diffusiophoresis (Figure 5.12c). As here a temperature gradient amounts to a gradient in the thermal energy of the air molecules, the average velocity of air molecules colliding with an aerosol particle is higher on its warm side than on its colder side. This leads to a net drift of the aerosol particle towards the cooler side. A temperature gradient can also be established due to the latent heat consumption of an evaporating hydrometeor: the thermophoretic force will push the aerosol particle along the negative gradient, i.e. towards the cool spot of the evaporating hydrometeor. This removal process of aerosol particles due to thermophoretic forces is called thermophoresis, from the Greek "carried by heat".

When a hydrometeor evaporates, the increase in total gas volume due to the evaporation leads to a net drift of water vapor molecules away from the hydrometeor. This drift is transferred (partly) to the aerosol particles in the vicinity of the hydrometeor and is called diffusiophoresis. For a growing hydrometeor, however, diffusiophoresis causes a net drift of water molecules towards the hydrometeor that also moves aerosol particles towards the hydrometeor. Here diffusiophoresis acts as a sink for aerosols. As the vapor and temperature gradient are always of opposite sign for either evaporating or growing hydrometeors, so also thermophoresis and diffusiophoresis always act against each other in such cases, with thermophoresis being the dominant force (see the arrows in Figure 5.12c).

On the whole the wet scavenging process by which aerosol particles are removed from the atmosphere depends strongly on the size of the aerosol particles. For small particles, with radii of less than approximately 0.1 μm, Brownian motion is the dominant process and can cause the collision of these particles with a falling hydrometeor. For aerosol particles with radii > 0.5 μm, Brownian motion is not important; here, inertial impaction and interception are the most important forces since particles in this size range do not follow the streamlines of the air around a falling hydrometeor but instead deviate from them towards the hydrometeor and may collide with it. Phoretic forces contribute to the overall collection efficiency mainly for aerosol particles of intermediate sizes, i.e. those in the accumulation mode, where both Brownian motion and inertial forces are small. The exact size ranges corresponding to the importance of the different forces depend on the size of the falling hydrometeor and, if applicable, on the evaporation or sublimation rate of the hydrometeor. The higher its evaporation or sublimation rate, the larger is the contribution from thermophoresis. For aerosol particles less than 10 nm, particle motion due to Brownian motion is sufficiently random that thermo- and diffusiophoresis do not induce an unidirectional movement.

Figure 5.13 shows the forces that contribute to the collection efficiency. As can be seen, the collection efficiency for wet deposition has a minimum for accumulation-mode particles. This minimum in wet deposition coincides with the minimum in dry deposition and is commonly referred to as the Greenfield gap (Greenfield, 1957).

The process of collecting a particle requires not only a collision but subsequent coagulation of the collision partners, i.e. that they do not separate again after the collision. The coagulation efficiency $\hat{E}$ of an aerosol particle with a liquid particle (water droplet or raindrop) is assumed to be unity, so that every particle coming into contact with a liquid particle becomes incorporated into it (Croft *et al.*, 2009). Therefore $\tilde{E} = E$ and thus the size dependence of $\tilde{E}$ shown in Figure 5.13 reflects the size dependence of E.

The removal of aerosols by hydrometeors is best described in terms of the collection kernel K [m^3 s^{-1}], which depends on the radii of the hydrometeor R and the aerosol particle r as well as the velocity difference between them:

$$K(R,r) = \pi(R+r)^2 E(R,r)|v_T(R) - v_T(r)|. \tag{5.19}$$

Here $E(R,r)$ is the collision efficiency (eq. 5.18), and $v_T(R)$ and $v_T(r)$ are the terminal velocities of the hydrometeor and the aerosol particle, respectively (Section 7.2.3).

In Figure 5.14 it can be seen that, for giant aerosol particles larger than 5 μm in radius, collision with an approaching raindrop is virtually inevitable whereas only one in 10^3–10^5 accumulation mode particles in the sweep-out volume of a raindrop ends up colliding with it. Also shown is the collection efficiency for the removal of aerosol particles by falling snowflakes. It is similar, and of the same order of magnitude, as that for removal by smaller raindrops. The removal of aerosol particles by snowflakes is, however, more

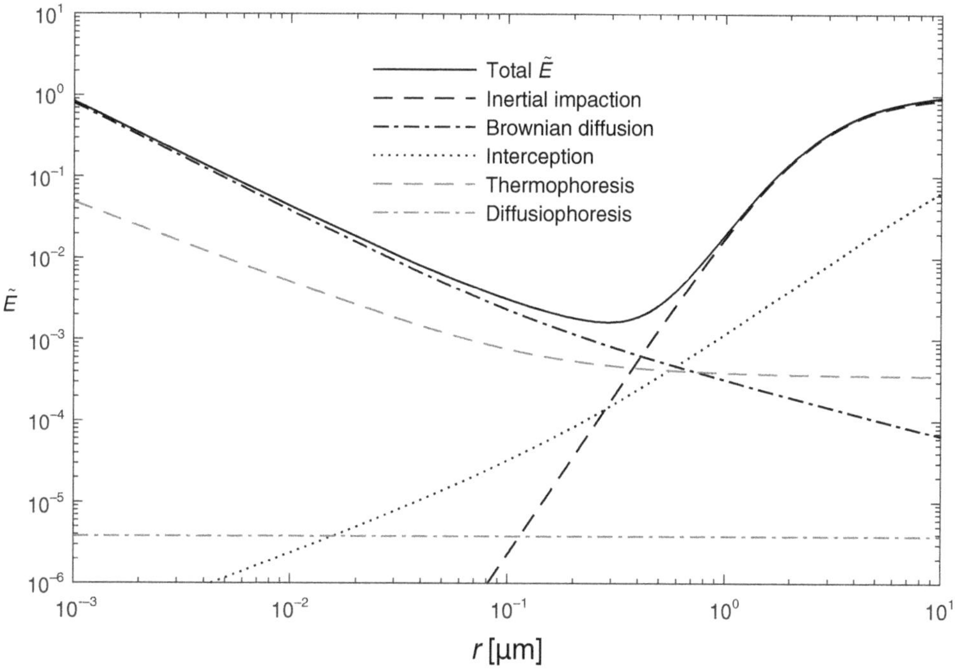

Fig. 5.13 Theoretical calculations of different forces contributing to the collection efficiency $\tilde{E}$ for hydrometeor–aerosol collisions as a function of aerosol radius r. The liquid hydrometeor has radius 40 μm at $T = 245$ K and is kept at ice saturation. The parameterizations are taken from Park *et al.* (2005) for Brownian diffusion, inertial impaction and interception and from Andronache *et al.* (2006) for thermophoresis and diffusiophoresis. Figure adapted from Nagare *et al.* (2015).

uncertain than removal by raindrops and depends on the morphology of the snowflake. In Figure 5.14 two estimates of the aerosol collection efficiency of snowflakes of differing shapes are shown. Snow-A refers to a plate with equivalent radius 500 μm and snow-B to a rimed crystal with characteristic dimension 100 μm. As with rain, the smaller the snowflake, the higher is the collection efficiency of nucleation and Aitken mode particles. The reason is that larger hydrometeors disturb the streamlines of the approaching particles more and therefore a larger fraction of the fine-mode particles has a chance to avoid them. For coarse-mode particles this effect becomes negligible or is even reversed because of their large inertia, which drives them towards the collector hydrometeor.

5.4.2.2 Nucleation scavenging

Nucleation scavenging refers to the ability of aerosols to act as CCN or INPs and to be incorporated into a hydrometeor that grows to precipitation size and finally reaches the ground as precipitation.

Given that accumulation-mode aerosol particles are poorly removed by either dry or collision-induced wet scavenging, there must be another way in which they are removed

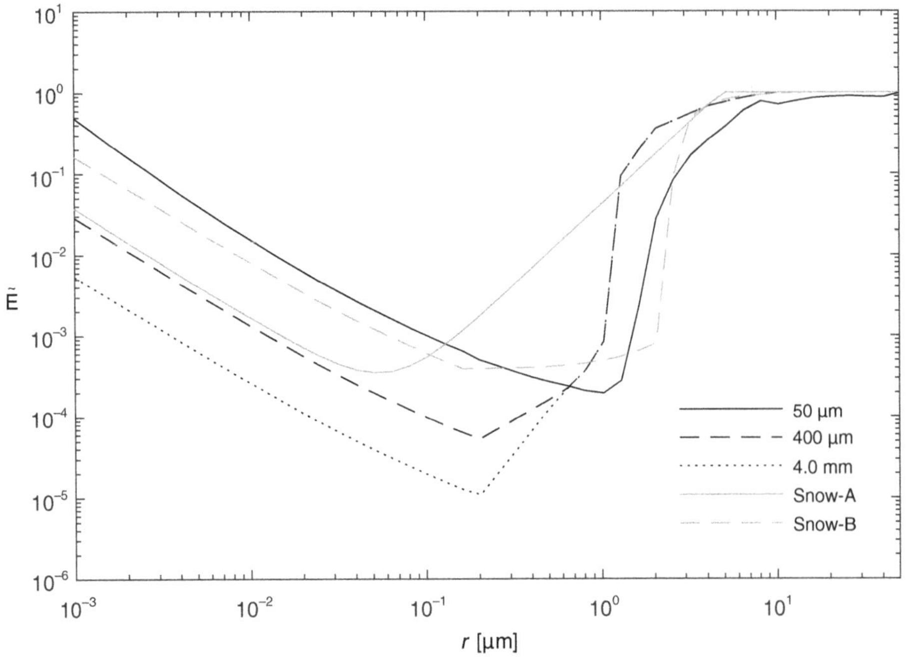

Fig. 5.14 Collection efficiency $\tilde{E}$ for raindrop–aerosol collisions as a function of aerosol radius r and collector raindrop size. In addition $\tilde{E}$ is shown for snowflake–aerosol collections: Snow-A refers to a plate with equivalent radius 500 μm and Snow-B to a rimed crystal with characteristic dimension 100 μm. Figure adapted from Croft *et al.* (2009).

from the atmosphere. In fact accumulation mode particles are most efficiently removed by nucleation scavenging. As will be discussed in Chapter 6, the ability of aerosol particles to act as CCN depends, besides other aerosol characteristics, strongly on size.

An example of how nucleation scavenging removes first the largest particles and subsequently smaller and smaller particles is shown in Figure 5.15. It mimics the way in which aerosol particles act as CCN in that the largest particles are activated first (Section 6.5). The degree of supersaturation in the cloud determines the size above which aerosol particles can be activated. The larger the aerosol particle, the lower the (critical) supersaturation needed to activate the aerosol particle into a cloud droplet. If the supersaturation is high enough, some Aitken mode particles can also be subject to nucleation scavenging.

At time t_0, the assumed initial aerosol number size distribution is bimodal, with a large number density of particles in the accumulation mode. At time $t_1 > t_0$ (Figure 5.15), the accumulation mode has mainly gone, showing the significant contribution of nucleation scavenging at the beginning of the cloud formation process. The accumulation mode subsequently decreases further until at time $t_2 > t_1$ nearly all aerosol particles with radii > 0.2 μm have been removed. Smaller particles act less readily as CCN for the supersaturation reached in this example and therefore are removed less efficiently by nucleation scavenging.

As the number of aerosols acting as INPs rather than as CCN is much fewer, nucleation scavenging is only of minor importance in ice clouds.

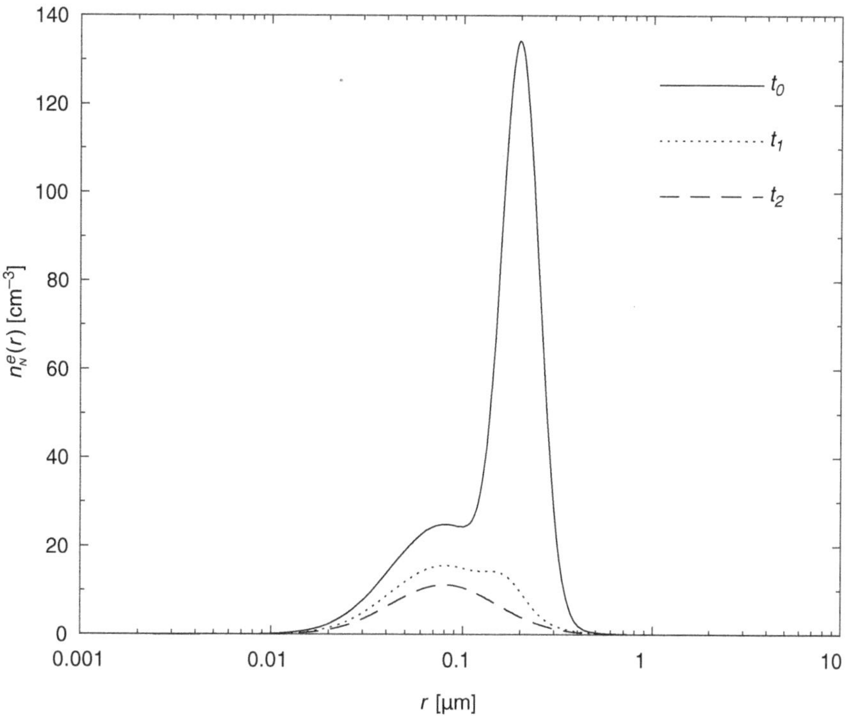

Fig. 5.15 Schematic of the nucleation scavenging of aerosol particles as a function of time: initial size distribution, at time t_0, and that at times t_1 and t_2 after nucleation scavenging has taken place.

5.4.3 Atmospheric processing of aerosol particles

Prior to being removed from the atmosphere, aerosols can change their properties within the atmosphere either by chemical conversions or by collision and coagulation with other aerosol particles. This change in size and/or chemical composition can transfer them into a different aerosol mode and thereby influence their rate of removal from the atmosphere.

An example of the chemical conversion of an aerosol particle is the uptake of ammonia gas, NH_3, by a sulfuric acid (H_2SO_4) particle, which results in either ammonium bisulfate NH_4HSO_4, or ammonium sulfate, $(NH_4)_2SO_4$, depending on the amount of ammonia gas available:

$$H_2SO_4(aq) + NH_3(g) \rightarrow NH_4HSO_4(aq), \tag{5.20}$$

$$H_2SO_4(aq) + 2NH_3(g) \rightarrow (NH_4)_2SO_4(aq). \tag{5.21}$$

In general any altering of the particle properties is referred to as the aging of aerosol particles. Aerosol particles can grow by the condensation of gaseous species onto pre-existing particles and so become secondary or aged particles; consider for instance the uptake of water vapor by a salt particle:

$$NaCl(s) + H_2O(g) \rightarrow NaCl(aq). \tag{5.22}$$

Water uptake is relevant for aerosol particles with an affinity for water vapor. The affinity of an aerosol particle to take up water depends on its chemical composition and mixing state. Depending on its state a particle may be either hydrophilic or hygroscopic. Hygroscopic aerosol particles are those that readily take up water from the atmosphere and retain it, examples of which are salts such as sulfates, nitrates and sodium chloride. Hydrophilic aerosols, however, are solid particles that can be wetted by water, such as mineral dust particles.

The uptake of water vapor by a solid hygroscopic aerosol particle is accompanied by a phase transition, as will be discussed in Section 6.3. If an aerosol particle is already liquid, then the uptake of water vapor just leads to further growth and dilution of the salt solution.

Some aerosol particles are hydrophobic upon emission, such as black carbon and partly also particulate organic matter. These hydrophobic aerosol particles can be removed only by dry deposition or wet impaction scavenging, not by nucleation scavenging. If they acquire soluble material either by the collisions with other particles that contain soluble material or by condensation of sulfuric acid or organics from the vapor phase, they can participate in nucleation scavenging as well. Growth by condensation and aging by gas-phase oxidation can transfer nucleation-mode particles to Aitken-mode particles and Aitken-mode particles further into the accumulation mode.

5.5 Burden and lifetime of aerosols

The estimates of the aerosol burden and lifetimes in Table 5.3 are taken from a model intercomparison involving 16 global aerosol climate models (Textor *et al.*, 2006). As an example of the geographical distribution of the various aerosol burdens, they are shown for the ECHAM6 global climate model (Neubauer *et al.*, 2014) in Figure 5.16.

Even though the mineral dust emissions are almost a factor 5 smaller than the sea salt fluxes, mineral dust dominates the aerosol burden (Table 5.3). This is caused by the different mechanisms by which these aerosol types are injected into the atmosphere and the associated different lifetimes of these aerosol types. Dust storms can transport mineral dust aerosols rather high into the atmosphere. For instance, particles from the Asian Taklamakan desert are entrained to altitudes > 5 km (Wiacek *et al.*, 2010) and thus enter the free troposphere. Sometimes mineral dust is lifted up in deep convective clouds and reaches even higher altitudes. As the lifetime of aerosol particles increases with altitude (Figure 5.17a), the higher the mineral dust particles are injected into the atmosphere, the longer their lifetime and thus the higher the burden. In contrast, sea salt has the largest emission flux but the particles remain near the surface in humid air, where the sea salt is removed very quickly by wet and dry deposition. Thus the lifetime of sea salt particles, half a day, is the shortest of the different aerosol species included in Table 5.3. This explains its rather small burden compared with that of mineral dust. It further elucidates why little sea salt is found over the continents (Figure 5.16e). Mineral dust, however, can be transported from the Sahara across the Atlantic all the way to North America.

Of the three anthropogenic species (sulfate particles, BC particles and POM), sulfate particles have the shortest lifetime, 4 days. It is shorter than that of BC particles and POM because BC particles and some fraction of the POM are hydrophobic when injected into the

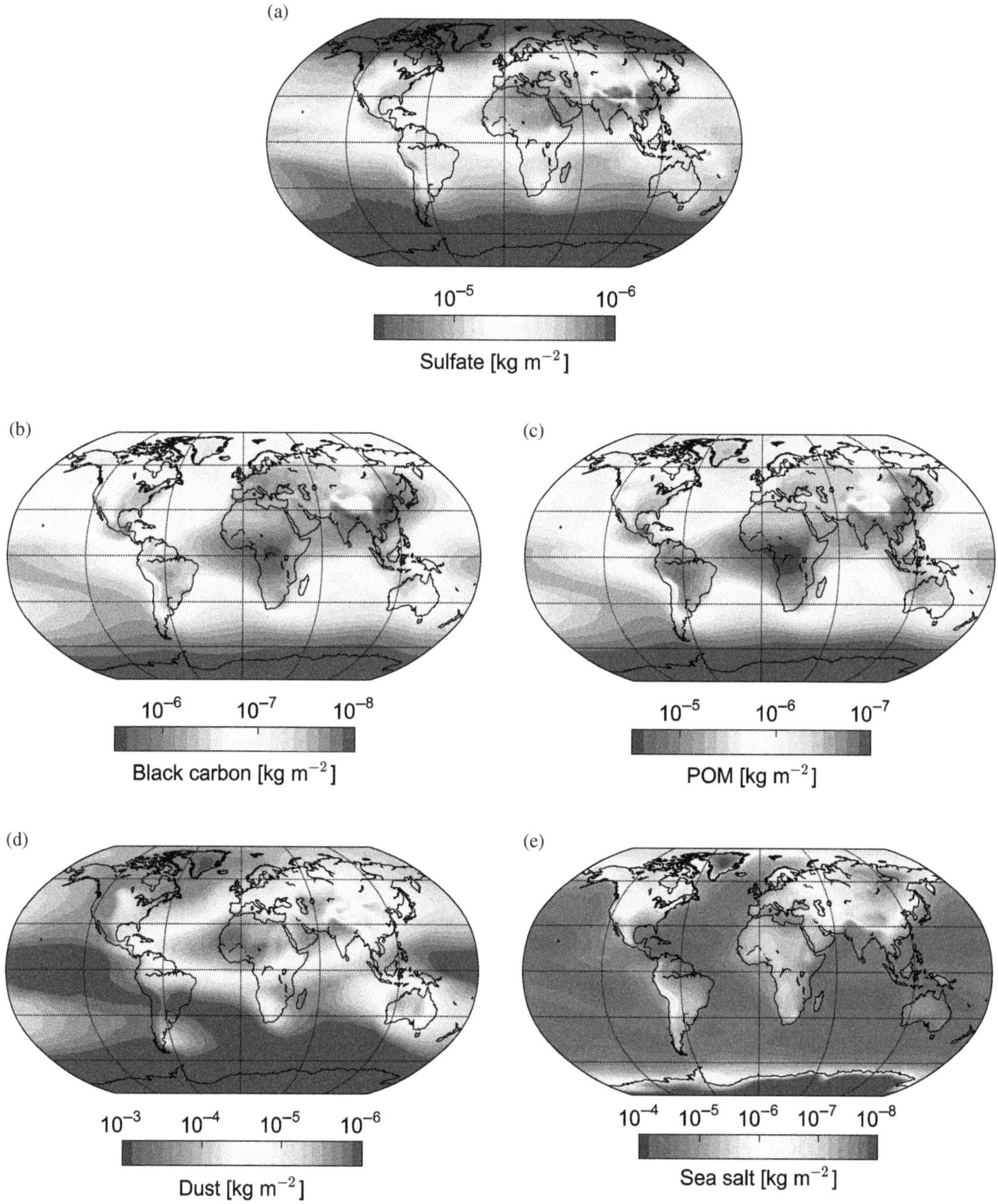

Fig. 5.16 Annual mean vertically integrated aerosol burden of (a) sulfate particles, (b) black carbon particles, (c) particulate organic matter (POM), (d) mineral dust particles and (e) sea salt particles from the ECHAM6-HAM global climate model (Neubauer *et al.*, 2014). Note that the scale is different for each aerosol component. A black and white version of this figure will appear in some formats. For the color version, please refer to the plate section.

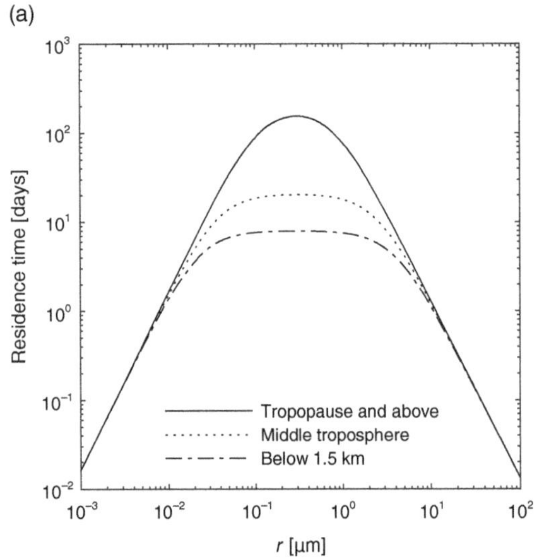

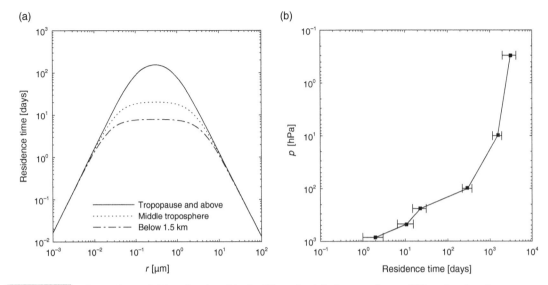

Fig. 5.17 Lifetime of aerosols (a) as a function of size for different levels in the atmosphere and (b) as a function of pressure. Figure adapted from Jaenicke (1988) with permission from Springer Science and Business Media.

atmosphere; consequently they are subject only to dry deposition, below-cloud scavenging and in-cloud collision scavenging, which all have a minimum for accumulation mode particles. It takes up to a few days to convert BC particles and POM into internally mixed aerosols (Figure 5.1) that contain some soluble material. Only thereafter can these particles be removed by nucleation scavenging. Thus the lifetimes of BC particles and POM are 2–3 days longer than that of sulfate particles (Table 5.3). The sulfate aerosol burden is highest in the northern hemisphere in proximity to the major industrial sources, whereas the POM burden is highest in regions with biomass burning in South America and Africa. The BC aerosol burden shows a similar pattern to that of POM, but it has a third maximum in southeast Asia.

Because both dry and wet scavenging efficiencies are smallest for the accumulation mode, the lifetime of aerosol particles is largest in this mode (Figure 5.17a). It is around one week in the lowest 1.5 km, two weeks in the mid-troposphere and more than 100 days in the tropopause region and above.

The lifetime decreases for smaller particles owing to Brownian motion, which can either result in a collision with a hydrometeor or in a coagulation with another aerosol particle that creates an aerosol particle of a larger size. The lifetime decreases for larger particles owing to gravitational settling with subsequent removal by dry deposition and increased probability of nucleation scavenging as CCN.

Major volcanic eruptions, such as the Mount Pinatubo eruption in 1991, inject large amounts of gaseous sulfur dioxide into the stratosphere; this is subsequently oxidized to gaseous sulfuric acid and further nucleates to form sulfate aerosols or condenses on pre-existing aerosol particles. These sulfate aerosols have a lifetime of a couple of years (Figure 5.17b) because the only sink in the stratosphere is gravitational settling. A typical mode radius for sulfate aerosols formed by gas-to-particle conversion with subsequent coagulation after a median volcanic eruption is 0.3 μm (Timmreck *et al.*, 1999). This radius lies in the Greenfield gap between removal by collisions due to Brownian motion and removal by inertial impaction. A particle with 0.3 μm radius has a sedimentation velocity on the order of 0.01 mm s^{-1}. Thus it would take more than one year for such a particle to settle one kilometer. In reality, however, some coagulation will occur and sulfate aerosols will be subject to wet deposition once they enter the troposphere. The long lifetime of stratospheric aerosols constitutes one reason behind the idea that part of the global warming that has been observed over the last 250 years could be offset by injecting aerosols into the stratosphere. This is referred to as climate engineering and will be discussed in Section 12.4.

5.6 Summary of aerosol processes

A summary of the main aerosol processes as a function of aerosol size is shown in Figure 5.18. Nucleation mode aerosols originate from gas-to-particle conversion and are commonly found over forests that emit organic precursors. They are also formed in the upper troposphere. Aitken mode particles are mainly emitted during combustion processes of which the most important sources are related to anthropogenic activities. These aerosol particles form predominantly by gas-to-particle conversion producing secondary particles. Aerosol particles such as BC that are emitted during biomass burning and forest fires are directly emitted as particles into the atmosphere and are primary particles. All particles can

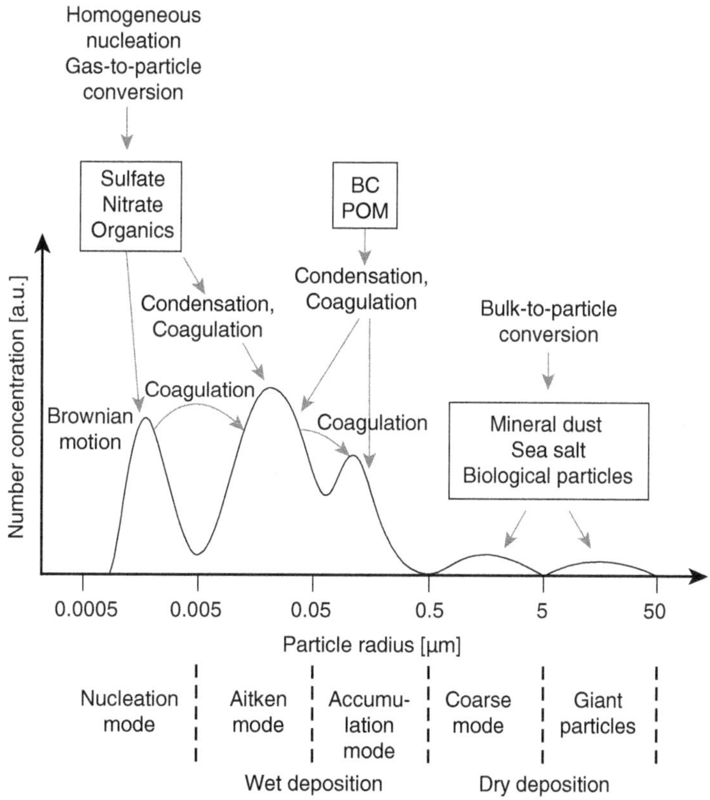

Fig. 5.18 Schematic of the main aerosol processes as a function of their size; BC refers to black carbon and POM to particulate organic matter.

grow by condensation of condensable material or by water uptake if they are hydrophilic or hygroscopic.

The collision and coagulation of aerosol particles shifts them to larger sizes. This internal redistribution is especially important as a sink for nucleation mode particles, transferring them to Aitken mode particles. It is still important for Aitken mode particles that are transferred to accumulation mode particles but is negligible as a growth process of accumulation mode particles. Collision and coagulation are linked to the mobility of aerosol particles. Their mobility rapidly decreases as their size increases, confining collision and coagulation due to Brownian motion to nucleation and Aitken mode particles.

The main sink of accumulation mode particles is by their acting as CCN and being removed by nucleation scavenging if the cloud precipitates. Nucleation scavenging also removes coarse mode particles and depending on the updraft velocity, also the larger Aitken mode particles.

The source of coarse mode particles is bulk-to-particle or liquid-to-particle conversion. Here dry deposition is the main removal mechanism, which increases in importance with increasing particle size. For comparison, the dry deposition velocity of an aerosol particle with 1 μm radius is $\sim$ 0.02 cm s^{-1} and that of a 10 μm aerosol particle is 2 cm s^{-1} (Figure 5.10).

5.7 Exercises

Single/multiple choice exercises

1. Number size distributions of ambient aerosol particles are often represented as a superposition of individual log-normal distributions. Consider a bimodal log-normal distribution shown in $dN/d\ln(r)$ versus r parameter space. Which of the following is/are true?
   (a) One advantage of the log-normal distribution over a normal distribution when describing aerosol populations is that the distributed quantity can only have positive values.
   (b) The geometric mean radius of the log-normal distribution equals the arithmetic mean of the normal distribution of $\ln(r)$.
   (c) In the given parameter space the area under the curve corresponds to the total aerosol number concentration.
   (d) Because the aerosol size distribution is log-normally distributed, the logarithm of the aerosol sizes is normally distributed.
2. Aerosol particles range in size from nanometers to hundreds of micrometers and are often classified in size bins. Which of the following is/are true?
   (a) The change in an aerosol size distribution with time depends on the lifetime of the aerosol particles.
   (b) Log-normal size distributions are symmetric.
   (c) Volume distributions usually have a large contribution from the nucleation mode.
   (d) Surface distributions can be used to estimate radiative properties of aerosol particles.

3. Sulfate is mainly a secondary particle. Which of the following is/are true?
   (a) Sulfate aerosols have no natural source.
   (b) Ship tracks are visible in sulfur emission maps.
   (c) Sulfate aerosols can grow by the condensation of water vapor.
   (d) Sulfate aerosols have a lifetime of around four days.
4. Aerosol particles are removed by dry and wet scavenging. Which of the following is/are true?
   (a) In the annual global mean, the majority of the Aitken and accumulation mode particles are removed by wet scavenging.
   (b) The dry deposition velocity steadily increases from smaller to larger particle sizes and is highest for coarse mode aerosol particles.
   (c) Only aerosol particles which can act as CCN or INPs can be removed by wet scavenging.
   (d) Wet scavenging is the dominant sink for coarse mode particles, as large particles are activated first.
5. Wet scavenging is divided into nucleation and impaction scavenging. Which of the following is/are true?
   (a) All particles in the sweep-out volume of a falling hydrometeor collide with the collector drop.
   (b) Turbulence increases the probability of a collision.
   (c) Small particles tend to follow the streamlines of the air and do not collide with the collector drop.
   (d) Impaction scavenging has a minimum in the size range of accumulation-mode particles.
   (e) A bare BC particle in the accumulation mode is directly removed by nucleation scavenging.
6. Phoretic forces can contribute to a collision between an aerosol particle and a hydrometeor. Which of the following is/are true?
   (a) The thermophoretic force moves aerosol particles towards the hydrometeor, whereas the diffusiophoretic force pushes them away.
   (b) Thermophoresis and diffusiophoresis always act in the opposite direction for a collision of an aerosol particle with a hydrometeor.
   (c) For an evaporating raindrop, diffusiophoresis is the dominant force whereas for a growing raindrop thermophoresis dominates.
   (d) Phoretic forces are especially important for aerosol particles in the nucleation and Aitken modes.
7. The residence time of aerosol particles in the atmosphere varies from hours to years. Which of the following is/are true?
   (a) Sea salt particles dominate aerosol emissions and thus also the aerosol burden.
   (b) Aerosol particles in the accumulation mode have the longest lifetime in the atmosphere.
   (c) The residence time of aerosol particles depends on the altitude at which they are injected into the atmosphere.
   (d) Aerosol particles that contain insoluble material usually have a longer lifetime than soluble particles of the same size.

8. Aerosol particles are classified into primary and secondary particles according to their source. Which of the following aerosol particles belong exclusively to the group of primary particles? (1) pollen, (2) sea salt, (3) POM, (4) nitrate, (5) sulfate, (6) BC, (7) sulfuric acid particles, (8) mineral dust.
   (a) (1), (2), (6), (8)
   (b) (1), (2), (4), (5), (8)
   (c) (1), (2), (3), (6), (8)
   (d) (1), (2), (3), (4), (5), (8)
   (e) (1), (2), (3), (6), (7), (8)
9. Why is the log-normal number distribution a good choice for plotting aerosol size distributions?
   (a) Because the size bins are in SI units.
   (b) Because there are generally many more small particles (e.g. between 5 and 10 nm) than large particles (e.g. between 200 and 205 nm).
   (c) Because its values do not depend on the length scale.
   (d) Because the area below the curve is proportional to the total particle number concentration.
10. Which of the following examples refer to secondary particles?
   (a) Sea salt particles originating from water droplets ejected into the air.
   (b) Most sulfuric acid particles.
   (c) An uncoated black carbon particle.
   (d) The oxidation products of alpha-pinene (an organic gas), which form a liquid phase.
11. Sea salt particles have a very short atmospheric lifetime compared with mineral dust particles (approximately 0.5 days compared with 4 days). Why?
   (a) Because sea salt particles are larger than mineral dust particles and therefore sediment more quickly.
   (b) Because sea salt particles are smaller than mineral dust particles and therefore deposit more quickly owing to their larger Brownian motion.
   (c) Because sea salt particles are better INPs than mineral dust particles and therefore the Wegener–Bergeron–Findeisen process occurs more often and the particles are more efficiently scavenged by wet deposition.
   (d) Because sea salt particles mainly occur over the ocean, where the relative humidity is high and therefore wet deposition is more efficient.
   (e) Because mineral dust particles are often ejected rather high into the atmosphere.

Long exercises

1. Consider a hypothetical aerosol population containing only aerosol particles with radii between 0.1 and 1.1 μm, where we have exactly 100 particles in each 0.1 μm size bin. Imagine that you have measured the aerosol size distribution with two instruments A and B which have different resolutions. The size distribution measured with instrument A is displayed in a histogram in Figure 5.19, where the dots on top of each bar denote

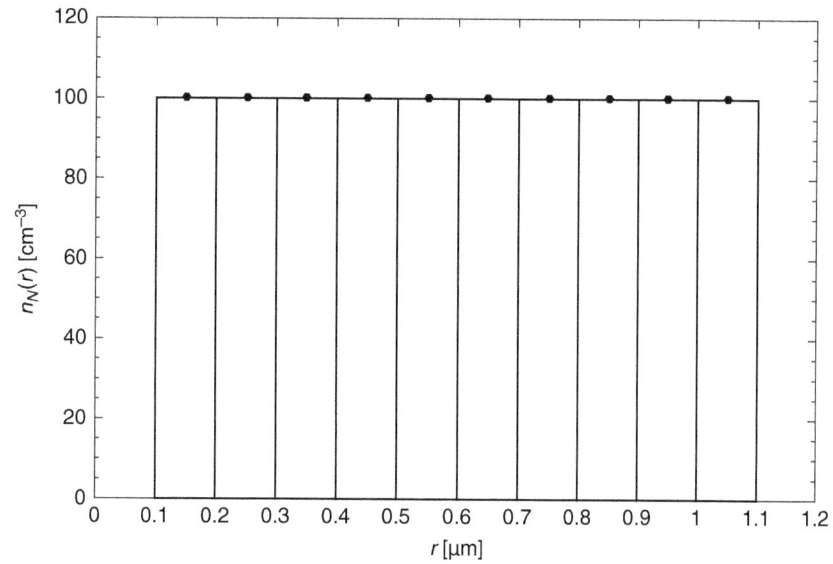

Fig. 5.19 Histogram of the size distribution measured with instrument A.

the mid-point radius. The size ranges of the individual size bins and the corresponding aerosol concentrations as measured by the second instrument, B, are given in Table 5.4.

(a) Is the aerosol population sampled by instrument A monodisperse or polydisperse?
(b) Plot the size distribution from instrument B in the same way as shown in Figure 5.19, using the information given in Table 5.4. Are more aerosol particles measured in the range with radii 0.1–0.3 μm or 0.8–1.0 μm?
(c) Plot the histogram of the size distribution measured with instrument A in the same figure and compare the two size distributions. Do they differ?
(d) Explain the limitations of discrete size distributions. What do you need to consider when comparing discrete size distributions from instruments with different size resolutions?
(e) Calculate the normalized size distributions of both instruments in μm cm^{-3} and plot them as histogram plots. Reconsider your results from (b).

2. The data in Table 5.5 show the values of an individual size distribution from Figure 5.20b, giving the particle radius r_i of the individual size bins i, the corresponding particle concentration N_i and the normalized particle concentration $n^e_{N,i}$. The complete dataset can be downloaded from www.cambridge.org/clouds.
(a) Calculate the arithmetic and geometric mean radii of the number size distribution as well as the geometric standard deviation.
(b) Derive the relation between $n_N(r)$ und $n^{10}_N(r)$. Looking at your result, how are $n^e_N(r)$ and $n^{10}_N(r)$ related?
(c) The aerosol concentration values in the third column of Table 5.5 correspond to the size distribution function $n^e_N(r)$. Transform them to $n^{10}_N(r)$. How do the particle concentrations within the individual size bins change with this transformation? Explain your results.
(d) Draw the size distribution in $n^e_N(r)$–r space using a logarithmic x-axis. Insert the arithmetic and geometric mean radii into it.

Table 5.4 Size distribution data as measured with instrument B.

Size range [μm]	$N(r_i)$ [cm^{-3}]
0.1–0.3	200
0.3–0.45	150
0.45–0.6	150
0.6–0.7	100
0.7–0.8	100
0.8–0.875	75
0.875–0.95	75
0.95–1.0	50
1.0–1.05	50
1.05–1.075	25
1.075–1.1	25

Table 5.5 Size distribution data for ambient aerosol particles as obtained with a scanning mobility particle sizer, showing the particle radius r_i of the individual size bins i, the corresponding particle concentration N_i and the normalized particle concentration $n^e_{N,i}$.

r_i [nm]	N_i [cm^{-3}]	$n^e_{N,i}$ [cm^{-3}]
4.74	0.00	0.00
4.91	0.00	0.00
5.10	0.00	0.00
5.30	0.06	2.18
5.45	0.00	0.00
5.65	0.28	6.35
5.90	0.23	6.87
6.10	0.20	6.04
6.30	0.42	10.87
6.55	0.49	13.15
6.80	0.64	17.76
7.05	0.58	16.74
7.30	0.83	24.74
⋮	⋮	⋮
192.70	2.19	60.89
199.75	1.99	55.19
207.10	2.08	57.67
214.70	1.96	54.66
222.55	2.00	55.55
230.70	2.12	58.84
239.15	1.94	53.94
247.90	2.27	62.95
⋮	⋮	⋮
318.90	0.00	0.00
330.60	0.00	0.00

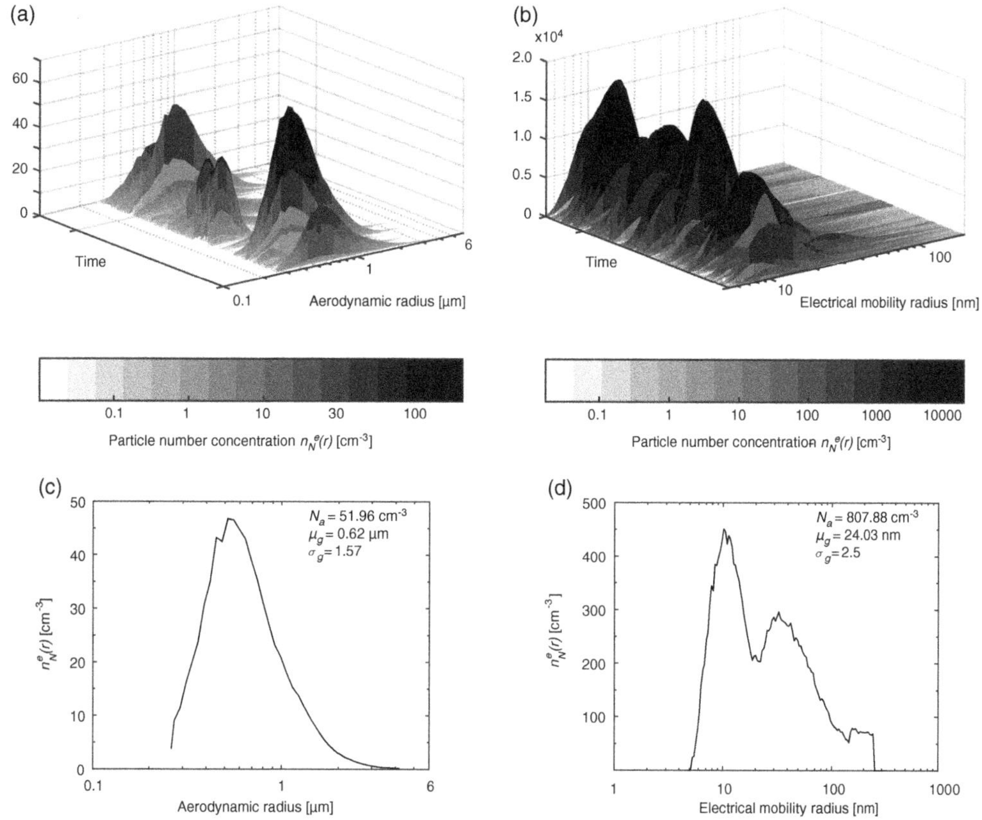

Fig. 5.20 Temporal evolution of the size distribution of ambient aerosol particles measured at Tenerife, Canary Islands, by (a) an aerodynamic particle sizer and (b) a scanning mobility particle sizer. The shading indicates the particle concentration as shown on the vertical-axis. (c), (d) Snapshots of the size distributions for August 19, 2014 in (a) and (b), respectively.

(e) In which size range lie 68% of all aerosol particles?
(f) In which size mode lies this type of aerosol?
(g) Plot the surface distribution $n_S^e(r)$, the volume distribution $n_V^e(r)$, and the mass distribution $n_m^e(r)$. Assume that all particles are spherical and that the aerosol particles consist mainly of mineral dust with density $\rho_a = 2.6$ g cm^{-3}.

3. A few times every summer, large amounts of Saharan dust are transported over the Canary Islands, carried within the so-called Saharan Dust Layer. Size measurements of the transported mineral dust particles show a bimodal size distribution. Figures 5.20a, b respectively show the temporal evolution of the aerosol number concentrations for these two modes obtained by two instruments measuring in different size ranges. Mineral dust events can clearly be identified as peaks in the supermicron size range in Figure 5.20a. Figures 5.20c, d give snapshots of the temporal evolution of the size distributions shown in Figures 5.20a, b. The total particle number concentration, the number geometric mean radius and the geometric standard deviation are denoted by N_a, μ_g and σ_g, respectively.

(a) On the basis of Figures 5.20c, d, name the two aerosol modes that were observed on August 19, 2014.

(b) Stokes' law states that the drag force F_D for a spherical body can be described by

$$F_D = 6\pi r \mu v,$$

where μ is the dynamic viscosity of air ($\approx 1.82 \times 10^{-5}$ kg m^{-1} s^{-1}), r is the particle radius and v the velocity of the particle. By equating F_D with the gravitational force F_g, the terminal velocity of a particle can be calculated. However, for small bodies such as aerosol particles, the formula is imprecise and a correction factor C_c (the so-called Cunningham correction factor) should be applied:

$$F_D = \frac{6\pi r \mu v}{C_c},$$

$$C_c = 1 + 2.34\frac{\lambda}{d} + 1.05\frac{\lambda}{d}\exp\left(\frac{-0.39d}{\lambda}\right),$$

where $\lambda \approx 66$ nm is the mean free path of air and d is the diameter of the aerosol particle. How long does it take for the mineral dust aerosols in the different modes given in Table 5.1 to settle out? Assume initial altitude $z_{t=0} = 200$ m and use the mass geometric mean radius for your calculation. The density of the particles is assumed to be $\rho_a = 2600$ kg m^{-3}.

(c) The scavenging coefficient Λ_B [s^{-1}] for Brownian diffusion is given by

$$\Lambda_B = \frac{1.35 M_l D_a}{\bar{r}_d^2},$$

where M_l is the cloud liquid-water content [g cm^{-3}], $\bar{r}_d$ is the radius of a drop with average mass [cm] and $D_a = kT/6\pi\mu\mu_g$ [cm^2 s^{-1}] is an approximation for the Brownian diffusivity, with Boltzmann constant k, dynamic viscosity of air μ and number geometric mean aerosol radius μ_g. Note that the factor 1.35 in the equation is in cm^3 g^{-1}. The scavenging rate is given by

$$\frac{1}{N_a}\frac{dN_a}{dt} = -\Lambda_B.$$

How long does it take for half the mineral dust aerosols in the different modes to be removed by scavenging? Assume the values $T = 275$ K, $M_l = 10^{-6}$ g cm^{-3}, $\mu = 1.82 \times 10^{-5}$ kg m^{-1} s^{-1} and $\bar{r}_d = 10$ µm.

(d) Interpret your results: Is removal by dry deposition (b) or by wet scavenging (c) faster?

4. (a) Where does the collection efficiency $\tilde{E}$ for wet deposition have a maximum and a minimum and why?

   (b) How does this change for dry deposition?

5. Can the global distribution of the main aerosol types be explained simply by their emission processes? Why (not)?

6 Cloud droplet formation and Köhler theory

In this chapter the details of phase transitions, nucleation and cloud droplet formation are discussed. Some phase transitions require a nucleation process, for instance, ice nucleation or the gas-to-particle conversion of aerosol particles. Other phase transitions, such as cloud droplet formation, do not require a nucleation process: cloud droplets form on soluble or hydrophilic aerosol particles such as sulfate, sea salt (soluble) and mineral dust (hydrophilic) particles. These particles acquire water by condensation and hence develop a liquid phase at conditions where the nucleation of pure water droplets from vapor cannot occur. Although no nucleation process initiates the existence of a cloud droplet, the concept of nucleation is a prerequisite for the understanding of cloud droplet formation and is therefore introduced in Section 6.1. The mathematical formulation of nucleation leads to the Kelvin equation (Section 6.2). Hygroscopic growth (Section 6.3) and Raoult's law (Section 6.4) consider the contribution of aerosol particles to cloud droplet formation. Combining Raoult's law with Kelvin's equation yields the Köhler theory describing the activation of a deliquesced aerosol particle – forming an aqueous solution droplet – into a cloud droplet (Section 6.5). The chapter ends with a discussion of measurements on cloud condensation nuclei (CCN), the subset of aerosol particles that can be activated into cloud droplets (Section 6.6).

6.1 Nucleation

From thermodynamics of phase transitions (Chapter 2), we do not obtain information about how a new phase is initiated. One possibility for the initiation of a new phase is a nucleation process; *nucleation* denotes a phase transition where a cluster of a thermodynamically stable phase forms and grows within the surrounding metastable parent phase. In the case of aerosol nucleation this refers to the formation of a liquid phase out of the gas phase. Ice crystal formation requires the initiation of the ice phase within the liquid or vapor phase.

The nucleation of a new phase requires the parent phase to be in a metastable state, which can be a supersaturated vapor or a supercooled liquid in the atmospheric context. The transition to the new, stable, phase is not spontaneous but is inhibited by an intermediate maximum in the Gibbs free energy, a so-called energy barrier. This energy barrier is of crucial importance, as it governs the kinetics and therefore the likelihood of a specific nucleation process under given atmospheric conditions.

The initiation of phase transitions is discussed in Section 6.1.1 and cluster formation in Section 6.1.2; this includes the calculation of the energy barrier for the hypothetical case of homogeneous cloud droplet nucleation from the vapor phase (i.e. with no aerosol particles involved). Analysis of this energy barrier yields the Kelvin equation, an integral part of Köhler theory (Section 6.5). Ice nucleation from the vapor or liquid phases is discussed in Section 8.1. For the homogeneous nucleation of aqueous aerosol particles from precursor gases, see Section 5.3.1.

6.1.1 Initiation of phase transitions

In Section 2.4 phase transitions were treated purely thermodynamically, i.e. assuming that a bulk quantity of all involved phases is already present in the system. However, in the atmosphere, phase transitions usually start with only the initial phase present. For simplicity we start with a one-component system. An example of such a system is one containing only water in different phases, for instance water vapor and liquid water. Considering a one-component system for the derivation of phase transitions and extending the findings to multi-component systems is the typical approach in thermodynamics. In a one-component system the partial pressure p_i of the gas phase equals the total pressure p.

For a system at constant pressure and temperature, the minimum in the Gibbs free energy indicates the state in which the system is in thermodynamic equilibrium (TDE). Figure 6.1 shows schematically the Gibbs free energy G of a representative substance as a function of the volume V occupied by the substance, for three different pressures. For each pressure the Gibbs free energy G exhibits two local minima, one corresponding to the substance in the gas phase and one corresponding to the liquid phase. The local minimum with higher G corresponds to a metastable state, meaning that this state will change into the stable state of TDE (i.e. the state of the global minimum of G) over a given time scale. For $p = p_3 > p_s$, with p_s the saturation vapor pressure, the gas is supersaturated and tends to condense into

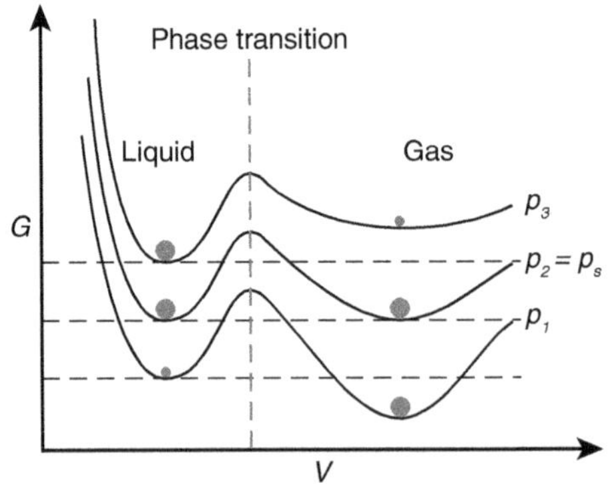

Fig. 6.1 Schematic of the Gibbs free energy G of a substance as a function of the volume V of the substance, for three different pressure levels ($p_1, p_2 = p_s$= saturation vapor pressure, and p_3) at a given T. The stable states of TDE are shown as large disks and the metastable states as small disks. The broken vertical line represents the energy barrier for the phase transition. The horizontal broken lines are intended for visual orientation.

the liquid state. For $p = p_2 = p_s$, both states have identical G values and neither the gaseous nor the liquid state can be identified as more favorable. For $p = p_1 < p_s$, the vapor phase is the preferred state and thus liquid water tends to evaporate.

The fact that metastable states are separated from stable states by an energy barrier leads to a kinetic inhibition of the phase transitions, i.e. phase transitions do not occur spontaneously. Regarding the minimum principle for G, there is thermodynamically no reason why a system in a metastable state should increase its Gibbs free energy and cross the energy barrier to reach the stable state. The reason why a system nevertheless temporarily increases its Gibbs free energy involves concepts from statistical mechanics. The time scale on which phase transitions occur increases with increasing height of the energy barrier, because the probability of a phase transition decreases with increasing height of the energy barrier. The existence of an energy barrier is common to first-order phase transitions, which are characterized by a jump in density between the two phases. The first-order phase transitions related to clouds are condensation, deposition, freezing, evaporation, sublimation and melting (Figure 2.10).

A purely theoretical pathway for a phase transition is provided by a uniform change in density, transferring the entire substance simultaneously from the metastable to the stable phase. However, as this involves all molecules of the system, the energy barrier impeding this pathway is very high. Thus the intermediate states on the path between the metastable and the stable state are so unlikely that they can be considered non-physical.

Nucleation of a new phase within the parent phase is an alternative pathway for a phase transition. It is by far the more likely pathway since the new phase is initiated locally, via the formation of a small cluster of the new phase, resulting in a much lower energy barrier. In fact, first-order phase transitions do generally occur via a nucleation process.

As pointed out in Section 2.1, thermodynamics describe the macrostate of a system, which is the temporal and spatial average of an infinite number of microstates that the system can run through by continuous fluctuations. However, the description of a nucleation process requires a much higher temporal resolution. Taking the thermodynamic concept of state variables and allowing them to fluctuate on the time scale of microstate fluctuations leads to a model that is a hybrid between a thermodynamic and an atomistic description of a system, allowing for some quantitative insight into the kinetics of the nucleation process. In that sense overcoming the energy barrier of nucleation means that a fluctuation in the microstate of the system forming a cluster of the new, stable, phase is associated with a fluctuation in G with an amplitude higher than or equal to the energy barrier.

Considering a cluster of n molecules to constitute the beginning of a nucleation process, we are interested in the difference in G between the system with the cluster formed and the system at its initial state. Compared to the total number of N molecules in the system, we regard n as sufficiently small that neither p nor T changes significantly upon the formation of a cluster of n molecules.

In the following we derive an expression for the energy barrier impeding the formation of cloud droplets within water vapor for the case of homogeneous nucleation. The term "homogeneous" refers to the case where no foreign surface is available to catalyze the nucleation process by lowering the energy barrier. The process of homogeneous droplet

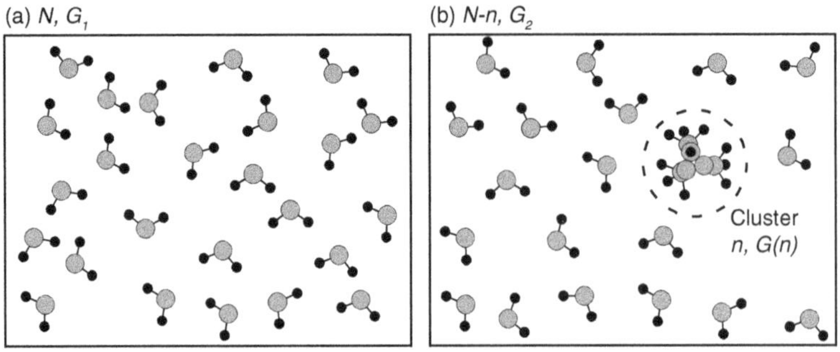

Fig. 6.2 Schematic of the states before and after homogeneous cluster formation. (a) State 1; the parent phase, with N molecules and Gibbs free energy G_1, before cluster formation. (b) State 2; the parent phase with $N - n$ molecules and Gibbs free energy G_2 along with a cluster of n molecules with Gibbs free energy $G(n)$.

nucleation provides a basis for the understanding of cloud droplet formation according to Köhler theory.

6.1.2 Cluster formation

Consider the transition from an initial state consisting of the uniform parent phase to a state where a cluster of n molecules of the new phase has formed within the parent phase. These two states are illustrated in Figure 6.2. State 1 has Gibbs free energy $G_1 = N\mu_{old}$ (eq. 2.44), where μ_{old} is the Gibbs free energy per molecule (or chemical potential) of the parent phase. State 2 has Gibbs free energy $G_2 = (N - n)\mu_{old} + G(n)$, where $G(n)$ is the total Gibbs free energy of the cluster; $G(n)$ can be written as $G(n) = n\mu_{new} + G_{ex}(n)$, where μ_{new} is the chemical potential of the new bulk phase and $G_{ex}(n)$ denotes the excess Gibbs energy. This arises because the condensed phase appears as a cluster, i.e. because of the formation of a new interface between two phases. Hence it takes into account the fact that, for a microscopic cluster, a considerable fraction of molecules is located at or close to the phase boundary and cannot be described by bulk properties. In contrast, for macroscopic quantities the number of molecules at the phase boundary is negligible compared with the number of molecules within the bulk.

The work needed for cluster formation is given by the change in G:

$$W(n) = \Delta G = G_2 - G_1 = n\Delta\mu + G_{ex}(n), \tag{6.1}$$

where $\Delta\mu = \mu_{new} - \mu_{old}$. For the transition from a vapor phase to a liquid phase, it can be shown that $\Delta\mu$ can be approximated as follows (eq. 2.47):

$$\Delta\mu(p, T) = -kT \ln\left(\frac{p}{p_s(T)}\right) = -kT \ln S, \tag{6.2}$$

where $p_s(T)$ is the saturation vapor pressure, k is the Boltzmann constant and

$$S = \frac{p}{p_s(T)} \tag{6.3}$$

defines the saturation ratio.

Furthermore, it can be shown that the "excess" energy for cluster formation, $G_{ex}(n)$, is given by

$$G_{ex}(n) = n(\mu_{new,n} - \mu_{new}) - (p_n - p)V_n + \varphi(V_n). \tag{6.4}$$

Here $\mu_{new,n}$ is the chemical potential of the molecules in the cluster, p_n the pressure within the cluster, V_n the volume of the cluster and $\varphi(V_n)$ the total Gibbs free energy of the interface between the cluster and the parent phase. The above expression shows that $G_{ex}(n)$ accounts for three effects: (i) the difference between the chemical potentials of the molecules within the cluster and that of the molecules in the bulk new phase, (ii) the difference between the pressures within the cluster and outside the cluster and (iii) the existence of an interface between the cluster and the parent phase. The pressure inside a spherical cluster with radius r is related to the pressure of the vapor surrounding the cluster via the Laplace equation,

$$p_n = p + \frac{2\sigma}{r}, \tag{6.5}$$

where σ is the surface tension between the parent phase and the new phase (Section 6.2). The Laplace equation states that the pressure is higher inside the cluster and that the pressure difference from that outside increases with decreasing radius of the cluster. Note that the Laplace equation can be used only under the assumption that the cluster is large enough that the concept of a surface tension makes sense. Because of the Laplace equation, $p_n \neq p$ and therefore also $\mu_{new,n} \neq \mu_{new}$; from eq. (2.53) it follows that

$$dg = d\mu = \alpha dp - sdT = \alpha dp = \frac{1}{n} V_n(p)dp$$

and therefore

$$\mu_{new,n} \equiv \mu_{new}(p_n) = \mu_{new}(p) + \frac{1}{n} \int_p^{p_n} V_n(p')dp'. \tag{6.6}$$

In the case of the condensation of vapor, the equation of state $V_n = V_n(p_n, T)$ of the cluster is simple. The liquid phase is approximately incompressible, i.e. $V_n(p_n) \approx nv_0$, where v_0 is the volume of an individual molecule in the bulk liquid phase. Inserting this into eqs. (6.4) and (6.6), we end up with a simple expression for G_{ex}:

$$G_{ex} = \varphi(V_n). \tag{6.7}$$

In other words the excess Gibbs free energy of the cluster is sensitive only to the surface energy associated with the new interface, not to the increased pressure within the cluster. For the homogeneous nucleation of a cluster of a condensed phase within the vapor phase, the change in G associated with cluster formation can be written as

$$\Delta G = n\Delta\mu + \varphi(V_n). \tag{6.8}$$

It can further be shown that, for spherical clusters of a condensed phase, we have $\varphi(V_n) = \varphi(r) = 4\pi\sigma r^2$, where r is the cluster radius and σ is the free energy or surface tension associated with the formation of a unit surface area of interface between the liquid and the vapor phase. Finally, we obtain the following expression for the change in Gibbs free energy associated with homogeneous nucleation of the liquid phase within the vapor phase:

$$\Delta G = -nkT \ln S + 4\pi\sigma r^2 = \underbrace{-\frac{4\pi r^3}{3\alpha_l} R_v T \ln S}_{\Delta G_v(\text{volume term})} + \underbrace{4\pi\sigma r^2}_{\Delta G_s(\text{surface term})} . \qquad (6.9)$$

In Section 6.2 we will discuss the dependence of ΔG on the cluster radius and see how ΔG forms an energy barrier for nucleation at $S > 1$. Note that no latent heat release has appeared in the formulation of nucleation because G, unlike the enthalpy H, is not sensitive to the release or absorption of latent heat (Section 2.4).

From a molecular point of view, the release of latent heat and the associated rise in temperature is not straightforward to understand. Macroscopically, latent heat release arises from the fact, that on average, molecules in the condensed phase are in a lower energetic state than molecules in the vapor phase. If an individual molecule in the vapor phase is to change to the condensed phase, its kinetic energy has to be low enough that it can enter a "bound state" with other molecules. If this is not the case, then it has to reduce its kinetic energy first, by collisions with other molecules in the vapor phase. If we remove a certain number of molecules from the lower-energy tail of the Maxwell–Boltzmann distribution (which describes the statistical distribution of the kinetic energies of molecules in an ideal gas) of the vapor phase and transfer them to the condensed phase, the Maxwell–Boltzmann distribution of the remaining vapor phase will shift to higher energies, i.e. to a higher temperature.

It is important to note that the above derivation of the work needed for cluster formation has been performed under the assumption of pure substances, i.e. for a cluster of the condensed phase forming within its pure vapor phase. It is obvious, however, that this is not the case for cloud droplets forming in the atmosphere, as water vapor is only a minor constituent of the atmosphere. Nevertheless, adding air at atmospheric pressure to our system does not change the result for ΔG, and eq. (6.9) is also valid for multi-component systems as an overall added partial pressure p_{atm} cancels out in the calculation of $G_{ex}(n)$. However, this generalization for multi-component systems holds only for the case of condensation. For the opposite case, that of evaporation, the work needed for the formation of a small "steam bubble" within the old liquid phase is more complicated and depends on the absolute pressure, an example of which is the boiling point of water, which depends on atmospheric pressure. Note that in the absence of an already existing liquid–vapor interface, evaporation is also a first-order phase transition with an associated energy barrier, and has to be initiated by the nucleation of small steam bubbles.

6.2 Kelvin equation

The height of the energy barrier for droplet nucleation critically depends on the saturation ratio S. The change in Gibbs free energy for cluster formation (eq. 6.9) is shown in

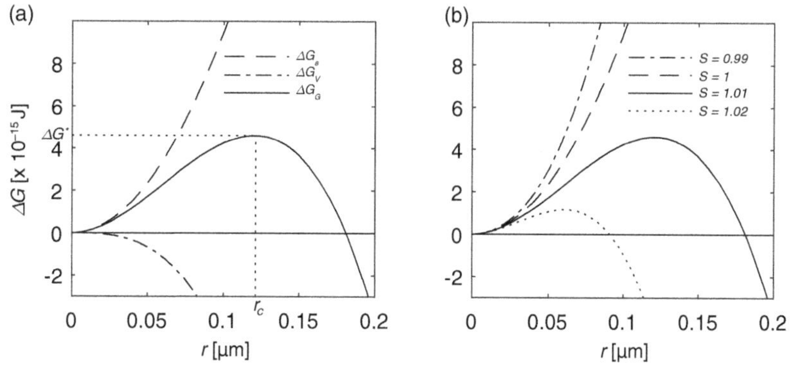

Fig. 6.3 Change in Gibbs free energy for the homogeneous nucleation of a water droplet of radius r from the vapor phase at $T = 273.15$ K with $\sigma_{w,v} = 0.0756$ N m^{-1}. (a) ΔG and its contributing terms ΔG_v and ΔG_s, according to eq. (6.9), for $S = 1.01$. The critical radius $r_c = 0.12$ μm for $S = 1.01$ corresponding to an energy barrier $\Delta G^* = 4.6 \times 10^{-15}$ J is shown as a dotted line. (b) ΔG for various saturation ratios S.

Figure 6.3b for various values of S. It shows that if $S < 1$, both terms on the right-hand side of eq. (6.9) are positive and ΔG increases monotonically with increasing r. If $S = 1$, the volume term is zero and ΔG still increases with r because of the positive contribution of the surface term. An increase in ΔG with increasing radius means that it is more likely for a cluster to evaporate than to grow. Only if $S > 1$ does ΔG contain both positive and negative contributions. At small values of r, the surface term dominates and the behavior of ΔG is similar to the $S \leq 1$ case. As r increases, the volume term becomes more important and ΔG reaches a maximum (the energy barrier) at the critical cluster radius r_c, beyond which it sharply decreases (hereafter we will write ΔG^* for the energy barrier to distinguish it from the radius-dependent change in Gibbs free energy $\Delta G(r)$). As long as $r < r_c$ the formed cluster will most likely disappear again, and no phase transition will be triggered by this specific cluster. However, if a fluctuation in the microstate brings together a sufficient number of molecules to form a cluster with $r > r_c$, this cluster will most likely continue to grow and eventually lower its G below the corresponding value in the initial, metastable gaseous state. In that case, the phase transition is triggered, and the cluster can grow to macroscopic size.

Figure 6.3b shows how strongly the height of the energy barrier depends on the saturation ratio in the vapor phase: increasing S by 1% from 1.01 to 1.02 decreases ΔG^* by a factor of almost 4 and halves the critical radius r_c. The latter can be calculated by differentiating eq. (6.9) with respect to r and setting the derivative equal to zero:

$$\left.\frac{\partial \Delta G}{\partial r}\right|_{r_c} = 0 = -\frac{4\pi r_c^2 R_v T}{\alpha_l} \ln S + 8\pi r_c \sigma_{w,v}. \tag{6.10}$$

For the specific case of the nucleation of a water drop, $\sigma_{w,v}$ is the surface tension between water vapor and liquid water. Solving for r_c yields

$$r_c = \frac{2\sigma_{w,v}\alpha_l}{R_v T \ln S}. \tag{6.11}$$

From eq. (6.11) the minimum saturation ratio S required for a droplet with radius r_c to exist can be calculated. "Existing" for a droplet in this context means that it most likely will not evaporate and vanish immediately. For the homogeneous nucleation of a cloud droplet, the saturation ratio S can be understood as the ratio of the saturation water vapor pressure $e_{s,w}(r)$ at which a droplet with radius r can exist and the saturation water vapor pressure over a flat and pure water surface.

Replacing S in eq. (6.11) by $e_{s,w}(r)/e_{s,w}(\infty)$ and replacing r_c by r leads to

$$\ln \frac{e_{s,w}(r)}{e_{s,w}(\infty)} = \frac{2\sigma_{w,v}\alpha_l}{R_v T r}. \tag{6.12}$$

Note that in the case of a water droplet in air the surface tension entering eq. (6.12) is that between pure water and moist air, $\sigma_{w,a}$; for simplicity we will refer to it as σ_w. At 273.15 K, σ_w is given as 0.0756 N m^{-1}. Replacing α_l by $1/\rho_l$ and solving for $e_{s,w}(r)$ yields the Kelvin equation,

$$e_{s,w}(r) = e_{s,w}(\infty) \exp\left(\frac{2\sigma_w}{\rho_l R_v T r}\right). \tag{6.13}$$

Equation (6.13) states that the equilibrium vapor pressure over a water droplet with a finite radius r is larger than that over a flat (or bulk) water surface. This can be understood as follows. At equilibrium, the cost in surface energy associated with an infinitesimal increase in droplet size is offset by a gain in energy associated with the lower chemical potential of the liquid phase. The former scales with the surface, the latter with the volume of the droplet. Since the surface-to-volume ratio increases for decreasing droplet radius, a larger difference in chemical potential between the liquid and the vapor phase (i.e. a larger vapor pressure) is required to offset the increase in surface energy for a small droplet as compared with that for a large droplet.

The surface tension has been introduced as the free energy per unit surface area of the interface between the liquid and vapor phases, i.e. the work per unit area that is required to extend the interface at constant temperature. It can also be understood in a phenomenological way: surface tension is produced by the asymmetry in interaction forces of the water molecules at the surface compared with those in the bulk. Since a molecule at the surface, flat or curved, has fewer neighboring molecules than a molecule in the bulk, it experiences a net force towards the bulk. In the case of a droplet, the surface molecules experience a net force towards the center of the droplet, which is illustrated schematically in Figure 6.4.

Some critical radii (eq. 6.11) for homogeneous droplet nucleation are given in Table 6.1 as a function of the saturation ratio S. Only for saturation ratios between 5 and 10 is the number of molecules necessary for droplet nucleation small enough that random cluster formation could lead to nucleation. However, under typical atmospheric conditions S rarely exceeds 1.01. This renders the homogeneous nucleation of water droplets directly from the vapor phase highly unlikely. Instead, cloud droplet formation in the atmosphere normally occurs in the presence of foreign substances, i.e. CCN. This case

Table 6.1 Critical radius based on eq. (6.11) and the corresponding number of water molecules for a given saturation ratio at temperature 273.15 K and surface tension σ_w 0.0756 N m^{-1}.

Saturation ratio, S	Critical radius, r_c[μm]	Number of molecules, n
1	∞	∞
1.01	0.12	2.47×10^8
1.1	0.0126	2.81×10^5
1.5	2.96×10^{-3}	3645
2	1.73×10^{-3}	730
5	7.47×10^{-4}	58
10	5.22×10^{-4}	20

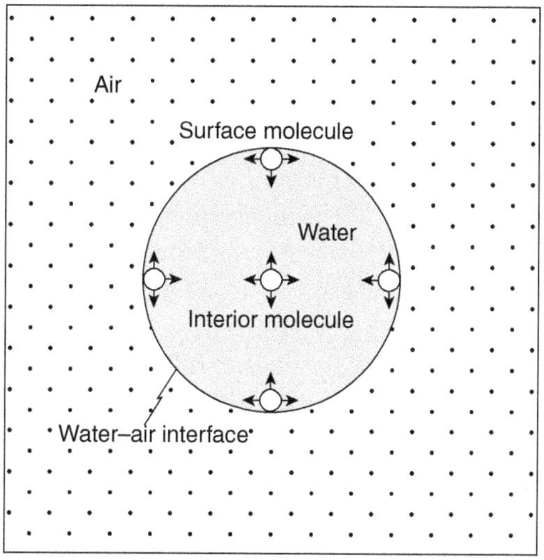

Fig. 6.4 A simplistic picture of surface tension for an air–water interface. Molecules at the interface experience a net force towards the interior.

is also referred to as heterogeneous droplet formation. This term is misleading, however, as nucleation of the liquid phase is not required, since the liquid phase is already present in soluble aerosol particles that have undergone deliquescence and hygroscopic growth in subsaturated conditions.

6.3 Hygroscopic growth

Aerosol particles occur in solid and liquid form. Some change their state under atmospheric conditions, as will be discussed in this section. The phase transition of a solid soluble aerosol to a liquid aerosol is called deliquescence and involves the dissolution of the particle in water. The opposite process, the formation of a solid aerosol from a liquid aerosol, is called efflorescence or crystallization.

In general, cloud droplets form on liquid aerosol particles that have taken up water before the ambient saturation ratio is high enough for them to be activated into cloud droplets. We will therefore first discuss the different phases of soluble aerosol particles (liquid and crystalline) before we study cloud droplet formation.

The hygroscopic growth of an aerosol particle refers to the change in its size when the relative humidity (RH) increases. It is described in terms of a growth factor, defined as the ratio of the actual radius of the particle at a specific RH and the radius at 0% RH. It is only defined for RHs below 100%. The actual size of an aqueous aerosol particle is also referred to as its wet size, whereas the size without any water uptake is denoted its dry size. The growth factor of a particle at a specific RH depends mainly on its chemistry. To a lesser extent it also depends on the aerosol particle size because of Kelvin's law (eq. 6.13), which reduces the growth factor for small particles as compared to large ones. Strong acids, such as nitric acid or sulfuric acid (H_2SO_4, shown in Figure 6.5), remain liquid over the full RH range at atmospheric temperatures and pressures. Hence they experience hygroscopic growth over the full RH range.

Other soluble aerosol particles, predominantly simple ones, i.e. small salts such as ammonium sulfate, $(NH_4)_2SO_4$, or sodium chloride, NaCl, change phase between liquid and crystalline with changing RH. At which RH this phase transition occurs depends on the direction of the change in RH. Let us take the example of sodium chloride, starting at 10% RH where an NaCl particle is solid. If we increase RH then its radius remains constant until its deliquescence RH (DRH) is reached at 75% RH, where the radius suddenly

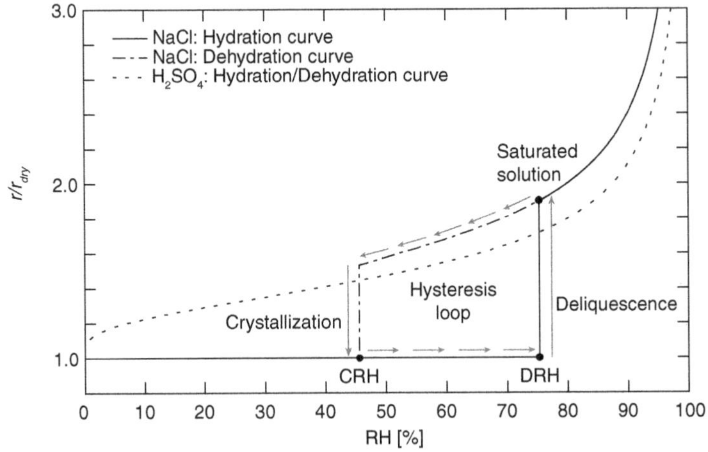

Fig. 6.5 Growth factor as a function of RH for sulfuric acid (H_2SO_4) and sodium chloride (NaCl) particles, expressed as the ratio of the actual radius r and the radius at 0% RH, r_{dry}. The solid line denotes stable equilibrium, the broken-and-dotted line metastable equilibrium, CRH the crystallization RH and DRH the deliquescence RH. The arrows indicate the hysteresis loop.

increases by 90% (Figure 6.5). At the DRH, the aerosol particle suddenly takes up water and changes from being solid to being liquid, i.e. the salt dissolves in water. At the DRH the number density of water molecules in the air is equal to that needed to dissolve the salt into a saturated solution. After deliquescence the solution becomes increasingly subsaturated with increasing RH and consequently the particle grows until it predominantly consists of water. The dependence of the saturation vapor pressure of an aqueous solution on the amount of dissolved solute, as will be discussed in Section 6.4. For a pure substance, the DRH is determined by the solubility of the salt in water: The higher the solubility, the lower the DRH. For a mixture of substances, the lowest DRH of the individual components determines the initial phase transition to liquid.

If the RH is decreased starting from a high value, NaCl remains in the liquid phase. At 75% RH the solution is saturated with respect to NaCl. At lower RHs, down to 45% RH, the solution becomes increasingly supersaturated. At 45% RH the solution reaches the critical supersaturation with respect to NaCl for crystallization to occur. This RH is called crystallization RH (CRH) or efflorescence RH.

Therefore the growth factor for NaCl is not unambiguously determined by the RH but depends also on the history of the particle (hysteresis). Between 45% and 75% RH the dissolved phase of NaCl is metastable. Here the aerosol particle would prefer to be in the solid state, because the solid state of NaCl has a lower chemical potential μ.

Figure 6.6 shows μ as a function of RH for an aerosol with DRH 75% such as NaCl. The arrows indicate a cycle of RH similar to that shown in Figure 6.5. Whereas μ is constant for a solid substance, μ for a solute in an aqueous solution decreases with increasing dilution. Above the DRH the solution is sufficiently dilute that its chemical potential lies below that of the crystalline salt. Hence, at RHs higher than the DRH the aqueous phase is the preferred state and for RHs below the DRH the solid is the preferred state. Nevertheless, the NaCl aerosol remains liquid down to the CRH as the RH decreases. For a supersaturated NaCl solution, transition to the crystalline phase is impeded by an energy

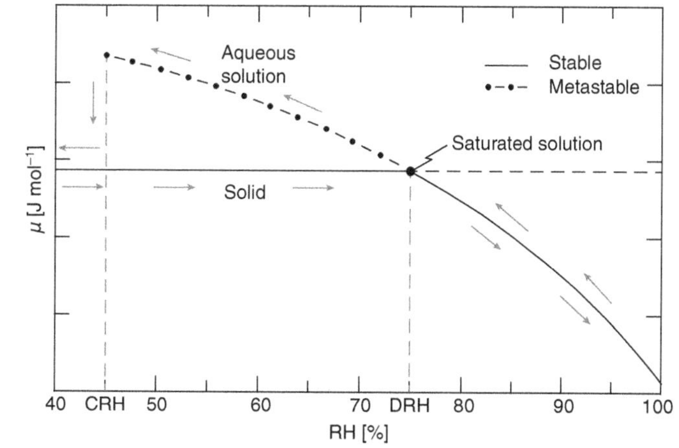

Fig. 6.6 Schematic of the chemical potential μ of sodium chloride (NaCl) in an aqueous solution (curved line) and in its solid state (solid horizontal line), as a function of RH. The arrows indicate how μ changes with increasing and decreasing RH, respectively. Temperature and pressure are held constant.

barrier and requires a nucleation process (Section 6.1). This deviation from the pathway of lowest possible chemical potential of the solute gives rise to the hysteresis loop visible in the growth factor in Figure 6.5. In the present example, the supersaturated state is maintained down to RH = 45%, the CRH for NaCl, where the solution reaches the critical supersaturation, i.e. the supersaturation at which the energy barrier for crystallization is sufficiently small for crystallization to occur. Note that we have so far discussed deliquescence and crystallization for bulk quantities, whereas in the context of aerosol particles, the CRH and DRH actually depend on particle radius. This is a direct consequence of the Kelvin effect: at bulk DRH of a specific salt, a small aerosol particle cannot deliquesce because the resulting small droplet would not be in equilibrium with the surrounding water vapor. Hence, the particle DRH increases for decreasing particle radius. The same argument holds for the CRH: if the RH decreases, a small solution droplet reaches saturation at an RH above the bulk CRH owing to the Kelvin effect. The particle size dependence of the CRH and DRH is illustrated in Figure 6.16.

Why is there no analogous energy barrier in the deliquescence process, i.e. why is there no nucleation process? Let us first consider the analogous case of the melting of ice. Ice melts immediately at 0 °C because a quasi-liquid layer (QLL) forms on ice crystals at temperatures higher than −25 °C (Li and Somorjai, 2007). The QLL consists of a thin layer on the surface of the ice where the water molecules are neither in the rigid solid structure of ice nor in the random order of liquid water. This is a consequence of the fact that the water molecules in the top layer of the ice have fewer neighboring molecules in the crystal lattice. Thus, they are not completely trapped within a crystal structure but can move around to some extent, forming the QLL. The QLL minimizes the surface energy because the surface tension between air and liquid water is smaller than that between air and solid ice. It is this interface with air at the edge of a piece of ice that explains the absence of an energy barrier for melting in all atmospherically relevant situations. If melting is considered from a pure thermodynamic point of view, where no edges are taken into account, the liquid phase has to form within an infinite block of ice, i.e. an interface between the liquid and solid phase has to be formed. This is equivalent to the existence of an energy barrier, as in every other first-order phase transition. The effect of an interface between ice and air in terms of the QLL is that the new liquid phase does not have to be formed as new but is already there when the melting phase transition becomes thermodynamically favorable. Hence it is not surprising that melting always starts at the surface. Similarly, crystalline aerosol particles form a quasi-liquid layer at RHs below their DRH by adsorbing a few monolayers of water molecules at their surface, so that deliquescence does not require a nucleation process either.

6.4 Raoult's law

Given that homogeneous nucleation from the vapor phase is highly unlikely for cloud droplet formation, we now discuss the role of CCN in cloud droplet activation. We refer to it as activation to emphasize that it is not a nucleation process. Whereas the curvature of a water surface increases its equilibrium vapor pressure, soluble material dissolved in water can reduce the equilibrium vapor pressure. In cloud droplet formation this effect is important, because a sufficient amount of soluble material is present in the atmosphere to substantially reduce the supersaturation. This reduction in vapor pressure due to the presence of soluble material is described by Raoult's law for ideal solutions, i.e. solutions with equal strengths of interaction between all molecules. Raoult's law is given for a flat surface over a solution as

$$\frac{e^*(\infty)}{e_{s,w}(\infty)} = a_w = \frac{n_w}{n_s + n_w}. \quad (6.14)$$

Here e^* is the equilibrium vapor pressure over a solution consisting of n_w water molecules and n_s solute molecules. The ratio of the equilibrium vapor pressure of the solution and that of pure water, i.e. the saturation water vapor pressure, is also referred to as water activity a_w. Equation (6.14) states that the equilibrium vapor pressure over the solution is lower than that of pure water and that the vapor pressure is reduced in proportion to the amount of solute present ($a_w \leq 1$).

The equilibrium vapor pressure reduction can be visualized by recalling that the saturation water vapor pressure is related to the exchange of water molecules between the liquid and the vapor phase. Assuming that we have an ideal solution of water and solute molecules, the exchange of water molecules between the surface of the solution and the overlying vapor is limited to those places where water molecules occupy the surface (Figure 6.7b). Although the probability per unit time for a single water molecule at the surface to evaporate is the same as that in pure water, the flux of water molecules per unit time from the surface towards the vapor phase is lower in the solution, since it has fewer water molecules at the surface. If the solution is in equilibrium with the vapor, this is equivalent to a lower density of water molecules in the vapor, i.e. to a lower saturation water vapor pressure.

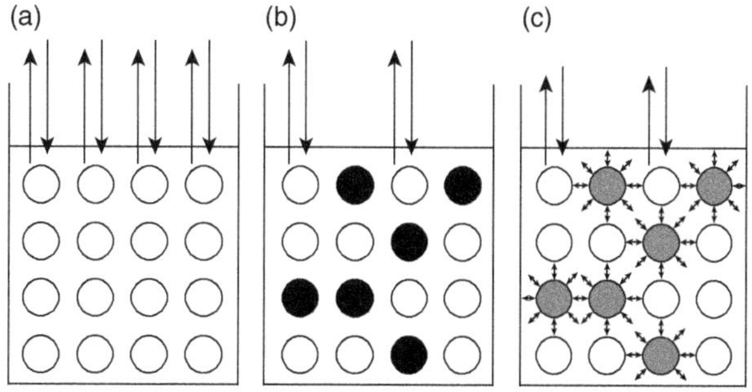

Fig. 6.7 Schematic of the saturation vapor pressure for (a) pure water, (b) an ideal solution and (c) a non-ideal solution. The white disks denote water molecules, the black disks dissolved molecules and the gray disks dissolved molecules forming hydrates with the water molecules. The reduction in vapor pressure from (a) to (b) corresponds to there being fewer sites from where water molecules can escape through the air–water interface, and the reduction from (b) to (c) is indicated by the shorter vertical arrows, and is due to increased bonding to the solute molecules. Figure adapted from Curry and Webster (1999) with permission of Elsevier.

For non-ideal solutions the exchange of water molecules between the water surface and the overlying vapor is further reduced. Here hydrates of the dissociated solute molecules occupy some places at the surface, at which the water molecules are more tightly bound due to the stronger interaction with the solute molecules. This interaction reduces the chance for the water molecules to leave the condensed phase, as indicated by the shorter vertical arrows in Figure 6.7c. In cases where the solute tends to break up strong interactions of the solvent with itself and replaces them by weaker ones, the vapor pressure of the solution would be higher than that expected from Raoult's law. However, these cases are not atmospherically relevant, and so in the remainder of this chapter we assume ideal solutions.

For ionic solutions n_s must be multiplied by the degree of ionic dissociation. This degree of ionic dissociation is frequently expressed by i, the van 't Hoff factor. For example, the van 't Hoff factor for NaCl is 2 because NaCl fully dissociates into two ions ($NaCl \leftrightarrow Na^+ + Cl^-$). Also, the dissociation of $(NH_4)_2SO_4$ is best described by $i = 2$ when compared with the measured water activities of aqueous $(NH_4)_2SO_4$ solutions, since it does not fully dissociate into two NH_4^+ and one SO_4^{2-} ion, which would result in $i = 3$. There are acid–base reactions in the presence of water ($SO_4^{2-} + H^+ \leftrightarrow HSO_4^-$), which reduce the number of dissolved ions for $(NH_4)_2SO_4$ to two. The van 't Hoff factor for non-electrolytes, e.g. most organic molecules, is $i = 1$ (Broekhuizen *et al.*, 2004) because they do not dissociate in solution.

As discussed above, Raoult's law describes the reduction in equilibrium vapor pressure over a solution consisting of n_w water molecules and n_s solute molecules compared with the saturation vapor pressure of pure water. For a mixture of two arbitrary liquid substances A and B, the equilibrium vapor pressure of the mixture relative to the one of the pure substance B is shown in Figure 6.8. If only substance B is present, the total vapor pressure over the solution is the same as the vapor pressure of B ($p_B = 1$). As the number fraction of substance A increases, the number fraction of substance B decreases because the sum of their mole fractions must equal unity for the mixture. As the vapor pressure of the pure substance A is smaller than the one of B, the increase in substance A causes the total vapor pressure over the mixture to decrease, and it is smallest if only substance A is present.

If we associate substance A with a salt and B with water, the p_B line represents a_w given in eq. (6.14). Raoult's law is satisfied best for small deviations from pure liquids, i.e. in dilute solutions with small amounts of solute. These conditions are met in the atmosphere near 100% relative humidity when deliquesced aerosol particles (i.e. non-activated solution droplets) have taken up large amounts of water and become strongly diluted, i.e. $n_s \ll n_w$. Again associating substance B in Figure 6.8 with water then only the curves in the right-hand side of the plot are relevant, where the mole fraction of B is much larger than the mole fraction of A and where it is justified to regard water as the solvent. In concentrated solutions deviations from Raoult's law are large. Here Henry's law can be used instead.

For dilute solutions, eq. (6.14) can be simplified using a Taylor series expansion:

$$\frac{e^*(\infty)}{e_{s,w}(\infty)} = \frac{n_w}{n_s + n_w} = \frac{1}{n_s/n_w + 1} \approx 1 - \frac{n_s}{n_w}. \tag{6.15}$$

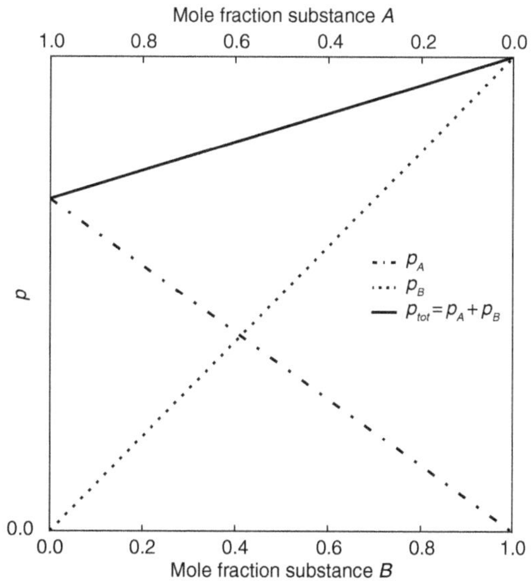

Fig. 6.8 Schematic illustration of Raoult's law for a system containing two liquid substances A and B. The upper x-axis shows the mole fraction of substance A and the lower x-axis of substance B. The total vapor pressure p_{tot} is the sum of the partial vapor pressures p_A and p_B of the substances A and B (eq. 2.76).

If the solute dissociates, n_s equals the number of effective ions in a solute of mass m_s:

$$n_s = \frac{iN_A m_s}{M_s}, \tag{6.16}$$

where N_A is Avogadro's number and M_s is the molecular weight of the solute. The number of water molecules in the solvent, of mass m_w, may likewise be written as

$$n_w = \frac{N_A m_w}{M_w}. \tag{6.17}$$

We now apply Raoult's law to a spherical solution droplet of radius r. Its size is related to the masses of water and the solute as follows:

$$m_{tot} = m_s + m_w = \frac{4}{3}\pi r^3 \rho_s, \tag{6.18}$$

where ρ_s is the density of the solution. For dilute solutions, we can neglect the mass of the solute m_s in eq. (6.18) and approximate the density of the solution droplet by that of water, so that

$$m_{tot} \approx m_w = \frac{4}{3}\pi r^3 \rho_l. \tag{6.19}$$

This approximation is the equivalent to the approximation in eq. (6.15). A more precise approach to calculating ρ_s would be to weight the density of each substance by its mass fraction $w_j = m_j/m_{tot}$ and so obtain the density of the solution:

$$\rho_s = \Sigma_j w_j \rho_j. \tag{6.20}$$

Inserting the expressions for n_s and n_w into the simplified form of eq. (6.15), replacing m_w by eq. (6.19) and evaluating the result for a droplet with radius r instead of a bulk ($r = \infty$) solution yields

$$\frac{e^*(r)}{e_{s,w}(r)} = 1 - \frac{3im_sM_w}{4\pi M_s\rho_l r^3} = 1 - \frac{b}{r^3}, \tag{6.21}$$

where

$$b = \frac{3im_sM_w}{4\pi\rho_l M_s} \approx 43 \times 10^{-6}\mathrm{m}^3\ \mathrm{mol}^{-1} \cdot \frac{im_s}{M_s}. \tag{6.22}$$

Writing Raoult's law in this way shows its dependence on the droplet radius for a given solute mass m_s and molecular weight M_s: the larger the droplet radius, the more dilute the solution droplet. The transition from a small, concentrated, droplet to a large, diluted, droplet is the key process in atmospheric cloud droplet formation, where each cloud droplet activates on a salt particle with mass m_s. On the one hand, the smaller the droplet, the larger the effect of the solute molecules in reducing the ambient water vapor pressure required for the droplet to be in equilibrium with the ambient air. On the other hand, the curvature of the droplet, via the Kelvin effect, progressively increases the equilibrium vapor pressure for decreasing droplet radius. The balance between the two effects is discussed next.

6.5 Köhler curve

In order to evaluate the competing effects of Raoult's law and the Kelvin equation, we will derive an expression describing the equilibrium saturation ratio of a solution droplet with radius r that takes into account both effects. Invoking eqs. (6.21) and (6.13) we obtain an expression for $e^*(r)/e_{s,w}(\infty)$:

$$\frac{e^*(r)}{e_{s,w}(\infty)} = \underbrace{\frac{e^*(r)}{e_{s,w}(r)}}_{\text{Raoult}} \underbrace{\frac{e_{s,w}(r)}{e_{s,w}(\infty)}}_{\text{Kelvin}} = \underbrace{\left(1 - \frac{b}{r^3}\right)}_{\text{Raoult}} \underbrace{\exp\left(\frac{a}{r}\right)}_{\text{Kelvin}}, \tag{6.23}$$

where

$$a = \frac{2\sigma_w}{\rho_l R_v T} \approx 3.3 \times 10^{-7}\mathrm{Km} \cdot \frac{1}{T}, \tag{6.24}$$

and σ_w is the surface tension of water in air. For the approximation in eq. (6.24), σ_w and ρ_l are evaluated at $T = 273.15$ K. Using a Taylor series, the exponential term can be linearized for small values of a/r:

$$\exp\left(\frac{a}{r}\right) \approx 1 + \frac{a}{r}.$$

For a droplet 1 μm in radius or larger, the error is less than 5%. We also ignore the term $-ab/r^4$ because $a/r \ll 1$ and $b/r^3 \ll 1$ for particles larger than 0.1 μm. This simplifies eq. (6.23) to:

$$\frac{e^*(r)}{e_{s,w}(\infty)} = S(r) = 1 + \frac{a}{r} - \frac{b}{r^3}. \tag{6.25}$$

Here $S(r)$ denotes the size-dependent equilibrium saturation ratio of a solution droplet with radius r containing a fixed amount of soluble mass (originating from a deliquesced CCN). Equation (6.25) is known as the approximated Köhler equation (Köhler, 1922).

The Köhler equation combines Kelvin's equation and Raoult's law. In the form of eq. (6.25) it is valid for ideal solutions of water-soluble substances and droplets that are not too small. The Köhler equation describes the ratio of the equilibrium vapor pressure over a solution droplet with radius r and the saturation water vapor pressure over a flat pure water surface. The vapor pressure over the solution is increased by the Kelvin effect and reduced by the Raoult effect, as discussed above.

Figure 6.9 displays the Köhler curves for droplets containing different amounts of sea salt. Each sea salt particle can be associated with an individual Köhler curve, given its dry radius r_{dry}. However, the transition from r_{dry} to the radius of the concentrated solution droplet, i.e. the deliquescence of the particle, is not described by Köhler theory but is assumed to have taken place already. Such solution droplets are present prior to cloud formation and appear as haze.

For aerosol particles, starting at small dry sizes a maximum in the saturation ratio $S(r)$ is clearly visible. Loosely speaking, this maximum separates the ascending branch of the Köhler curve in Figure 6.9, where Raoult's effect dominates, from the descending branch, where the dilution of the solution droplets increases and the Kelvin effect dominates. The so-called activation or critical saturation ratio S_{act} and the corresponding activation radius r_{act} can be obtained by differentiating the Köhler equation with respect to r and setting the derivative to zero. They are then given by

$$r_{act} = \sqrt{\frac{3b}{a}}, \quad S_{act} = 1 + \sqrt{\frac{4a^3}{27b}}. \tag{6.26}$$

The activation saturation ratio S_{act} of a specific Köhler curve is the minimum saturation ratio that is required for the corresponding droplet to grow theoretically to infinity. For aerosol particles with large dry radius the Kelvin effect becomes negligible. Therefore S_{act} approaches 1 for large particles and corresponds to a large r_{act}. Once a droplet has grown to $r > r_{act}$ it is called an activated droplet. In practice, activated droplets can grow to macroscopic sizes, i.e. to cloud droplet size (tens of micrometers) or even drizzle-drop size (around 0.1 mm). All droplets that have an activation saturation ratio S_{act} below the ambient saturation ratio S can become activated. If $S < S_{act}$, a non-activated solution droplet can grow only to the radius at which the ascending branch of the Köhler curve reaches the value S. Note that the Köhler curve does not represent an energy barrier. The correlation

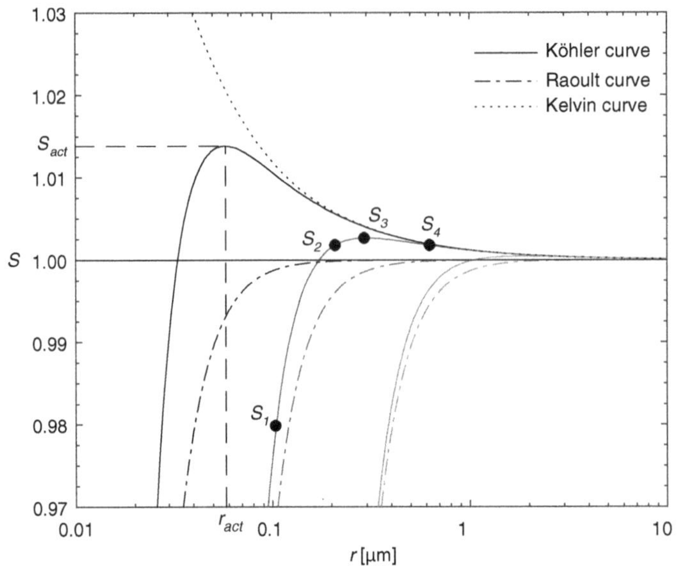

Fig. 6.9 Köhler curves $S(r)$ (solid curves) and Raoult's law (broken-and-dotted curves) for NaCl particles of different dry radii r_{dry} (left to right, $r_{dry} = 0.01\,\mu m$, $0.03\,\mu m$ and $0.1\,\mu m$), van 't Hoff factor 2, dry density 2200 kg m^{-3}, $T = 273.15$ K and the surface tension of water in air $\sigma_w = 0.0756$ N m^{-1}. The Kelvin curve is the same for all particles. The activation radius r_{act} and the corresponding activation saturation ratio S_{act} are shown for $r_{dry} = 0.01\,\mu m$. The saturation ratios S_1—S_4 refer to Figure 6.10.

between the Köhler curve and the energy barrier for droplet formation in the presence of a soluble aerosol particle is elucidated in the following.

6.5.1 Stable and unstable equilibrium

So far we have derived the Köhler curve as a superposition of two distinct effects, the Kelvin and the Raoult effects. However, it is illustrative to discuss directly the Gibbs free energy of a deliquesced aerosol particle as a function of the droplet radius, in a similar way to the the derivation of the Kelvin effect (Section 6.2). This discussion readily explains the formation of the two branches of the Köhler curve, i.e. the ascending branch of the (meta)stable equilibrium and the descending branch of the unstable equilibrium.

For every given ambient saturation ratio $S > 1$, the Kelvin effect describes the radius at which a pure water droplet, in unstable equilibrium with the vapor phase, has its maximum Gibbs free energy. This maximum is the result of the competing surface term $\Delta G_{s,hom}$ and volume term $\Delta G_{v,hom}$ in eq. (6.9), where the negative contribution of the volume term becomes dominant for increasing droplet radius. We will now analyze the change in ΔG for a solution droplet neglecting the dependence of the surface tension on the solute concentration in ΔG_s, i.e. we set $\Delta G_{s,sol} = \Delta G_{s,hom} = 4\pi\sigma_w r^2$. In order to formulate the volume term $\Delta G_{v,sol}$ for the activation of a liquid aerosol particle (a solution droplet) we have to add two contributions due to the presence of the solute (Seinfeld and Pandis, 2006). First, the volume term has to be corrected by the water activity a_w, which is reduced in the presence of the solute. Second, the activity of the solute has to be considered. Thus $\Delta G_{v,sol}$ is a combination of the water volume term and the solute volume term:

$$\Delta G_{v,sol} = \underbrace{-n_w kT \ln \frac{S}{a_w}}_{\text{water volume term}} + \underbrace{n_s kT \ln a_s}_{\text{solute volume term}}, \tag{6.27}$$

where n_w and n_s are the numbers of water and solute molecules in the solution droplet, respectively, and a_w and a_s are the respective activities. The activities are equal to the mole fractions under the assumption of an ideal solution. Both activity values are ≤ 1 by definition, and their relative contributions decrease for increasing droplet radius, as expected when one is approaching the homogeneous case. For decreasing droplet radius, i.e. an increasingly concentrated solution, a_w decreases but at the same time the number of water molecules n_w in the solution droplet decreases, leading to a weak radius dependence of the first term in eq. (6.27) for small droplet radii.

The second term represents the change in chemical potential of the solute for a changing dilution level. Its negative contribution decreases significantly for decreasing droplet radius, since a_s increases and n_s is constant in the solution droplet. Hence, the radius dependence of $n_s kT \ln a_s$ introduces a local minimum in ΔG_{sol}, which represents a metastable or stable equilibrium. Note that eq. (6.27) relies on the assumption of a sufficiently dilute solution. This condition is not met, and therefore eq. (6.27) is not valid, for very small droplet radii. Nevertheless this approximation is good enough to illustrate the origin of the two branches of the Köhler curve.

For the case $S < 1$, the minimum in ΔG_{sol} is a global minimum and hence corresponds to the stable equilibrium (Figure 6.10a) found on the ascending branch of the Köhler curve S_1 in Figure 6.9. Therefore, activation into cloud droplets is impossible for $S < 1$ but nevertheless solution droplets can exist in a stable equilibrium with the subsaturated vapor phase due to Raoult's effect.

For a saturation ratio $S = 1.002$ that lies between 1 and S_{act} (Figure 6.10b), there are two solutions to the equation $d\Delta G_{sol}/dr = 0$, one corresponding to the metastable equilibrium on the ascending branch of the Köhler curve (S_2 in Figure 6.9) and one corresponding to the unstable equilibrium on the descending branch for radius 0.64 μm (S_4 in Figure 6.9). This is in contrast with the pure water case where according to the Kelvin equation (eq. 6.13) only an unstable equilibrium exists.

For $S = S_{act} = 1.0027$ (Figure 6.10c), the two solutions of the equation $d\Delta G_{sol}/dr = 0$ merge into a single one. For $S > S_{act}$ no equilibrium exists, i.e. the solution droplet grows spontaneously into a cloud droplet immediately after deliquescence of a initial aerosol particle. Point S_3 in Figure 6.9 corresponds to S_{act} for the particle with r_{dry} = 0.03 μm. Whereas homogeneous nucleation is always impeded by an energy barrier, because the surface energy term dominates for small cluster radius, a deliquesced aerosol particle only encounters an energy barrier for activation into a cloud droplet at very low supersaturation values.

Let us now visualize the stable and unstable equilibria directly using the Köhler curve in Figure 6.11. Consider a solution droplet at point A and assume that it is in equilibrium with its environment, i.e. that the ambient water vapor pressure e is equal to $e^*(r)$. Let a small perturbation cause a few molecules of water to be added to the solution droplet

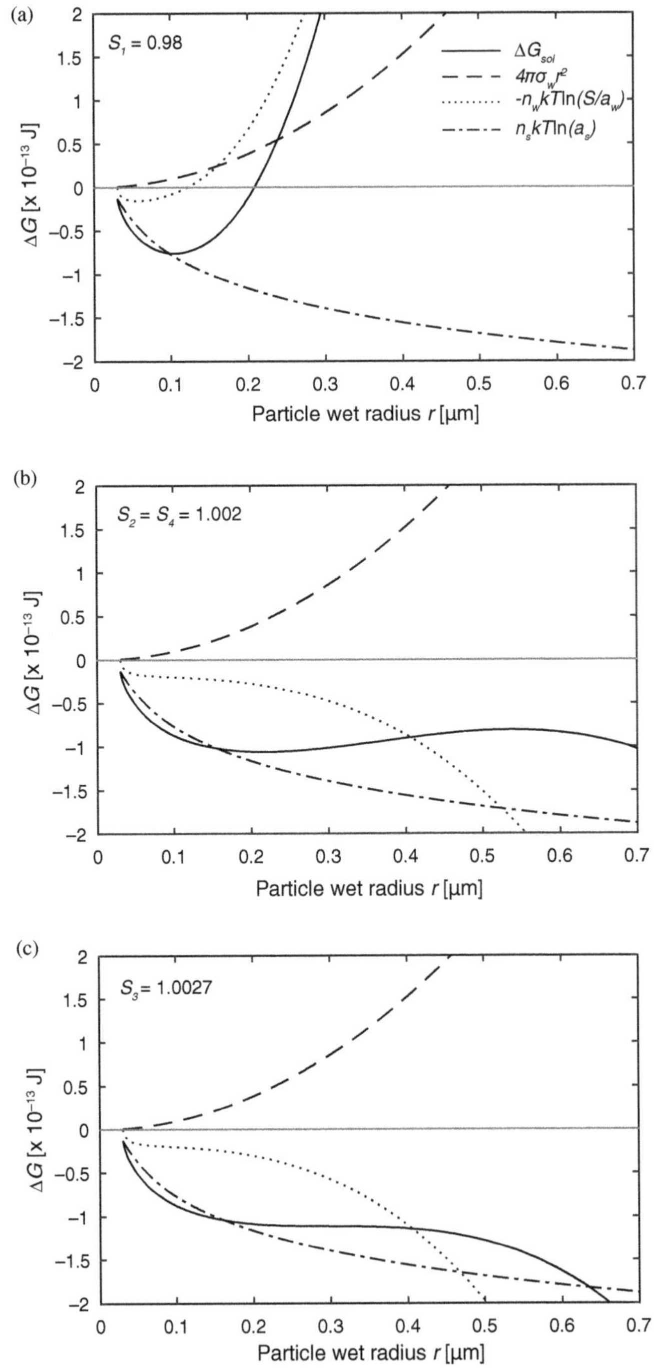

Fig. 6.10 The change in Gibbs free energy ΔG_{sol} with its contributions $\Delta G_{s,sol}$ and $\Delta G_{v,sol}$ according to eq. (6.27) for the aerosol activation of an NaCl particle with dry radius $r_{dry} = 0.03\ \mu$m and $S_{act} = 1.0027$ at $T = 273.15$ K for different ambient saturation ratios S: (a) $S_1 = 0.98$, (b) $S_2 = S_4 = 1.002$ and (c) $S_3 = S_{act} = 1.0027$, where the saturation values S_1–S_4 are indicated on the Köhler curve in Figure 6.9.

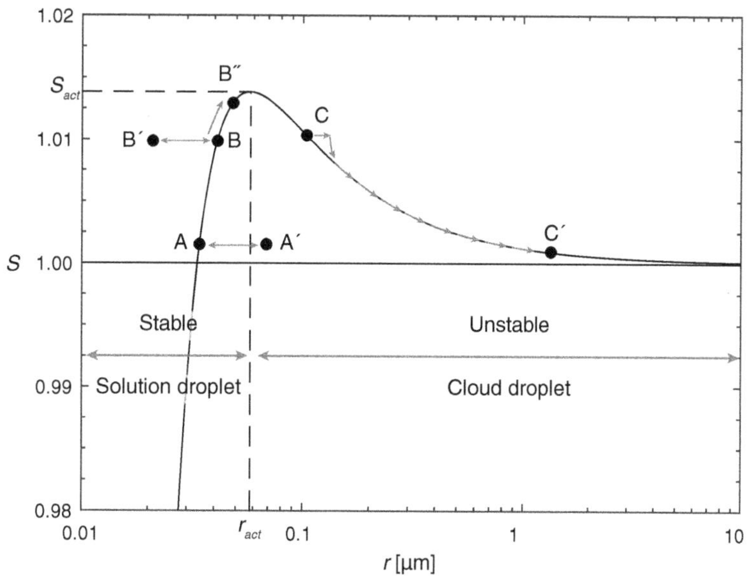

Fig. 6.11 Köhler curve as a function of droplet radius r and saturation ratio S for an NaCl particle with dry radius $r_{dry} = 0.01\ \mu m$ and van 't Hoff factor $i = 2$ at $T = 273.15$ K and $\sigma_w = 0.0756$ N m^{-1}. Also indicated are different starting points A–C. Perturbations of droplets starting from A, B or C are indicated by the small arrows and the primed letters. The activation radius r_{act} divides the Köhler curve into a stable and an unstable branch and separates solution droplets from cloud droplets.

without changing the ambient saturation ratio. This moves the solution droplet to point A′ at a slightly larger new radius $r + \delta$, where δ represents an infinitesimal change. The equilibrium vapor pressure of the solution droplet $e^*(r + \delta)$ described by the Köhler curve is higher than $e^*(r)$, i.e. the atmosphere is subsaturated with respect to the solution droplet at point A′. This will cause water molecules to evaporate until the solution droplet is back in equilibrium with the environment (point A). The corresponding mechanism also applies for a perturbation that removes molecules from the solution droplet at point B, moving it to B′, where $e^*(r - \delta) < e^*(r)$. Now the atmosphere is supersaturated with respect to the solution droplet at B′ and water vapor molecules condense on the solution droplet until it is back at point B, in equilibrium with the environment. Hence, as long as $r < r_{act}$ a solution droplet changes its size only as a response to a change in the ambient saturation ratio S. Upon a small increase in S the droplet initially at point B moves to point B″ corresponding to a new equilibrium at a new, larger, droplet radius. These observations are consistent with the fact that every point on the ascending branch of a Köhler curve corresponds to a local minimum in ΔG_{het} (Figure 6.10a, b).

If the ambient saturation ratio S exceeds S_{act} for an individual solution droplet, there is no equilibrium for that droplet. The droplet quickly grows beyond r_{act}, and the environment becomes increasingly supersaturated for a cloud droplet, because the difference $S - S(r)$ increases with increasing radius for $r > r_{act}$.

A solution droplet which has been activated into a cloud droplet can be in unstable equilibrium with the atmosphere at a specific saturation ratio $1 < S < S_{act}$ (i.e. after the particle has been activated and S has decreased below S_{act} again owing to the condensational growth of all activated cloud droplets, as discussed in Chapter 7). Consider a droplet starting at point C in Figure 6.11. If a few random molecules condense on it and slightly increase its radius, the environment suddenly becomes supersaturated with respect to the droplet at its new radius, $e^*(r + \delta) < e^*(r)$. Hence it will continue to grow as long as $S > S(r)$ and no other processes take over.

If the cloud was advected in a drier environment where the ambient vapor pressure is below $e^*(r)$ or the air parcel locally experiences these conditions due to the entrainment of dry air, the droplet will shrink. This could even be the case for RH $\geq$ 100% provided that $e < e^*(r)$. A shrinking droplet will regain a stable equilibrium with the environment when it reaches the ascending branch of the Köhler curve at the new environmental RH. Note that, for a pure water droplet, an infinitesimal reduction in the radius causing e to be smaller than $e_{s,w}(r)$ inevitably leads to complete evaporation, since the Kelvin curve has no ascending branch of stable equilibrium.

6.5.2 The role of particle size and chemistry for Köhler activation

Figure 6.9 shows that, for higher S, more and smaller aerosol particles can be activated. The figure also shows that larger aerosol particles can more readily act as CCN because they require smaller supersaturation values to reach the unstable regime in which unlimited condensational growth is theoretically possible. However, there are fewer of them available. Also, because large aerosol particles need a large amount of water to condense on them in order to reach r_{act}, such large particles may not reach r_{act} by condensational growth within a time scale relevant for cloud formation (e.g. in the case of short-lived convective clouds). However, nucleation-mode aerosol particles would need supersaturations higher than the ones existing in the atmosphere to become activated. So these aerosol particles are not activated as cloud droplets either. It is mainly accumulation-mode and some Aitken-mode aerosol particles that act as CCN and get activated to form cloud droplets.

Particle chemistry also influences Köhler theory. As shown in Figure 6.12, S_{act} is smaller for sodium chloride (NaCl) and for ammonium nitrate (NH_4NO_3) than for ammonium sulfate, $(NH_4)_2SO_4$. This is due to the smaller molecular weight of NaCl, with $M_s = 58$ g mol^{-1}, and of NH_4NO_3, with $M_s = 80$ g mol^{-1} compared with that of $(NH_4)_2SO_4$, with $M_s = 132$ g mol^{-1}. Thus the Raoult term (eq. 6.21) is largest for NaCl and smallest for $(NH_4)_2SO_4$.

An alternative way to describe cloud droplet activation is by lumping all chemical parameters in the Köhler theory into a single parameter, the so-called hygroscopicity parameter κ (Petters and Kreidenweis, 2007). The parameter κ is zero for completely insoluble particles, which can only gain water by adsorption on the surface, and ranges over 0.003–0.023 for mineral dust particles composed of clay (Herich *et al.*, 2009), 0.1–0.2 for organic aerosols, 0.6 for ammonium sulfate to 1.2 for sodium chloride. As a rough guidance, a κ value of 0.3 can be considered as representative for aged continental aerosol particles and a κ value of 0.7 as representative for marine aerosol particles (Andreae and Rosenfeld, 2008). However, large variations in κ at different locations have been found, which suggest that one still has to take these variations into account rather than using a single κ value for a given aerosol type. For example, the continental aerosol

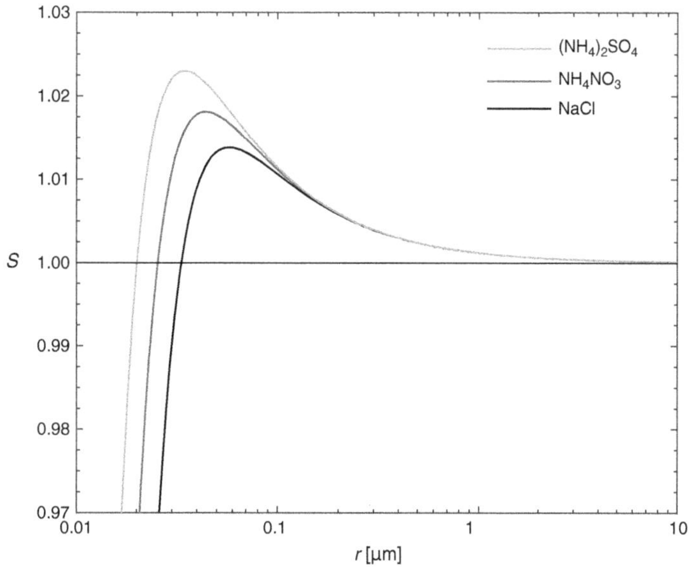

Fig. 6.12 Köhler curves $S(r)$ for (right to left) sodium chloride (NaCl), with molecular weight $M_s = 58\ \mathrm{g\ mol^{-1}}$, ammonium nitrate ($NH_4NO_3$), with $M_s = 80\ \mathrm{g\ mol^{-1}}$ and ammonium sulfate ($(NH_4)_2SO_4$), with $M_s = 132\ \mathrm{g\ mol^{-1}}$. For these salts a van 't Hoff factor equal to 2, $T = 273.15$ K, $\sigma_w = 0.0756\ \mathrm{N\ m^{-1}}$ and a dry radius $r_{dry} = 0.01\ \mu$m were used.

observed in Finland has a lower κ value than that observed in England (Paramonov *et al.*, 2015) although both can be categorized as aged continental aerosol. Particle size thus remains the most important parameter for the activation of aerosol particles into cloud droplets.

6.6 Measurements of cloud condensation nuclei

The most common way to measure the activation of aerosol particles is using a so-called CCN counter (CCNC). A CCNC is a chamber in which the saturation ratio S can be varied. As S is increased, more and more particles are activated. The most popular CCNC nowadays is a continuous-flow thermal gradient device, as described in detail by Roberts and Nenes (2005).

The CCNC consists of a cylindrical column. The inner walls of this column are kept wet and thus are at 100% RH with respect to water. A temperature gradient is applied along the length of the column, with temperature increasing in the direction of the air flow (see Figure 6.13). The operating principle of this type of CCNC uses the fact that the diffusion of water vapor (and therefore mass) in air is faster than diffusion of heat (and therefore temperature), because water molecules are lighter than N_2 and O_2 (the most common air molecules).

While the diffusion of water vapor only involves the small H_2O molecules, the diffusion of heat occurs mainly via collisions between air molecules, which on average move and thus diffuse less quickly than the water molecules. The desired supersaturation is then

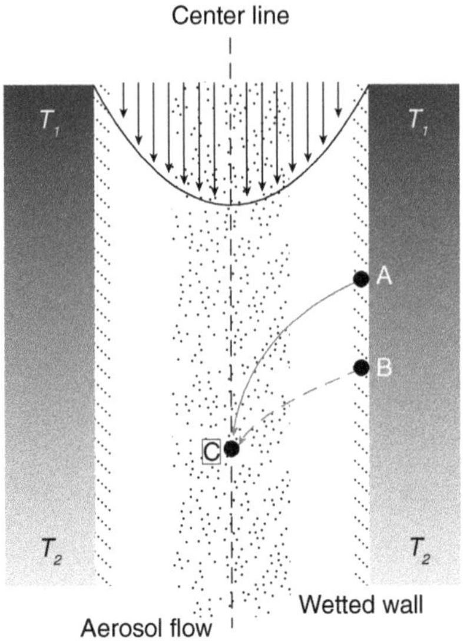

Fig. 6.13 The working principle of a CCNC that establishes a supersaturation inside the flow chamber which depends on the temperature gradient between the top T_1 and the bottom T_2 of the column. Figure adapted from Roberts and Nenes (2005).

established, as shown in Figure 6.13. Water vapor diffuses from the warm, wet, column walls toward the center line at a faster rate than the heat. Therefore the heat diffusing to point C on the center line of the instrument originates from a point on the wall further away (solid line, point A) than the diffusing water vapor (broken line, point B). Because of the increase in temperature from point A to point B, assuming 100% RH, it follows that $e_{s,w}$ also increases from point A to point B. Thus, at point C we have $e(\mathrm{C}) = e_{s,w}(\mathrm{B})$ and $T(\mathrm{C}) = T(\mathrm{A}) < T(\mathrm{B})$. This results in a supersaturated environment inside the chamber with a maximum along the center line.

This supersaturation can be controlled by the temperature gradient along the chamber walls. A sheath flow keeps the sample flow in the center of the column at maximum supersaturation. Here, the supersaturated water vapor condenses on the CCN-active particles to form droplets, which are subsequently counted and sized using a specially designed optical particle counter. Activated cloud droplets are distinguished from unactivated particles by their size. The instrument can operate at supersaturations between 0.1% and 2% RH, which are representative of atmospheric values.

In typical CCNC measurements the supersaturation is scanned over a given range of set points, and the activated fraction ($f_{act} = N_{CCN}/N_a$, where N_{CCN} is the number concentration of CCN and N_a is the total aerosol number concentration) of a polydisperse aerosol is measured downstream of the CCNC column. However, measurements with size-selected aerosol particles are also performed in the field, or for calibration purposes, to provide the CCN-activated fraction as a function of particle dry size and supersaturation. An example of the activation of ammonium sulfate particles is shown in Figure 6.14. From Köhler theory we would expect that, for size-selected particles, activation proceeds as a step function,

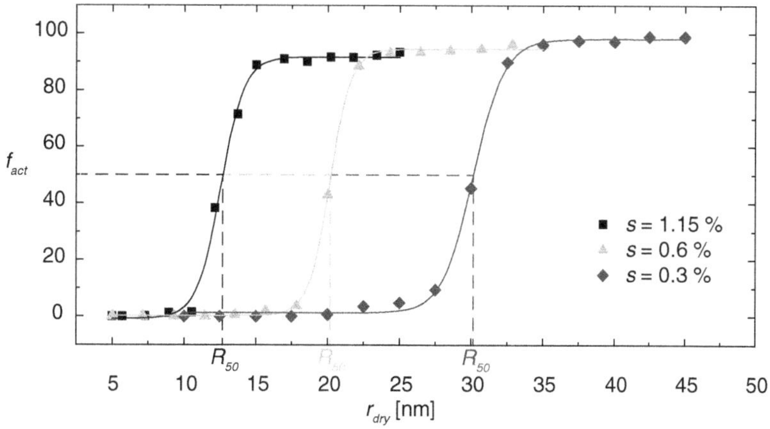

Fig. 6.14 Ratio of the number concentration of CCN to the total number concentration of aerosol particles (the activated fraction f_{act}) plotted as a function of dry particle radius r_{dry} for $(NH_4)_2SO_4$ particles at different supersaturations s. The best fits through the data in form of a sigmoidal curve are shown as solid lines. From these the radii at which 50% of all particles have been activated R_{50} are deduced.

because if the dry particle size increases above a certain value, then the critical saturation ratio S_{act} decreases below the saturation ratio in the instrument and suddenly all particles should activate. However, CCNCs usually show a sigmoidal curve instead, as indicated by the fitted curves, because (i) a size-selected aerosol still exhibits a size distribution, albeit with a narrow but finite width and (ii) the finite width of the aerosol flow within the chamber leads to a finite range of supersaturation experienced by the particles. Therefore, instead of r_{act}, the radius R_{50} at which 50% of all aerosol particles have been activated (Figure 6.14) is used to characterize aerosol activation in a CCNC. The R_{50} values are obtained by fitting the data with a sigmoidal curve. As expected from Köhler theory, the larger the supersaturation, the smaller the minimum size of aerosol particles that activate.

The relationship between CCN concentration and supersaturation is sometimes approximated by an empirical power law:

$$\mathrm{CCN} = Cs^k. \tag{6.28}$$

Here C denotes the CCN concentration in cm^{-3} at 1% supersaturation with respect to water and k is an empirical coefficient. The CCN concentration is given in cm^{-3} and the supersaturation s in % by

$$s = (S - 1)100. \tag{6.29}$$

Figure 6.15 shows an activation spectrum obtained from a CCNC divided into marine and continental aerosol particles along with a fit curve having the form of eq. (6.28). The values of C are 3172 and 87 cm^{-3} and those of k are 0.7 and 0.48 for the continental and marine data, respectively. Generally these values need to be interpreted with care, as they vary strongly, depending on the air mass and thus the aerosol concentration and type. Consequently they can vary strongly even within individual categories such as continental and marine air masses. The beauty of eq. (6.28) stems from the fact that the

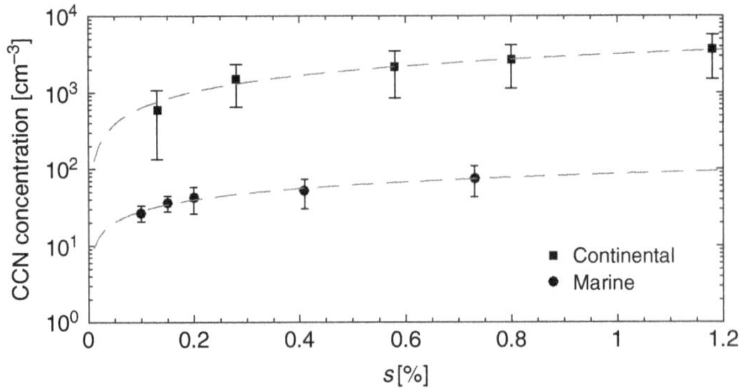

Fig. 6.15 Example of CCN measurements with error bars for marine (disks) and continental (squares) conditions, as a function of supersaturation s. The data are taken from surface measurements carried out in Zurich, Switzerland (urban site; Madonna, 2009) and in the Arctic (clean marine site; Martin *et al.* 2011). The broken curves correspond to eq. (6.28) and yield values for C of 3172 and 87 cm^{-3} and for k of 0.7 and 0.48 for the continental and marine data, respectively.

CCN concentration can be estimated even for supersaturations not directly measured by the CCNC. From Figure 6.15 we see that the CCN concentrations of the marine aerosol on average remain below 100 cm^{-3} at all supersaturations measured. The continental CCN concentrations, however, are 1–2 orders of magnitude larger and on average exceed 1000 cm^{-3} even at supersaturations of 0.25%, mainly owing to the higher aerosol concentrations in continental air masses.

6.7 Summary of cloud droplet formation by Köhler activation

The pathway from a dry particle of soluble material to a cloud droplet involves several different theoretical concepts. The deliquescence of a soluble particle leads to the point where Köhler theory starts. Köhler theory describes the transformation of solution droplets into activated cloud droplets and combines the concept of homogeneous droplet nucleation (leading to the Kelvin effect) and that of the effect of a dissolved compound on the equilibrium between water and its vapor phase (Raoult's law). Although deliquescence circumvents the necessity to form a new liquid phase, so that no nucleation process is required to activate cloud droplets from aerosol particles, the concept of an energy barrier does apply to Köhler activation. The main difference from homogeneous droplet nucleation is that whether cloud droplet activation is impeded by an energy barrier depends on the saturation ratio. For a sufficient supersaturation, the energy barrier vanishes and cloud droplet activation occurs spontaneously. This makes Köhler activation the only pathway for cloud droplet formation occurring in the atmosphere and underlines the importance of the presence of soluble aerosol particles for atmospheric cloud formation.

Figure 6.16 shows the full pathway of a water-soluble aerosol particle that starts with a dry radius $r_{dry} = 0.03\,\mu\text{m}$ at 0% relative humidity and is exposed to higher and higher relative humidities, e.g. in an updraft. The particle undergoes deliquescence at

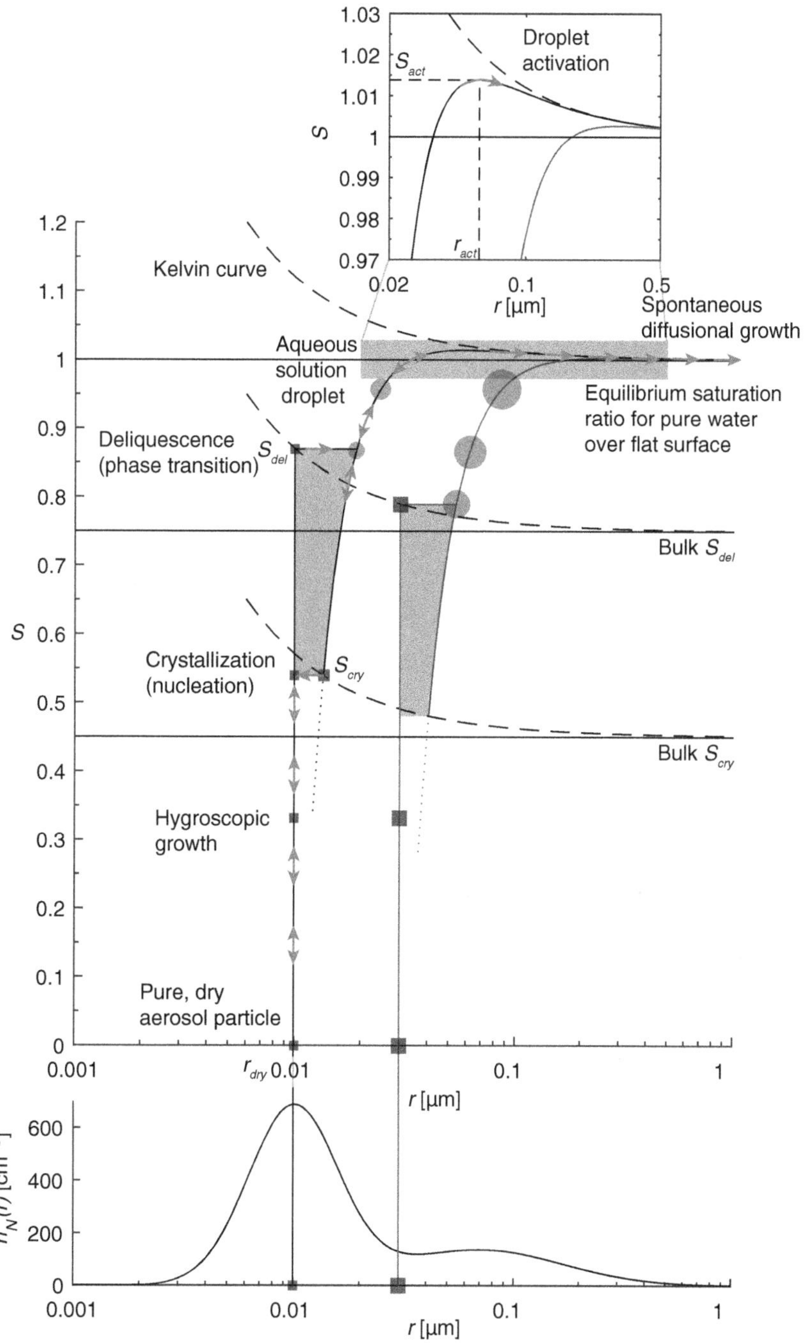

Fig. 6.16 Schematic of the entire growth process, as a function of r, of an NaCl aerosol particle forming a cloud droplet in response to an increasing ambient saturation ratio S. S_{del} and S_{cry} denote the particle size dependent deliquescence saturation ratio and crystallization saturation ratio, respectively. The solid horizontal line at $S \approx 0.75$ denotes the bulk deliquescence saturation ratio. The shaded areas highlight the hysteresis in the cycle of deliquescence and crystallization shown for bulk quantities in Figure 6.5.

approximately 80% RH, where its radius suddenly increases and it turns into an aqueous solution droplet. Note that the deliquescence saturation ratio S_{del} differs progressively from the one of the bulk NaCl salt ($S_{del} \approx 0.75$), with decreasing particle radius, owing to the Kelvin effect, as explained in Section 6.3. The condensation of water vapor onto the deliquesced particle below S_{act} is referred to as hygroscopic growth (Section 6.3) and can be described by Köhler theory. A stable equilibrium between the droplet and the surrounding water vapor is maintained for all radii ($r < r_{act}$). When the saturation ratio exceeds S_{act}, the solution droplet is activated as a cloud droplet. Beyond this radius the cloud droplet is no longer in equilibrium with the surrounding vapor and spontaneously grows further, by the condensation of water vapor, into a micrometer-sized cloud droplet.

This growth only stops when the ambient supersaturation decays due to depletion of water vapor or quenching of the updraft. The pathway sketched in Figure 6.16 continues until cloud droplets with sizes of 10 μm (Table 1.2) are formed. For the formation of raindrops other, more efficient processes as discussed in Chapter 7 take over.

6.8 Exercises

Single/multiple choice exercises

1. Which of the following phase changes require(s) a nucleation process?
   (a) Cloud droplet formation in the atmosphere.
   (b) The formation of a solid aerosol particle from a liquid aerosol particle (crystallization).
   (c) The phase change of a crystalline aerosol particle to a liquid aerosol particle (deliquescence).
   (d) The formation of new aerosol particles from precursor gases.
   (e) The homogeneous freezing of water droplets.
2. Assume that a cloud droplet was activated on a 0.05 μm sodium chloride particle and has grown to radius 5 μm. Imagine that now it is transported to a region where the relative humidity is 100%. What will happen to the cloud droplet?
   (a) The droplet will continue to grow because once larger than the activation radius r_{act} it can theoretically grow to infinity.
   (b) The droplet will retain its size since the environment needs to be supersaturated with respect to water to allow further growth.
   (c) The droplet will shrink to a smaller radius because the atmosphere is subsaturated with respect to the given droplet radius.
3. Which of the following is/are true?
   (a) From the Köhler equation it can be concluded that larger particles are more likely to act as CCN under atmospheric conditions.

(b) The Kelvin curve is equal for particles of different dry radii but depends on the chemical composition of the particles.
(c) An increasing concentration of the solvent decreases the equilibrium vapor pressure of an ideal solution.
(d) Of the two competing terms in the Köhler equation, the Raoult term dominates for smaller droplet radii.

4. The Kelvin equation, Raoult's law and the Köhler equation are important for describing cloud droplet formation in the atmosphere. Which of the following is/are true?
   (a) The Kelvin equation states that the equilibrium vapor pressure over a solution droplet (containing for example NaCl) is larger than over a plane water surface, provided that the surface tension of the solution is unaffected by the salt.
   (b) Raoult's law implies that the equilibrium vapor pressure is reduced over a solution containing soluble material.
   (c) The Köhler curve levels off at saturation ratio of one at large radii and never intersects the Kelvin curve.
5. In the atmosphere, cloud droplet formation does not occur via homogeneous nucleation since the saturation ratios present in the atmosphere are too small. Instead, cloud droplets form in the presence of foreign surfaces, i.e. on cloud condensation nuclei (CCN). Which of the following is/are true?
   (a) Accumulation mode aerosol particles are more common CCN in the atmosphere than nucleation mode aerosol particles.
   (b) The number of CCN does not depend on the supersaturation present in the atmosphere.
   (c) The concentrations of CCN are generally larger over the ocean than over land, owing to the emissions of sea salt aerosols.
   (d) Ammonium sulfate $(NH_4)_2SO_4$ cannot act as a CCN because it is soluble.
6. How large must the Gibbs free energy barrier be for nucleation not to occur?
   (a) Larger than zero.
   (b) Larger than the energy gain associated with the phase transition to the stable phase.
   (c) Infinite.
7. Which of the following is/are true?
   (a) A system in a metastable state will always increase its Gibbs free energy to reach the stable state because this is thermodynamically more favorable.
   (b) In an atmosphere without CCN, supersaturations between 50% and 100% (with respect to water) would be necessary to form a water droplet.
   (c) The measured H_2SO_4 gas concentration in the atmosphere is generally very small because of gas-to-particle conversion and/or condensation.
8. The Köhler equation written in the form of eq. (6.25) is only valid if the radius of the particle is not too small, because ...
   (a) ... the term abr^{-4} is neglected.
   (b) ... the exponential function is approximated by its Taylor series.
   (c) ... aerosol particles that are too small would never act as CCN.
   (d) ... for very small aerosol particles the surface tension of the pure salt needs to be used in the equation.

Long exercises

1. Consider two boxes at equal temperatures filled with a mixture of air and water vapor. Box A is at a total pressure p_1 and partial water vapor pressure e_1; box B is at $p_2 > p_1$ and $e_2 = e_1$. In which box is homogeneous nucleation of liquid water droplets more likely?
2. Consider a steam bubble with radius 1 mm in a lake at depth 1 m. Temperature and pressure of the atmosphere are 285 K and 1005 hPa, respectively.
   (a) Use the Laplace equation to calculate the pressure p_b of the bubble.
   (b) Since the bubble consists of pure water vapor, $p_b = e_{s,w}(T)K$ in equilibrium, where K is smaller than 1. Can the bubble exist under equilibrium conditions?
3. A pure water droplet of radius $r_d = 0.1$ μm is put into a supersaturated environment with RH = 101% and $T_{env} = 278$ K.
   (a) Compute the theoretical equilibrium vapor pressure $e_{s,w}$ just above the droplet.
   (b) Will it shrink or grow? Will it be in a stable or unstable equilibrium? Explain briefly.
   (c) How would the previous answer change if the same-sized droplet contains 5×10^{-20} kg of NaCl?
4. A solution is made by adding 50 g of $NaHCO_3$ to half a liter of pure water.
   (a) Compute the equilibrium vapor pressure e^* above the solution if the environmental temperature is $T_{env} = 276$ K. Assume that the salt dissociates completely.
   (b) At what relative humidity is the solution in equilibrium with its environment?
5. Sea salt particles are emitted into the atmosphere by high winds and breaking waves in the ocean.
   (a) Assuming equal masses, order the following marine salts with respect to their activation saturation ratio (from low to high): $MgCl_2$, $CaCl_2$, NaCl, $CaSO_4$.
   (b) Assume a spherical NaCl particle of dry radius $r_{dry} = 0.1$ μm. To which size does it need to grow to become activated at $T = 278$ K? The density of NaCl is 2.17 g cm^{-3}.
   (c) At what supersaturation would this occur?
6. Consider a cloud where cloud droplets are beginning to form on a population of cloud condensation nuclei (CCN) as the saturation ratio is gradually increased, with temperature held constant. The CCN each have the same chemical composition but a broad range of sizes. Droplets with radii $r_{act} = 0.22$ μm are activated when the supersaturation reaches 0.35%.
   (a) Determine the temperature T of the cloud.
   (b) Other CCN are activated at supersaturation 0.9% at the same temperature. What is the activation radius r_{act} in that case?
   (c) Assume that the CCN in (a) consist of ammonium sulfate particles. How does the activation of your aerosol population change if it consists instead of dust particles coated by the same amount of ammonium sulfate? Explain qualitatively.
7. A cloud condensation nuclei counter (CCNC), as described in Section 6.6, is a useful tool to study the nucleating properties of aerosol particles. Consider a 2 cm thick CCNC chamber with a parallel plate design. The temperatures of the two wetted plates

are $T_1 = 5$ °C (left-hand side) and $T_2 = 10$ °C (right-hand side) and the air is saturated with respect to water close to the plates. The specific humidity profile can be regarded as linear.

(a) Plot the supersaturation inside the chamber.

(b) At what distance from the cold plate is the maximum supersaturation observed? How large is the latter?

7 Microphysical processes in warm clouds

In Chapter 6 we discussed how cloud droplets are activated. Both before and after a droplet reaches its critical size, it grows by the uptake of water molecules from the vapor phase. Water vapor molecules are transported by diffusion towards the droplet where condensation takes place at the droplet surface. This is called diffusional or condensational growth (Section 7.1). Additional processes involving collisions with other droplets are required for the droplets to reach precipitation size, since diffusion of water vapor alone can be too slow to produce precipitation-sized drops on the time scales observed in the atmosphere. This growth is referred to as growth by collision–coalescence. It is induced by the difference in terminal velocity of the larger and smaller drops. Diffusional growth (and evaporation) and growth by collision–coalescence (Section 7.2), along with evaporation and the break-up of raindrops, which limits raindrops to a maximum equivalent radius of around 5 mm (Section 7.3), are discussed in this chapter.

Figure 7.1 shows typical sizes for a CCN, a cloud droplet, a drizzle drop and a raindrop, along with their typical concentrations in the marine atmosphere: typical marine CCN concentrations range between 10^2 and 10^3 cm^{-3} while that for cloud droplets is only around 10^2 cm^{-3} (Kubar *et al.*, 2009). This shows that for typical supersaturations (0.1%–1.6%) in the atmosphere not all CCN are activated to become cloud droplets (Chapter 6), i.e. the number of cloud droplets is less than the number of CCN.

After activation a cloud droplet must grow by two orders of magnitude in radius in order to reach the size of a raindrop, with typical radius 1 mm. Put differently, it requires a million cloud droplets to form one raindrop. Raindrops are large enough to overcome the updraft velocities within a cloud and leave the cloud as precipitation. Their large sizes explain their typical concentration, of only 10^{-4} cm^{-3} (100 m^{-3}), which is six orders of magnitude smaller than that of cloud droplets. Drizzle drops have sizes in between cloud droplets and raindrops, with radii between 25 μm and 0.25 mm (Table 1.2). Their typical concentration in marine clouds is 0.1 cm^{-3}, which is three orders of magnitude smaller than that of cloud droplets (Wood *et al.*, 2009) and three orders of magnitude larger than that of raindrops. Only one in ten clouds forms precipitation. Nine of ten times, the cloud evaporates.

7.1 Droplet growth by diffusion and condensation

As discussed in Chapter 6 and shown in Figure 7.2, a solution droplet can spontaneously grow into a cloud droplet in an environment in which $S > S_{act}$. Here we will discuss this

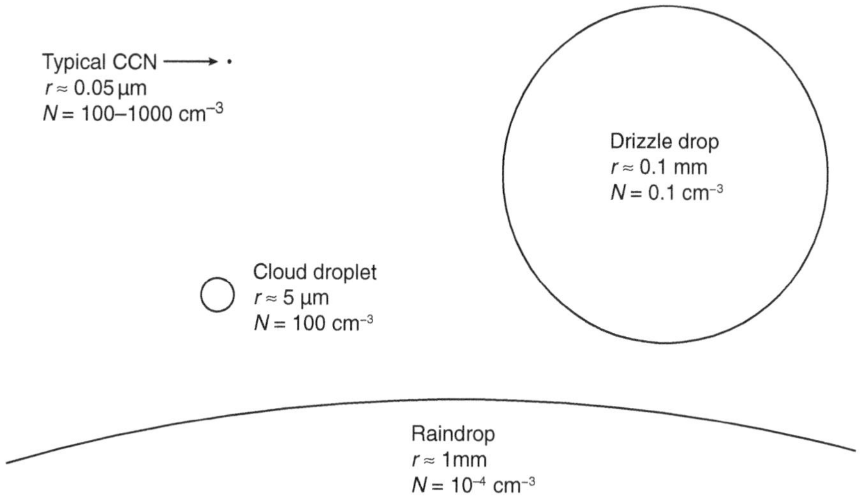

Fig. 7.1 Comparative sizes and concentrations of the particles involved in cloud and precipitation processes (Table 1.2). Typical values for marine clouds are given for the radii r and concentrations N.

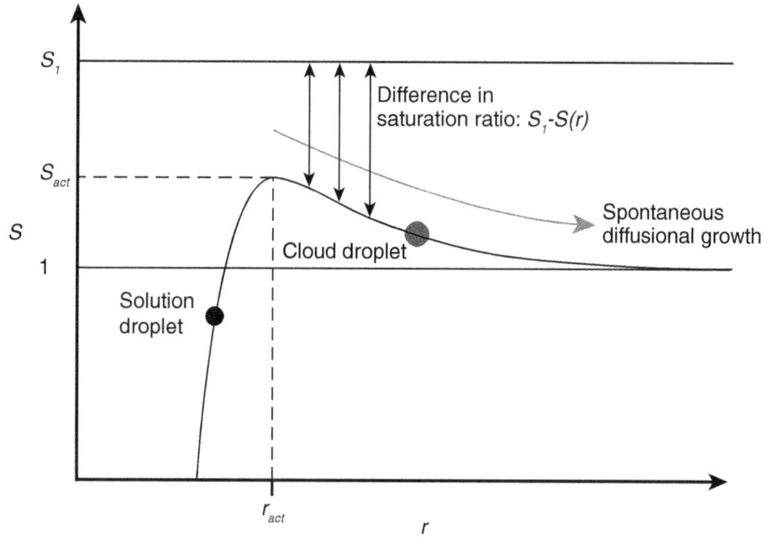

Fig. 7.2 Schematic of the diffusional growth process for a cloud droplet at a given ambient saturation ratio S_1 after Köhler activation at r_{act}. The diffusional growth process is driven by the difference $S_1 - S(r)$ in saturation ratios.

growth in more detail. Before and after the droplet reaches the activation radius r_{act}, it grows by the diffusion and condensation of water molecules from the surrounding vapor onto its surface. Before reaching r_{act} the solution droplet grows in equilibrium with the ambient saturation ratio. Once r_{act} is reached, the droplet grows spontaneously and continues even if $S < S_{act}$ as long as there is supersaturation. The droplet growth equation can be obtained from a combination of the diffusion equations for water vapor and heat, as will be discussed in the following.

7.1.1 Diffusion equation for water vapor

The diffusion equation for water vapor is given as

$$\frac{\partial n_v}{\partial t} = D_v \nabla^2 n_v = D_v \Delta n_v, \tag{7.1}$$

where n_v is the number concentration of water vapor molecules and D_v is the diffusion coefficient of water vapor in air. It is also called Fick's second law of diffusion. Fick's second law is based on his first law, which postulates that in the presence of a vapor gradient the flux is from regions of high concentration to regions of low concentration, with magnitude proportional to the concentration gradient.

We now place a drop with radius r_d inside the vapor field and assume that the concentration of water molecules at a radial distance r away from the drop is $n_v(r)$. We will assume that the water vapor field is isotropic around the drop. In such a case, it is advantageous to use spherical coordinates (r, θ, φ) because then the derivatives with respect to θ and φ are zero. Furthermore, we assume steady-state, or stationary, conditions, so that $\partial n_v/\partial t = 0$. Then eq. (7.1) can be written in spherical coordinates as follows:

$$D_v \Delta n_v = D_v \frac{1}{r^2}\frac{\partial}{\partial r}\left(r^2 \frac{\partial n}{\partial r}\right) = 0, \tag{7.2}$$

with general solution

$$n_v(r) = C_1 - \frac{C_2}{r}, \tag{7.3}$$

where C_1 and C_2 are constants of integration. The water vapor field has the following boundary conditions. For $r \to \infty$ the vapor concentration n_v approaches $n_{v,\infty}$, the undisturbed water vapor concentration far from the droplet surface. For $r \to r_d$ the vapor concentration n approaches n_{v,r_d}, the water vapor concentration at the droplet's surface.

This yields the special solution

$$n_v(r) = n_{v,\infty} - \frac{r_d}{r}(n_{v,\infty} - n_{v,r_d}), \tag{7.4}$$

where r_d is the droplet radius. The gradient of the vapor concentration is given as

$$\frac{\partial n_v}{\partial r} = \frac{r_d}{r^2}(n_{v,\infty} - n_{v,r_d}). \tag{7.5}$$

The gradient in the water vapor concentration between the environment far from the droplet and near the droplet's surface causes a diffusive flux of water vapor. In a supersaturated environment, the flux is directed towards the droplet, where the molecules condense on the surface and cause the droplet to grow. From eq. (7.5) we see that the gradient scales with $1/r^2$; hence it is small at large radial distances from the droplet and large close to the droplet surface, where it maximizes. This can be seen when evaluating eq. (7.5) at the droplet surface:

$$\left(\frac{\partial n_v}{\partial r}\right)_{r=r_d} = \frac{1}{r_d}(n_{v,\infty} - n_{v,r_d}). \tag{7.6}$$

Similarly, we obtain the flux density of water vapor molecules onto the surface of the droplet as $D_v\,(\partial n_v/\partial r)_{r=r_d}$. This yields the following equation for the rate of mass increase of the droplet due to diffusional growth:

$$\frac{dm_l}{dt} = 4\pi r_d^2 D_v \left(\frac{\partial n}{\partial r}\right)_{r=r_d} m_0, \tag{7.7}$$

where m_0 denotes the mass of one water molecule (3×10^{-25} kg). Inserting eq. (7.6) into eq. (7.7) yields

$$\frac{dm_l}{dt} = 4\pi r_d D_v (n_{v,\infty} - n_{v,r_d}) m_0, \tag{7.8}$$

where we see that the rate of mass increase depends linearly on the droplet radius. Replacing $n_{v,r_d} m_0$ by the vapor density ρ_{v,r_d} at the droplet surface (where $r = r_d$) and $n_{v,\infty} m_0$ by the ambient vapor density ρ_v yields

$$\frac{dm_l}{dt} = 4\pi r_d D_v (\rho_v - \rho_{v,r_d}). \tag{7.9}$$

Equation (7.9) is the diffusional growth equation for an isolated droplet at rest. It shows that the droplet grows if $\rho_v > \rho_{v,r_d}$ and evaporates if $\rho_v < \rho_{v,r_d}$. Whereas ρ_v is available from ambient measurements, ρ_{v,r_d} depends on the droplet size, chemical composition and temperature. The temperature at the droplet surface is usually not the same as the ambient temperature, so the heat transfer between the droplet and the environment needs to be considered as well.

7.1.2 Heat conduction equation

The description of diffusional growth requires not only the diffusion of water vapor to be taken into account but equally the release of latent heat and the associated transport of heat away from the droplet surface. This has large implications for the parameters which determine droplet growth. Condensation leads to an increase in droplet temperature owing to the release of latent heat and this causes the saturation vapor pressure at the droplet surface to increase. Since the vapor pressure at the droplet surface is equilibrated at saturation, the vapor pressure gradient between the supersaturated environment and the droplet surface decreases. This means that neglecting the latent heat release would cause an overestimation of the growth rate of cloud droplets due to condensation.

The latent heat can be taken into account by looking at the rate of heat transfer dQ/dt, to a droplet, which is given by the heat conduction equation. This is analogous to the vapor diffusion equation (eq. 7.9):

$$\frac{dQ}{dt} = 4\pi r K (T_{r_d} - T), \tag{7.10}$$

where T is the ambient temperature, T_{r_d} the temperature at the droplet surface and K the coefficient of thermal conductivity in air. The heat conduction equation is based on the Fourier law of conduction, which states that in the presence of a temperature gradient an energy flux is transported down the gradient (Lamb and Verlinde, 2011).

7.1.3 Droplet growth equation

In order to obtain an expression for the rate of change of the temperature at the surface of a water droplet, both vapor diffusion and heat conduction need to be taken into account. From the first law of thermodynamics the change in droplet surface temperature (here evaluated at the droplet's surface) is given by

$$m_l c_l \frac{dT_{r_d}}{dt} = \frac{dq}{dt}, \tag{7.11}$$

where m_l is the mass of the droplet and c_l is the specific heat capacity of liquid water. Note that dq is the total heat transfer while dQ (eq. 7.10) refers just to heat conduction. Thus dq/dt comprises the heat transfer due to latent heat release and also that due to the conduction of heat, so that

$$m_l c_l \frac{dT_{r_d}}{dt} = L_v \frac{dm}{dt} - \frac{dQ}{dt}. \tag{7.12}$$

We will assume that the diffusion of water vapor to the droplet with associated latent heat release upon condensation onto the droplet surface exactly balances the heat conduction away from the droplet, so that $dT_{r_d}/dt = 0$. Replacing the expressions for dm/dt and dQ/dt with, respectively, eqs. (7.9) and (7.10) yields

$$4\pi r K (T_{r_d} - T) = 4\pi r D_v (\rho_v - \rho_{v,r_d}) L_v; \tag{7.13}$$

after rearranging terms we obtain

$$\frac{K}{D_v L_v} = \frac{\rho_v - \rho_{v,r_d}}{T_{r_d} - T}. \tag{7.14}$$

In eq. (7.14) ρ_{v,r_d} and T_{r_d} are unknown. They are related by the Köhler equation (6.25). Using the ideal gas law for water vapor (eq. 2.77) allows us to express the Köhler equation in terms of ρ_{v,r_d}:

$$\rho_{v,r_d} = \frac{e^*(r)}{R_v T_{r_d}} = \left(1 + \frac{a}{r} - \frac{b}{r^3}\right) \frac{e_{s,w}(T_{r_d})}{R_v T_{r_d}}, \tag{7.15}$$

where e^* refers to the equilibrium vapor pressure over a solution droplet (Chapter 6); $e_{s,w}(T_{r_d})$ can be obtained from the Clausius–Clapeyron equation (2.59) given the temperature at the droplet surface T_{r_d}. Equations (7.14) and (7.15) comprise a coupled set of equations, which can only be solved numerically.

An analytical approximation of the droplet growth equation is possible if the solute and curvature terms of the Köhler equation are ignored, as described in Mason (1971). Starting from the Clausius–Clapeyron equation together with the ideal gas law for water vapor yields the change in saturation water vapor density ρ_{vs} with temperature:

$$\frac{d\rho_{vs}}{\rho_{vs}} = \frac{L_v}{R_v} \frac{dT}{T^2} - \frac{dT}{T}. \tag{7.16}$$

Integrating from T_{r_d} to T gives

$$\ln \frac{\rho_{vs}}{\rho_{vs,r_d}} = (T - T_{r_d}) \left(\frac{L_v}{R_v T_{r_d} T}\right) - \ln \frac{T}{T_{r_d}}. \tag{7.17}$$

Assume that $\rho_{vs}/\rho_{vs,r_d} \approx 1$ and $T/T_{r_d} \approx 1$ to use a Taylor series to approximate $\ln \rho_{vs}/\rho_{vs,r_d}$ by $\rho_{vs}/\rho_{vs,r_d} - 1$ and $\ln T/T_{r_d}$ by $T/T_{r_d} - 1$, and assume that $TT_{r_d} \approx T^2$: thus

$$\frac{\rho_{vs} - \rho_{vs,r_d}}{\rho_{vs,r_d}} \approx \left(\frac{T - T_{r_d}}{T}\right)\left(\frac{L_v}{R_v T} - 1\right). \tag{7.18}$$

Then use eq. (7.10) to replace $T - T_{r_d}$ and eq. (7.12) in the steady state to replace dQ/dt by $L_v dm_l/dt$:

$$\frac{\rho_{vs} - \rho_{vs,r_d}}{\rho_{vs,r_d}} = -\frac{dm_l}{dt}\left(\frac{L_v}{4\pi r K T}\right)\left(\frac{L_v}{R_v T} - 1\right). \tag{7.19}$$

Now divide eq. (7.9) by ρ_{vs,r_d}, obtaining

$$\frac{\rho_v - \rho_{v,r_d}}{\rho_{vs,r_d}} = \frac{dm_l}{dt}\left(\frac{1}{4\pi r D_v \rho_{vs,r_d}}\right). \tag{7.20}$$

Finally, assume that the vapor density at the droplet's surface equals the saturated water vapor density ($\rho_{v,r_d} = \rho_{vs,r_d}$), approximate ρ_{vs,r_d} in the denominator by ρ_{vs} and subtract eq. (7.19) from it:

$$\frac{\rho_v - \rho_{vs}}{\rho_{vs}} = \frac{dm_l}{dt}\left[\left(\frac{1}{4\pi r D_v \rho_{vs}}\right) + \frac{L_v}{4\pi r K T}\left(\frac{L_v}{R_v T} - 1\right)\right]. \tag{7.21}$$

Replace ρ_v/ρ_{vs} by the saturation ratio S and rearrange terms to obtain the mass growth equation for a cloud droplet:

$$\frac{dm_l}{dt} = 4\pi r \frac{S - 1}{\frac{L_v}{KT}\left(\frac{L_v}{R_v T} - 1\right) + \frac{1}{D_v \rho_{vs}}} = 4\pi r \frac{S - 1}{F_k^l + F_d^l}. \tag{7.22}$$

Here F_k^l is a thermodynamic term related to the latent heat release due to the condensation and diffusion of heat away from the droplet and F_d^l is a vapor diffusion term related to the diffusion of water vapor to the growing droplet.

The term F_k^l depends on the ambient temperature T, the latent heat of vaporization L_v and the thermal conductivity coefficient K and is given as

$$F_k^l = \left(\frac{L_v}{R_v T} - 1\right)\frac{L_v}{KT} \approx \left(\frac{L_v^2}{K R_v T^2}\right); \tag{7.23}$$

the coefficient K increases with increasing temperature since heat conduction at higher temperatures is faster owing to the higher kinetic energy of the air molecules. For dry air the temperature dependence of K in W m^{-1} K^{-1} can be approximated by an empirical relation based on Beard and Pruppacher (1971):

$$K = 4.1868 \times 10^{-3}[5.69 + 0.017(T - T_0)], \tag{7.24}$$

where $T_0 = 273.15$ K. For typical atmospheric conditions the molar fraction of water vapor in moist air is small, allowing us to approximate the thermal conductivity of moist air by eq. (7.24). Besides, the pressure dependence of K is very weak at atmospheric pressures and can therefore be neglected (Kadoya *et al.*, 1985).

Table 7.1 Values for the thermal conductivity K in W m^{-1} K^{-1} according to eq. (7.24), and for the diffusion coefficient for water vapor in air D_v in 10^{-5} m^2 s^{-1} according to eq. (7.26) as a function of temperature in °C for the pressure values indicated.

T	K	D_v (1000 hPa)	D_v (800 hPa)	D_v (600 hPa)	D_v (400 hPa)	D_v (200 hPa)
−40	0.021	1.57	1.97	2.62	3.93	7.86
−20	0.022	1.84	2.31	3.07	4.61	9.22
0	0.024	2.14	2.67	3.56	5.34	10.69
20	0.025	2.45	3.07	4.09	6.13	12.26
40	0.027	2.79	3.48	4.64	6.97	13.94

The term F_d^l depends on the saturation vapor pressure, the temperature and the water vapor diffusion coefficient in air, D_v:

$$F_d^l = \frac{1}{D_v \rho_{vs}} = \frac{R_v T}{D_v e_{s,w}(T)}. \tag{7.25}$$

The coefficient D_v increases with increasing temperature but decreases with increasing pressure: an increase in pressure means a decrease in the mean free path of the molecules so that collisions with other molecules and particles occur more frequently. The increase in D_v with increasing temperature represents the faster diffusion at higher temperatures. This quantity can be approximated for temperatures between −40 and +40 °C as follows (Hall and Pruppacher, 1976):

$$D_v = 2.11 \times 10^{-5} \left(\frac{T}{T_0}\right)^{1.94} \left(\frac{p_0}{p}\right) \text{ m}^2 \text{ s}^{-1}, \tag{7.26}$$

where $p_0 = 101\,325$ Pa. Values of D_v and K for various temperatures, and for different pressures in the case of D_v, are listed in Table 7.1.

For the growth of cloud droplets it is more common to write the droplet growth equation in terms of the growth in radius. This version of the equation can be obtained on replacing dm_l/dt by $4\pi r^2 \rho_l dr/dt$ in eq. (7.22):

$$r\frac{dr}{dt} = \frac{S-1}{F_k + F_d}. \tag{7.27}$$

where $F_k = \rho_l F_k^l$, $F_d = \rho_l F_d^l$ and ρ_l denotes the water density. Equation (7.27) is known as the simplified droplet growth equation. It is valid for sufficiently large droplets when neglect of the solution and curvature effects from the Köhler equation on the droplet's equilibrium vapor pressure is justified. If these are included then the full droplet growth equation is given as

$$r\frac{dr}{dt} = \frac{(S-1) - a/r + b/r^3}{F_k + F_d}. \tag{7.28}$$

Equation (7.28) cannot be solved analytically. However, for most applications the simple droplet growth equation (eq. 7.27) is sufficient.

7.1.4 Solution of the droplet growth equation

Equation (7.27) can be integrated to show that the cloud droplet radius varies with time according to

$$r(t) = \sqrt{r_0^2 + 2\left(\frac{S-1}{F_k + F_d}\right)t}. \tag{7.29}$$

As can be seen in eqs. (7.23) and (7.25), F_k depends on temperature and F_d depends on temperature and pressure. The higher the temperature, the smaller is F_k. The term F_d has temperature dependent terms in both the nominator and denominator. Here the temperature dependence in the denominator dominates so that F_d also decreases with increasing temperature. Consequently the denominator in eq. (7.29) is smaller for higher temperatures, i.e. at higher temperatures droplets grow or shrink faster for a given ambient saturation ratio S. This has two reasons: diffusion is more efficient at higher temperatures and the absolute difference in water vapor pressure between the droplet surface and the surroundings is larger for a given ambient saturation ratio at higher temperatures. Therefore the water vapor gradient at the droplet surface, which drives the diffusional growth or evaporation of the droplet, is larger at higher temperatures.

The solution of the droplet growth equation is shown in Figure 7.3 for droplets starting at different initial sizes r_0 and at two different temperatures. It can be seen that the diffusional growth in terms of radius slows down as the droplets become larger. This can be explained by the decrease in surface-to-volume ratio of the larger droplets. As diffusional growth is proportional to r^2 and mass growth is proportional to r^3, the small droplets, with a larger surface-to-volume ratio, catch up with the larger droplets. This leads to the characteristic parabolic form of eq. (7.29) and to a narrowing of the droplet size distribution as growth proceeds.

Figure 7.3 further illustrates the temperature dependence of droplet growth. It shows that raindrops would not form after 20 minutes at an ambient temperature of 253 K, because the droplets would not yet have reached precipitation size (Figure 7.1). However, a radius of 15 μm, which is considered the minimum size for the growth by collision–coalescence (Section 7.2) and thus for the development of rain, has been reached after 3–5 minutes at an ambient temperature of 293.15 K. At these high temperatures, with a constant saturation ratio $S = 1.005$ diffusional growth is fast enough to produce drizzle drops of 40 μm in 20 minutes.

7.1.5 Growth of a droplet population

In reality, droplets within a cloud are not of uniform size and thus start their condensational growth from different sizes. Besides, the supersaturation does not remain constant

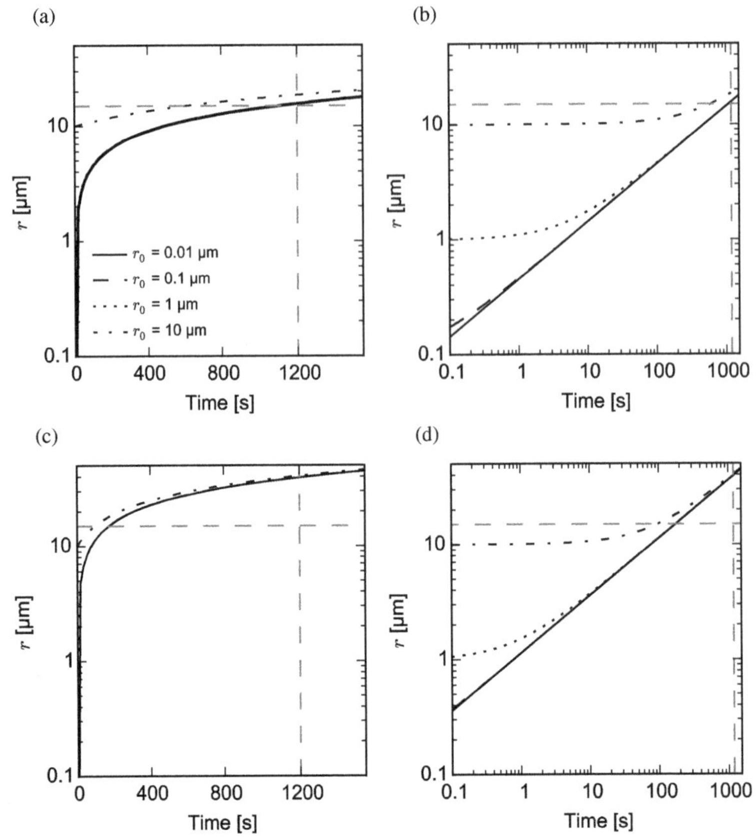

Fig. 7.3 Solution of the droplet growth equation (7.29) for different r_0 values plotted (a), (c) on a linear and (b), (d) on a logarithmic time scale for $T = 253$ K (a), (b) and for $T = 293$ K (c), (d). The curves were calculated for $S = 1.005$ and pressure 800 hPa. The broken horizontal lines correspond to a droplet with $r_d = 15\ \mu$m, beyond which collision–coalescence becomes important. Twenty minutes (broken vertical line) is the time elapsed when newly formed cumuli are first observed to precipitate.

but instead will be depleted by condensational growth. Thus we need to take the interactions of cloud droplets with the ambient water vapor field and with each other into account. In the early development of a cloud the droplets are too small for sedimentation and coalescence to be important; their initial growth is governed purely by condensational growth. Therefore we need to examine the vapor budget of a developing cloud. Supersaturation is established in the ascending air of a forming cloud owing to adiabatic expansion and cooling of the air ($Q_1 dz/dt$), and is consumed by the condensation of water vapor on the growing cloud droplets ($Q_2 dq_l/dt$). The rate of change of the saturation ratio S in an ascending air parcel is thus given by

$$\frac{dS}{dt} = Q_1 \frac{dz}{dt} - Q_2 \frac{dq_l}{dt}. \tag{7.30}$$

Here dz/dt is the vertical velocity w and dq_l/dt is the condensation rate, i.e. the mass condensate per unit mass of air and unit time; Q_1 and Q_2 are thermodynamic variables given by

$$Q_1 = \frac{1}{T}\left(\frac{\epsilon L_v g}{R_d c_p T} - \frac{g}{R_d}\right), \tag{7.31}$$

$$Q_2 = \rho\left(\frac{R_d T}{\epsilon e_{s,w}} + \frac{\epsilon L_v^2}{p T c_p}\right), \tag{7.32}$$

Q_1 has the unit m^{-1} and Q_2 is dimensionless. A derivation of Q_1 and Q_2 is given in Rogers and Yau (1989). Equation (7.30) shows that the higher the vertical velocity, the larger the production of supersaturation.

7.1.6 Application of the droplet growth equation

A Lagrangian parcel model can simulate the evolution of droplet sizes, droplet number concentrations and saturation ratio as a function of height above cloud base. The model is considered Lagrangian because it describes an air parcel and follows it, i.e. the observer moves with the air parcel. A Lagrangian parcel model can be initialized with a population of aerosol particles with different sizes. Examples of such simulations are shown in Figures 7.4 and 7.5. The parcel model used for these simulations numerically solves the droplet growth equation (7.28) (Neubauer, 2009). It is suitable for simulating the early stages of cloud formation.

In the example shown in Figure 7.4, the aerosol particles start from a monomodal log-normal distribution of ammonium sulfate particles with $N_a = 1000$ cm^{-3}, $\mu_g = 75$ nm and $\sigma_g = 1.5$, as indicated in terms of dry radii in Figure 7.4b. Before they reach cloud base in Figure 7.4a they have deliquesced and grown hygroscopically (Section 6.3); this explains the size shift between panels 7.4a and 7.4b.

The supersaturation reaches its peak at 20–25 m above cloud base. Up to this altitude all particles are growing and thereby reducing the ambient water vapor. After the peak supersaturation S_{max} has been passed, the droplet population diverges. The non-activated solution droplets with dry radii of 0.03 and 0.04 µm shrink, according to Köhler theory (Figure 6.11), in order to remain in equilibrium with the decreasing ambient saturation ratio. The separation in size between activated cloud droplets and non-activated solution droplets depends on the peak supersaturation, which again is governed by Köhler theory. According to eq. (7.30) the peak supersaturation in turn is determined by the updraft velocity: a higher updraft velocity activates more aerosol particles. In addition to its dependence on the updraft velocity, the number of cloud droplets formed depends on the aerosol size, number concentration and chemical composition. The cloud droplets experience rapid growth even at decreasing saturation ratios. As they continue to grow, their spread in size becomes narrower because of the parabolic form of the growth equation (7.29).

The dependence of some cloud properties on the height above cloud base for the early stages of cloud formation is shown in Figure 7.5 for the lowest 100 m in a cloud for two different updraft velocities, 0.5 and 3 m s^{-1}. As in Figure 7.4, one can

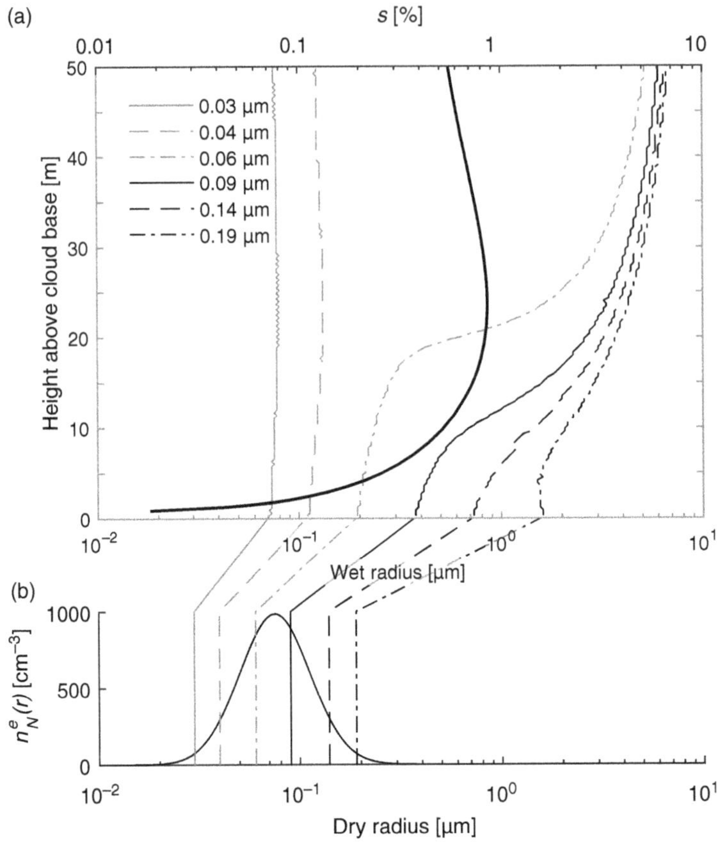

Fig. 7.4 (a) Simulated growth of a droplet population as a function of height above cloud base in a rising air parcel without entrainment and with constant updraft velocity 3 m s^{-1} (Neubauer, 2009). Droplet size (thin lines) corresponds to the lower x-axis and ambient saturation ratio (thick solid line) to the upper x-axis; the bold curve corresponds to s. The initial dry particle size distribution is a monomodal log-normal distribution for ammonium sulfate particles with $N_a = 1000$ cm^{-3}, $\mu_g = 75$ nm and $\sigma_g = 1.5$, as indicated by the solid curve in panel (b). Data by courtesy of David Neubauer.

see the sharp rise and gentle settling down of the supersaturation as one goes upward from the cloud base (Figure 7.5a). The peak supersaturation lies within about 30 m above cloud base. The supersaturation approaches a constant value higher up in the cloud when the rate of condensation approaches the production of supersaturation in the updraft. Within the 100 m shown in Figure 7.5, Q_1 and Q_2 (eqs. 7.31, 7.32) can be regarded as constant because T, ρ and $e_{s,w}$ vary only a little over this small height range. This also implies that the production rate of supersaturation is constant (eq. 7.30). In fact the shape of the supersaturation profile in Figure 7.5a is determined by the profile of the condensation rate. The condensation rate is small initially when only a few cloud droplets are activated and increases steadily with the increasing number of activated droplets and thus the larger surface area available for condensation. Once the

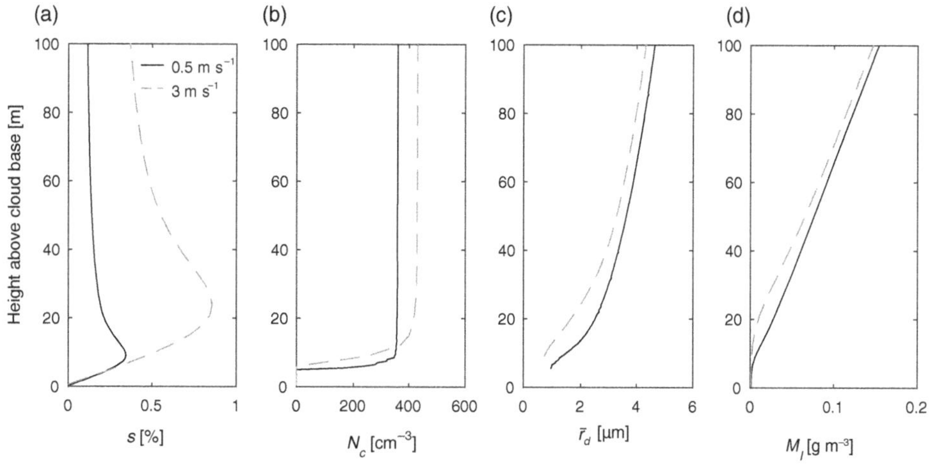

Fig. 7.5 Simulated development of cloud properties in the lowest 100 m of a cloud for air ascending at a constant updraft velocity of 0.5 m s^{-1} or 3 m s^{-1} starting at 850 hPa and 15 °C with the same initial aerosol size distribution as in Figure 7.4. (a) Supersaturation s, (b) cloud droplet number concentration N_c, (c) mean volume radius $\bar{r}_d$ and (d) liquid water content M_l. Data by courtesy of David Neubauer.

condensation rate exceeds the production rate, it reduces the supersaturation and therefore limits itself because of the decreasing water vapor gradient between the ambient air and the droplets.

The cloud droplet number concentration N_c increases rapidly initially but then settles to a constant value at the height where S_{max} is reached. The reason is that S_{max} determines the number (and thus indirectly the size) of the activated particles. Thus, the cloud droplet number concentration is determined in the lowest layers of a cloud. It is smaller for a slower updraft velocity because a smaller updraft velocity leads to a lower production of supersaturation and thus to a smaller S_{max}, causing fewer droplets to become activated (Figure 7.5b). Because fewer droplets compete for the available water vapor, the droplets are on average larger in these clouds (Figure 7.5c).

Both the liquid water content M_l (Figure 7.5d) as well as the volume mean droplet radius $\bar{r}_d$ increase with height above cloud base. In the first 100 m, mixing with the environment is negligible and M_l is approximately the adiabatic liquid water content w_l (eq. 2.106); w_l increases almost linearly with height in shallow clouds, such as stratocumulus clouds with a vertical extent smaller than 1 km (Brenguier *et al.*, 2000). Note that M_l is slightly smaller at a given altitude for a larger updraft velocity, when the cloud droplets have less time to grow while rising. In the example of Figure 7.5, the factor 3 increase in S_{max} causes N_c to increase by roughly 25%. The condensation rate, which depletes the supersaturation (eq. 7.30), is proportional to N_c and therefore increases only by 25%, i.e. much less than the production rate of supersaturation. On the whole, the supersaturation remains larger for larger updraft velocities and less water is found in the condensed phase in the lowest 100 m of the cloud.

7.2 Droplet growth by collision–coalescence

Higher up in the cloud than is shown in Figure 7.5 other processes, such as the entrainment of environmental air and the formation of precipitation, become important. The challenge of microphysics in warm clouds is to explain how raindrops can be formed in 20–30 minutes, which is the time between the first appearance of a cumulus and the earliest appearance of rain. Owing to the decreasing growth rate for increasing cloud droplet radius, condensational growth is usually too slow to be responsible for the observed speed of rain formation.

Rain formation by collision–coalescence is the dominant precipitation formation process in warm clouds whose tops may extend to subzero temperatures. Growth by collision–coalescence refers to the collision of droplets of the same or different sizes that then stick together (coalescence) to form a larger drop. This process is also referred to as growth by accretion. It starts once some droplets exceed a radius of 15 μm, after which they can achieve a 100-fold increase in radius (from a 10 μm cloud droplet with number concentration 10^2 cm^{-3} to a 1 mm raindrop with number concentration 10^{-4} cm^{-3}) by collision–coalescence.

As shown in Figure 7.3, cloud droplets can grow by condensation to $r = 15$ μm in less than 10 minutes at warm temperatures if $s \geq 0.5\%$ or $w \geq 5$ m s^{-1}. Such conditions can be found in developing cumuli. The 15 μm threshold is not cast in stone but is, rather, an empirical value; different authors use thresholds varying between 14 μm (Rosenfeld *et al.*, 2001) and 20 μm (Wallace and Hobbs, 2006). Droplets of that size start to settle (sediment) because gravity leads to an appreciable fall velocity at this droplet size. The settling of droplets is important because it causes them to collide with smaller droplets in their paths (Section 5.4.2). In clouds with a variety of different cloud droplet sizes, among which some are larger than $\sim$ 15 μm, growth by collision–coalescence becomes the dominant precipitation-formation process. This variety of droplet sizes cannot be produced by condensational growth alone, because the latter leads to a narrowing of the droplet size spectrum instead of a broadening and thus to a reduction in the velocity difference between the largest and smallest droplets. A broadening of the cloud droplet size distribution, with associated increase in the velocity difference between the largest and smallest droplets, is needed for the onset of precipitation formation and can be achieved either by turbulence or by giant CCN as will be discussed next.

7.2.1 Initiation of the collision–coalescence process

The collision–coalescence process is rather efficient, given that a raindrop of 1 mm results from $\sim 10^5$ collisions (Rogers and Yau, 1989). As shown schematically in Figure 7.6 the condensational growth of droplets tends to level off at larger droplet sizes owing to the parabolic form of eq. (7.29). Then growth by collision–coalescence takes over.

One possible way in which the collision–coalescence process is initiated is by the broadening of the drop size distribution in the presence of giant CCN (GCCN, Table 5.1). These could for instance be sea salt nuclei or giant insoluble aerosol particles such as

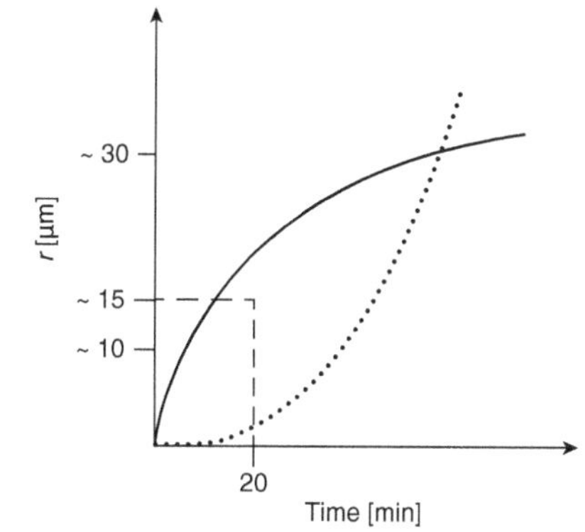

Fig. 7.6 Schematic change in droplet radius versus time due to growth by condensation from the vapor phase (eq. 7.29) (solid line) and due to growth by collision–coalescence (dotted line). Adapted from Wallace and Hobbs (2006) with permission from Elsevier.

mineral dust. As shown in Figure 7.3, a droplet that has grown from a 10 μm salt aerosol remains significantly larger than those grown from smaller particles over the calculated 30 minutes, for a temperature of 253 K but not at 293 K. Thus such large cloud droplets can initiate the collision–coalescence process by settling (sedimentation) due to gravity. This effect of GCCN is important at colder temperatures where diffusional growth is slower.

As shown in the modeling study of Feingold *et al.* (1999) the small observed concentrations of GCCN over the ocean, between 10^{-4} and 10^{-2} cm^{-3} (depending on the ambient wind speed and state of the ocean), are sufficient to transform a non-precipitating stratocumulus into a precipitating state. In their calculations, with CCN concentrations of 50 to 250 cm^{-3}, higher concentrations of GCCN were required at higher CCN concentrations where the average cloud droplet size is smaller and the droplet size distribution tends to be narrower. In these clouds the collision–coalescence process would not be initiated without the help of some larger drops. Relatively low CCN concentrations, however, yield on average larger cloud droplets and a broader size distribution (Figure 1.11) and the addition of GCCN has little impact.

An example of the impact of GCCN on the surface precipitation for different CCN concentrations and GCCN sizes is shown in Figure 7.7. It shows results from a GCM simulation where the GCCN number concentration was varied between 10^{-4} and 1 cm^{-3}, while the number concentrations of the background CCN were taken as 100, 250 and 500 cm^{-3} and the activated GCCN-formed raindrops as having initial sizes 12, 25 or 55 μm. Some precipitation is formed through condensational growth and collision–coalescence regardless of the GCCN concentration. The smaller precipitation rate in clouds with higher CCN concentrations is caused by a less efficient collision–coalescence process due to the smaller size of the individual droplets. The impact of GCCN on precipitation is smallest for the cleanest cloud, with 100 CCN cm^{-3}, and largest for the most polluted cloud, with 500 CCN cm^{-3}. Thus, the addition of GCCN matters most for continental clouds but is negligible for marine clouds, which consist of fewer cloud droplets (Table 1.3 and Figure 1.10).

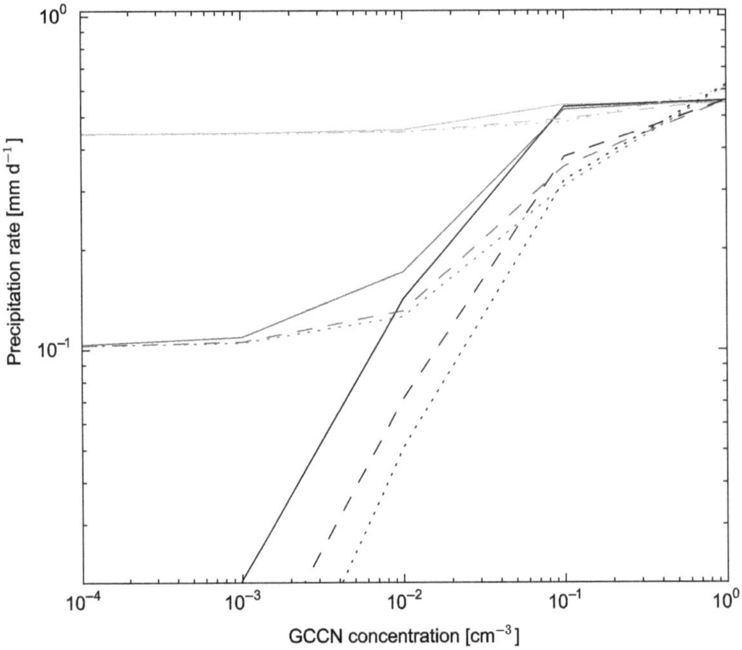

Fig. 7.7 Precipitation rate as a function of the GCCN concentration, illustrating the impact of GCCN on precipitation formation. The (background) CCN number concentrations N_{CCN} are indicated in the plot as follows: top group of lines, $N_{CCN} = 100\ \text{cm}^{-3}$; middle group, $N_{CCN} = 250\ \text{cm}^{-3}$; bottom group, $N_{CCN} = 500\ \text{cm}^{-3}$. For each value of N_{CCN}, three minimum radii of the GCCN, r_{GCCN}, were used: $r_{GCCN} = 55\ \mu\text{m}$ (solid lines), $r_{GCCN} = 25\ \mu\text{m}$ (broken lines) and $r_{GCCN} = 12\ \mu\text{m}$ (dotted lines). The simulations were conducted with the ECHAM-GCM using single-column mode. Data by courtesy of Vivek Sant.

The size of the initially formed raindrops, however, is of minor importance, especially for clean conditions.

If no GCCN are present, turbulence is needed to explain the broadening of the cloud droplet size distribution. Turbulence is always present in clouds to some degree. Turbulent fluctuations promote coalescence by increasing the chance of droplet collisions. In particular the presence of turbulence causes droplets to have a relative velocity not only in the vertical direction due to gravitational settling but additionally in all directions due to the turbulent air flow. The effective collision kernel, a measure of the collision efficiency, increased owing to turbulence by up to 47%, as estimated from direct numerical simulations of droplet collisions (Ayala *et al.*, 2008).

7.2.2 Collision and coalescence efficiencies

Collisions may occur due to gravitational effects once some drops become larger than others. The probability of collisions is expressed in terms of the collision efficiency E (eq. 5.18). Collision efficiencies from theoretical calculations (Schlamp *et al.*, 1976) for collector drops with radii R between 11 and 74 μm and smaller droplets of radius r are shown in

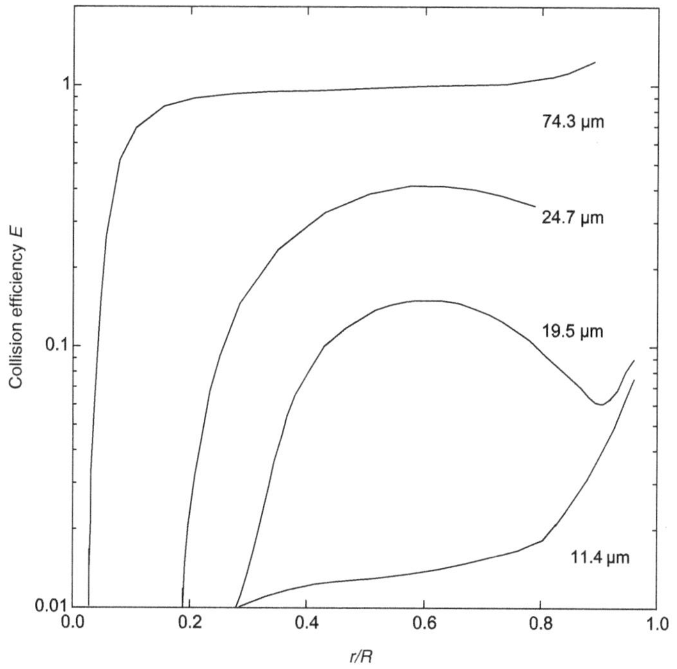

Fig. 7.8 Computed collision efficiencies E for pairs of drops as a function of the ratio of their radii, r/R. The curves are labeled according to the radius R of the larger collector drop. Figure adapted from Schlamp *et al.* (1976) with permission of the American Meteorological Society.

Figure 7.8 as a function of the ratio of their radii, r/R. The collision efficiency E is small for small values of r/R: then the collected droplets are small, have little inertia and are thus easily deflected by the flow around the collector drop (Figure 5.11). The inertia increases with increasing mass of the small droplets and thus with increasing values of r/R. The value of E increases up to approximately $r/R = 0.6$. Beyond this ratio, two counteracting effects influence E. On the one hand the difference in size between the collector drop and the collected droplet is getting smaller. Hence the inertia of the collected drop increases, which reduces its ability to be deflected around the collector drop. On the other hand the forces of deflection have a longer time in which to operate, owing to the reduced velocity difference of the two droplets, leading to a decrease in E. Overall, the increase in inertia is the more important effect and therefore E generally increases with increasing values of r/R.

There is also the possibility that a trailing drop will be attracted into the wake of a drop falling close by at nearly the same speed. This can cause a capture from behind, called wake capture. For a given r/R ratio, wake capture is more important for larger collector drops (Figure 5.11) because they have a higher Reynolds number and thus induce more turbulence, which pulls the trailing drops into their wake. Wake capture can be compared to cycling in the lee of another bicycle. Cycling with the same power in the lee as in front would cause a collision because the person cycling in front experiences the full aerodynamic resistance whereas the person behind experiences a much reduced aerodynamic

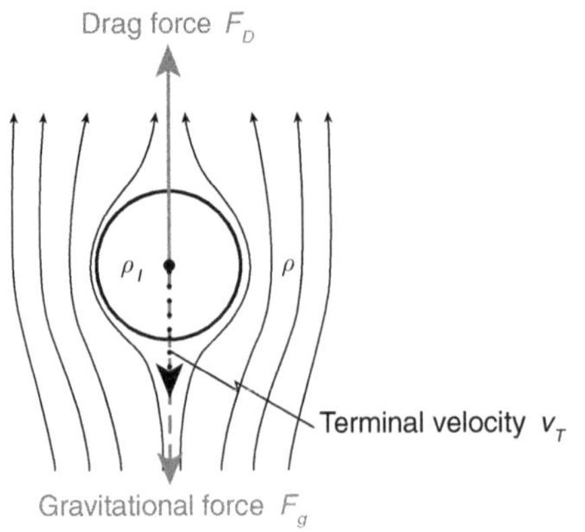

Fig. 7.9 Schematic showing the terminal velocity v_T, defined as the velocity when a balance between the drag force F_D and the gravitational force F_g has been established.

resistance. Likewise, the lower drop experiences the full aerodynamic resistance as it falls whereas the drop falling in its wake experiences a much reduced aerodynamic resistance, causing it to fall at a higher velocity, which might induce a collision with the drop below. Because the definition of E does not take wake capture into account, the latter effect can cause E to exceed 100% for large collector drops (Figure 7.8).

So far we have assumed that colliding drops stick together, i.e coalesce. This, however, is only one of three possibilities. If a pair of drops collides, the following things can happen: (i) they coalesce and remain permanently united, (ii) they coalesce temporarily and separate, retaining their initial identities, or (iii) they coalesce temporarily and break into a number of small drops.

For drops with $r < 100\ \mu m$, (i) is the most important process. The coalescence efficiency $\hat{E}$ is defined as the number of coalescence events divided by the total number of collisions. The growth of droplets by collision–coalescence is governed by the collection efficiency $\tilde{E}$, which is the product of the coalescence efficiency and the collision efficiency:

$$\tilde{E} = \hat{E} \times E. \tag{7.33}$$

A good assumption for the collision–coalescence process for droplets with $r < 100\ \mu m$ is that $\hat{E} \sim 1$, so that $\tilde{E} = E$. A temporary coalescence as in case (ii) is rather unimportant in the atmosphere. Case (iii) is important for raindrops ($r > 100\ \mu m$) and will be discussed in Section 7.3.

7.2.3 Terminal velocity of cloud droplets and raindrops

In order to derive the droplet growth equation due to collision–coalescence, we need to know at what velocities the droplets fall. A droplet reaches its sedimentation velocity when the gravitational force is balanced by the drag force, i.e. the aerodynamic resistance exerted on the droplet by the air as a viscous medium, as shown in Figure 7.9.

The gravity force on a spherical water droplet with radius r is given by

$$F_g = m_l g = \frac{4}{3}\pi r^3 g \rho_l, \tag{7.34}$$

and the drag force exerted on this droplet in a viscous fluid is

$$F_D = \frac{\pi}{2} r^2 v^2 \rho C_D = 6\pi \mu r v \left(\frac{C_D Re}{24}\right), \tag{7.35}$$

where μ is the dynamic viscosity of air, C_D is the drag coefficient, an empirical factor accounting for all non-geometrical components of the drag force, and the Reynolds number is given by

$$Re = \frac{2\rho v r}{\mu}. \tag{7.36}$$

The sedimentation velocity or fall speed is also known as the terminal velocity v_T, which emphasizes that a given droplet does not accelerate beyond v_T. Assuming equality of the drag force and the gravitational force, we can calculate the terminal velocity from

$$6\pi \mu r v_T \left(\frac{C_D Re}{24}\right) = \frac{4}{3}\pi r^3 g \rho_l. \tag{7.37}$$

Rearranging the factors yields the general form of v_T for spherical particles:

$$v_T = \frac{2}{9} \frac{r^2 g \rho_l}{\mu C_D Re/24}. \tag{7.38}$$

Since Re and C_D depend on v, eq. (7.37) cannot be solved analytically for v_T. The calculation of v_T depends on a calculation of Re and C_D and thus the drag force F_D, since F_g is a simple function of drop size and density. Both Re and C_D vary with the type of flow and the drop size, making it impossible to give a general expression that is valid for all drop sizes. In order to reduce complexity we can divide the range of possible drop sizes into regimes; this allows us to define empirical expressions for the terminal velocities in each individual regime.

For droplets with radii $< 30\ \mu$m, i.e. cloud droplets and small drizzle drops (Figure 7.1), where $Re \ll 1$, the factor $C_D Re/24$ is approximately 1 (Rogers and Yau, 1989), so that we have for the terminal velocity

$$v_T = \frac{2}{9}\frac{r^2 g \rho_l}{\mu} = k_1 r^2, \qquad r < 0.03\,\text{mm}. \tag{7.39}$$

Using the values of μ and ρ_l at 20 °C we can approximate k_1 as 1.2×10^6 cm^{-1} s^{-1}. Equation (7.39) is known as Stokes's law. It is characterized by the quadratic dependence of v_T on the drop size.

For drops of intermediate size, in the range $30\ \mu\text{m} < r < 0.6$ mm, an empirical formula for the terminal velocity is given by Rogers and Yau (1989) as

$$v_T = k_2 r, \qquad 0.03\,\text{mm} < r < 0.6\,\text{mm}, \tag{7.40}$$

with $k_2 = 8000\ \text{s}^{-1}$, showing that v_T varies approximately linearly with droplet radius in this regime.

For large drops ($r > 0.6$ mm), for which $Re > 100$, C_D is independent of Re and has the value 0.45. With that value eq. (7.38) becomes

$$v_T = \frac{2}{9}\frac{r^2 g\rho_l}{0.45\mu Re/24} = \frac{2}{9}\frac{r^2 g\rho_l}{0.45\mu 2\rho v_T r/24\mu}. \tag{7.41}$$

Here the Reynolds number has been replaced by its definition. Rearranging factors yields

$$v_T^2 = \frac{rg\rho_l}{\rho}\frac{160}{27} \tag{7.42}$$

or

$$v_T = k_3\sqrt{r}, \qquad 0.6\,\text{mm} < r < 2\,\text{mm}. \tag{7.43}$$

Here k_3 is approximately 2010 $\text{cm}^{1/2}\ \text{s}^{-1}$ for a raindrop size between 0.6 mm and 2 mm (Rogers and Yau, 1989). The upper limit corresponds to the size up to which raindrops can be considered spherical.

The terminal velocity is shown in Figure 7.10, where the broken lines mark the separations between the different regimes. Because of their empirical character, the transition between the different formulas is not smooth.

Figure 7.10 also illustrates how raindrops become increasingly non-spherical with increasing size, as indicated by the insets (Pruppacher and Beard, 1970). Beyond radius 2 mm the bottom of the raindrops flattens considerably. Because of this the raindrop sizes in Figure 7.10 are given in terms of the equivalent spherical radius, i.e. the radius which one would obtain if a raindrop of that given mass was spherical. A raindrop with equivalent size 2.5 mm can be considered as an oblate spheroid (Figure 7.10). Because of its large surface area, its drag force approaches the same radius dependence as the gravitational force acting on it. Its terminal velocity is thus constant for equivalent raindrop radii > 2.5 mm. The maximum equivalent raindrop radius is limited by break-up (Section 7.3). Depending on the pressure and temperature, the maximum equivalent raindrop radius varies between 4.3 mm and 6.2 mm (Pruppacher and Klett, 1997).

7.2.4 Growth model for continuous collection

For drops to collide they have to have different velocities and thus sizes. Consider a large raindrop with radius R falling with terminal velocity $v_T(R)$ through a population of smaller droplets of uniform sizes r, falling at terminal velocities $v_T(r)$. During settling, the large drop can collide with smaller droplets in its path if the smaller drops are within the impact parameter of the large drop (eq. 5.18). As in the case of the wet deposition of aerosol particles discussed in Section 5.4.2, we can define the cross-sectional area as $\pi(R + r)^2$, and hence the volume swept out by the collector drop per unit time is:

$$\frac{dV_{sweep}}{dt} = \pi(R + r)^2[v_T(R) - v_T(r)]. \tag{7.44}$$

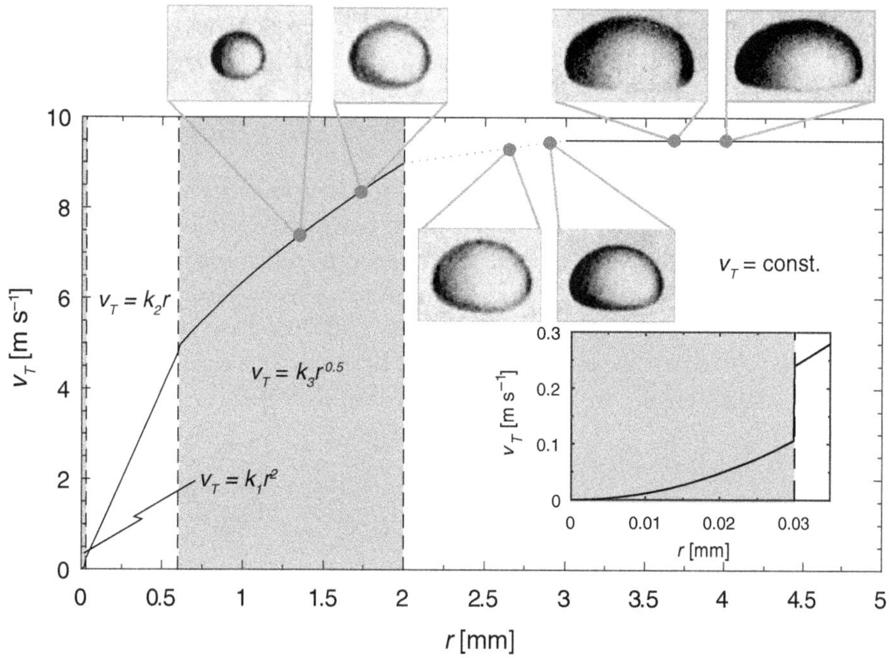

Fig. 7.10 Terminal velocities of cloud droplets, drizzle and raindrops up to 2 mm radius using eqs. (7.39), (7.43) and (7.40). The broken lines mark the separations between the different terminal velocity regimes. The v_T of drops between 2 and 3 mm is shown qualitatively to indicate the transition to a constant terminal velocity for raindrops larger than 3 mm. The inset photographs are of raindrops with equivalent radii of 1.35, 1.73, 2.65, 2.9 and 4 mm. The inset graph shows a zoom to the region $r < 0.035$ mm, where v_T is given by $v_T = k_1 r^2$. Photographs taken from Pruppacher and Beard (1970) with permission from Wiley.

We can express the mass growth rate of the large drop as the product of the rate of change of sweep out volume times the collection efficiency $\tilde{E}(R, r)$ and the (cloud) liquid water content M_l in kg m^{-3} of the smaller droplets:

$$\frac{dm_R}{dt} = \frac{dV_{sweep}}{dt}\tilde{E}(R,r)M_l = \underbrace{\pi(R+r)^2[v_T(R) - v_T(r)]\tilde{E}(R,r)}_{K(R,r)}M_l, \tag{7.45}$$

where $K(R, r)$ denotes the gravitational collection kernel (eq. 5.19), and the cloud liquid water content M_l is given as

$$M_l = \int_0^R \frac{4}{3}\pi\rho_l r^3 n_N(r)dr. \tag{7.46}$$

For the derivation so far we have assumed that the cloud consists of droplets of uniform size, continuously distributed in space. We can apply eq. (7.45) to a large droplet falling through a more realistic polydisperse cloud droplet population, where each droplet of radius r_i has its own liquid water content $M_{l,i}$ and collection efficiency $\tilde{E}(R, r_i)$. Treating the cloud droplet population as a continuum we can generalize eq. (7.45) by integrating over the size range where collection by the collector drop with radius R can take place. The mass growth rate of the collector drop can then be written as:

$$\frac{dm_R}{dt} = M_l \frac{dV_{sweep}}{dt} = \int_0^R \pi(R+r)^2[v_T(R) - v_T(r)]\tilde{E}(R,r)\frac{4}{3}\pi\rho_l r^3 n_N(r)dr. \quad (7.47)$$

From eq. (7.47) we see that the mass growth rate for growth by collection is proportional to r^3, whereas the increase in mass per unit time for diffusional growth is only proportional to r (eq. 7.9). This explains the fundamental difference between these two growth mechanisms (Figure 7.6) and why growth by collision–coalescence is needed to explain the formation of rain in warm clouds within 20–30 minutes of cloud formation.

In order to express the growth rate in terms of the change in radius of the large collector drop R in time, we replace dm_R with $4\pi R^2 \rho_l dR$ in eq. (7.47) and rearrange the equation:

$$\frac{dR}{dt} = \frac{\pi}{3}\int_0^R \left(\frac{R+r}{R}\right)^2 [v_T(R) - v_T(r)]n_N(r)r^3\tilde{E}(R,r)dr. \quad (7.48)$$

If the droplets are much smaller than the collector drop, $R + r$ can be approximated as R and their terminal velocity can be neglected, so that $v_T(R)$ can be moved in front of the integral:

$$\frac{dR}{dt} \approx \frac{\pi}{3} v_T(R) \int_0^R n_N(r)r^3\tilde{E}(R,r)dr. \quad (7.49)$$

In order to integrate this equation we assume an average collection efficiency for the droplet population, $\bar{\tilde{E}}$. This simplification is justified when the size distribution of the cloud droplets is narrow, and it leads to

$$\frac{dR}{dt} \approx \frac{\pi}{3}\bar{\tilde{E}} v_T(R) \int_0^R n_N(r)r^3 dr. \quad (7.50)$$

Making use of eq. (7.46) we obtain the continuous growth rate of the collector drop in terms of radius as

$$\frac{dR}{dt} \approx \frac{\bar{\tilde{E}} M_l v_T(R)}{4\rho_l}. \quad (7.51)$$

From eq. (7.48) we can deduce that the growth rate of the collector drop is extremely sensitive to the droplet number distribution $n_N(r)$, as follows. The broader the distribution, the faster the growth because the relative velocities between the droplets are larger. Growth by collision–coalescence is an accelerating process, since the terminal velocity of the collector drop $v_T(R)$ increases with increasing R and the growth rate dR/dt also increases with increasing R. This is in strong contrast with growth by condensation, where the increase in radius over time is proportional to $t^{1/2}$ (eq. 7.29 and Figure 7.6). Nevertheless, diffusional growth still takes place; it provides the small droplets that feed the growth of the large collector drop by collision–coalescence.

Equation (7.51) describes droplet growth as a continuous collection process. However, in reality growth by collision–coalescence is not purely continuous but occurs stepwise by random collisions; in a probabilistic sense these are very rare for small droplet sizes and increasingly likely for larger droplet sizes. Approximating droplet growth as a continuous collection process cannot capture the non-linear increase in collision efficiency for increasing r/R ratio, shown in Figure 7.8. Moreover the continuous collection model predicts time scales for rain formation, or more precisely the formation of precipitation sized drops, which are larger than those observed.

7.2.5 Growth model for stochastic collection

A better approach to the collection process is found by describing it stochastically with the stochastic coalescence equation (SCE). The SCE expresses the stochastic growth of cloud droplets in terms of the probability that a larger drop will collect another droplet of any smaller size. It also takes into account the probability for every possible combination of drops to coalesce and how these probabilities change after each coalescence event (Berry, 1967). In this stochastic approach, the collision efficiency is the probability that a drop of radius R collects a droplet of radius r located at a random position in its swept-out volume. The SCE can be written in terms of the temporal evolution of the droplet size distribution function $f(m_x, t)$, where m_x is the drop mass and t the time:

$$\frac{\partial f(m_x,t)}{\partial t} = \frac{1}{2}\int_0^{m_x} f(m_x - m_y, t) f(m_y, t) K(m_x - m_y, m_y) dm_y - \int_0^{\infty} f(m_x, t) f(m_y, t) K(m_x, m_y) dm_y. \tag{7.52}$$

Here $m_x - m_y$ and m_y refer to the pair of drops colliding in the collision–coalescence process with collection kernel $K(m_x - m_y, m_y)$ (eq. 5.19). Here, the droplet mass is used in the SCE because mass is conserved during the collision–coalescence process, but the hydrometer number concentration decreases. The first term corresponds to a source of drops arriving at mass m_x after collision–coalescence. The second term corresponds to a sink of drops of mass m_x due to collision–coalescence with drops of all other sizes. The factor $1/2$ in the first term avoids the double counting of collections. Solving the SCE requires a knowledge of the collection kernel.

An example of the evolution of a droplet mass distribution when collision–coalescence is the only growth mechanism is shown in Figure 7.11. Here the droplet growth is calculated using an approach that divides the hydrometeor mass distribution into three size classes (cloud droplets, drizzle drops and raindrops), describing each class with a modified gamma distribution (eq. 1.1). In Figure 7.11 the growth starts from a droplet population consisting purely of cloud droplets. The cloud droplet mass distribution was initiated with total cloud droplet number concentration $N_c = 75$ cm^{-3} and liquid water content $M_l = 0.75$ g m^{-3}, corresponding to a mean radius $\sim$ 13.4 μm. In this clean cloud drizzle is produced early on, as indicated in Figure 7.11 by the gray-shaded bimodal mass distribution at $t = 5$ minutes. With ongoing time, the initial narrow distribution becomes broader as cloud droplets collide and form larger hydrometeors. At the same time the number concentration of the cloud droplets decreases. Raindrops are formed after around 15 minutes. Although relatively few in number initially, these large drops will grow rapidly and dominate the mass distribution. Once rain is formed the cloud droplet number concentration decreases noticeably, starting at around 50 minutes. At the end of the simulation, after 70 minutes, hardly any cloud droplets remain and the drizzle has been entirely converted to

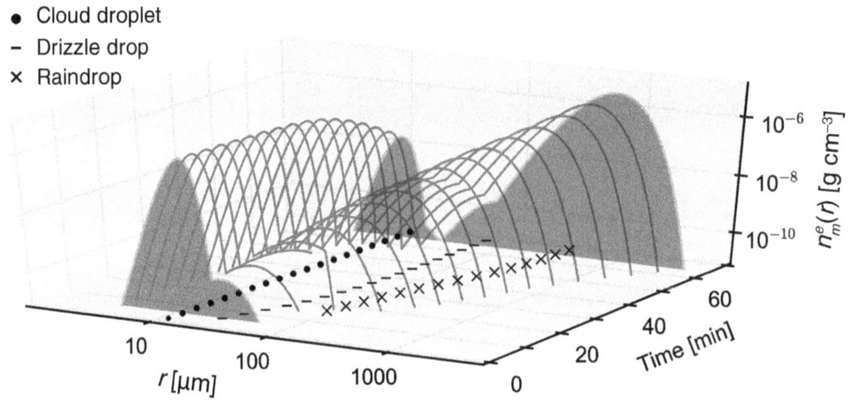

Fig. 7.11 Three-dimensional illustration of the temporal evolution of a droplet size distribution (cloud droplets, drizzle and rain) in an idealized simulation over 70 minutes in steps of 200 s. The cloud class was initialized with a total number concentration $N_c = 75\ \text{cm}^{-3}$ and a cloud water content $M_l = 0.75\ \text{g m}^{-3}$, indicated by the gray shaded mass distribution at $t = 0$. Figure adapted from Sant *et al.* (2013).

rain. Since this simulation comprises only the collision–coalescence process, the raindrops neither sediment nor evaporate.

7.3 Evaporation and break-up of raindrops

7.3.1 Evaporation of cloud droplets and raindrops

When the relative humidity drops below 100%, the droplet growth equation can be used to calculate the evaporation of cloud droplets and raindrops. Since raindrops are much larger than cloud droplets, the flow around them is turbulent. This causes the evaporation to be larger than that calculated from eq. (7.27). The turbulent enhancement of the evaporation rate for raindrops is obtained by multiplying the right-hand side of eq. (7.27) by the mean ventilation coefficient $\bar{f}_v$, defined as the ratio of the water mass flux to or from the drop for a moving drop dm/dt and that for a motionless drop $(dm/dt)_0$:

$$\bar{f}_v \equiv \frac{dm/dt}{(dm/dt)_0}. \tag{7.53}$$

Turbulent air motions enhance the transport of mass and heat to and away from a droplet. The coefficient $\bar{f}_v$ can be obtained from empirical expressions as a function of the Reynolds number Re and the Schmidt number Sc (eq. 5.17) which in turn has been converted into an equivalent drop radius r_{eq} in Figure 7.12. The figure shows that $\bar{f}_v$ is approximately 1 for drops with $r_{eq} < 60\ \mu\text{m}$, 9.5 for a raindrop with $r_{eq} = 1$ mm and 15 for a raindrop with $r_{eq} = 2$ mm. Thus, the evaporation of a 1 mm raindrop, falling at its terminal velocity in air, proceeds approximately 10 times faster than the evaporation of a drop of the same size at rest. Consequently, calculating the evaporation rate of raindrops on the basis of eq. (7.27) gives an underestimation.

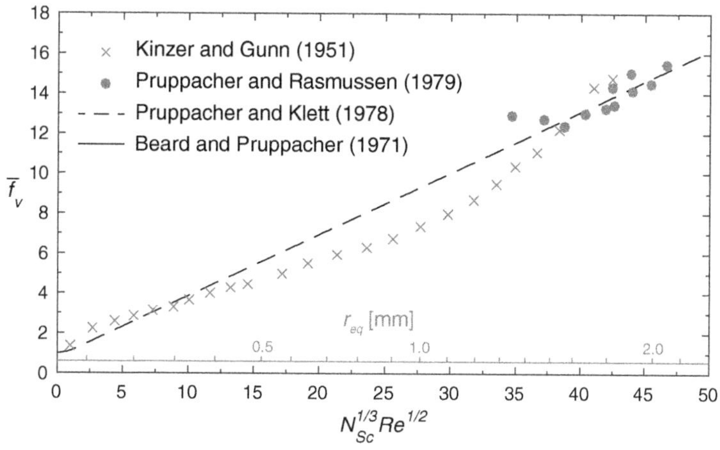

Fig. 7.12 Mean ventilation coefficient $\bar{f}_v$ as a function of the Reynolds number *Re* and the Schmidt number *Sc*, converted into an equivalent raindrop radius r_{eq}. The crosses and disks denote measurements and the solid and broken lines correspond to empirical expressions for $\bar{f}_v$. Figure adapted from Pruppacher and Rasmussen (1979) with permission of the American Meteorological Society.

The evaporation of raindrops depends on the relative humidity. Before rain can reach the surface, the relative humidity below a cloud needs to be high enough that the raindrops do not evaporate completely on their way to the surface. In fact, not all precipitation reaches the ground; some evaporates as it falls to the ground. The evaporation of raindrops can be observed in the form of virga. An example of a stratocumulus cloud with virga is shown in Figure 7.13.

7.3.2 Maximum raindrop size

Raindrops are limited in size to between 4.3 and 6.2 mm in equivalent radius, because the chance of disruption increases with size. Break-up can occur because the aerodynamically induced circulation that develops inside the drop is large enough to overcome the surface tension force. In most cases, however, raindrop break-up is induced by the collisions of a raindrop with smaller droplets.

The maximum raindrop radius r_{max} can be calculated by equating the drag force (eq. 7.35) and the surface tension force:

$$\frac{\pi}{2} r_{max}^2 v_T^2 \rho C_D = 2\pi r_{max} \sigma_w. \tag{7.54}$$

Solving for r_{max} yields

$$r_{max} = \frac{4\sigma_w}{C_D \rho v_T^2}. \tag{7.55}$$

Taking maximum terminal velocity $v_T = 9$ m s^{-1}, drag coefficient $C_D = 0.825$ (which lies between that for a sphere, 0.46, and that for a disk, 1.19, owing to the oblate shape of raindrops (Figure 7.10)), pressure = 850 hPa and temperature 298 K, so that the surface

Fig. 7.13 Stratocumulus with virga. Photograph taken by Luisa Ickes.

tension for the air–water interface $\sigma_w = 0.072$ N m^{-1}, and air density $\rho = 1$ kg m^{-3}, we arrive at $r_{max} = 4.3$ mm.

7.3.3 Energy transformation during collision–coalescence

In a collision where both the droplet and the collector drop (raindrop) are smaller than 100 μm, the coalescence efficiency $\hat{E}$ can be assumed to be unity. As shown in Figure 7.14b, $\hat{E}$ decreases with increasing raindrop size and is smallest for similar sized raindrops. If two drops collide and coalescence occurs, the total energy of coalescence E_{coal} in J, which the resulting drop obtains due to collision–coalescence, must be either dissipated or used to form new surfaces. It is given as

$$E_{coal} = \text{CKE} + \Delta S, \tag{7.56}$$

where CKE is the collision kinetic energy in J and ΔS is the change in surface energy. The CKE is the amount of energy that the newly formed drop obtains from the kinetic energy of the colliding drop pairs. Since a collision between droplets is not an elastic collision, in which kinetic energy would be conserved, some of the kinetic energy of the colliding droplets must be dissipated by circulation inside the raindrop. If the dissipation of energy is insufficient, break-up will occur.

The probability of whether break-up occurs after collision–coalescence is determined by the CKE, where CKE depends on the kinetic energies of the two drops before the collision. It is given as follows (Low and List, 1982):

$$\text{CKE} = \left(\frac{2}{3}\pi\rho_l\right)\frac{R^3}{1+\gamma^3}v_r^2, \tag{7.57}$$

where v_r is the relative velocity of impact and is given by the difference in velocity between the collector drop and the droplet with which it collides. The water density is ρ_l and $\gamma = R/r$ is the ratio of the radius of the larger raindrop R and the radius of the smaller droplet r.

The surface energy term describes the energy that the newly formed droplets obtain due to the reduction in the total surface (which releases energy). It is given by

$$\Delta S = 4\pi\sigma_w(r^2 + R^2) - S_C, \tag{7.58}$$

where S_C in J is the surface energy of the equivalent sphere for the united drop mass:

$$S_C = 4\pi\sigma_w(r^3 + R^3)^{2/3}. \tag{7.59}$$

The first term in eq. (7.58) represents the total surface energy of the incident drops and S_C is the excess energy from the decrease in the total surface area due to coalescence.

An empirical coalescence efficiency $\hat{E}$ based on the total energy of coalescence E_{coal}, was derived by Low and List (1982) from experimental data. It is based on the concept that E_{coal} must be either dissipated or used up by the formation of new surfaces.

$$\begin{aligned}\hat{E} &= a\left(\frac{\gamma}{\gamma+1}\right)^2 \exp\left(-\frac{b\sigma_w E_{coal}^2}{S_C}\right), \quad \text{for } E_{coal} < 5\mu\text{J}, \\ &= 0 \text{ otherwise,}\end{aligned} \tag{7.60}$$

where σ_w is the surface tension of the air–water interface, $a = 0.778$ and $b = 2.61 \times 10^6$ J^{-2} m^2. The CKE and the empirical coalescence efficiency are shown for four different values of γ in Figure 7.14. As can be seen, CKE increases rapidly with raindrop size owing to the increase in the colliding masses. At the same time $\hat{E}$ decreases for sufficiently high values of the CKE, thus increasing the chance of break-up. As shown in Figure 7.14b, $\hat{E}$ drops to

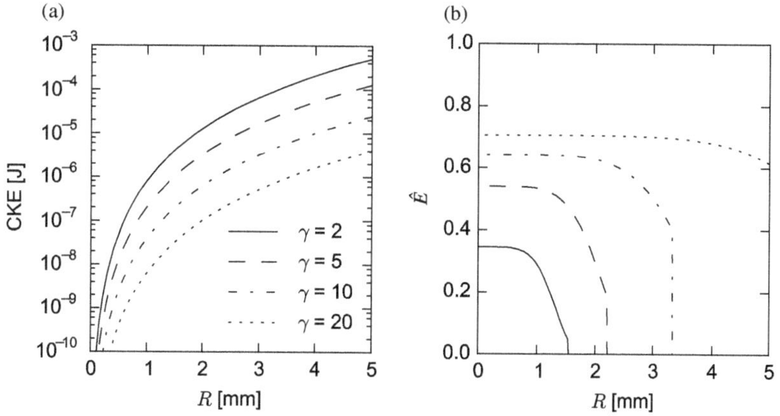

Fig. 7.14 (a) Collision kinetic energy (CKE) as a function of the larger raindrop radius R for four different values of the radius ratios of colliding hydrometeors, $\gamma = R/r$, using eq. (7.57). (b) Empirical coalescence efficiency $\hat{E}$ as a function of R for four different values of γ, using eq. (7.60). When E_{coal} exceeds 5 μJ and hence $\hat{E} = 0$ (eq. 7.60), the curves turn into vertical lines.

zero ensuring break-up when E_{coal} exceeds 5 μJ. This occurs for instance when the radius of the large raindrop is 1.5 mm and that of the smaller raindrop is 0.75 mm or when the larger raindrop is 5 mm in radius (the upper limit of the raindrop size) and the colliding drop is larger than 0.5 mm.

In summary, for cloud droplets the collision efficiency E is found to be ≤ 1 due to hydrodynamic interactions of the surrounding air with the different drops, but the coalescence efficiency $\hat{E}$ can be approximated as unity. Therefore the collection efficiency $\tilde{E}$ is approximated by E (eq. 7.33). For raindrops, the opposite is true: E can be assumed to be unity but $\hat{E} < 1$ due to break-up, so that $\tilde{E} \approx \hat{E}$.

7.3.4 Types of raindrop break-up

Collisions that do not lead to permanent coalescence were found to produce 1–10 small so-called satellite drops with radii 20–220 μm (Brazier-Smith *et al.*, 1973). Low and List (1982) found three different types of raindrop break-up: neck or filament break-up, sheet break-up and disk break-up, as shown in Figure 7.15.

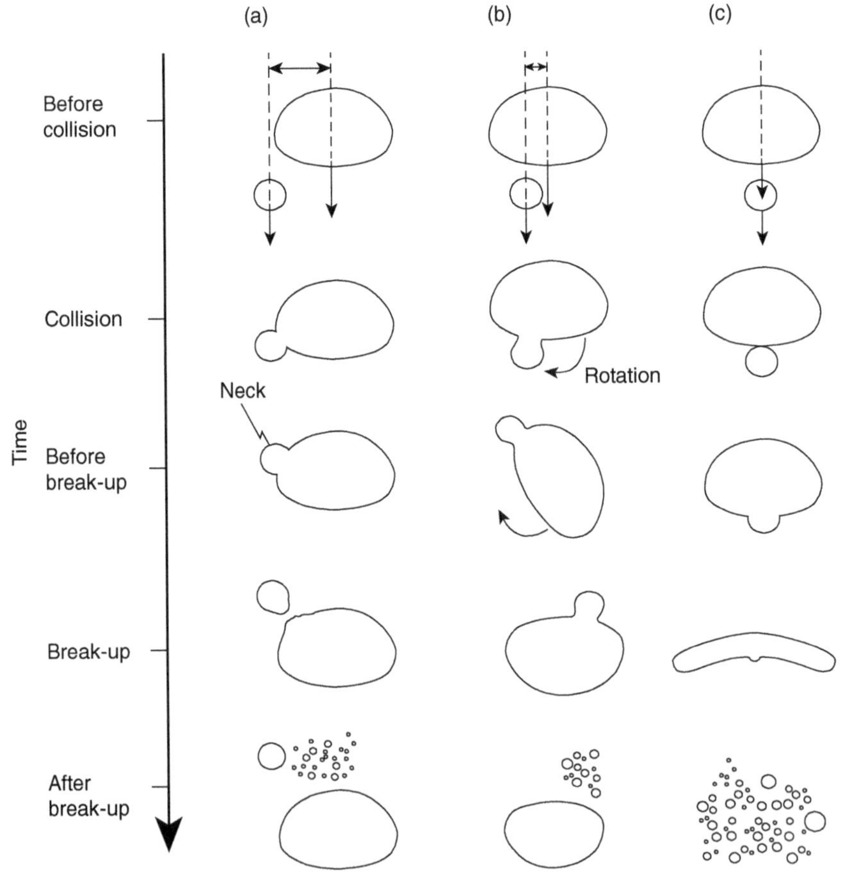

Fig. 7.15 Schematic illustration of the three common types of raindrop break-up: (a) neck or filament break-up, (b) sheet break-up and (c) disk break-up.

Neck or filament break-up is caused by grazing collisions, i.e. the point of impact is at the edge of the larger drop. It produces a spinning, elongated drop which may quickly fly apart. Satellite droplets are created by disintegration of the connecting neck. The identities of the two colliding drops are preserved (Figure 7.15a).

Sheet break-up occurs when drops collide in such a way that one side of the large drop is torn off, with the point of impact between the edge and the center of the large drop. The bulk of the large drop then rotates about the point of impact, issuing a sheet of water that breaks apart into satellite drops. The identity of only the large drop is preserved (Figure 7.15b).

For disk break-up the point of impact is close to the center of the bigger drop. Coalescence occurs temporarily, but the disk that forms upon impact disintegrates into a large number of medium-sized drops. The original identities of the drops are lost (Figure 7.15c).

Filament or neck break-up needs the least amount of CKE and disk break-up the most, because the disk break-up produces the largest number of small drops. This requires the formation of a large surface area and hence needs the most energy. To summarize, break-up is important in limiting the growth of the largest drops and is the reason why raindrops rarely exceed 5 mm in radius. The consequences of raindrop break-up for the raindrop size distribution will be discussed in Chapter 9.

7.4 Exercises

Single/multiple choice exercises

1. The sizes and concentrations of CCN, cloud droplets and raindrops differ. Which of the following is/are true?
   (a) The CCN size is always smaller than 1 μm.
   (b) The CCN concentration depends on the temperature and saturation ratio prevailing in the ambient air.
   (c) The concentration and volume of raindrops are in general a factor of one million larger than those of cloud droplets.
   (d) Higher supersaturation implies a larger cloud droplet concentration and thus increased droplet size.
2. Cloud droplets grow initially by diffusional growth. Which of the following is/are true?
   (a) The droplet growth equation takes the diffusion of water vapor and heat into account.
   (b) Raindrops grow in size by diffusional growth more efficiently than droplets because of the larger particle surface.
   (c) In warm-phase clouds, growth by diffusion is the most efficient process owing to the high temperatures.
   (d) Cloud droplet growth depends on both the environmental temperature and the relative humidity.

3. Assume two droplets, with $r = 0.5$ μm and $R = 5$ μm respectively, which are exposed to supersaturated conditions and grow by water vapor diffusion. Which of the following is/are true?
   (a) The difference in size between the droplets, $R - r$, gets smaller with increasing condensational growth time.
   (b) The ambient supersaturation remains constant in an updraft.
   (c) The saturation vapor pressure over the droplet surface is larger for r than for R.
   (d) It is likely that r and R collide and stick together, i.e. grow by collision–coalescence.
4. For cloud droplets larger than 15 μm, collision–coalescence is the dominant precipitation-formation process. Which of the following is/are true?
   (a) Sedimentation and a broad cloud droplet size distribution are important for growth by collision–coalescence.
   (b) The presence of turbulence is favorable for initiating collision–coalescence.
   (c) The addition of giant CCN is especially important in clouds with low cloud droplet concentrations.
   (d) The collision of two cloud droplets usually implies coalescence.
5. Consider a collected drop r and a collector drop R which may undergo collision–coalescence. Which of the following is/are true?
   (a) The collision efficiency generally increases for larger r/R ratios owing to the increasing relative velocity between them.
   (b) The coalescence efficiency depends on the inertia of the collected drops.
   (c) The collision efficiency can exceed unity owing to wake capture.
   (d) The collision efficiency increases for larger R with constant r/R.
6. Consider two droplets of 40 μm and 70 μm in diameter. What is the absolute difference in terminal fall velocity between the droplets?
   (a) 9 cm s^{-1}.
   (b) 24 cm s^{-1}.
   (c) 28 cm s^{-1}.
   (d) 37 cm s^{-1}.
7. Raindrops are limited in size. Which of the following is/are true?
   (a) The maximum raindrop size depends on temperature, pressure and the terminal velocity of the drop.
   (b) The raindrop radius is limited to 1 cm.
   (c) Raindrop break-up requires a collision with smaller droplets.
8. In comparison with a drop of 3.7 mm equivalent radius, the terminal fall velocity of a drop with 2.5 mm equivalent radius is ...
   (a) ... smaller than that of the larger drop because it experiences a smaller gravitational force.
   (b) ... equal to that of the larger drop.
   (c) ... larger than that of the larger drop because it experiences less aerodynamic resistance from the surrounding air.

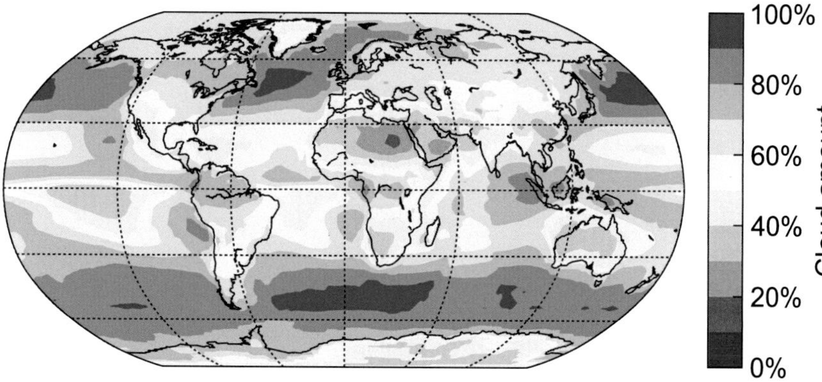

Fig. 1.1 Annual mean total cloud cover [%] averaged over 1983–2009. Data from the ISCCP web site in December 2014.

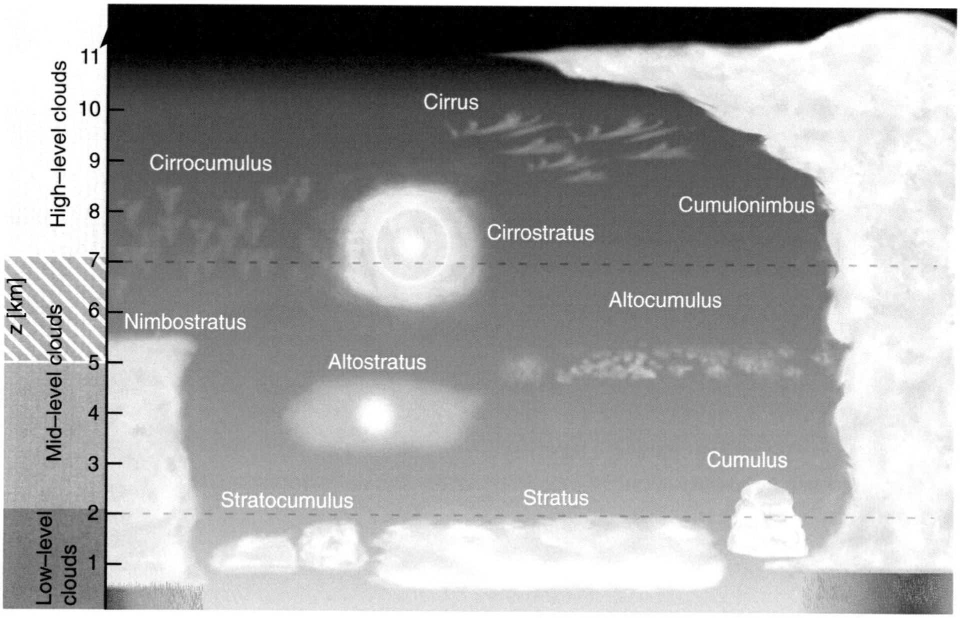

Fig. 1.3 Schematic figure of the ten cloud types according to WMO and as described in Table 1.1.

Fig. 1.4 (a) Cumulus humilis, Cu hum, (b) cumulonimbus, Cb, (c) nimbostratus, Ns, (d) stratus, St, (e) stratocumulus, Sc, as viewed from above, and (f) satellite image of ship tracks off Europe's Atlantic coast (MODIS Land Rapid Response Team). Photographs taken by Fabian Mahrt (a), (b), (d), (e) and Larissa Lacher (c).

Fig. 1.6 (a) Altostratus, As, (b) altocumulus, Ac, (c) cirrus uncinus, Ci, (d) cirrostratus, Cs, with a 22° halo, and (e) cirrocumulus, Cc. Photographs taken by Robert David (a), Fabian Mahrt (b), (e) and Blaz Gasparini (c), (d).

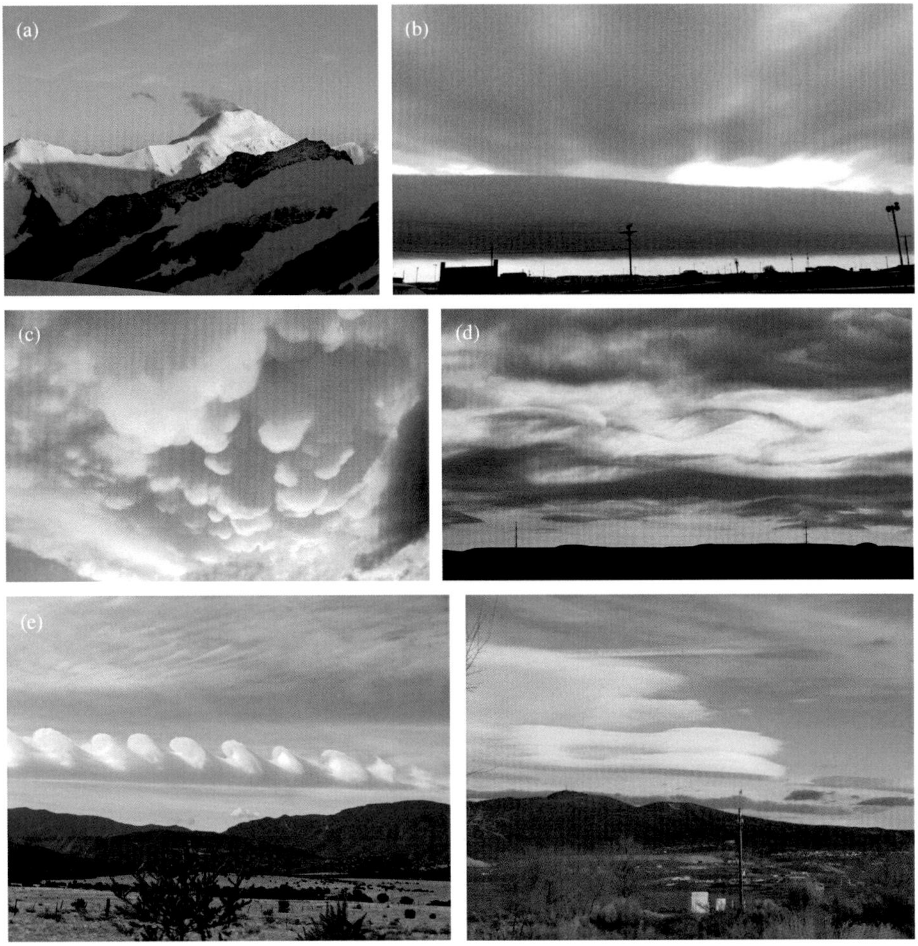

Fig. 1.8 (a) Banner cloud, (b) roll cloud (the dark shape above a light horizon), (c) mammatus cloud, (d) asperatus cloud and (e) Kelvin–Helmholtz wave cloud. Photographs taken by Larissa Lacher (a), Sandra LaCorte (b), Robert David (c), Christian Grams (d) and Laurie Krall (e). Bottom right: Altocumulus (Ac) lenticularis cloud. Photograph taken by Jeremy Michael.

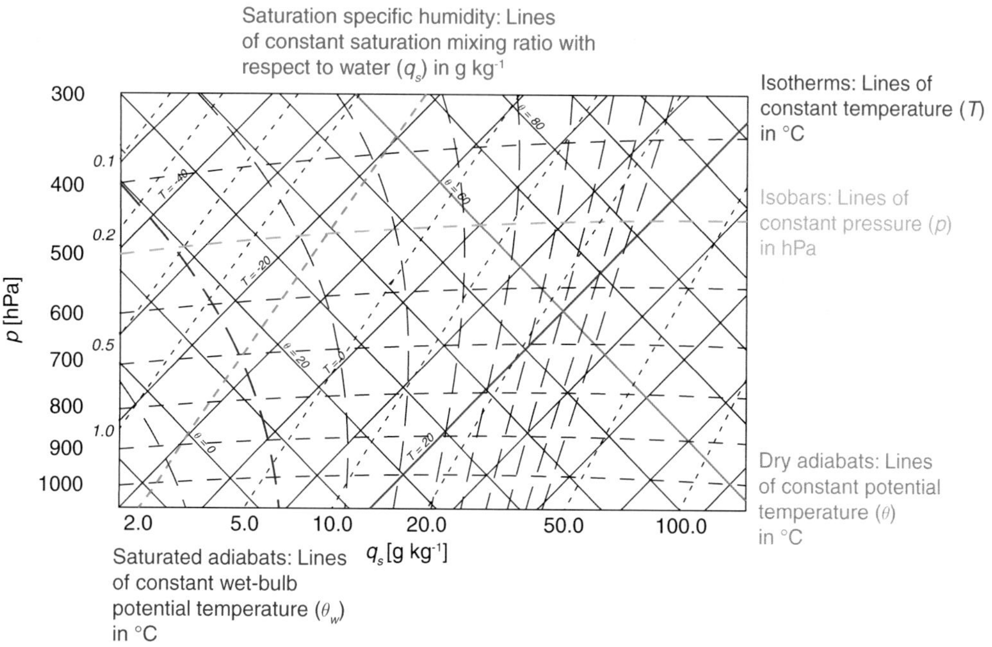

Fig. 2.7 Tephigram in which examples of the various types of lines are highlighted; p is the atmospheric pressure and q_s is the saturation specific humidity (Section 2.5.3).

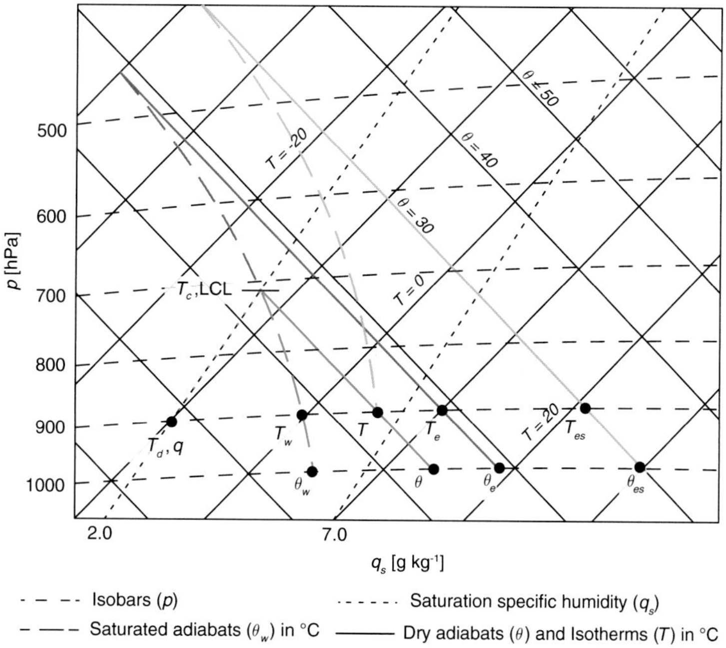

Fig. 2.15 This tephigram depicts the following temperatures for an air parcel at 900 hPa with $T = 5\,°\mathrm{C}$ and $q_v = 2\ \mathrm{g\ kg}^{-1}$, and the pathways by which they can be obtained: T_d is the dew point temperature, T_w is the wet-bulb temperature, T_c is the isentropic condensation temperature, i.e. the temperature at the lifting condensation level (LCL), T_e is the equivalent temperature and T_{es} the saturated equivalent temperature, θ the potential temperature (Section 2.5.8), θ_w the wet-bulb potential temperature, θ_e the equivalent potential temperature and θ_{es} the saturated equivalent potential temperature.

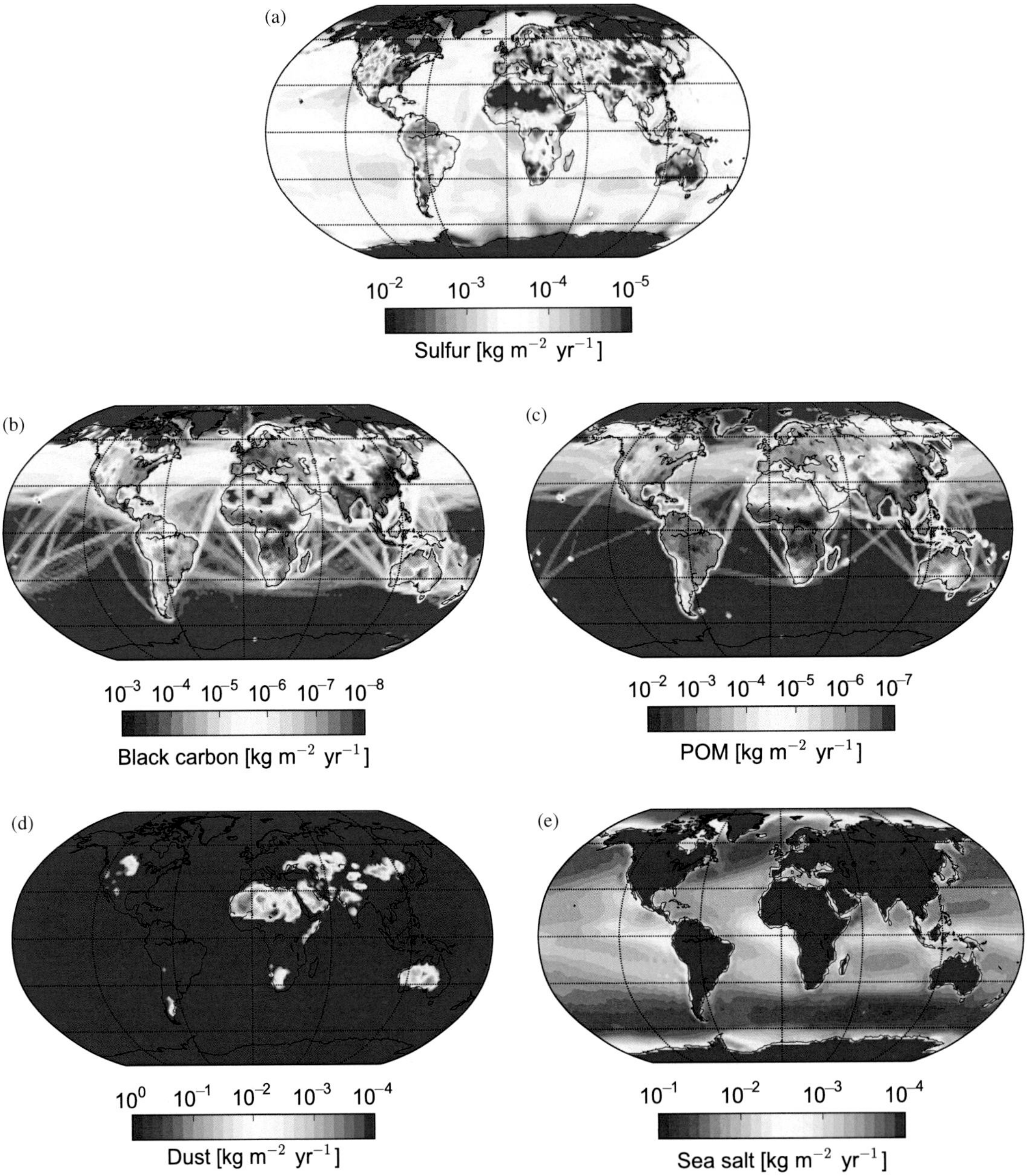

Fig. 5.9 Annual mean emissions of (a) sulfur, (b) black carbon particles, (c) particulate organic matter (POM), (d) mineral dust particles and (e) sea salt particles obtained from the ECHAM6-HAM global climate model (Neubauer *et al.*, 2014). Note that the scale is different for each aerosol type.

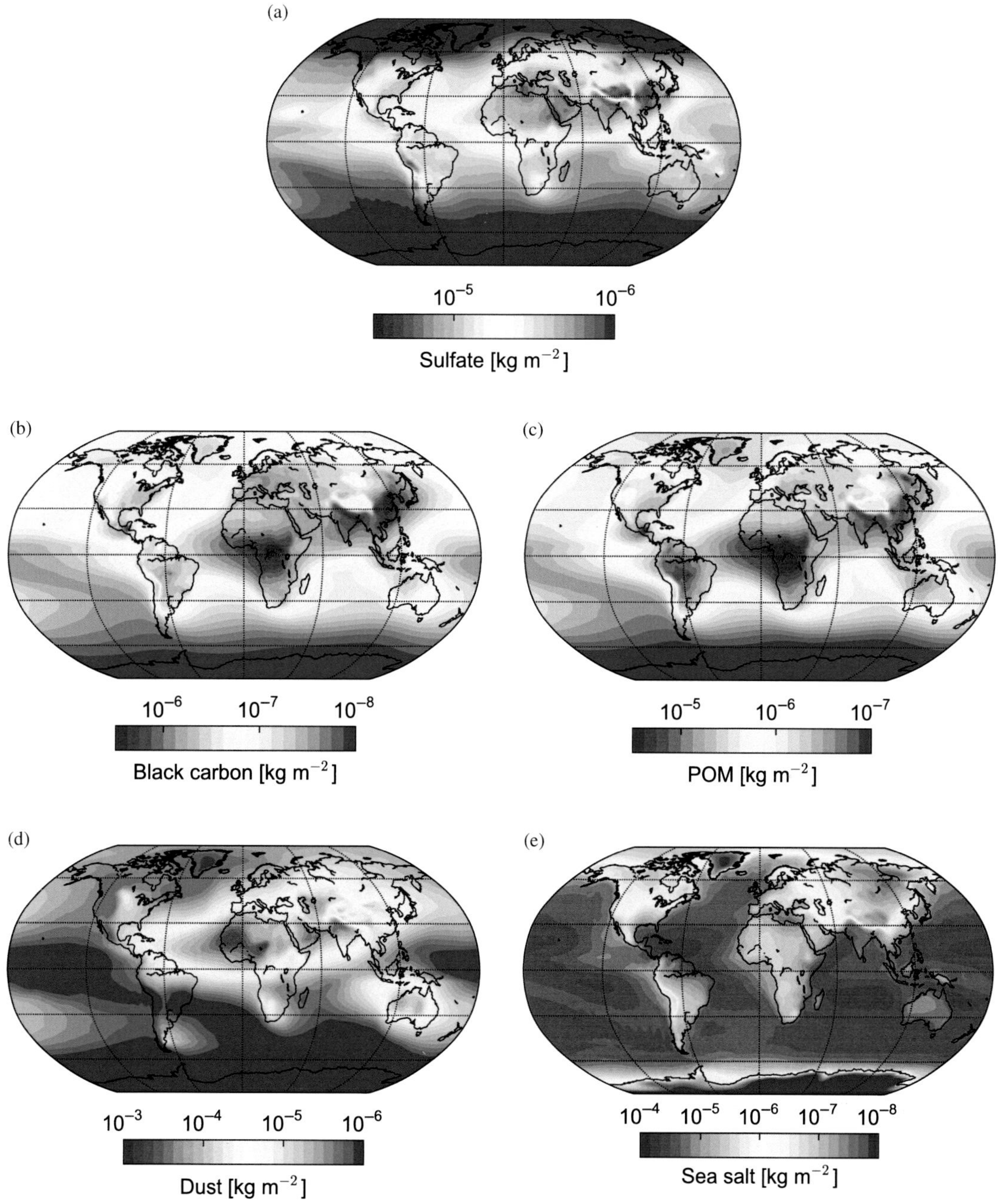

Fig. 5.16 Annual mean vertically integrated aerosol burden of (a) sulfate particles, (b) black carbon particles, (c) particulate organic matter (POM), (d) mineral dust particles and (e) sea salt particles from the ECHAM6-HAM global climate model (Neubauer *et al.*, 2014). Note that the scale is different for each aerosol component.

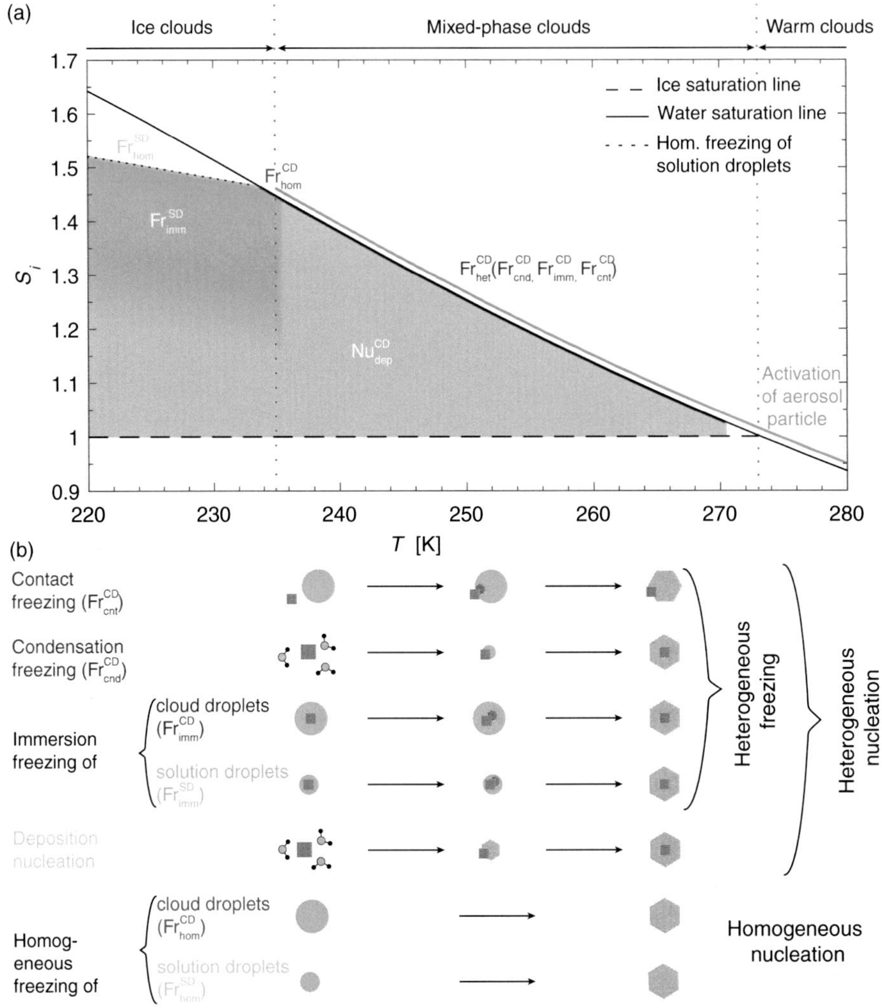

Fig. 8.4 (a) Schematic of the main freezing pathways as a function of temperature T and saturation ratio S_i with respect to ice [%], along with the cloud types in which these processes occur. The sloping dotted line that starts at 235 K denotes the minimum saturation ratio for solution droplets to freeze homogeneously, according to Koop *et al.* (2000). Colored areas denote the regions where the individual nucleation types occur. The superscripts CD and SD refer to cloud droplets and solution droplets respectively. (b) Schematic of the mechanisms associated with the individual processes. The cubes represent INPs, the spheres droplets, the hexagons ice crystals.

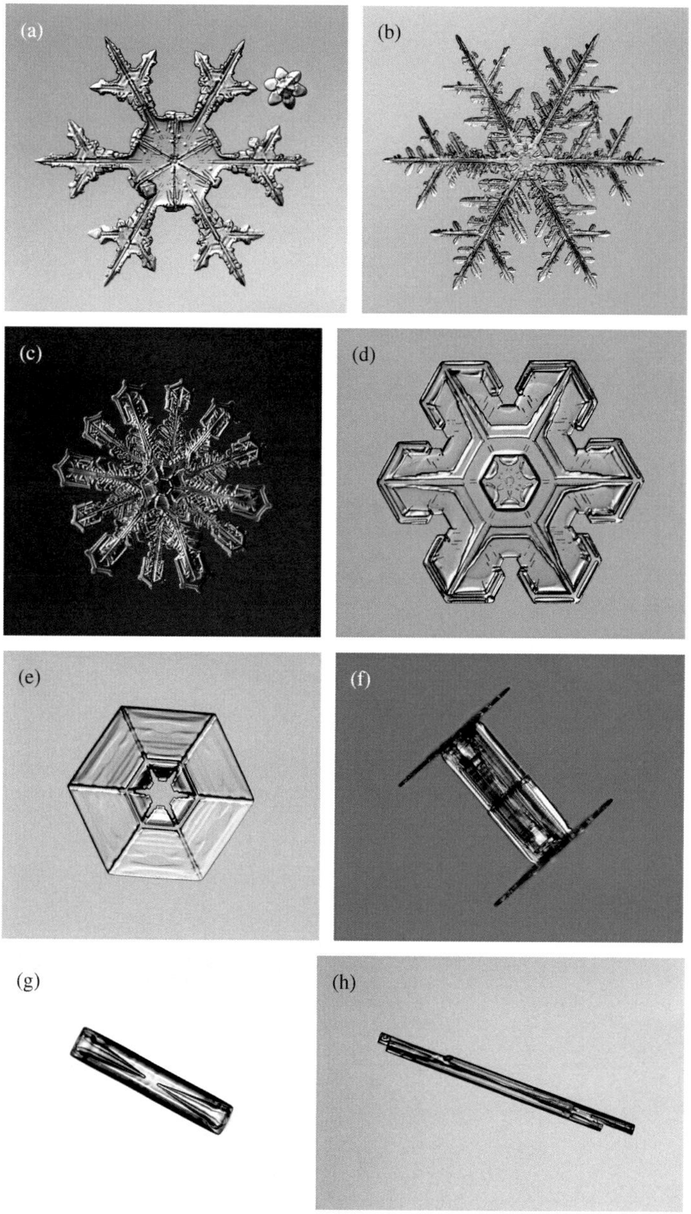

Fig. 8.14 Photographs showing (a) a stellar dendrite, (b) a fern-like stellar dendrite, (c) a 12-sided snowflake, (d) a stellar plate, (e) a plate, (f) a capped column, (g) a hollow column and (h) a needle. Photographs taken from www.snowcrystals.com and used by permission and courtesy of Kenneth Libbrecht.

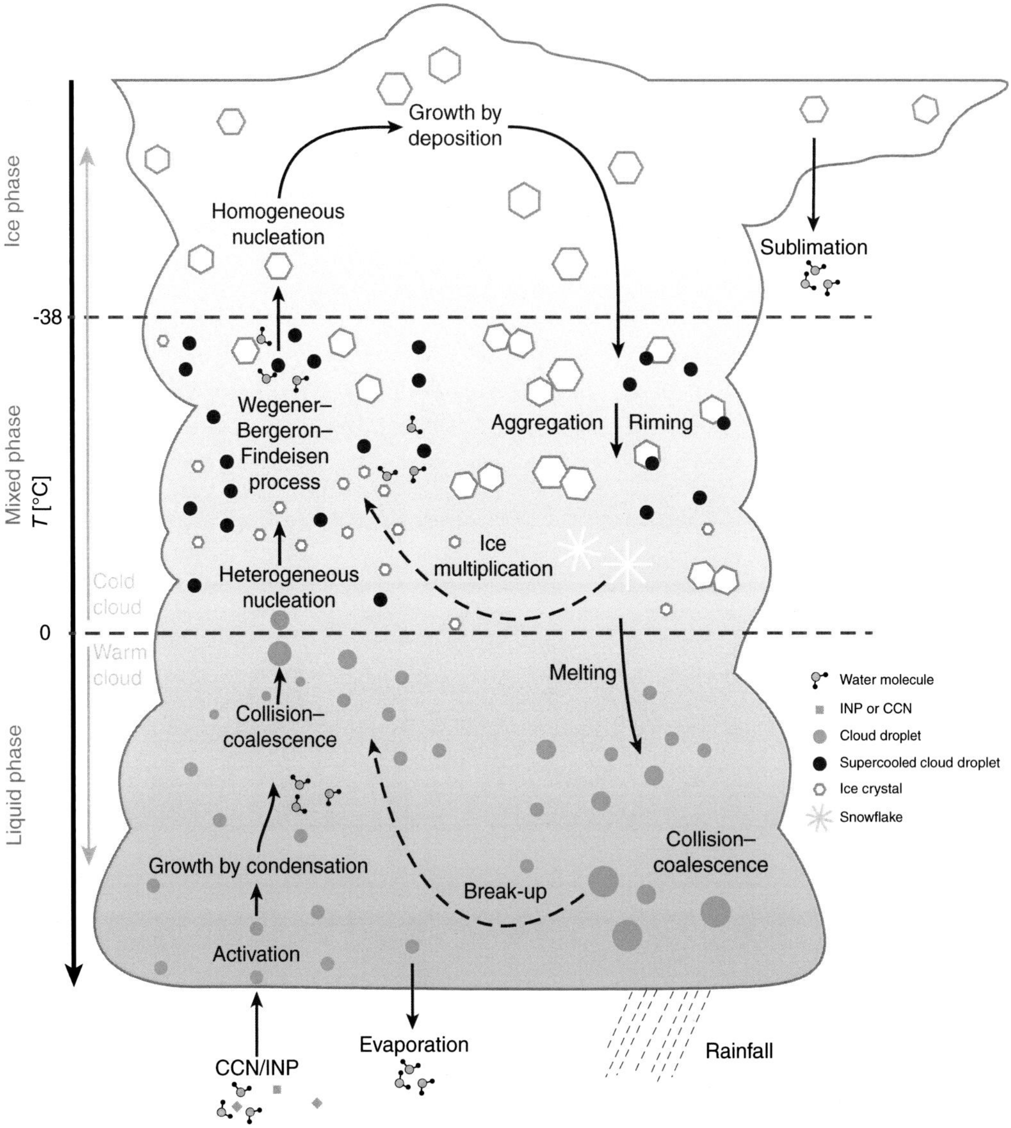

Fig. 8.19 Microphysical processes occurring in a convective cloud with a cloud base temperature higher than 0 °C and a cloud top temperature lower than −38 °C, so that warm-, mixed- and ice-phase processes take place. The small arrows show typical trajectories of a cloud particle.

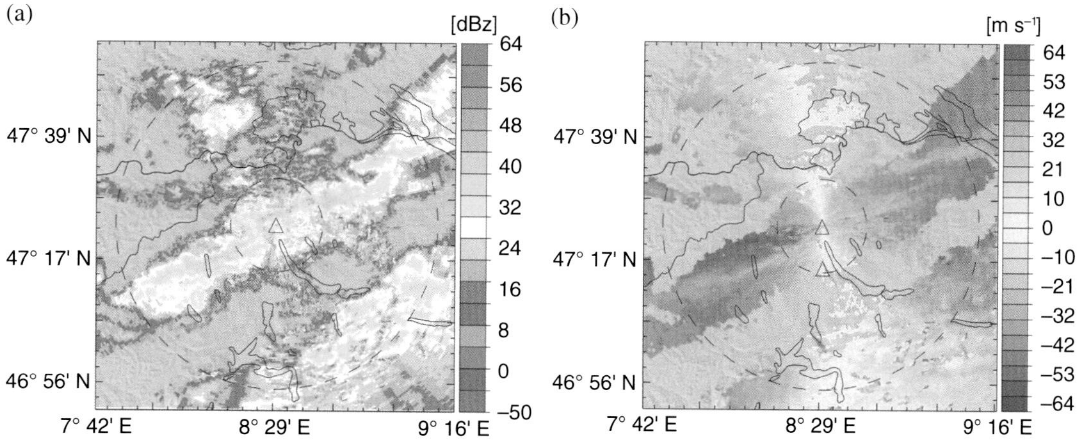

Fig. 9.7 Plan position indicator images of (a) radar reflectivity and (b) Doppler velocity, scanned by the ETH radar in Zurich (red triangle in the center) on December 26, 1999. The radar antenna was installed at 600 m above sea level and the scan was performed with a 1.5° elevation angle (Figure 9.4). The broken circles mark PPI heights of 1 km (smaller circle) and 2 km above sea level (larger circle). The those directed brown colors indicate velocities directed away from the radar and bluish colors those directed towards the radar. Figure adapted from Wüest (2001).

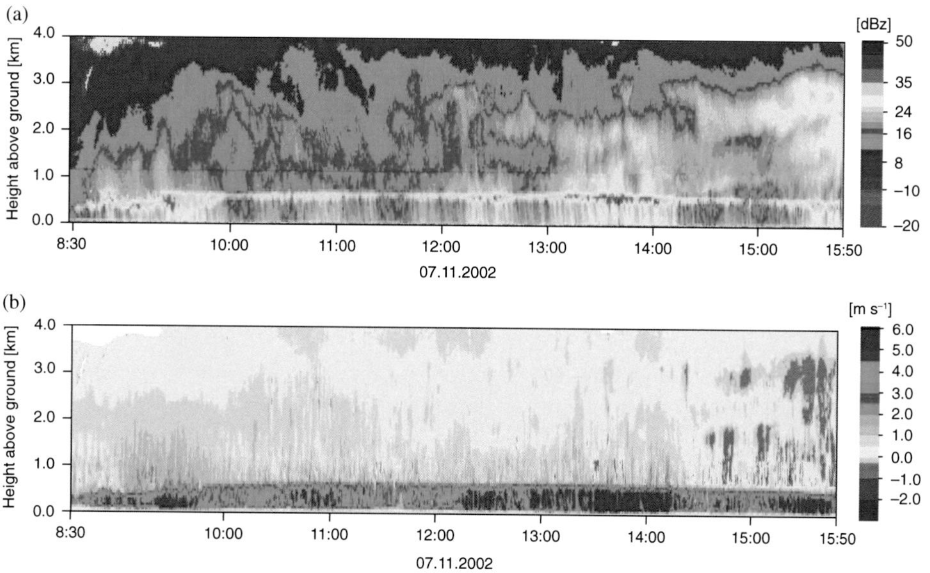

Fig. 9.9 Example of an HTI scan, produced by a vertically pointing radar of a stratiform precipitation event on November 7, 2002. (a) Radar reflectivities and (b) net vertical velocities, where negative values indicate net upward motion. Figure adapted from Baschek (2005).

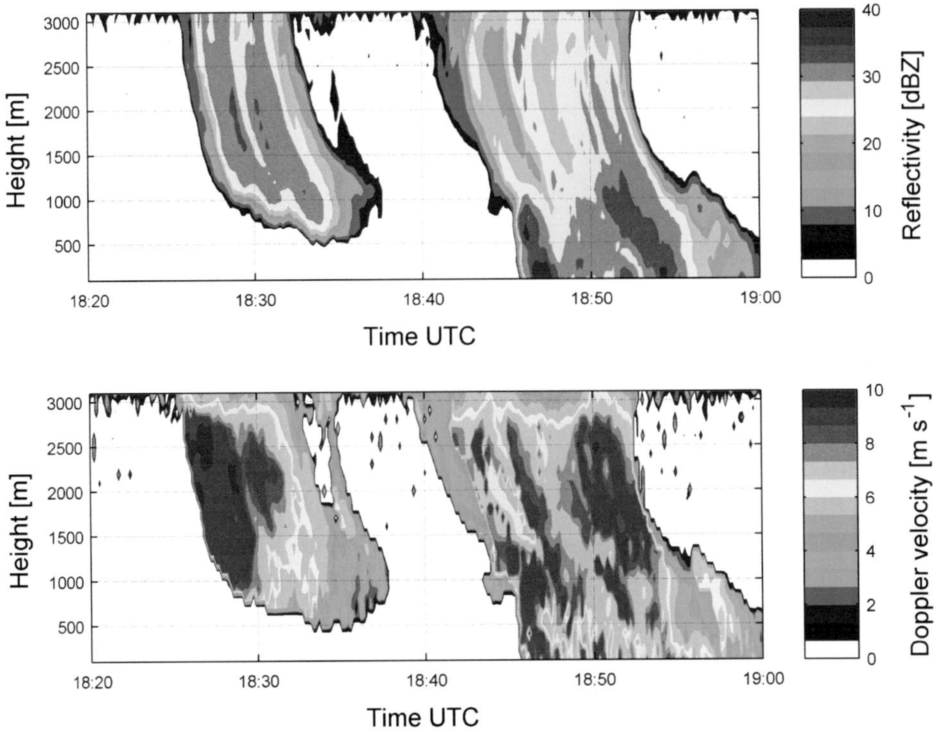

Fig. 9.11 Height–time indicator radar image of convective cloud showers as observed in Zurich, Switzerland on August 13, 2015: (a) reflectivity and (b) Doppler velocity. Data by courtesy of Pascal Graf.

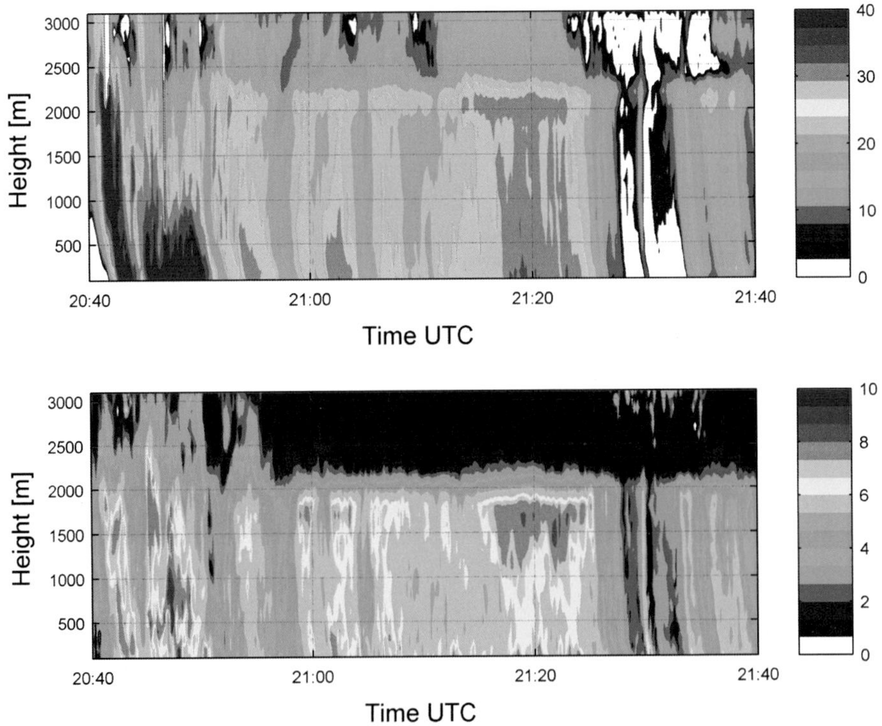

Fig. 9.20 Height–time Indicator (HTI) radar images as observed in Zurich, Switzerland on June 21, 2015. Data by courtesy of Pascal Graf.

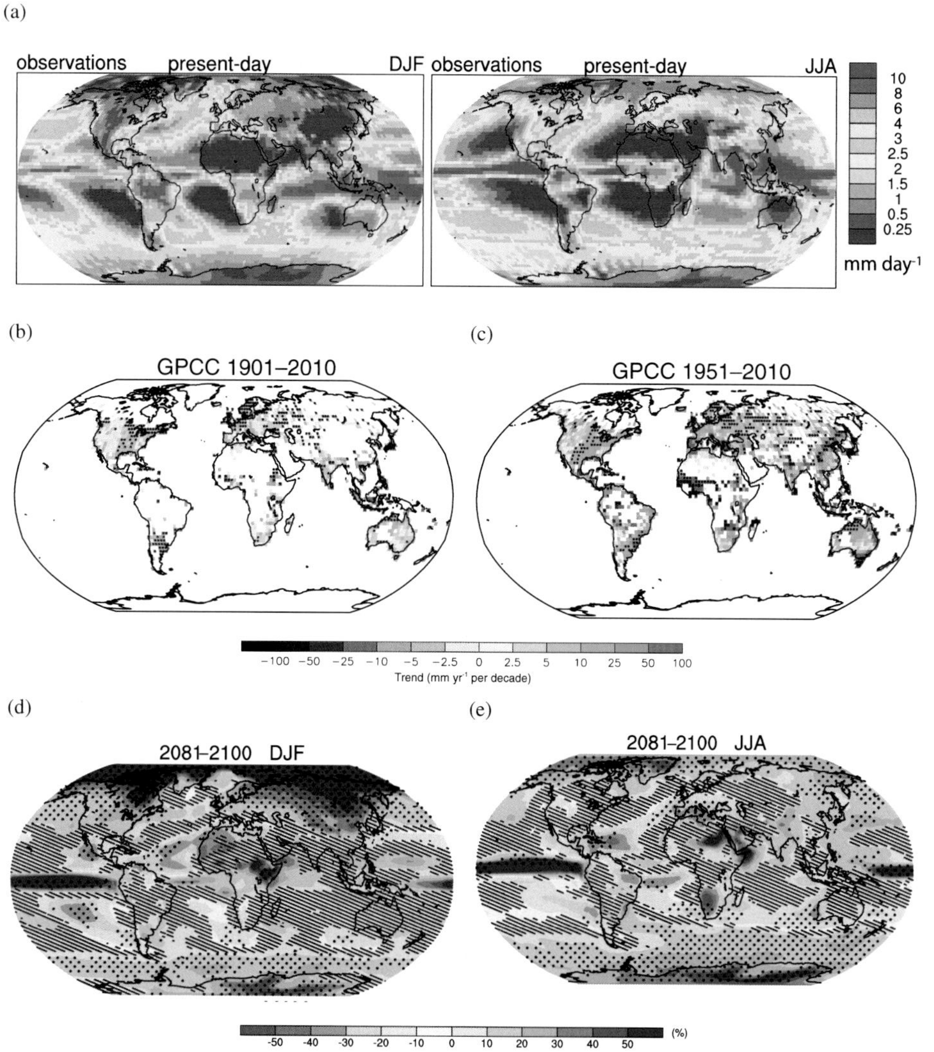

Fig. 9.19 (a) Seasonal mean precipitation rates over the period 1979–1993 in DJF and in JJA. Trends in annual precipitation over land (b) 1901–2010 and (c) 1951–2010. The trends were calculated for grid boxes with $>$ 70% complete records and for $>$ 20% data availability in the first and last deciles of the period. The white areas in (b) and (c) indicate incomplete or missing data and the black plus signs indicate grid boxes where trends are significant. (d), (e) projected percentage changes of seasonal mean precipitation for the period 2081–2100 relative to the reference period 1986–2005 following the RCP8.5 forcing scenario for DJF and JJA. The hatching indicates regions where the multimodel mean change is less than one standard deviation from the natural internal variability in 20-year means. the stippling indicates regions where the multimodel mean change is greater than two standard deviations from the natural internal variability in 20-year means and where $>$ 90% of the models agree on the sign of change. Data sources: (a), Figure TS.30 of IPCC AR4 (Solomon *et al.*, 2007); (b), (c), Figure 2.22 of IPCC AR5 (Hartmann *et al.*, 2013); (d), (e), Figure 12.22 of IPCC AR5 (Collins *et al.*, 2013).

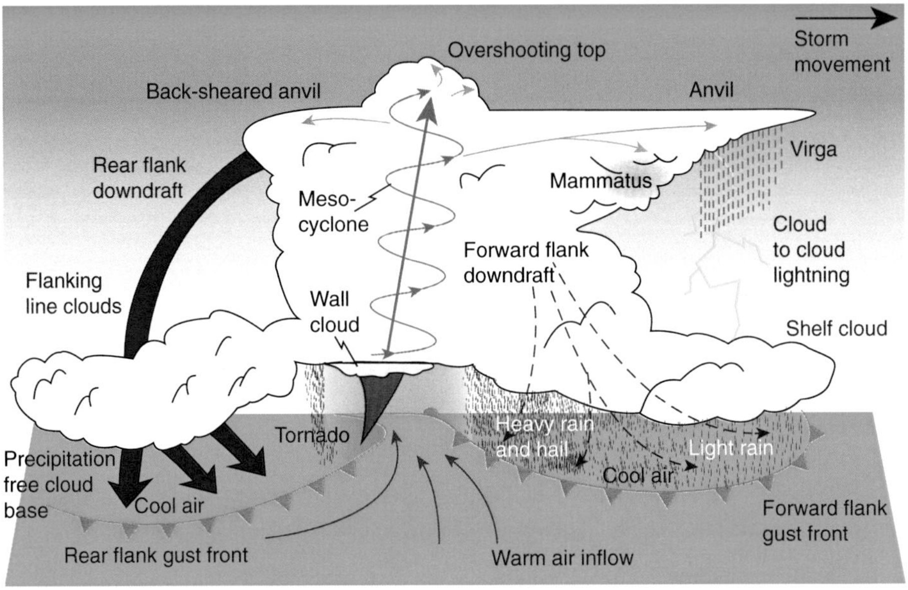

Fig. 10.13 Schematic of the main features, including the updraft and downdrafts and surface air flow, associated with a tornado-breeding supercell thunderstorm.

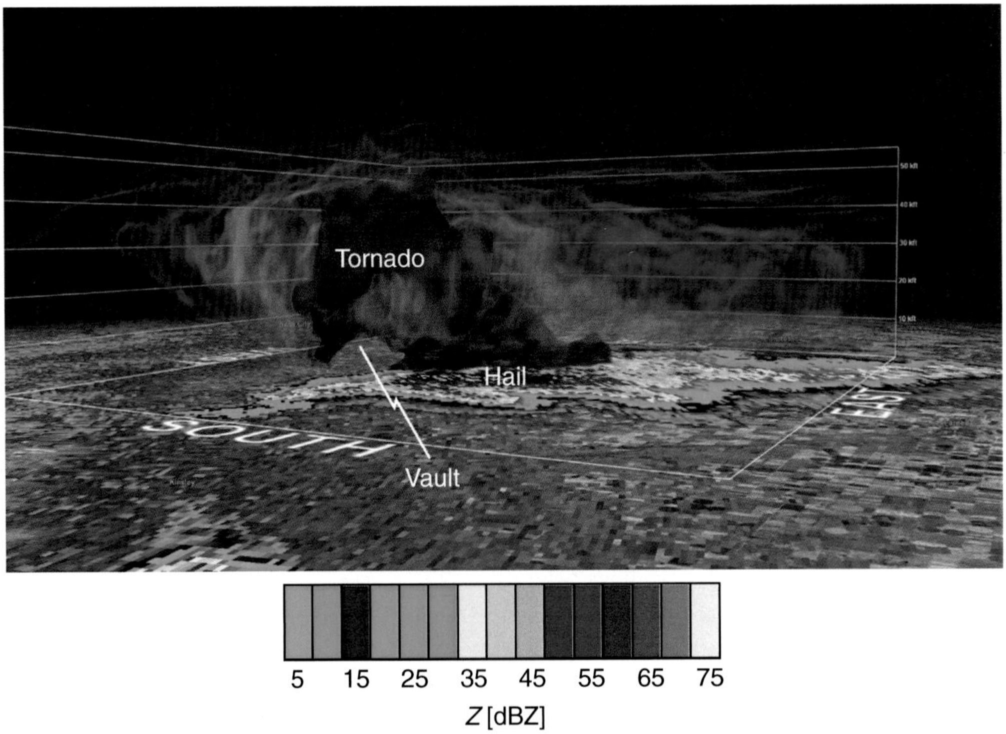

Fig. 10.17 3D radar image of a tornadic supercell over Kansas on May 18, 2013 at 7:15 pm local time. The radar data were obtained from the National Climate Data Center in the United States. The picture was produced with GR2Analyst and kindly provided by Thomas Winesett.

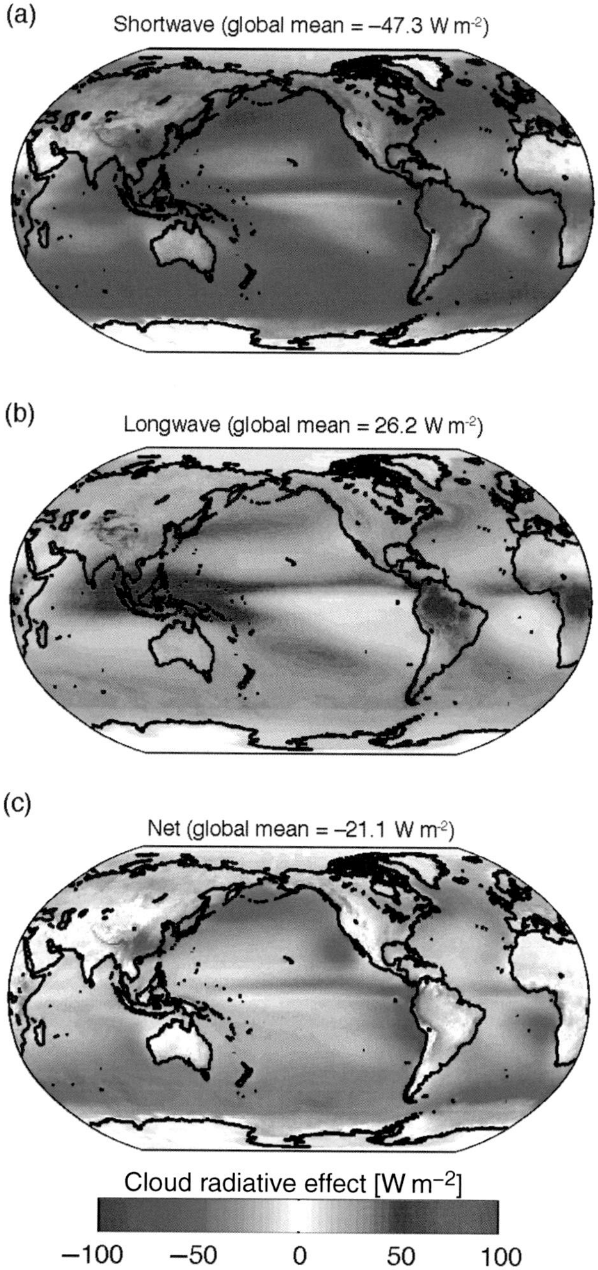

Fig. 11.6 Annual mean (a) shortwave (SCRE), (b) longwave (LCRE) and (c) net cloud radiative effect (CRE) averaged over the years 2001 to 2011, as obtained from the CERES satellite data. Data source: Figure 7.7 from IPCC AR5 (Boucher *et al.*, 2013).

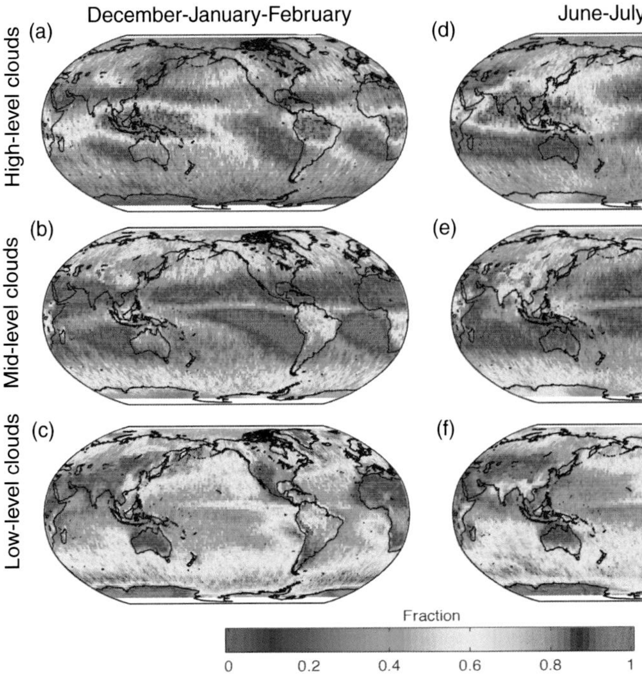

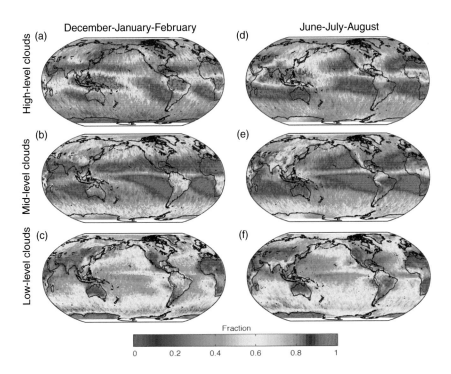

Fig. 12.6 (a)–(c) DJF and (d)–(f) JJA averaged high-level, mid-level and low-level cloud coverage based on data from CloudSat and CALIPSO for the period 2006–2011 (Mace *et al.*, 2009).

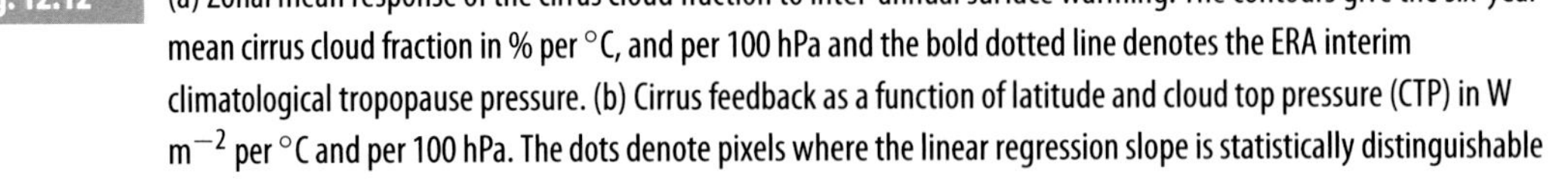

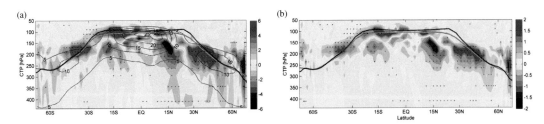

Fig. 12.12 (a) Zonal mean response of the cirrus cloud fraction to inter-annual surface warming. The contours give the six-year mean cirrus cloud fraction in % per °C, and per 100 hPa and the bold dotted line denotes the ERA interim climatological tropopause pressure. (b) Cirrus feedback as a function of latitude and cloud top pressure (CTP) in W m^{-2} per °C and per 100 hPa. The dots denote pixels where the linear regression slope is statistically distinguishable from zero.

9. Using the droplet growth equation, the evaporation of cloud droplets and raindrops can also be calculated. The ventilation coefficient is introduced to account for turbulent flow. Which of the following is/are true?
   (a) The turbulent flow around raindrops causes them to evaporate faster than calculated with the droplet growth equation.
   (b) The turbulent flow around cloud droplets is weaker that that around raindrops. Therefore, the evaporation of cloud droplets proceeds more slowly than as calculated with the droplet growth equation.
   (c) The flow around small cloud droplets is not turbulent and therefore the ventilation coefficient is zero.
   (d) The ventilation coefficient is larger for raindrops with increasing equivalent droplet radius.
10. Which of the following is/are true regarding the growth of a droplet population?
    (a) Aerosol particles that experience steady growth until peak supersaturation will inevitably become activated to cloud droplets.
    (b) Activated particles can be defined as those aerosol particles that still grow when the relative humidity decreases below 100%.
    (c) With a critical supersaturation lower than the peak supersaturation, aerosol particles become activated to cloud droplets.
    (d) Which particles become activated depends on the vertical velocity in the atmosphere and on the properties of the aerosol population.
11. During the early stages of cloud formation, the peak supersaturation that can be reached depends on the updraft velocity, which consequently determines some of the properties of the cloud that is forming. Which of the following is/are true?
    (a) The supersaturation within the cloud rises sharply in the lower part of the cloud and reduces gradually in the upper parts after peak supersaturation is surpassed.
    (b) To keep the supersaturation constant in the upper part of the cloud, the rate of condensation has to equal the production rate of supersaturation.
    (c) Higher updraft velocities cause more supersaturation and higher liquid water content.
    (d) The relative humidity at cloud base determines the number of activated particles.
12. The size of raindrops is limited owing to collisions with smaller droplets and aerodynamically induced circulation within the drop. Which of the following is/are true?
    (a) The maximum radius of raindrops is limited to around 4 mm.
    (b) If the collision of two drops occurs at the edge of the larger drop, both drops will lose their original identity.
    (c) The maximum radius of a raindrop can be determined by equating the drag force and the surface tension.
13. Collision–coalescence is one mechanism by which hydrometeors can grow. The droplets collide and stay together permanently. Which of the following is/are true?

(a) The collection efficiency approaches unity for small values of the collision kinetic energy.
(b) When two drops collide and coalesce, some of the total energy of coalescence is used to increase the total surface of the newly formed drop.
(c) The larger the difference in two colliding droplets' velocities at a given collector drop size R, the smaller the coalescence efficiency.

Long exercises

1. (a) Derive the diffusion equation for water vapor from the continuity equation and Fick's first law; in them n_v is the number concentration of water vapor molecules, J is the flux of these molecules per unit area and D_v is the diffusion coefficient of water vapor. The continuity equation is

$$\frac{\partial n_v(r,t)}{\partial t} = \nabla J$$

and Fick's first law is

$$J = -D_v \, \nabla n_v.$$

(b) What happens when $n_{v,r_d} - n_{v,\infty} > 0$ and when $n_{v,r_d} - n_{v,\infty} < 0$? Here n_{v,r_d} is the vapor concentration above the droplet's surface and $n_{v,\infty}$ is the vapor concentration far from the droplet.
(c) What are the limitations of the diffusion equation for water vapor when one is describing the growth of cloud droplets?

2. Explain three mechanisms which can lead to a broadening of a cloud droplet size distribution.
3. Explain qualitatively why the collision efficiency E is largest for $r/R \approx 0.6$ in the case $R = 20$ μm, where r and R are the radii of the smaller and the larger collision partners, respectively.
4. A cloud has a thickness of 3 km with uniform liquid water content $M_l = 0.4$ g m^{-3}. A drop with radius $r_d = 0.04$ mm and density $\rho_l = 1000$ kg m^{-3} falls from the cloud top through the cloud.
(a) How large is the drop when it leaves the cloud? Assume that the drop is larger than the drops with which it is colliding, and that the collection efficiency is 1. Vertical air motions in the cloud and condensational growth can be neglected. *Hint: Substitute dt by dz.*
(b) Calculate the time required for the drop to reach the cloud base.
(c) In view of the fact that the growth of droplets via collisions does not happen continuously but in discrete steps, would the estimate of the time from part (b) still be satisfactory?
5. To determine the terminal fall velocity of particles, it can be helpful to distinguish between different flow regimes.
(a) In which flow regime is Stokes's law valid?
(b) Calculate the terminal fall velocities for the following cloud droplets:

1. $r_d = 15$ μm.
2. $r_d = 100$ μm.
3. $r_d = 700$ μm.

6. Break-up limits the observed raindrop sizes in the atmosphere.
   (a) Name two ways in which the break-up of a raindrop can occur.
   (b) Describe briefly the difference between filament or neck break-up and disk break-up. Which break-up type needs more CKE and why?

8 Microphysical processes in cold clouds

As discussed in Chapter 7, warm rain formation is not sufficient to explain most global precipitation. Globally 50% of all precipitation events of more than 1 mm d^{-1} involve the formation and growth of ice crystals somewhere in the cloud (Field and Heymsfield, 2015). In the tropics, precipitation via the ice phase accounts for 69% of the total precipitation (Lau and Wu, 2003). In mid-latitudes, warm rain is even less prevalent, especially over land. Here it accounts for less than 10% of the total precipitation because of the smaller cloud droplets in continental clouds as compared with marine clouds (Mülmenstädt *et al.*, 2015).

Once a cloud extends to altitudes where the temperature is below 0 °C, ice crystals may form by the homogeneous freezing of cloud droplets or by the heterogeneous ice nucleation, as discussed below in Section 8.1. Heterogeneous ice nucleation takes place with the help of ice nucleating particles (INPs) and can occur via different pathways depending on the temperature and supersaturation (Section 8.1.2).

Once nucleated, the ice crystals can grow by diffusion from the vapor phase (Section 8.3). As long as they keep their identity, they are called pristine crystals. They can also grow by aggregation to form snowflakes or by riming to form graupel and hail. The counterpart of raindrop break-up is ice multiplication, which also occurs upon collision and drastically enhances the number concentration of ice crystals in the cloud. The melting of ice and snow and the sublimation of snow (Section 8.4.2) complete our discussion of these microphysical processes; a summary is given in Section 8.5.

8.1 Ice nucleation

As discussed in Section 6.1, nucleation denotes a phase transition where a cluster of a thermodynamically stable phase forms and grows within the surrounding metastable parent phase. Ice crystals can form by the direct deposition of water vapor on an INP (deposition nucleation) or by the freezing of a cloud droplet or solution droplet (Section 6.4). The term solution droplet refers to a liquid aerosol, such as sulfuric acid or sodium chloride, that has undergone hygroscopic growth (Section 6.3). A cloud droplet, however, is sufficiently large that the amount of solute is negligible and it can be regarded as a pure water droplet. Both the freezing of cloud and solution droplets and the deposition of vapor to the solid phase are nucleation processes because a stable cluster of the new ice phase has to form within the parent phase (vapor or liquid).

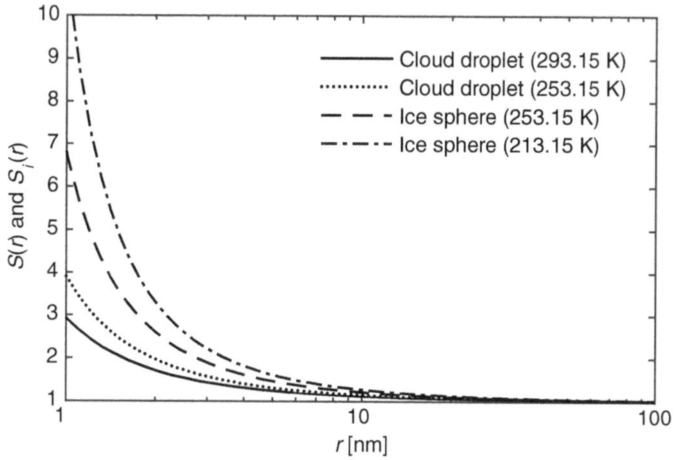

Fig. 8.1 Kelvin's law (giving the equilibrium saturation ratio as a function of radius) for a cloud droplet at 293.15 K and 253.15 K and for an ice sphere at 253.15 K and 213.15 K. The vertical axis represents the equilibrium saturation ratio with respect to water, $S(r)$, for the water droplet and with respect to ice, $S_i(r)$, for the ice sphere. The temperature dependence of the ice density is taken from Pruppacher and Klett (1997) and that of the water density from Wagner and Pruss (2002). The values of the surface tension follow the parameterization of Pruppacher and Klett (1997) and Vargaftik *et al.* (1983) for the cases of supercooled water and water vapor, respectively, as well as the parameterization given in Hale and Plummer (1974) for the ice–vapor interface.

Unlike the formation of liquid droplets, which does not occur homogeneously in the atmosphere, the homogeneous nucleation of ice crystals from the liquid phase (homogeneous freezing) and the heterogeneous nucleation of ice crystals from the vapor and liquid phases both occur in the atmosphere. The homogeneous freezing of solution droplets is a prominent freezing process in cirrus clouds. Heterogeneous ice nucleation, which requires the presence of an INP, is the dominant ice nucleation process in mixed-phase clouds but also occurs in cirrus clouds (Cziczo *et al.*, 2013).

8.1.1 Homogeneous ice nucleation

The homogeneous deposition nucleation of vapor to form ice crystals can be described analogously to homogeneous cloud droplet nucleation on replacing the phase transition vapor–liquid by the phase transition vapor–ice and the parameters related to liquid water by those related to ice. Thus, for homogeneous deposition nucleation also, a Kelvin effect (eq. 6.13) exists, i.e. the vapor pressure on the surface of a small ice sphere is enhanced compared with that for bulk ice and Kelvin's law can be formulated accordingly, as shown in Figure 8.1.

Below 0 °C the chemical potential and therefore the Gibbs free energy (Section 6.1) of the bulk ice phase are lower than those of the vapor phase. However, the formation of a new surface between the two phases increases the Gibbs free energy of the system. The energy barrier that has to be overcome is:

$$\Delta G_{i,v} = -\frac{4\pi r^3 R_v T}{3\alpha_i} \ln S_i + 4\pi r^2 \sigma_{i,a}, \qquad (8.1)$$

where $\Delta G_{i,v}$ is the Gibbs free energy for the phase transition from vapor to ice, α_i is the specific volume of ice and $\sigma_{i,a}$ is the surface tension between ice and humid air. Because ice occupies a larger volume than water for all $T < 0\,°C$ ($\alpha_i > \alpha_l$) and the surface tension between ice and humid air, $\sigma_{i,a}$, is larger than that between water and humid air, $\sigma_{w,a}$, the energy barrier for the homogeneous nucleation of ice from the vapor phase is larger than that for homogeneous cloud droplet nucleation for the same saturation ratio. This indicates that the homogeneous nucleation of ice from the vapor phase is even less likely than the homogeneous nucleation of liquid droplets. For example, in order for a water droplet with $r = 2$ nm to be stable at 253 K, a saturation ratio 2 with respect to water is needed. For an ice crystal of the equivalent size, a saturation ratio 2.7 with respect to ice is required, as shown in Figure 8.1. This comparison is, however, not entirely correct because in the atmosphere much higher supersaturations with respect to ice are found, owing to the scarcity of INPs, than supersaturations with respect to water.

Therefore, in order to compare the likeliness of homogeneous droplet and ice nucleation from the vapor phase with that for homogeneous droplet nucleation, it is better to compare the saturation ratios necessary to generate a typical nucleation rate J. J describes the number of nucleation events per unit volume and unit time in the parent phase. A nucleation rate on the order of $J = 1\ \text{cm}^{-3}\ \text{s}^{-1}$ is required for nucleation from the vapor phase to occur on an atmospherically relevant time scale. The nucleation rate J is given as:

$$J = K \exp\left(-\frac{\Delta G^*}{kT}\right), \qquad (8.2)$$

where ΔG^* is the energy barrier for the phase transition under consideration, k is the Boltzmann constant and K is the kinetic prefactor. The height of the Gibbs free energy barrier depends on substance-related quantities such as the surface tension between the two phases and on the conditions in the parent phase, such as temperature and supersaturation. These quantities are combined in the thermodynamic factor $\exp(-\Delta G^*/kT)$. For the homogeneous nucleation of ice from the vapor phase, $\Delta G^*_{i,v}$ can be determined from the maximum of eq. (8.1) and is given by (Ickes *et al.*, 2015):

$$\Delta G^*_{i,v} = \frac{16\pi}{3} \frac{v_i^2 \sigma_{i,v}^3}{(kT \ln S_i)^2}. \qquad (8.3)$$

This expression is also valid for homogeneous droplet nucleation if $\Delta G^*_{i,v}$ is replaced by $\Delta G^*_{w,v}$, $\sigma_{i,v}$ by $\sigma_{w,v}$, v_i by v_w and S_i by S.

Besides the thermodynamic parameters, the kinetic parameters such as the viscosity of the parent phase or – in the case of ice nucleation from the supercooled liquid phase – the rate at which molecules can rearrange their orientation and join

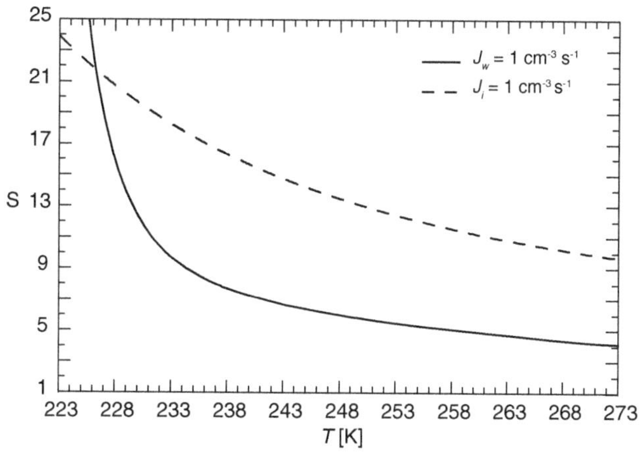

Fig. 8.2 Saturation ratio with respect to water, S, as a function of the temperature required to obtain a homogeneous nucleation rate of 1 cm^{-3} s^{-1} for water from vapor, J_w, and for ice from vapor, J_i.

the ice lattice determine the rate at which critical ice embryos form in the parent phase. It is intuitive that, with increasing viscosity for decreasing temperature, K also decreases with temperature. In extreme cases, the kinetics can inhibit nucleation even under thermodynamically favorable conditions. For instance, very concentrated solution droplets enter a near-solid amorphous glassy state at very low temperatures which makes it impossible for the system to transform into the thermodynamically favored ice structure.

Figure 8.2 shows the saturation ratio with respect to water S that is required to obtain a homogeneous nucleation rate for water in vapor of $J_w = 1$ cm^{-3} s^{-1} and for ice in vapor of $J_i = 1$ cm^{-3} s^{-1}. Both nucleation rates require the same S at 226 K. At higher temperatures the homogeneous nucleation of water is more favorable and at lower temperatures the homogeneous nucleation of ice from the vapor phase. However, the saturation ratios needed for both processes are much higher than those observed in the atmosphere, causing both the homogeneous nucleation of water and ice in vapor to be negligible in the atmosphere.

In contrast, the homogeneous nucleation of ice within a liquid drop is indeed observed in the atmosphere. The homogeneous freezing of a water drop requires the random motion of water molecules in order to bring a sufficient number together that forms a cluster. This cluster of an ice-like structure is called an ice embryo. If the ice embryo exceeds the critical radius at the prevailing temperature, freezing of the drop is initiated. The embryo is then called an ice germ or critical ice embryo. The Gibbs free energy for homogeneous freezing, $\Delta G_{i,w}$, is given as

$$\Delta G_{i,w} = -\frac{4\pi r^3 R_v T}{3\alpha_i} \ln \frac{e_{s,w}}{e_{s,i}} + 4\pi r^2 \sigma_{i,w}, \tag{8.4}$$

and the corresponding energy barrier is given by

$$\Delta G^*_{i,w} = \frac{16\pi}{3} \frac{v_{ice}^2 \sigma_{i,w}^3}{\left[kT \ln(e_{s,w}/e_{s,i})\right]^2}. \tag{8.5}$$

The dependence of $\Delta G^*_{i,w}$ on the saturation vapor pressures $e_{s,w}$ and $e_{s,i}$ can be understood by expressing $\ln(e_{s,w}/e_{s,i})$ as a function of the supercooling $T_0 - T$ according to eq. (2.89):

$$\ln \frac{e_{s,w}}{e_{s,i}} = \frac{L_f}{R_v T_0 T}(T_0 - T)$$

and therefore

$$\begin{aligned} \Delta G_{i,w} &= -\frac{4\pi r^3 R_v T}{3\alpha_i} \frac{L_f}{R_v T_0 T}(T_0 - T) + 4\pi r^2 \sigma_{i,w} \\ &= \underbrace{-\frac{4\pi r^3}{3\alpha_i} \frac{L_f}{T_0}(T_0 - T)}_{\text{volume term}} + \underbrace{4\pi r^2 \sigma_{i,w}}_{\text{surface term}} . \end{aligned} \tag{8.6}$$

In the case of droplet nucleation, overcoming the energy barrier refers to the transition from a nanometer-size fluctuating cluster to a critical cluster growing spontaneously to a micrometer-size cloud droplet. For homogeneous freezing, the spontaneous growth of an ice germ within a droplet leads to instantaneous freezing of the entire droplet.

The energy barriers ΔG^* that need to be overcome, for the homogeneous freezing of a cloud droplet, homogeneous cloud droplet nucleation from the vapor phase and homogeneous ice crystal nucleation from the vapor phase, as functions of temperature and saturation ratio with respect to water, are shown for comparison in Figure 8.3. Homogeneous droplet nucleation depends only slightly on temperature but strongly on supersaturation (eq. 6.9). The temperature dependences of $\Delta G^*_{i,v}$ and $\Delta G^*_{w,v}$ at constant S_i and S, respectively, are dominated by the increase in the respective surface tension for decreasing temperature (eq. 8.3). This manifests itself in a moderate increase in $\Delta G^*_{w,v}$ for decreasing temperature (Figure 8.3). However, $\Delta G^*_{i,v}$ increases with temperature for constant S_i but decreases for constant S, since S_i increases with decreasing temperature for constant S. In contrast with nucleation from the vapor phase, the surface tension between ice and liquid water $\sigma_{i,w}$ decreases with decreasing temperature (Ickes *et al.*, 2015). Hence, $\Delta G^*_{i,w}$ strongly decreases with decreasing temperature and thus makes homogeneous freezing the much more favorable process at cold temperatures as compared with homogeneous droplet nucleation and homogeneous ice nucleation from the vapor phase.

Comparing the critical radii for ice nucleation from the vapor and the liquid phases also emphasizes that freezing is the preferred process. The critical radius that needs to be overcome at −20 °C at water saturation in the first case is ~ 10 nm, according to eq. (8.1), but is only 2 nm for freezing (see Figure 6.7 from Pruppacher and Klett, 1997). This finding is consistent with Ostwald's rule of stages (Ostwald, 1897) stating that a supersaturated phase (in our case water vapor) does not directly transform into the most stable phase (ice) but rather into the next most stable or metastable phase (supercooled water).

Experimental data on the freezing of pure water show that pure water droplets below a radius of 5 μm will freeze spontaneously only at temperatures below −38 °C (Pruppacher

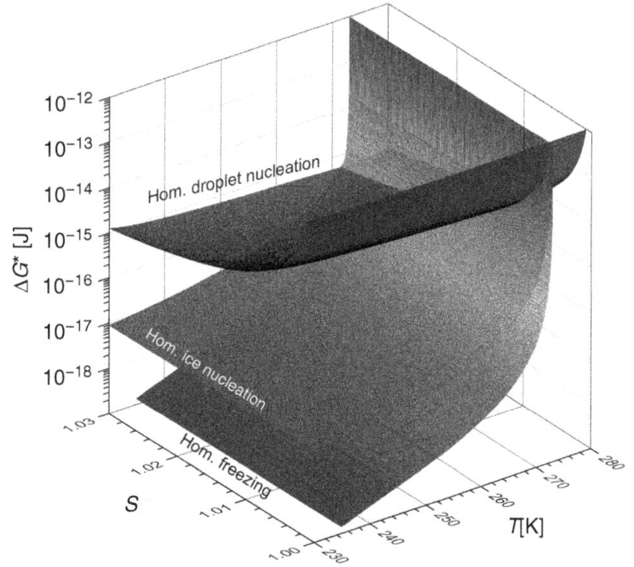

Fig. 8.3 Energy barriers ΔG^*, in J, that need to be overcome, for the homogeneous freezing of a water droplet, homogeneous droplet nucleation and homogeneous ice nucleation from the vapor phase, as functions of temperature and saturation ratio S with respect to water.

and Klett, 1997). At −38 °C the nucleation rate J of ice in pure water becomes large enough that at least one nucleation event is likely to occur in a droplet of that size within an atmospherically relevant time period. The freezing probability P of a droplet with volume V in a time interval Δt is given by

$$P = 1 - \exp(-JV\Delta t). \tag{8.7}$$

Equation (8.7) shows that, for constant J and Δt, P increases with increasing droplet volume, i.e. with increasing radius. Therefore the freezing temperature also increases with increasing droplet radius. One reason why in everyday life we experience the freezing of water at temperatures very close to 0 °C is that everyday life involves sufficiently large volumes of water that one nucleation event within seconds or minutes is likely to occur even at a low J value. Another reason is the presence of INPs, as will be discussed in Section 8.1.2.

The various freezing pathways are shown in Figure 8.4. Pure water droplets exist in equilibrium only at or above water saturation. Cloud droplets contain some salt from the CCN on which they formed, but the salt concentration is usually very dilute and can be neglected (as was also assumed in deriving the droplet growth equation (7.28)). Thus the freezing of cloud droplets can be approximated to that of pure water droplets, as discussed above.

Ice crystals in cirrus clouds that form in situ, i.e. in the absence of a pre-existing cloud, will do so at relative humidities below water saturation even in the absence of INPs. Owing to the freezing point depression of dissolved ions, solution droplets require lower temperatures to freeze compared to pure water droplets. Below the onset of homogeneous

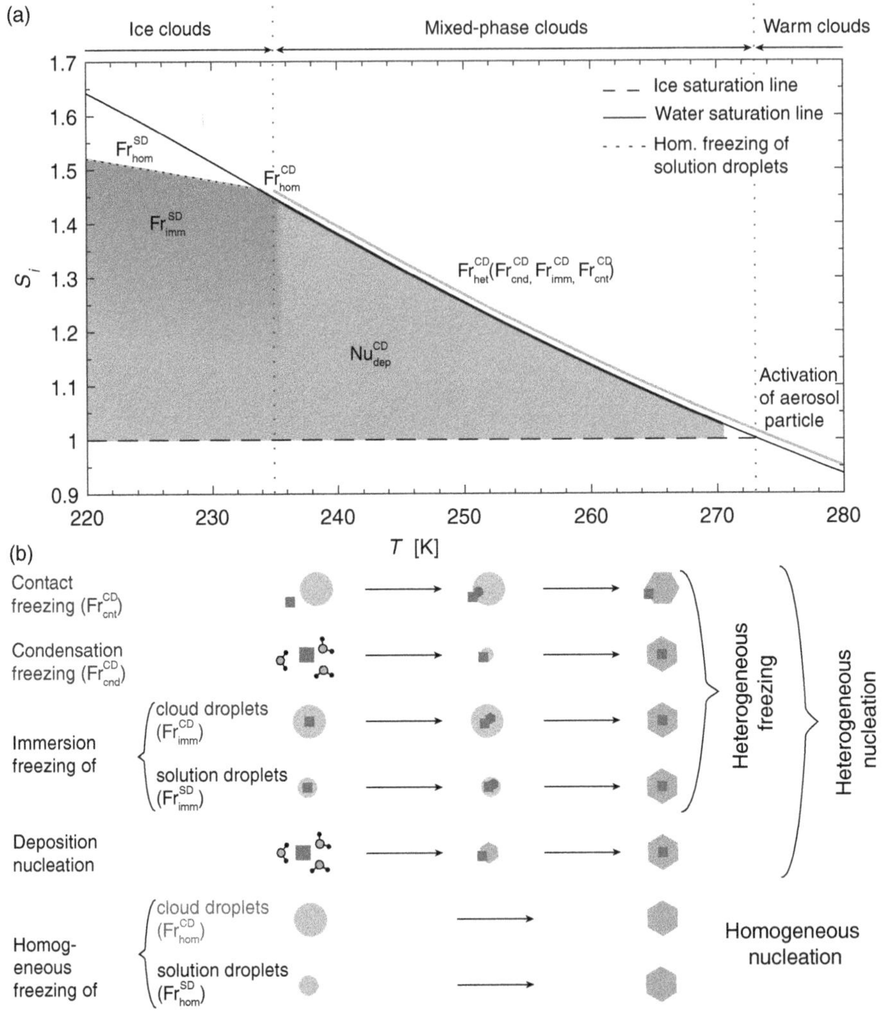

Fig. 8.4 (a) Schematic of the main freezing pathways as a function of temperature T and saturation ratio S_i with respect to ice [%], along with the cloud types in which these processes occur. The sloping dotted line that starts at 235 K denotes the minimum saturation ratio for solution droplets to freeze homogeneously, according to Koop *et al.* (2000). The shaded areas denote the regions where the individual nucleation types occur. The superscripts CD and SD refer to the cloud droplets and solution droplets respectively. (b) Schematic of the mechanisms associated with the individual processes. The cubes represent INPs, the spheres droplets, the hexagons ice crystals. A black and white version of this figure will appear in some formats. For the color version, please refer to the plate section.

freezing at −38 °C, increasingly concentrated solution droplets can freeze with decreasing temperature. Owing to Raoult's effect (Section 6.4), this extends the range of homogeneous freezing below the water saturation line, down to the sloping dotted line at the top left of Figure 8.4a (Koop *et al.*, 2000). Hence, for decreasing temperature, an increasing range of saturation ratios below 100% RH exists in which the homogeneous freezing of solution droplets can be found in the atmosphere.

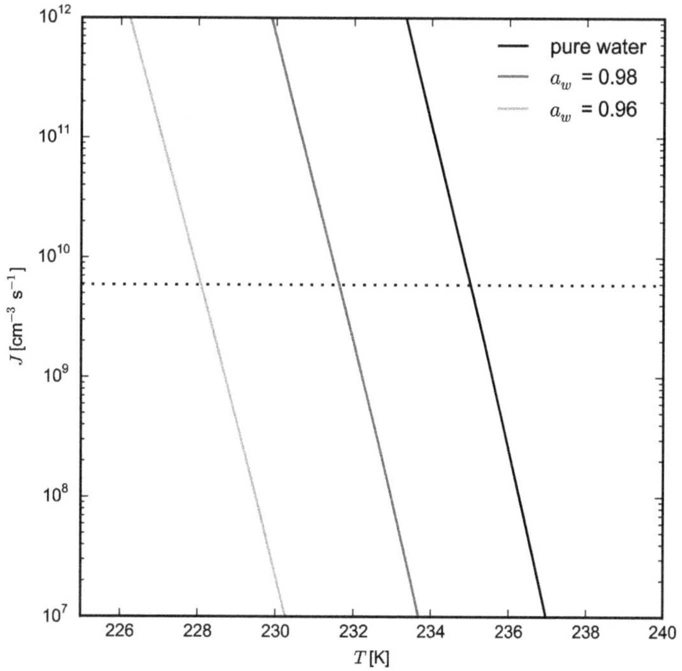

Fig. 8.5 Nucleation rates J for the homogeneous freezing of pure water droplets and aqueous solution droplets (in equilibrium at the corresponding relative humidity), represented by two different water activities, $a_w = 0.98$ and $a_w = 0.96$, according to Koop *et al.* (2000). The nucleation rate $J = 6 \times 10^9$ cm^{-3} s^{-1} is shown as a dotted line.

Freezing point depression is the reason for putting salt on roads in winter to avoid the formation of black ice. This is presented in Figure 8.5 for two different water activities, $a_w = 0.98$ and $a_w = 0.96$ (eq. 6.14), indicative of an increasing fraction of salt ions in the solution. Figure 8.5 highlights the nucleation rate, $J = 6 \times 10^9$ cm^{-3} s^{-1}, necessary for the homogeneous freezing at 235 K of pure water droplets with the size of typical cloud droplets. In order to reach the same nucleation rate when the solution consists of 2% salt ions ($a_w = 0.98$), the temperature needs to be lowered to 231.5 K, and to 228 K when 4% of the solution consists of salt ions ($a_w = 0.96$). This is of importance for the homogeneous freezing of solution droplets in in-situ-formed cirrus clouds, where values of a_w down to 0.8 can be found at temperatures below 200 K (Koop *et al.*, 2000).

In summary, the homogeneous freezing of cloud droplets occurs in deep convective clouds and the homogeneous freezing of solution droplets occurs in in-situ-formed cirrus clouds. It is, however, currently a matter of debate whether in-situ-formed cirrus clouds are mainly formed homogeneously (Spichtinger and Krämer, 2012) or whether sufficient INPs exist to cause widespread heterogeneous nucleation (Cziczo *et al.*, 2013). The former would imply that the relative humidity in in-situ-formed cirrus clouds may exceed the threshold for the homogeneous freezing of solution droplets but generally does not reach water saturation (Krämer *et al.*, 2009).

8.1.2 Heterogeneous ice nucleation

Freezing with the involvement of INPs is termed heterogeneous ice nucleation. Ice nucleating particles are aerosol particles that provide a surface onto which water molecules are likely to adsorb, bond together and form aggregates with an ice-like structure. In contrast with CCN, only a small fraction (one in 10^5 to 10^6) of all aerosol particles can serve as INPs at temperatures higher than −38 °C. The reason is that heterogeneous freezing strongly depends on distinct surface features of INPs. For CCN, however, surface features do not matter because they change into solution droplets when the DRH is reached. Therefore the criteria for aerosol particles to act as INPs are less well understood than their ability to act as CCN.

An INP promotes the freezing process because it reduces the energy barrier for the formation of an ice germ by offering potential ice embryos a solid surface to form upon. This concept is often visualized by a "spherical-cap model", according to which an ice embryo with the shape of a spherical cap forms on the surface of an INP (Figure 8.6). The interaction between the ice embryo, the INP surface and the surrounding parent phase is incorporated in a contact angle, which denotes the angle between the INP and the spherical ice cap. A schematic showing the influence of the contact angle on ice nucleation is shown in Figure 8.7b.

In the context of water droplets forming on surfaces, the contact angle θ is usually understood as a measure of the "wettability" of a surface. A small contact angle represents a hydrophilic surface. For ice nucleation, a small θ value represents good "compatibility" of the ice embryo with the INP's surface and significantly reduces the energy barrier of nucleation by strongly reducing the number of water molecules required to join the embryo before it reaches its critical radius. A reduced number of molecules in turn reduces both the volume term and the surface term of the Gibbs free energy in eq. (8.1) as compared with the homogeneous case. However, because of the spherical cap geometry, the reduction is more important in the volume term. For the energy barrier to be smaller in the presence of the INP requires that the surface tension between the INP and the ice embryo $\sigma_{INP,i}$ is lower than that between the ice embryo and the parent phase ($\sigma_{i,v}$ or $\sigma_{i,w}$). The relation between the surface tensions and θ is mathematically described by

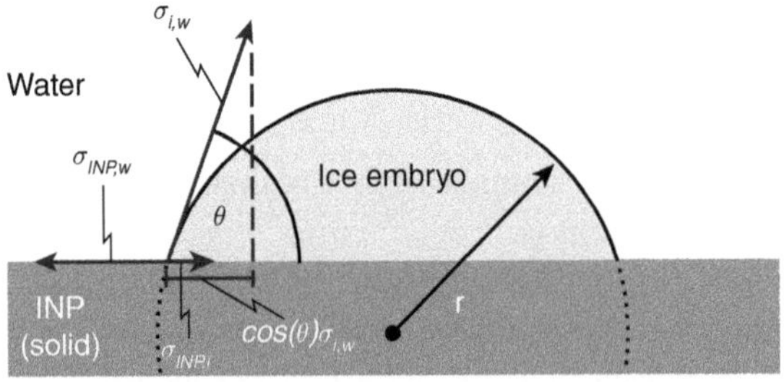

Fig. 8.6 Schematic of the contact angle θ of a cap-like ice embryo that is forming on a solid INP; $\sigma_{i,w}$ is the surface tension between ice and water, $\sigma_{INP,w}$ that between the INP and water and $\sigma_{INP,i}$ that between the INP and ice.

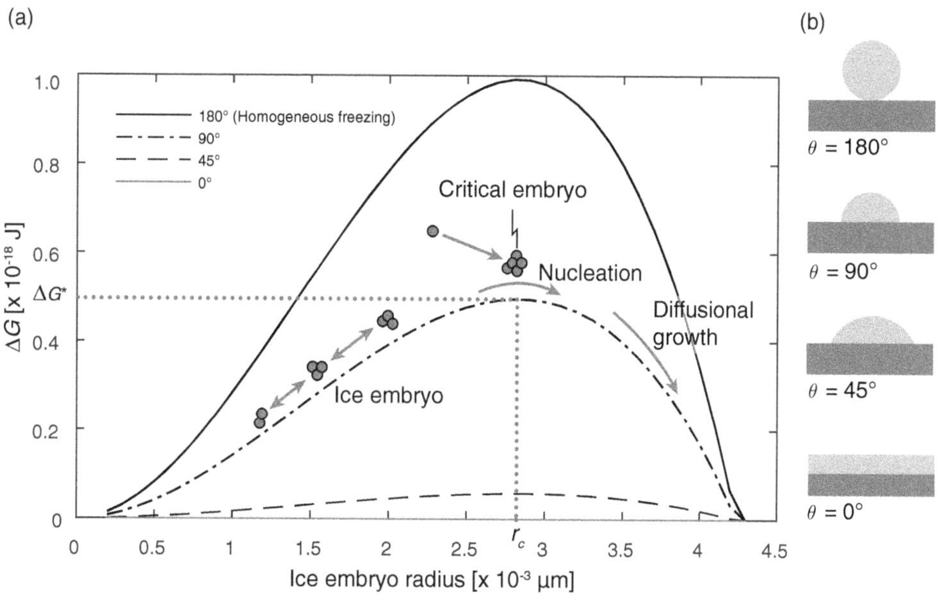

Fig. 8.7 (a) Gibbs free energy ΔG for homogeneous and heterogeneous freezing at $T = 253$ K as a function of the embryo radius r. Heterogeneous freezing is presented for three contact angles, $\theta = 0°, 45°$ and $90°$. The dotted horizontal and vertical lines indicate the Gibbs free energy barrier ΔG^* and the critical radius r_c. (b) Schematic illustration of the contact angles shown in (a); the dark gray areas show the INP and the light gray areas the ice embryo.

Young's equation and is illustrated in Figure 8.6. For the case of heterogeneous freezing it is given by

$$\sigma_{i,w} \cos\theta = \sigma_{INP,w} - \sigma_{INP,i}. \tag{8.8}$$

Figure 8.7a illustrates the Gibbs free energy barrier for homogeneous freezing and heterogeneous freezing using two different contact angles at $T = 253$ K. It can be seen that ice nucleation is much more likely to occur for an efficient INP, i.e. one with a contact angle smaller than 45° than for homogeneous freezing. Even a moderate INP with a larger contact angle of 90° can reduce the energy barrier to about half the value required for homogeneous freezing ($\theta = 180°$). The critical radius remains the same for homogeneous and heterogeneous freezing, but fewer water molecules are needed to reach critical size if the ice embryo forms a spherical cap on the INP, as shown in Figure 8.7b. This reduction in the energy barrier for heterogeneous freezing, ΔG^*_{het}, is taken into account by multiplying ΔG^* for homogeneous freezing by a compatibility parameter f:

$$\Delta G^*_{het} = f \Delta G^*, \tag{8.9}$$

where f depends on the contact angle (Pruppacher and Klett, 1997); it is unity for homogeneous freezing ($\theta = 180°$, Figure 8.7b) and 0 for $\theta = 0°$.

As discussed in the following, four heterogeneous ice nucleation modes are distinguished in the literature (Figure 8.4).

8.1.2.1 Immersion freezing

Immersion freezing refers to freezing that is initiated from within a cloud or solution droplet. It requires that the INP is already immersed in the cloud or solution droplet at higher temperatures. Upon cooling of the cloud or solution droplet, freezing is initiated (Figure 8.4). The immersion freezing of solution droplets can occur at all supersaturations with respect to ice where they exist. In contrast, the immersion freezing of cloud droplets requires the prior existence of cloud droplets and thus relative humidities near water saturation.

8.1.2.2 Condensation freezing

Condensation freezing refers to a different pathway, where an air parcel containing INPs starts from subsaturated conditions (Figure 8.4). When water saturation is approached, embryos of condensed water form on the INP surface and can grow to sizes large enough to contain an ice germ. At water saturation the critical radius for ice nucleation in liquid water is smaller than that for droplet nucleation. Therefore it is possible that a stable ice germ forms within a water embryo and the freezing therefore "stabilizes" the water embryo. As INPs in general contain at least a small amount of soluble compounds in addition to the insoluble part, they can act as CCN at supersaturations of only a fraction of a percent. Therefore, condensation freezing can occur only in the narrow range of S values before the CCN is activated and immersion freezing in cloud droplets takes place.

8.1.2.3 Contact freezing

Contact freezing refers to the collision of an INP with a supercooled cloud droplet, with subsequent freezing. Contact freezing requires the presence of a cloud droplet and is therefore shown on the water saturation line in Figure 8.4. However, a particular kind of contact freezing can also take place for RH $< 100\%$: an evaporating cloud droplet at subsaturated conditions could shrink until the INP within the droplet comes in contact with its surface. This freezing pathway has been termed "contact nucleation inside out" (Durant and Shaw, 2005). Because contact freezing depends on collisions between the supercooled droplet and aerosol particles, it is limited by the collision rate. The collision rate in turn is determined by the collision efficiency between a droplet and an aerosol particle and by the number density of aerosol particles available for collisions with a given supercooled cloud droplet.

8.1.2.4 Deposition nucleation

Deposition nucleation refers to the direct deposition of vapor onto an INP and requires that the ambient saturation ratio with respect to ice, S_i, is greater than unity. Deposition nucleation is found in cirrus clouds, when vapor is deposited for instance onto mineral dust particles that act as INPs (Cziczo *et al.*, 2013). Some early studies of heterogeneous freezing suggested that deposition nucleation only occurs below -18 °C and at $S_i \geq 1.2$

(Roberts and Hallett, 1968; Schaller and Fukuta, 1979). However, newer studies show that deposition nucleation on mineral dust particles can occur at lower saturation ratios with respect to ice (Kanji and Abbatt, 2006; Möhler *et al.*, 2006; Welti *et al.*, 2009).

It is not clear whether deposition nucleation continues at $S > 1$ or whether it is then replaced by other heterogeneous nucleation mechanisms such as condensation or immersion freezing, as discussed by Welti *et al.* (2014) and Marcolli (2014). It may even be that deposition nucleation does not exist at all because a phase transition from vapor to a solid is not favorable according to Ostwald's rule of stages. Instead, what appears to be deposition nucleation may in reality be immersion or condensation freezing taking place within the pores of an INP, where a liquid phase with an inversely curved surface can already be in equilibrium with the vapor phase at $S < 1$ (Marcolli, 2014). For a negative-curvature of the surface of the liquid phase, the Kelvin effect (eq. 6.13) lowers the equilibrium vapor pressure as compared with a flat surface.

Overall, deposition nucleation and condensation freezing do not seem to be important for the formation of mixed-phase clouds, as lidar observations have revealed that liquid cloud droplets are required before ice crystals can form via heterogeneous freezing mechanisms in mixed-phase clouds (Ansmann *et al.*, 2008).

8.1.2.5 Deterministic versus stochastic process

One question, regardless of the mode of heterogeneous nucleation, is whether the latter is a deterministic (singular) or a stochastic process. Two controversial hypotheses have been advanced.

The "stochastic hypothesis" assumes that although the surface of an INP reduces the energy barrier for nucleation as compared with the homogeneous case, the freezing process remains stochastic in nature and is described by a nucleation rate (eq. 8.2). In a sample of identical INPs (i.e. equal surface areas and contact angles), all will have the same probability per unit time that an ice embryo on their surface reaches its critical size by random fluctuations. For an individual INP it is impossible to predict if and when a nucleation event occurs. According to the stochastic hypothesis, the fraction of frozen droplets in a population of supercooled cloud droplets increases with time at constant temperature and S_i. Furthermore, in a freezing experiment with a given set of supercooled droplets held at constant temperature or supersaturation, the sequence in which the droplets freeze is completely random.

The "singular hypothesis", however, assumes that each INP has a distinct temperature or supersaturation at which it will trigger a nucleation event. Accordingly, heterogeneous ice nucleation is fully determined by the ambient conditions. If these are held constant, the fraction of frozen droplets in a population of supercooled droplets does not change with time. If a freezing experiment is repeated several times with the same set of droplets, the sequence in which the droplets freeze will always be the same.

An established theory to describe ice nucleation is classical nucleation theory (CNT). According to CNT heterogeneous nucleation is a stochastic process. Experimental studies in which the time dependence of immersion freezing was investigated provide support for a stochastic component of the ice nucleation process (Murray *et al.*, 2011; Niedermeier *et al.*, 2011; Pinti *et al.*, 2012; Welti *et al.*, 2012). At the same time, CNT for immersion freezing has a much lower sensitivity to time than it has to temperature, INP size or contact angle (Ervens and Feingold, 2013).

In fact, the scatter in freezing temperatures among individual INPs that is observed for a single particle type and size cannot be explained solely by the stochastic nature of the nucleation process. Instead one needs to assume that not all INPs are the same but that some are more efficient and some are less efficient. One possibility to account for this deterministic aspect of heterogeneous freezing is to consider disturbances on the INP surface, so-called active sites where ice nucleation is preferentially initiated (Section 8.1.3.2). In conclusion, this means that heterogeneous nucleation has both a deterministic and a stochastic component. There have recently been several attempts to modify CNT towards a hybrid model also incorporating differences in the surface properties between individual INPs; see for example Niedermeier *et al.* (2011).

8.1.3 Ice nucleating particles

8.1.3.1 Types of ice nucleating particles

A compilation of the onset temperatures and ice supersaturations S_i at which ice nucleation is first observed in laboratory studies is shown in Figure 8.8 (Hoose and Möhler, 2012). It includes deposition nucleation, condensation freezing and immersion freezing for bioaerosols (bacteria, fungal spores and pollen), mineral dust particles, organic aerosols, ammonium sulfate particles and soot particles.

Condensation and immersion freezing refer to those symbols in Figure 8.8 that lie at or above the water saturation line. It can be seen that there is a considerable spread between the different data, caused by the variety of aerosol types within one category (e.g. montmorillonite, kaolinite, or Arizona Test Dust in the mineral dust category), the different experimental methods used to detect nucleation and their different detection thresholds. The symbols that lie far above water saturation are not atmospherically relevant.

The onset temperature and S_i value at which a certain INP initiates ice nucleation depend on the requirements for an INP, discussed below in Section 8.1.3.2. While, for most mineral dust, bioparticles and crystalline ammonium sulfate particles, heterogeneous ice nucleation occurs at considerably higher temperatures and lower supersaturations than homogeneous freezing, this is not the case for black carbon and organic aerosol particles. Figure 8.8 shows that the highest onset temperatures for immersion and/or condensation freezing occur for bioaerosols, followed by mineral dust and then organic aerosols. Deposition nucleation (the symbols below the water saturation line) on mineral dust commences at lower S_i values and/or higher temperatures than on crystalline ammonium sulfate, organic aerosols and black carbon. The heterogeneous nucleation of mineral dust is further distinguished into sub- and supermicron mineral dust particles. The lower-onset relative humidities of the supermicron dust particles reveal that the size of the particle is an important factor in whether it acts as an INP, as discussed next.

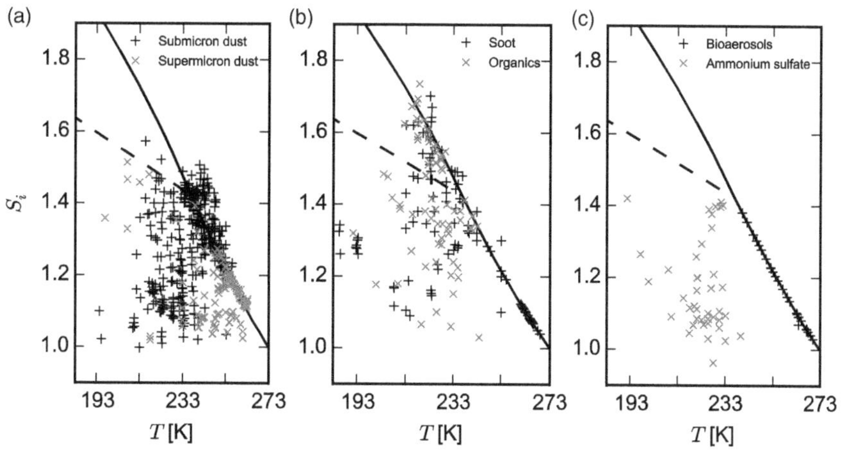

Fig. 8.8 Onset temperatures T and ambient saturation ratios with respect to ice, S_i for heterogeneous nucleation (deposition nucleation, condensation freezing and immersion freezing) for (a) mineral dust, divided into sub- and supermicron aerosol particles, (b) organic particles and soot and (c) bioaerosols and ammonium sulfate. The solid lines denote saturation with respect to water. The sloping broken lines refer to the onset conditions for the homogeneous freezing of solution droplets according to Koop *et al.* (2000). Figure adapted from Hoose and Möhler (2012).

8.1.3.2 Requirements of INP

The criteria for an aerosol particle to act as an INP are thought to be: (i) whether it is an insoluble or a soluble substance in the crystalline phase, (ii) its size, (iii) its lattice structure, (iv) its molecular bindings with water and (v) its active sites. Criteria (i), (iii) and (iv) are determined by the chemical composition of the aerosol particle. There is some dispute about the importance of each criterion and about how many are required for an aerosol particle to act as an INP.

Only aerosol particles providing a solid surface can significantly reduce the energy barrier for ice nucleation, as the solid surface enables an ice embryo on it to reach the critical radius with a lower number of water molecules in the ice phase than if the ice germ were to form without a solid surface. Hence, the most commonly occurring INPs, such as mineral dust, biological particles and soot, are solid insoluble substances. Aerosol particles that are crystalline at certain conditions of temperature and relative humidity can also act as INPs. Among them are crystalline ammonium sulfate and certain organic particles as found in laboratory studies (Abbatt *et al.*, 2006; Zobrist *et al.*, 2006; Wise *et al.*, 2012).

For INPs, as for CCN, size matters but for a different reason. The larger an INP's surface, the larger the probability that a critical cluster forms on it within a given time. Therefore data for all ice nucleation modes show higher freezing-onset temperatures for larger INPs (Figure 8.8). According to classical nucleation theory, the probability P for an ice nucleation event to occur in a given time span Δt and for a given nucleation rate J and surface area A of the INP is given by:

$$P = 1 - \exp(-JA\Delta t). \qquad (8.10)$$

Unlike in homogeneous nucleation (eq. 8.2), the heterogeneous nucleation rate is defined as the number of nucleation events per unit time and unit INP surface.

A crystalline structure similar to an ice lattice is preferable for an INP because then the water molecules can most easily form an ice lattice on the substrate (Evans, 1965). This is the case for silver iodide (AgI) because its lattice dimensions are very similar to those of ice. Indeed, AgI particles have been found to nucleate ice at temperatures as high as −6 °C (Detwiler and Vonnegut, 1981; Demott, 1995), which is much higher than those for freezing on clay minerals (Figure 8.10). Thus AgI has been used as an INP in cloud seeding experiments such as the Schweizer Grossversuche II–IV, which were designed to prevent hail (Federer *et al.*, 1986) and in shallow precipitating orographic cumuli (Pokharel *et al.*, 2014).

While in the case of bacteria a whole bacterium does not have a lattice structure resembling ice, their ice nucleating proteins (Kajava and Lindow, 1993) do show such a resemblance. However, for certain organic aerosols and mineral dust particles that nucleate ice, the lattice match is completely missing. Thus it is questionable how important this criterion is. Alternatively it has been suggested that for certain mineral dust types their ability to form hydrogen bonds with water matters for ice nucleation. For instance montmorillonite has more hydroxyl groups (-OH) and oxygen atoms in its silicates than kaolinite and showed a larger fraction of INPs at −30 °C at the same S_i (Salam *et al.*, 2006) and also a lower onset S_i at this temperature (Welti *et al.*, 2009).

While CNT usually treats the INP surface as uniform and defines one parameter (the contact angle) to quantify its ability to initiate ice nucleation, recent studies have attempted to take inhomogeneities on the INP surface into account (Murray *et al.*, 2012; Atkinson *et al.*, 2013; Kanji *et al.*, 2013). It is assumed that heterogeneous ice nucleation takes place on preferred sites on the INP surface, called active sites, which can be imperfections such as crevasses or steps on its surface or within pores of the INP (Marcolli, 2014). In the framework of CNT, these active sites can be interpreted as spots with a locally significantly reduced contact angle compared with the surrounding INP surface. Active sites are also used as an explanation why at a given temperature, supersaturation and particle size of a given aerosol type, some individual aerosol particles act as INPs but others do not. In order to explain the results of immersion freezing experiments, Marcolli *et al.* (2007) concluded that INPs need to have a radius of at least 0.05 μm in order to have on average one active site.

In summary, the efficiency of an INP is determined by a combination of size, solid state, lattice structure, molecular binding, low interfacial energy between the INP and ice and the presence of active sites.

8.1.4 Dependence of ice nucleation on temperature and supersaturation

For deposition nucleation it can be observed that as the supersaturation increases, so does the activated fraction $f_{act} = N_i/N_a$, which is defined as the ratio of the number concentration of ice crystals N_i and the number concentration of aerosol particles N_a before activation. This dependence on supersaturation is expected, in the framework of CNT, since

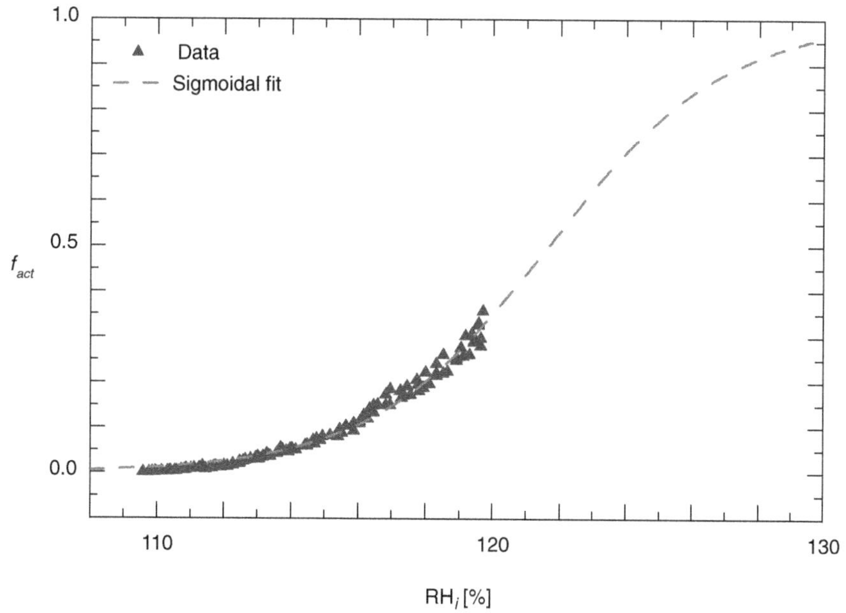

Fig. 8.9 Activated fraction $f_{act} = N_i/N_a$ in deposition nucleation mode as a function of relative humidity with respect to ice (RH_i) measured at a temperature of $-55\,^{\circ}C$ for illite particles of 200 nm in radius (triangles). The curve represents a least square fit to the data. Figure adapted from Welti *et al.* (2009).

the energy barrier of ice nucleation from the vapor phase decreases with increasing supersaturation (eq. 8.3). An example of deposition nucleation on mineral dust particles (illite) in a laboratory study together with a fit to the data is shown in Figure 8.9 (Welti *et al.*, 2009).

In-situ measurements in clouds have enabled assessment of the frozen fraction

$$f_f = \frac{N_i}{N_d + N_i},$$

defined as the ratio between the number of ice crystals N_i to the sum of N_i and the number of cloud droplets N_d as a function of supercooling. Measurements have shown that f_f increases with increasing supercooling (Korolev *et al.*, 2003), which is attributed to increasing INP concentrations with decreasing temperature. An example of immersion freezing experiments with ammonium sulfate ($(NH_4)_2SO_4$) particles, mineral dust particles (kaolinite) and silver iodide (AgI) particles is shown in Figure 8.10.

It shows that AgI is an excellent INP, initiating freezing even at 264 K, whereas kaolinite initiates freezing only at 243 K. This suggests that kaolinite is not an efficient INP, at least not in the immersion freezing mode (Hoose and Möhler, 2012). Ammonium sulfate is included as a reference experiment for homogeneous freezing. As expected, it does not show any significant freezing above 235 K. All cloud droplets activated on soluble ammonium sulfate particles freeze homogeneously at 233 K in the absence of a solid surface for heterogeneous freezing.

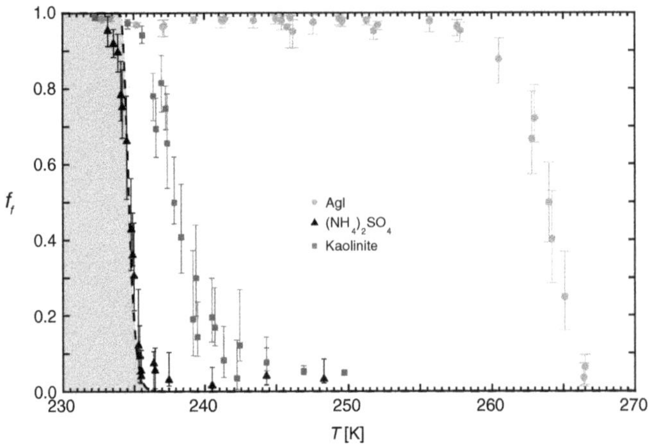

Fig. 8.10 Frozen fraction $f_f = N_i/(N_d + N_i)$, resulting from immersion freezing of droplets containing single immersed aerosol particles with radius $r = 100$ nm, as a function of temperature, for ammonium sulfate ($(NH_4)_2SO_4$) particles, mineral dust (kaolinite) particles and silver iodide (AgI) particles. The temperature uncertainty is $\pm$ 0.4 K. The shaded area indicates the temperature regime for the homogeneous freezing of solution droplets. Data taken from Hoyle *et al.* (2011), Lüönd *et al.* (2010) and André Welti.

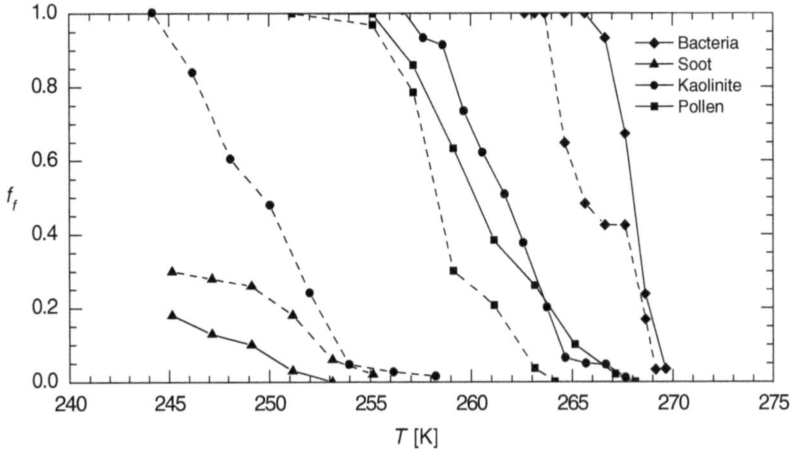

Fig. 8.11 Comparison of the frozen fraction f_f as a function of temperature T for kaolinite mineral dust particles, birch pollen, bacteria and soot, due to contact freezing (solid lines) and to immersion freezing (broken lines) from wind tunnel studies. Figure adapted from Ladino Moreno *et al.* (2013).

A comparison between the contact and immersion freezing of mineral dust (kaolinite), birch pollen, bacteria and soot from wind tunnel studies with drizzle-size droplets having radii of 100–200 μm can be made from Figure 8.11, in terms of the frozen fraction f_f versus temperature. This figure shows that contact freezing is initiated at higher temperatures than immersion freezing for all substances except soot (Ladino Moreno *et al.*, 2013). The difference between immersion and contact freezing was found to be largest for kaolinite and smallest for bacteria. The same qualitative conclusion about contact versus immersion freezing was reached for kaolinite particles using a different experimental set-up (Ladino *et al.*, 2011). In that study, contact freezing occurred at higher temperatures for the larger particles, because they have a larger surface area and a higher probability of containing active sites (Sections 8.1.2.5 and 8.1.3.2). However, contact freezing is limited by the collision efficiency and therefore its relevance in the atmosphere is not known. This limitation of contact freezing by the collision efficiency suggests that immersion freezing is the dominant freezing pathway at higher altitudes, i.e. at lower temperatures, where aerosol concentrations are lower than at lower altitudes (Section 5.3), at least according to studies conducted with global climate models (Lohmann and Diehl, 2006; Hoose *et al.*, 2010).

A rule of thumb for the dependence of INP concentration on temperature was obtained by Fletcher (1962). Based on laboratory data available at that time, he found one INP per liter of air at $T = -20$ °C, increasing by a factor 10 for each 4 °C of additional cooling. An exponential dependence of the observed INP concentration on supercooling is accepted as typical, but it is recognized that the INP number concentration at any given location and time can be at least an order of magnitude smaller or larger than observed by Fletcher (1962).

An example of N_i observed at various locations as a function of temperature is shown in Figure 8.12 (Vali, 1985). In this example N_i is taken to be the number concentration of INPs, i.e. secondary ice formation processes (Section 8.4.1) are deemed to be unimportant. Figure 8.12 shows a close to exponential increase of N_i with decreasing temperature, which originates from an increase in the nucleation rate with decreasing temperature (Ickes *et al.*, 2015). This is partly offset by the decrease in the aerosol, and hence INP, concentration with increasing altitude (and therefore decreasing temperature). Due to the large variability

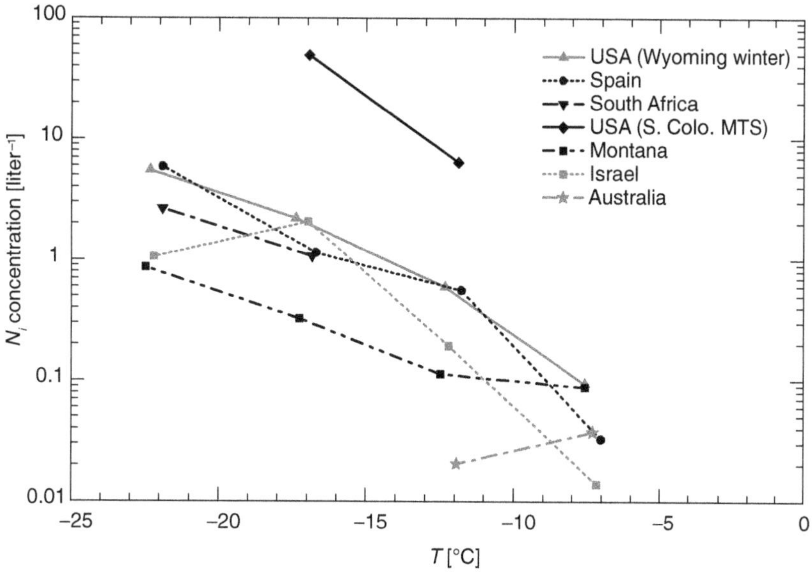

Fig. 8.12 Observed INP number concentrations N_i as a function of temperature at various locations. Secondary ice formation processes are assumed to be unimportant. Figure adapted from Vali (1985).

in aerosol concentration and composition depending on the geographical location, N_i at a given temperature varies up to three orders of magnitude.

8.2 Ice crystal habits

Ambient conditions, such as temperature and supersaturation, determine the shape, called the *habit*, that a growing crystal takes. The most common ice crystal habits are shown schematically in Figure 8.13 along with photographs showing these and other crystal shapes in Figure 8.14. Water molecules form hydrogen bonds and therefore arrange in a tetrahedral manner. This causes the ice lattice to consist of stacks of hexagons (Figure 8.13d) and explains why all ice crystal habits are comprised of hexagons.

As work is needed to create a new surface in a growing ice crystal, the equilibrium shape of an ice crystal must minimize the amount of surface energy for a given volume. Whereas CNT assumes that ice crystals nucleate in a spherical shape the reality is more complex owing to the ice lattice structure, which leads to anisotropic surface tensions. A spherical

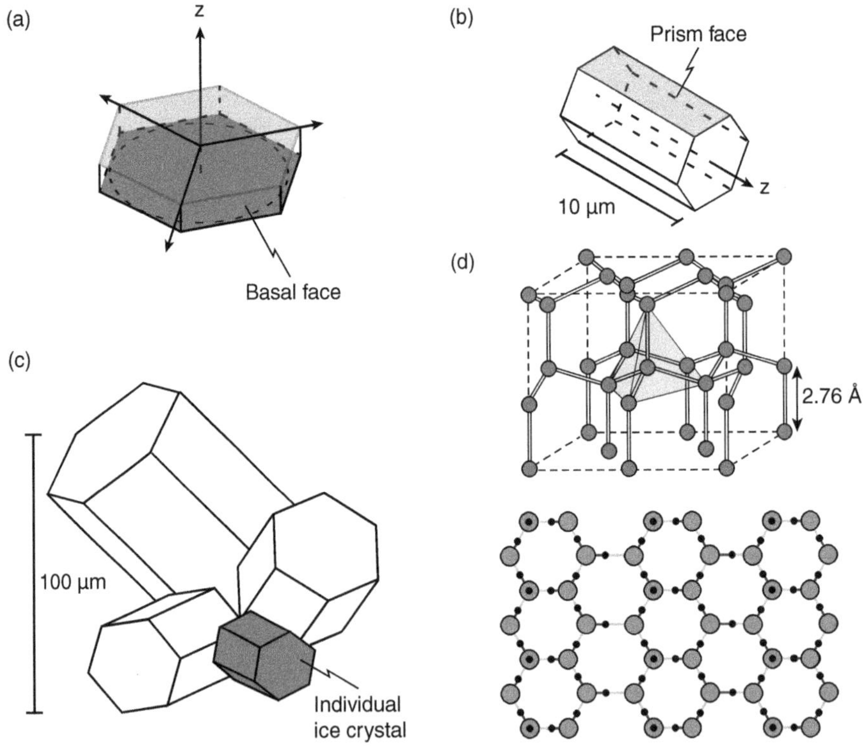

Fig. 8.13 Schematic of the most common ice crystal habits: (a) plate, (b) column, (c) aggregate and (d) schematic of the tetrahedral arrangement of the water molecules, along with the hexagonal structure of the ice lattice; gray disks denote oxygen atoms, the small black disks denote hydrogen atoms in the paper plane, black disks on gray disks denote hydrogen atoms pointing out of the paper plane and gray lines denote hydrogen bonds.

structure minimizes the surface energy in the case of an isotropic surface tension but this is not so for a real ice crystal, where the surface tension σ at a crystal face depends on the orientation of that face with respect to the crystal lattice. For example, the surface tension between a basal face of an ice crystal, $\sigma^B_{i,v}$, and air differs from the surface tension of a prism face $\sigma^P_{i,v}$ by a factor of approximately 0.9 (Pruppacher and Klett, 1997). Under this assumption, the equilibrium shape of an ice crystal is a hexagon of comparable width and height.

The requirement of minimum surface energy per volume shifts the equilibrium shape of a crystal between different crystal habits if the ambient conditions of T and S change, e.g. owing to the different temperature dependences of the different surface tensions (Pruppacher and Klett, 1997). If the growth of the hexagon preferentially takes place along the basal face, it results in a plate (Figures 8.13a and 8.14d, e). If growth along the prism face prevails, this leads to a column (Figures 8.13b and 8.14f, g). Aggregates (Figure 8.13c) form owing to collisions of ice crystals with each other at rather high temperatures (Section 8.3.2).

Deviations from the equilibrium shape occur because molecules do not have sufficient time to arrange themselves in the thermodynamically favored way. In this case growth is kinetically limited, leading for instance to dendrites (Figure 8.14a–c), needles (Figure 8.14h) or stellar plates (Figure 8.14d).

Ice crystal habits as a function of temperature and S_i, as shown in Figure 8.15, originate from laboratory experiments summarized by Libbrecht (2005). At constant S_i the growth is controlled by temperature. Between 0 and −5 °C and between −10 and −22 °C the growth rate of the prism face exceeds that of the basal face, and vice versa between −5 and −10 °C and below −22 °C (Pruppacher and Klett, 1997). These findings roughly agree with the observed morphology changes from plates at temperatures higher than −3 °C, to columns between −3 and −10 °C, to plates between −10 and −22 °C and again predominantly to columns at temperatures below −22 °C, as shown in Figure 8.15. If the supersaturation is increased, needles and more complex structures such as hollow columns or dendrites are formed.

Larger snowflakes (aggregates made up of 10 to 100 or more individual crystals) often consist of dendrites and thin plates indicative of diffusional growth conditions near water saturation. Far less common are columns and thick-plate aggregates, which point to growth conditions close to ice saturation but below water saturation. Rimed crystal forms, indicating growth by collisions of hydrometeors in different phases (Section 8.3.3), are common at rather mild temperatures. They tend to be produced in convective clouds, with a high liquid water content, as will be discussed further in Chapter 10.

8.3 Ice crystal growth

Once an ice embryo has formed, it will rapidly grow into a macroscopic pristine ice crystal by vapor diffusion. This is called growth by deposition and is shown in Figure 8.18b. A newly formed ice crystal grows more quickly than a freshly activated cloud droplet: if the

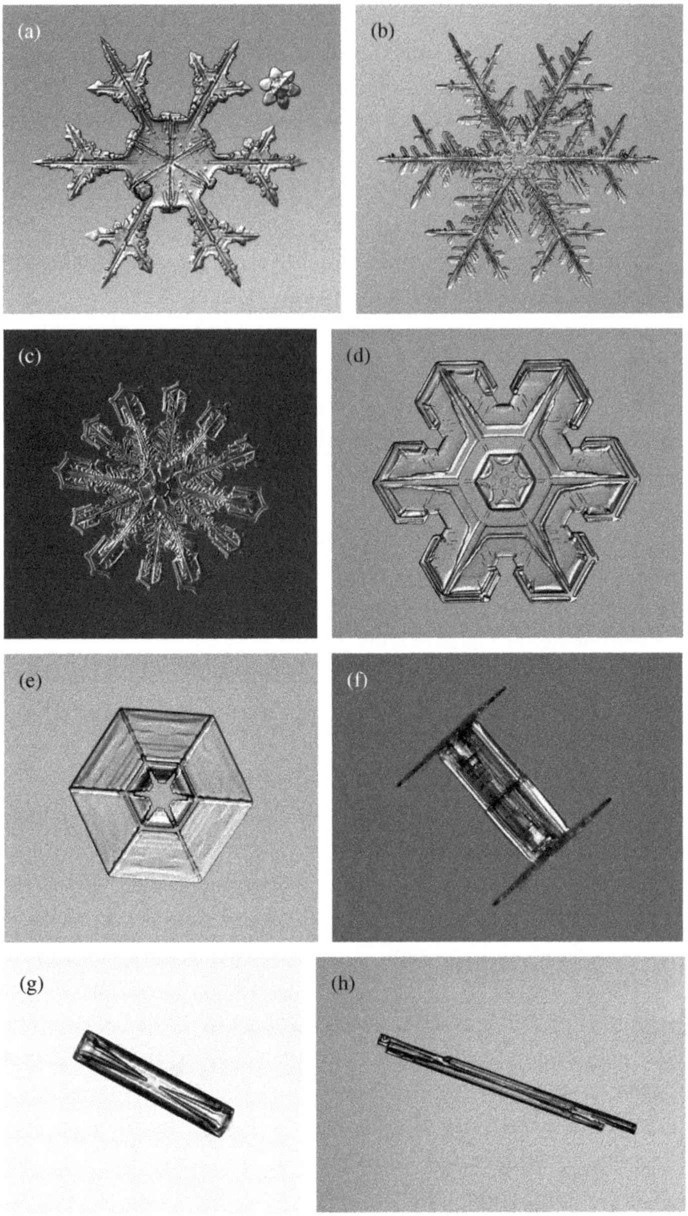

Fig. 8.14 Photographs showing (a) a stellar dendrite, (b) a fern-like stellar dendrite, (c) a 12-sided snowflake, (d) a stellar plate, (e) a plate, (f) a capped column, (g) a hollow column and (h) a needle. Photographs taken from www.snowcrystals.com and used by permission and courtesy of Kenneth Libbrecht. A black and white version of this figure will appear in some formats. For the color version, please refer to the plate section.

ice phase is nucleated within a cloud droplet then an RH value close to water saturation prevails, corresponding for instance to an ice saturation ratio $S_i = 1.22$ at -20 °C and 1.47 at -40 °C (Figure 2.11). Ice nucleation from the vapor phase also requires S_i to be significantly higher than the water saturation ratio S required for the Köhler activation of

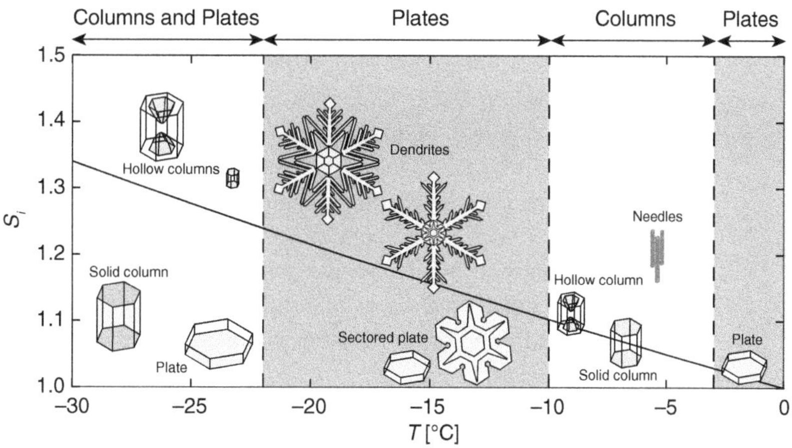

Fig. 8.15 An ice crystal habit diagram, showing the various types of ice crystals that grow in air at atmospheric pressure, as a function of T and the saturation ratio with respect to ice, S_i. The solid line denotes saturation with respect to water.

a cloud droplet. Either way, a newly nucleated ice crystal encounters a vapor phase at a significantly higher saturation ratio than that encountered by a newly activated cloud droplet. Hence, the water vapor gradient at the surface driving diffusional growth is larger for an ice crystal than for a cloud droplet.

The environment will remain favorable for ice crystal growth as long as supersaturation with respect to ice is maintained. As discussed in Chapter 2, the growth of ice crystals at the expense of evaporating cloud droplets at vapor pressures between saturation with respect to water and to ice is called the Wegener–Bergeron–Findeisen process (Figures 2.12 and 8.18h). This process is more efficient if a mixed-phase cloud is mixed homogeneously than if the cloud consists of individual pockets of ice and water. In a homogeneous mixed-phase cloud, the cloud droplets will evaporate more rapidly owing to their on-average larger proximity to the ice crystals. Eventually all cloud droplets will have disappeared and the saturation ratio will have decreased to the value with respect to ice. At this point the Wegener–Bergeron–Findeisen process stops, if it has not been interrupted already by the freezing of the remaining water droplets. The growth processes for ice crystals as described below are the same as for cloud droplets, namely diffusional growth followed by collisions in which crystals stick together.

8.3.1 Growth by diffusion

In only a few minutes an ice crystal can grow by diffusion to many tens of micrometers, if its fall velocity is on the order of 0.1 m s^{-1}. It may reach the ground as an individual ice crystal, highlighting again the importance of diffusional growth for ice crystals. In the Arctic, so-called diamond dust has been observed, which refers to ice crystal precipitation falling from a clear sky (Intrieri and Shupe, 2004). This clear-sky precipitation is attributed to very low INP concentrations: the individual ice crystals grow quickly because they do not need to compete for the available water vapor and are so sparse that no visible cloud is formed. Moreover, they sediment and reach the ground before colliding with other particles. Furthermore, some drizzle or light rain has its origin in unaggregated crystals which melted before reaching the ground.

The diffusional growth of an ice crystal (Figure 8.18b) is analogous to that of a cloud droplet (Figure 8.18a) but requires the non-spherical shape of the ice crystal to be taken into account. Therefore its growth is expressed in terms of a mass growth rather than a growth in radius. Also, water vapor molecules cannot join the ice crystal in any random orientation but must be oriented in such a way that the crystal structure is maintained; thus not every attempt of a water vapor molecule to join the crystal surface is successful. This effect is expressed by an accommodation coefficient α_m, so that the mass growth rate of an ice crystal is given as

$$\frac{dm_i}{dt} = \alpha_m 4\pi C(S_i - 1)\left[\left(\frac{L_s}{R_v T} - 1\right)\frac{L_s}{KT} + \frac{R_v T}{e_{s,i}(T) D_v}\right]^{-1}$$
$$= \alpha_m \frac{4\pi C(S_i - 1)}{F_k^i + F_d^i}, \tag{8.11}$$

where m_i is the mass of the ice crystal, D_v is the molecular diffusion coefficient of water vapor in air, K is the thermal conductivity of air and C is referred to as the capacitance. The mass growth rate given in eq. (8.11) results from integration of the flux of water vapor to the ice crystal over its surface. It can be shown that the flux of water vapor to or from a liquid droplet or ice crystal is governed by equations analogous to those describing the electrical field around a conducting surface held at a given electrostatic potential. In the electrostatic case the capacitance C, depending on the size and shape of the conductor, relates the electrical field strength to the conductor's electrostatic potential. Owing to this mathematical equivalence, we can see that eq. (8.11) also contains a "capacitance" C which incorporates the size and shape of the crystal. For a sphere, C equals its radius ($C = r$). Circular disks that are used as approximations for plate-type crystals have capacitance $C = 2r/\pi$. Capacitances for other shapes can be found in Lamb and Verlinde (2011).

Equation (8.11) is analogous to eq. (7.22) for cloud droplets if the constants are summarized as F_k^i and F_d^i, as follows:

$$F_k^i = \left(\frac{L_s}{R_v T} - 1\right)\frac{L_s}{KT} \quad \text{and} \quad F_d^i = \frac{R_v T}{D_v e_{s,i}(T)}.$$

However, in eq. (8.11) we use the saturation ratio with respect to ice, S_i, the latent heat of sublimation, L_s, instead of that for vaporization in the definition of the thermodynamic term F_k^l and the saturation vapor pressure with respect to ice, $e_{s,i}$, instead of saturation with respect to water in the vapor diffusion term F_d^l. In mixed-phase clouds, where the initial growth of ice crystals usually occurs at or close to water saturation, S_i can be approximated by $e_{s,w}/e_{s,i}$. As discussed in Chapter 7 the growth rate is a function of temperature and pressure.

The mass accommodation coefficient α_m was found to be 0.5 for a small ice crystal at $T > 263$ K (Rogers and Yau, 1989), but may be considerably less at lower temperatures.

While one study suggested an α_m value as small as 0.005 at 223 K (Magee *et al.*, 2006) implying that only 1 in 200 attempts of a water vapor molecule to join the ice crystal lattice is successful, a newer study suggests that α_m is independent of temperature, with values between 0.2 and 1 in the investigated temperature range 190–235 K (Skrotzki *et al.*, 2013).

8.3.2 Snow formation by aggregation

The collision–coalescence process in cold clouds extends beyond that in warm clouds because ice crystals, in addition to colliding with other ice particles (ice crystals or snowflakes), can collide with cloud droplets or raindrops.

An ice crystal can collide or clump together with another ice crystal as indicated in Figure 8.18d. This process is called aggregation and will lead to an aggregated snowflake (Figure 8.16e). Aggregation in cold clouds is thus the counterpart to growth by collision–coalescence of cloud droplets in warm clouds. However, the collection efficiency for aggregation is less well understood than that for the collision–coalescence process of cloud droplets (eq. 7.33). For ice crystals the coalescence efficiency $\hat{E}$ is referred to as the sticking efficiency. It depends on temperature because at higher temperatures ice crystals are more likely to stick together than at lower temperatures. Ice crystals have a quasi-liquid layer (QLL) that forms on their surface (Section 6.3). It is thicker at warmer temperatures and causes the sticking efficiency to be larger. The process whereby two pieces of a solid substance stick together is called sintering. It is effective only within a few degrees of the melting point. Sintering can explain observations showing that the aggregation process is mainly limited to temperatures higher than $-10\,°C$, where it can result in huge snowflakes. It also explains why snow remains powdery at cold temperatures but easily compacts to snowballs at higher (subzero) temperatures.

8.3.3 Growth by accretion and terminal velocity of snowflakes

Snowflakes can grow considerably larger than raindrops as they tend to be rather two-dimensional and have a much lower density. The dendritic structure of snowflakes allows rapid diffusional growth at the edges and branches. Their lower density and basically two-dimensionality result in a lower fall speed, which prolongs the time for interactions. Snowflakes thus experience more collisions than raindrops for a given distance fallen despite the lower sticking efficiency of ice crystals with each other.

8.3.3.1 Terminal velocities of snowflakes and graupel particles

The determination of fall speeds of ice particles is more complex that that of liquid hydrometeors due to the variety of shapes and densities of ice particles. Many empirical formulas for the fall velocity v_T in m s^{-1} are power laws of the form

$$v_T = ar_h^b, \tag{8.12}$$

Fig. 8.16 Images of different solid precipitation particles: (a) a lightly rimed ice crystal, (b) a moderately rimed ice crystal (c) a densely rimed ice crystal, (d) two graupel particles, (e) an aggregate and (f) hailstones. Photographs (a–d) are taken from Baschek (2005); Photograph (e) is by Jan Henneberger and (f) is by Thomas Winesett.

with coefficients a and b that describe the dependence of v_T on the hydrometeor size r_h (maximum dimension). As discussed in Section 7.2, v_T depends on the mass-to-area ratio of the falling hydrometeors, where area refers to the area projected to the flow. Both area and mass can be expressed in terms of the radius r_h as laws of the form $A(r) = \alpha r_h^{\beta}, m(r) = \gamma r_h^{\delta}$, where $\alpha, \beta, \gamma, \delta$ are empirical constants.

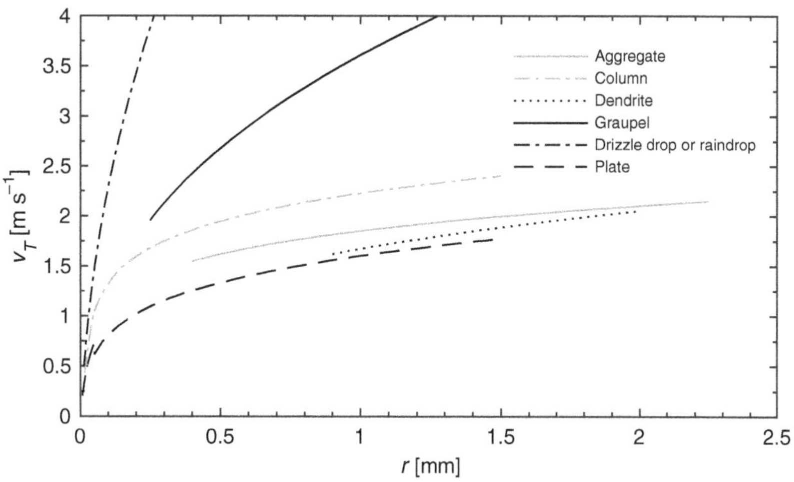

Fig. 8.17 Terminal velocities of different hydrometeors based on the parameterizations given in Mitchell and Heymsfield (2005).

For very flat snowflakes the area and mass and thus the drag force F_D introduced in eq. (7.35) and the gravitational force F_g are proportional to r^2. Therefore the terminal velocity does not increase with increasing size. Snowflakes of all sizes can be observed to fall with nearly the same velocity in a snow storm (Figure 8.17). Similarly, raindrops with equivalent radii 2.5 mm and larger become oblate spheroids and hence their terminal velocity levels off at that size (Figure 7.10). While the fall velocity of snowflakes approaches at most 2 m s^{-1}, that of raindrops levels off at 9–13 m s^{-1}, depending on the ambient pressure (Beard, 1976).

The increase in fall velocity of frozen hydrometeors with increasing size is largest for graupel, because the gravitational force on spherical graupel is proportional to r^3. Its increase with increasing r is not compensated by the increase in drag force and therefore its terminal velocity continues to rise with increasing equivalent size. The terminal velocity of a graupel particle is smaller than that for a raindrop for the same size because graupel particles are less dense due to the cavities they contain. The empirical fall velocities shown in Figure 8.17 are based on Mitchell and Heymsfield (2005), who consider a more realistic power law dependence of v_T on the hydrometeor size (maximum dimension) by allowing the coefficients to depend on particle size and shape (Mitchell, 1996).

8.3.3.2 Growth by accretion

The term "growth by accretion" is used for a collision process of two different hydrometeors that results in a permanent union of the two particles, which can either be of different size and/or in different phases. Various accretion processes are shown in Figure 8.18 and include the following:

- the collision–coalescence of differently sized droplets with each other (Figure 8.18c);
- the collision of differently sized ice crystals if they stick together. This process is also referred to as aggregation (Figure 8.18d);

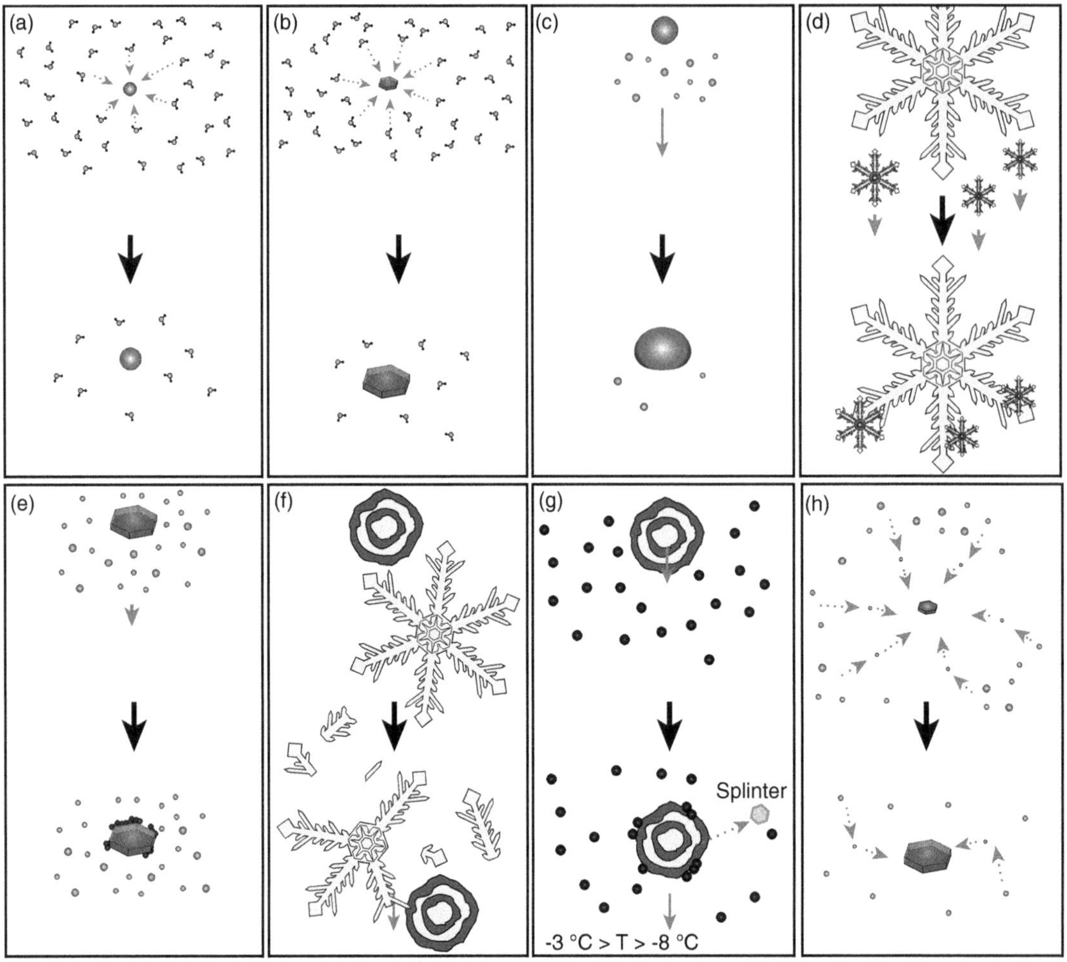

Fig. 8.18 Schematic of microphysical processes in an earlier and later stage: (a) condensational growth, (b) depositional growth, (c) collision–coalescence, (d) aggregation, (e) riming, (f, g) ice multiplication, (h) Wegener–Bergeron–Findeisen process. The processes shown in (c), (d) and (e) are also referred to as growth by accretion.

- the collision of an ice particle (ice crystal or snowflake) with a liquid particle (supercooled cloud droplet or raindrop) that freezes upon contact (Figure 8.18e); this process is also referred to as riming. Photographs of differently rimed ice crystals are shown in (Figure 8.16a–c).

Growth by accretion is caused either by the difference in fall velocities of the involved particles or by turbulence, which enables collisions between them. Whether the colliding particles remain together depends on the coalescence efficiency for liquid hydrometeors and on the sticking efficiency $\hat{E}$ when ice particles are involved. $\hat{E}$ of a cloud droplet or raindrop with an ice particle is approximately unity, because freezing is likely to occur upon contact with the supercooled cloud droplet or raindrop. Riming leads to a hydrometeor whose density is between the density of a liquid hydrometeor and the density of a snowflake or ice crystal. Typical end products of the riming process are graupel (Figure 8.16d) if the particles remain below 2.5 mm in melted radius or else hailstones (Figure 8.16f).

For hailstones to form convective clouds with high vertical velocities are necessary, so the rimed particle is caught in the updrafts repeatedly. There it experiences alternating growth by accretion with ice crystals at lower temperatures (the dry growth regime) and growth by accretion with cloud droplets when falling through regions of higher temperatures (the wet growth regime) as will be discussed in Chapter 10. A hailstone is denser than graupel because it can take up more liquid water prior to freezing. In graupel formation, the collected cloud droplets freeze more or less in their original shape, which leads to cavities within graupel and therefore to a mass density well below that of bulk ice.

The equation for growth by accretion in which the collector particle is an ice particle of equivalent radius R is analogous to that leading to a raindrop (eq. 7.51):

$$\frac{dm_i}{dt} = \bar{\bar{E}} M_l \pi R^2 [v_T(R) - v_T(r)], \tag{8.13}$$

where m_i is the mass of the frozen collector particle, M_l is the liquid water content of the cloud in the case of graupel and hail or the frozen water content in the case of aggregates, $v_T(R)$ is the terminal velocity of the collector particle, $v_T(r)$ is the terminal velocity of the particle being collected and $\bar{\bar{E}}$ is the mean collection efficiency over the full size range of collected particles.

8.4 Collapse of ice particles

Snowflakes and rimed crystals can disappear either because they melt or because they sublimate. Also, upon collision with other hydrometeors they could completely or partially collapse and may eject splinters. These splinters are a secondary source of ice crystals as will be discussed next.

8.4.1 Ice multiplication

The ice crystal number concentration in mixed-phase clouds has been observed to occasionally exceed the INP concentration by more than an order of magnitude (Crosier *et al.*, 2011; Henneberger *et al.*, 2013). The first ice crystals have to form on INPs, but additional ice crystals may be produced by secondary processes in which the primary ice crystals are "multiplied".

Secondary ice crystal production can occur by fracturing when graupel encounters slower-falling dendritic crystals (Figure 8.18g). During these collisions the dendritic crystals can break and leave a multitude of small ice particles behind. Another mechanism of ice multiplication is the production of secondary particles when supercooled drops of an appropriate size and temperature are captured by graupel (Figure 8.18g). This is called the Hallett–Mossop or rime-splintering process (Hallett and Mossop, 1974) and has been found to occur at temperatures between −3 and −8 °C when cloud droplets larger than 12.5 μm in radius collide with graupel of at least 0.25 mm in radius (Lamb and Verlinde, 2011).

It was observed that at $-5\,°C$ approximately 250 droplets need to accrete on a graupel particle for one splinter to be released (Pruppacher and Klett, 1997). As shown in Figure 8.12, the ice crystal concentration without secondary processes varies between 0.01 to 0.1 per liter at temperatures $> -8\,°C$. For a typical cloud droplet number concentration of 100 cm^{-3} (Figure 7.1 and Table 1.3) this yields an addition of 0.4 ice crystals per cm^3, which corresponds to an increase of up to 3–4 orders of magnitude at temperatures $> -8\,°C$.

The reason for the production of ice splinters during riming is the build-up of pressure within a freezing droplet. This means that the droplet freezes from the surface, so that freezing of the inner part then leads to a volume increase of the inner part, which attempts to expand against the rigid frozen surface. This pressure may be relieved by the cracking of the shell and the ejection of ice splinters. The narrow temperature range and the minimum size of the droplets are explained as follows (Mossop, 1985). Between -3 and $-8\,°C$, droplets larger than 12.5 μm in radius accrete on the ice particle by a narrow ice bridge, keeping their identities. At temperatures higher than $-3\,°C$, droplets tend to spread over the ice surface rather than remaining as discrete entities. The same argument holds for droplets below 12.5 μm in radius. At temperatures below $-8\,°C$, the freezing is thought to proceed so rapidly that the resulting ice shell is too thick to be disrupted.

8.4.2 Melting and sublimation of ice and snow

Our discussion of cloud microphysical processes closes with a brief consideration of the sublimation and melting of ice and snow. The melting of ice and snow does not require a nucleation process for its initiation. As the ice crystal structure at the surface starts to change into a quasi-liquid layer even at subzero temperatures, thermodynamically the new stable liquid phase is already present when the temperature approaches 0 °C. Therefore, there is no energy barrier at the melting temperature that needs to be overcome (Section 6.3). Note, however, that this holds only for finite ice crystals exposing a surface to air or water vapor, where the quasi-liquid layer can form. If melting were to be initiated from within the bulk, it would be associated with a nucleation process and an energy barrier, like a conventional first-order phase transition.

Because melting consumes energy, the temperature needs to exceed 0 °C for the melting of ice or snow to commence. Like the evaporation of raindrops the sublimation of snowflakes depends on the relative humidity, which needs to be high enough for the snowflakes, graupel particles or hailstones not to sublimate completely on their way to the ground.

8.5 Summary of microphysical processes in warm and cold clouds

The growth processes discussed in this and the previous chapter were illustrated schematically in Figure 8.18. Figure 8.19 shows where they would occur in a convective cloud with a warm cloud base and a cold cloud top.

In a cloud, droplets initially form by the activation of aerosol particles (Chapter 6). The droplets increase in size by diffusional growth until some droplets are large enough for the

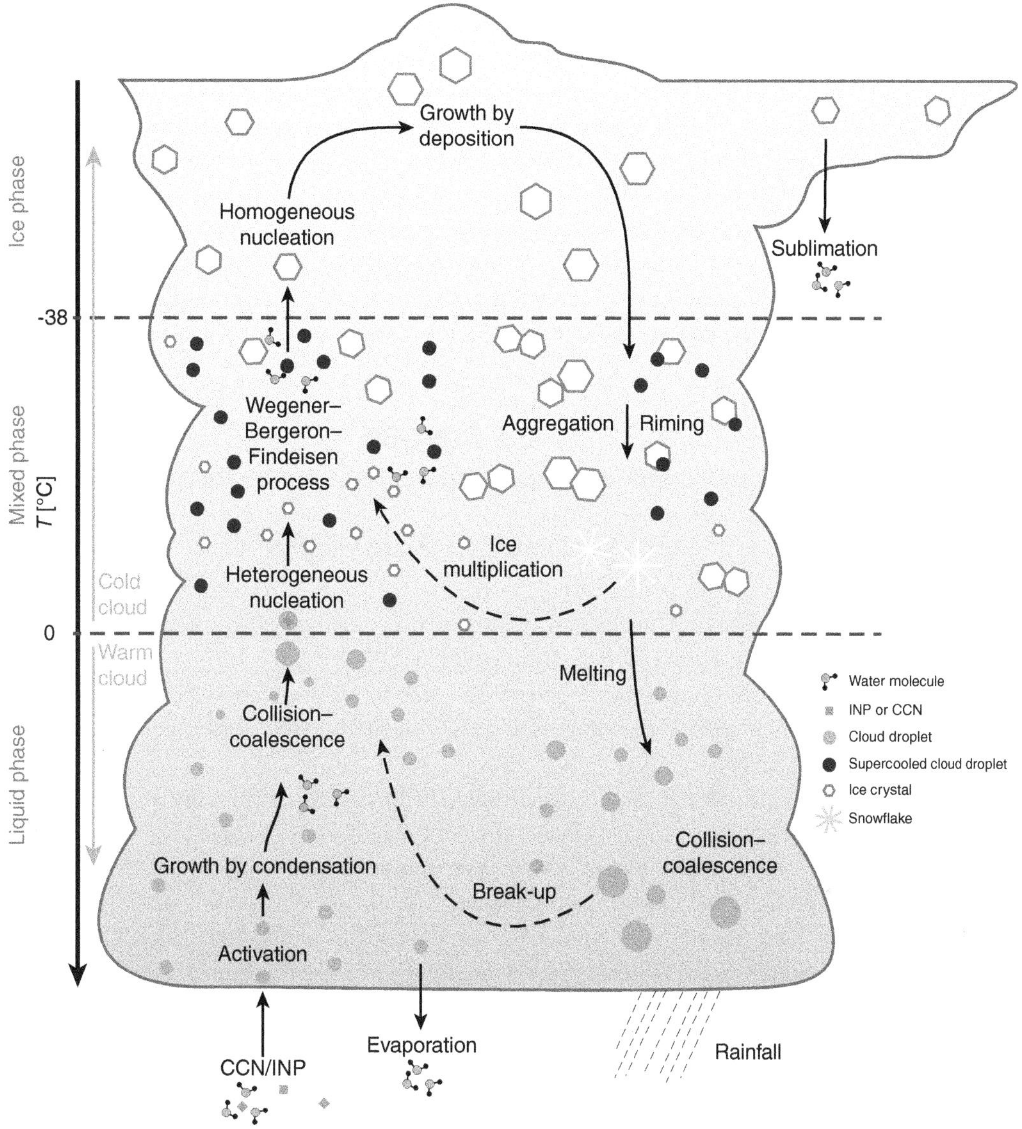

Fig. 8.19 Microphysical processes occurring in a convective cloud with a cloud base temperature higher than 0 °C and a cloud top temperature lower than −38 °C, so that warm-, mixed- and ice-phase processes take place. The small arrows show typical trajectories of a cloud particle. A black and white version of this figure will appear in some formats. For the color version, please refer to the plate section.

collision–coalescence process to commence. If the cloud top extends beyond −10 °C or so, the first ice crystals will nucleate to form primary ice. Then, because of the difference in vapor pressure between water and ice, these ice crystals will grow by deposition at the expense of the cloud droplets (the Wegener–Bergeron–Findeisen process, Figures 8.18h and 2.12) if the vapor pressure lies between saturation with respect to water and ice. Then this part of the cloud will be completely glaciated. Ice crystals grow further by deposition until they reach a sufficiently large terminal velocity to fall against the updraft. They predominantly grow by deposition at temperatures below −10 °C. At higher subzero temperatures the collection efficiencies involving ice crystals are higher because of their quasi-liquid layer. This favors growth by aggregation. Ice crystals and snowflakes can also grow by riming if they encounter supercooled cloud droplets. During riming, secondary ice crystals can be produced that are small enough to be carried upward in the cloud by the updraft. Aggregates and rimed structures falling into regions of the cloud that are at temperatures above 0 °C will melt to raindrops. These raindrops can grow further by collision–coalescence and leave the cloud. If the raindrops are larger than 5 mm, they will be unstable and break up into smaller droplets (Section 7.3). Below cloud base, raindrops may partly or fully evaporate depending on the relative humidity below cloud base and on their size. This increases the relative humidity below cloud base until it eventually reaches 100%.

8.6 Exercises

Single/multiple choice exercises

1. Heterogeneous ice nucleation is favored over homogeneous freezing. Which of the following is/are true? Ice nucleating particles (INPs) . . .
   (a) . . . reduce the energy barrier that needs to be overcome when water molecules group together randomly to form a sufficiently large cluster that can serve as an ice embryo.
   (b) . . . that have a large contact angle with an ice embryo are more efficient in nucleating ice crystals than those with a small contact angle.
   (c) . . . initiate freezing at lower supersaturation and temperature than would be possible by homogeneous freezing.
   (d) . . . are much less common in the atmosphere compared with CCN, but their concentration increases exponentially with increasing temperature.
2. The freezing mode in which water vapor is deposited directly onto an INP is called deposition nucleation. Which of the following is/are true?
   (a) The activated fraction INPs increases as the temperature decreases.
   (b) The frozen fraction INPs decreases as the temperature decreases.
   (c) The frozen fraction INPs increases but the activated fraction decreases.
   (d) The activated fraction INPs and the frozen fraction INPs are always equal.
3. Once an ice crystal has formed, it generally grows until it reaches precipitation size. Which of the following is/are true?
   (a) Growth by diffusion is more efficient for ice crystals than for cloud droplets. Ice crystals can even reach precipitation size by diffusional growth.
   (b) The diffusional mass growth of an ice crystal does not depend on its size and shape.
   (c) All water vapor molecules colliding with an ice crystal are incorporated into the ice crystal surface since the sticking coefficient is close to unity.
   (d) High supersaturations with respect to ice lead to the formation of needles and plates.

4. Different freezing modes exist in the atmosphere. In cirrus clouds formed in-situ, ...
   (a) ... ice crystals can be formed by homogeneous freezing of pure water droplets.
   (b) ... ice crystal formation cannot occur by heterogeneous nucleation as no INPs are present at these altitudes.
   (c) ... ice crystals cannot be formed if the saturation is below water saturation because water droplets are not in equilibrium under these conditions.
   (d) ... solution droplets play an important role in ice formation.
5. Consider a cloud which consists of ice crystals at the cloud top and supercooled droplets further below. The ice crystals can collide with these droplets when falling as (solid) precipitation. Which of the following is/are true?
   (a) Riming is an aggregation process.
   (b) During riming, additional ice crystals can be produced; thus riming can increase the ice crystal concentration of a cloud significantly.
   (c) In an environment of high supercooling, large INP concentration and low updraft velocities very large precipitation particles can be formed due to riming.
   (d) For riming, the sticking efficiency is close to one due to the quasi-liquid layer.
6. The fractions of pure ice clouds, pure liquid clouds and mixed-phase clouds from different observations show a large variability, even if measured at the same temperature. Which of the following could be direct reason(s) for this?
   (a) Differing number concentrations of CCN and INPs.
   (b) Differing updraft velocities.
   (c) Measurement uncertainties and detection limit of ice crystal concentration measurements.
   (d) Differing Coriolis parameters.
7. Crystalline ammonium sulfate can act as INPs at certain temperatures and relative humidities. Under which conditions may these INPs be important?
   (a) Where no better INPs (like mineral dust and biological particles) exist.
   (b) At temperatures between 0 °C and −20 °C.
   (c) In cirrus clouds.
8. Assume that you want to compare the INP ability in the immersion mode of two mineral dust particle types (e.g. a clay mineral and a quartz particle) in the laboratory. What do you need to ensure in your experiments?
   (a) A large number of data points for better statistics.
   (b) The same temperature in both experiments.
   (c) High temporal resolution.
   (d) Particles should originate from the same soil.
9. Which of the following process(es) is/are referred to as accretion?
   (a) The collision of two cloud droplets of the same size, which leads to the formation of a larger cloud droplet.
   (b) The collision of a cloud droplet with a raindrop, which leads to the formation of a larger raindrop.
   (c) The collision of an aerosol particle with a raindrop.
10. Precipitation can form via the liquid or ice phase. Which of the following statements is/are true?

(a) Because most precipitation in mid-latitudes occurs as rain, ice-phase processes are of secondary importance.
(b) Secondary ice crystal production requires the presence of graupel particles.
(c) Diffusional growth is the dominant growth mechanism in stratus clouds and growth by accretion is the dominant growth mechanism in convective clouds.
(d) The depositional growth of ice crystals is similarly described to the condensational growth of cloud droplets, taking the latent heat of sublimation and saturation vapor pressure with respect to ice, among other factors, into account.

Long exercises

1. (a) Explain the two freezing modes "contact freezing" and "deposition nucleation".
   (b) Name four factors that determine the efficiency of contact freezing.
2. (a) Explain the difference between a stochastic and a deterministic process. Which process is described by classical nucleation theory?
   (b) Name three parameters that influence the heterogeneous nucleation rate according to classical nucleation theory.
3. How many water molecules would be required to form a stable ice embryo within the vapor phase at $-15\,°\mathrm{C}$ and saturation ratio with respect to ice $S_i = 1.2$? The Gibbs free energy for cluster formation $\Delta G_{i,v}$ is given by eq. (8.1) and the volume of one H_2O molecule in ice $v_0 = 3.25 \times 10^{-29}$ m^3. The density of ice $\rho_i \approx 0.92$ g cm^{-3} and the surface tension between ice and vapor is 0.109 N m^{-1}.
4. Mixed-phase clouds contain ice crystals as well as liquid droplets. Consider such a cloud at a temperature $T = -4\,°\mathrm{C}$, pressure $p = 800$ hPa and a humidity which corresponds to supersaturations with respect to ice of 5% and with respect to water of 1%. In the cloud, an ice crystal and a droplet both grow by diffusion, each starting from mass $m_0 = 10^{-8}$ g. The ice crystal is a thin hexagonal plate, so that its capacitance can be approximated by $C = 2r_i/\pi$, where r_i is the radius of the ice crystal.
   (a) Determine the times it takes for the droplet and the ice crystal to grow to a total mass $m_1 = 1.1 \times 10^{-8}$ g. You can neglect solution and curvature effects when calculating the droplet growth. For the ice crystal, you can assume that its mass m_i and diameter d_i are related by $m_i = \alpha d_i^3$, with $\alpha = 1.9 \times 10^{-2}$ g cm^{-3}.
   (b) Which of the two cloud particles grows faster? Explain the main reason for the difference in growth speed.
   (c) How would the situation in the mixed-phase cloud change for supersaturation with respect to ice but subsaturation with respect to water? Explain qualitatively in a few sentences.
5. Consider a spherical graupel particle with radius 2 mm. This graupel particle falls through a cloud consisting uniformly of cloud droplets with radii of 5 μm. The mean collection efficiency is 0.85 and the liquid water content of the cloud M_l = 1 g m^{-3}. Approximately how much will the graupel particle grow in 10 seconds, i.e. what is the difference in mass Δm and its final melted radius r_h? Use the empirical relationship $v_T = a r_h^b$ for the terminal velocity of the graupel particle with $a = 82.4$ m$^{0.4}$ s^{-1} and $b = 0.6$. You can assume that the density of the graupel particle is constant at 0.5 g cm^{-3}.

9 Precipitation

In the previous three chapters we discussed on the microscale how cloud droplets and ice crystals form and how they can grow and reach precipitation size. In this chapter we take a more macroscopic perspective and start with an overview of observed precipitation rates (Section 9.1). Raindrop and snowflake size distributions have been observed to be exponential, as discussed in Section 9.2. At Earth's surface, rainfall and snowfall rates are measured with rain gauges. In the atmosphere they are estimated from radar reflectivity. In fact, radars are irreplaceable for the short-term forecast, called "nowcast", of precipitation events. We discuss how estimates of precipitation rates can be obtained from radar measurements and how radar images can be interpreted (Section 9.3). Next we discuss how precipitation can be classified into stratiform and convective precipitation (Section 9.4) and explain its mesoscale structure (Section 9.5). This chapter concludes with a discussion of the geographical distribution of precipitation in the present climate, how it has changed since the 1950s and how it is projected to change until the end of this century (Section 9.6).

9.1 Precipitation rates

As discussed in Chapter 7, precipitation in warm clouds involving only the liquid phase forms via collision–coalescence. Collision–coalescence is favored in clouds which have a large liquid water content with relatively few cloud droplets. Therefore clouds which form precipitation involving only the liquid phase are mainly convective clouds over the tropical oceans. However, even in the tropics only 31% of the total precipitation is "warm rain" (Lau and Wu, 2003). Warm-phase precipitation often produces light precipitation (drizzle) because it is less efficient than precipitation formation involving the ice phase. An exception are tropical islands such as Hawaii, where significant warm rain has been observed due to orographic forcing.

After hydrometeors are formed, they can continue to grow by collisions with other hydrometeors such as cloud droplets (Section 7.2) or ice crystals (Section 8.3) as illustrated in Figures 8.18 and 8.19. All collision processes that involve hydrometeors of either different sizes or different phases are summarized as growth by accretion (Section 8.3.3). Depending on the phases of the hydrometeors involved, accretion leads to raindrops, snowflakes, graupel or hailstones. Regardless of how precipitation is initiated inside clouds, over a large part of Earth's surface it reaches the ground as rain because the average surface temperature of Earth is +15 °C.

The intensity of precipitation is expressed in terms of a precipitation rate R, which is the volume flux of precipitation through a horizontal surface per unit time interval [m^3 m^{-2} s^{-1} = m s^{-1}], but is usually expressed either in mm h^{-1} or in mm d^{-1}:

$$R = \frac{4}{3}\pi \int_0^\infty n_N(r_h) r_h^3 v_T(r_h) dr_h, \tag{9.1}$$

where $v_T(r_h)$ is the terminal velocity of a hydrometeor with radius r_h and $n_N(r_h)$ is the number distribution of hydrometeors. This equation can be applied to both rain and frozen precipitation (snow, graupel and hail). For snowflakes, r_h refers to the radius of the melted snowflakes and R to the equivalent rainfall rate, expressed in mm h^{-1} of melted water. As discussed in Chapter 8, the fall velocity of snow is considerably smaller than the fall velocity of raindrops. Equation (9.1) can also be applied to graupel or hail, in which case the quantities correspond to the fall velocity and melted diameter of the graupel or hail.

Values of R vary from trace amounts up to several hundred mm h^{-1}. Rainfall rates in excess of about 25 mm h^{-1} are always associated with convective clouds. At most locations, annual snowfall rates tend to be at least an order of magnitude lower than rainfall rates. First, the terminal velocities of snowflakes are an order of magnitude lower than those of raindrops, decreasing snowfall rates in a given precipitation event as compared to rainfall rates. Second, snow is limited to temperatures < 0 °C, which in most locations occurs far less often than warmer temperatures. This applies only to overall climatological values. In a single precipitation event snowfall rates can exceed rainfall rates.

Globally, precipitation varies greatly. On the one hand Quillagua, located along the west coast of Chile, is the driest place on Earth with 0.2 mm observed annually. Arica, also in Chile, went a record 173 months without receiving any precipitation, from October 1903 to January 1918 (http://wmo.asu.edu/#global). On the other hand, the highest average annual precipitation rate can be found in Mawsynram, north-east India, with 11 872 mm per year (http://wmo.asu.edu/#global). Cherrapunji, in eastern India, holds the record for the most precipitation ever observed in one year when 26 470 mm fell between August 1860 and July 1861 (http://wmo.asu.edu/#global) owing to monsoonal rainfall associated with orographic uplift.

9.2 Size distributions of hydrometeors

9.2.1 Raindrop size distribution

Raindrop size distributions (Figure 9.1) show a rapid decrease in drop number concentration with increasing size for raindrop radii r_r larger than 0.5 mm, owing to gravitational settling. At Earth's surface the distributions have been observed to be approximately exponential, especially in rain that is fairly steady. They are given by the so-called Marshall–Palmer distribution (Marshall and Palmer, 1948):

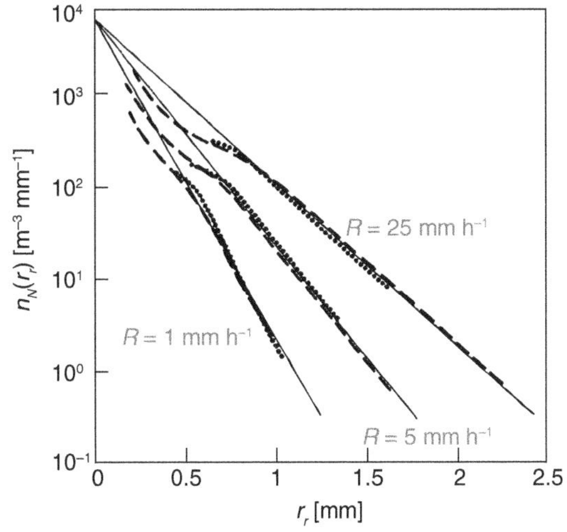

Fig. 9.1 Measured raindrop size distributions $n_N(r_r)$ (broken and dotted lines) for different rain rates R as a function of the radius of the raindrops r_r, together with best fit exponential curves (solid lines). Figure adapted from Marshall and Palmer (1948) with permission of the American Meteorological Society.

$$n_N(r_r) = N_0 \exp(-\Lambda r_r), \tag{9.2}$$

where N_0 is the intercept parameter in cm^{-4} and Λ is the slope of the raindrop size distribution in m^{-1} or cm^{-1}. In order to obtain the total number of raindrops, eq. (9.2) needs to be integrated over the number of raindrops per unit volume of air with radii between r_r and $r_r + dr_r$, which is $n_N(r_r)dr_r$. Marshall and Palmer (1948) found the following empirical dependence of Λ on the rainfall rate R:

$$\Lambda(R) = c_r R^{-0.21}, \tag{9.3}$$

where R is in mm h^{-1}, Λ is in cm^{-1} and c_r = 82 cm^{-1} (mm h^{-1})$^{0.21}$. The dependence of Λ on the rainfall rate indicates that the raindrop size distribution changes with rainfall intensity in such a way that the higher is R, the smaller is Λ. This means that $n_N(r_r)$ decreases less quickly towards larger sizes. Higher rainfall intensities are caused by higher updraft velocities in clouds and hence higher liquid water contents, which produce larger raindrops. Marshall and Palmer (1948) also observed a rather constant intercept parameter N_0 of 0.08 cm^{-4}, as shown in Figure 9.1. It has been found that the Marshall–Palmer distribution does not hold for small raindrops: the observations show deviations from the exponential distribution at radii below 0.8 mm (Figure 9.1). The Marshall–Palmer distribution holds when averaging over many rainfall events but cannot be applied for individual rainfall events.

Raindrops leaving a cloud hardly interact at all with other raindrops because their number density is much lower than that of cloud droplets. On average there is only one raindrop in a million cloud droplets (Figure 7.1). Raindrops may experience some shrinking due to evaporation when the RH below cloud is less than 100%.

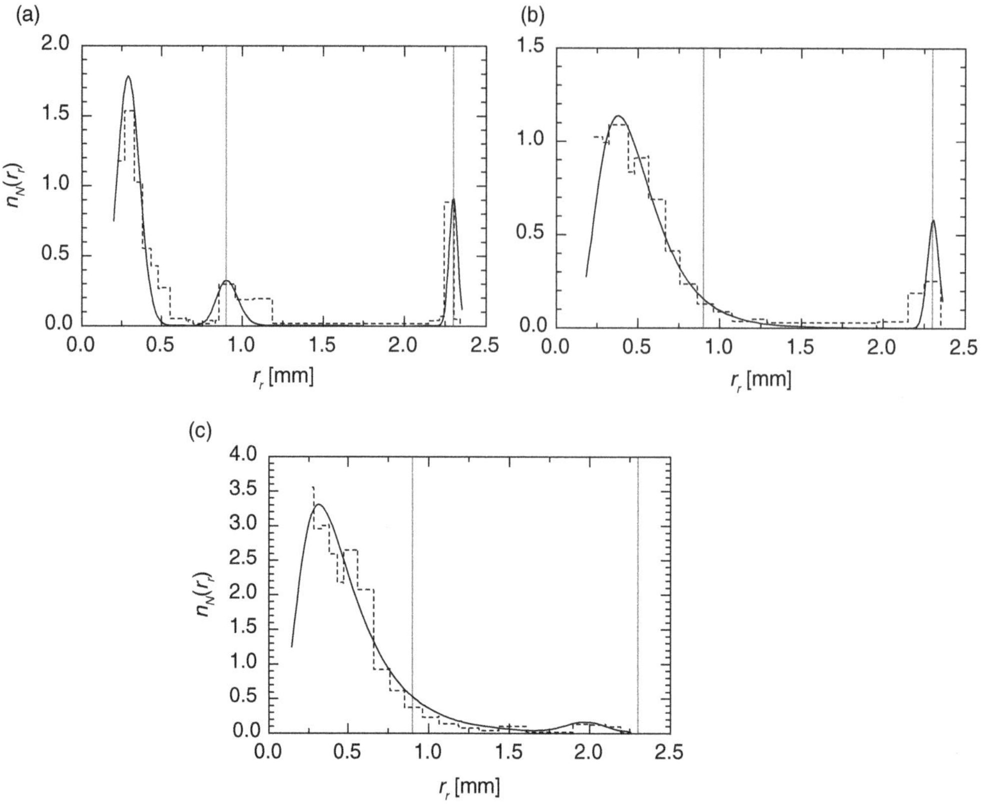

Fig. 9.2 Examples from laboratory experiments of average size distributions of raindrop fragments $n_N(r_r)$, in numbers per size interval (0.1 mm) as a function of raindrop radius r_r. The fragments were produced by (a) neck or filament break-up, (b) sheet break-up and (c) disk break-up (Figure 7.15). The measurements are shown as broken lines and distributions fitted to the data as solid lines. The original radii of the raindrops (0.9 and 2.3 mm) are indicated by the vertical lines. Note the different vertical scales. Figure adapted from Low and List (1982) with permission of the American Meteorological Society.

Size distributions of raindrops can, for instance, be observed acoustically with the Joss–Waldvogel distrometer (Joss and Waldvogel, 1967), which converts the kinetic energy of a raindrop impacting on a membrane into an electric pulse whose amplitude is a function of the raindrop size. The electric pulses are then converted into a raindrop size distribution.

Figure 9.2 shows an example of averaged measured size distributions after drop break-up, in numbers per fragment size interval (0.1 mm), obtained from laboratory experiments. The collisions resulted from a raindrop of 2.3 mm radius with a raindrop of 0.9 mm radius. For neck or filament break-up, a third peak consisting of satellite drops (Figure 9.2a) appears at around 0.3 mm in radius in addition to the two peaks centered at the original raindrop sizes. Sheet break-up is effective in removing the smaller drop size from the distribution whereas the larger drop keeps its identity (Figure 9.2b). Disk break-up creates the largest number of small drops by eliminating the identities of both colliding drops. Here the concentration of small satellite drops is largest (Figure 9.2c).

9.2.2 Snowflake size distribution

Unlike rain, which always originates via the collision–coalescence process, individual ice crystals can grow large enough by diffusional growth to sediment and reach the ground. This phenomenon has been observed in otherwise clear or nearly clear skies and is called "diamond dust" or "cloudless ice precipitation" (Section 8.3.1). Diamond dust can occur anywhere in the world where the temperatures are well below freezing but is most common in the Arctic and in Antarctica. In the Arctic, it has been observed to occur 13% of the time during winter but not during summer (Intrieri and Shupe, 2004).

Most snow originates from snowflakes that formed by aggregation (Section 8.3.1). Because snowflakes are either single dendrites or irregular aggregates of crystals, there is no easy way to measure their linear dimension. Thus, data on snowflakes are usually expressed in terms of particle mass or equivalent radius r_{eq} of the drop formed when the snowflake melts.

An example of a snowflake size distribution is shown in Figure 9.3. As in the case of rain (eq. 9.2), the snowflake size distribution has been found to be exponential (Gunn and Marshall, 1958):

$$n_N(r_{eq}) = N_0 \exp(-\Lambda r_{eq}). \tag{9.4}$$

For snow the intercept parameter N_0 [cm^{-4}] is not constant as in the case of rain but depends on the precipitation rate R [mm h^{-1}] expressed in the water-equivalent depth of the accumulated snow:

$$N_0 = c_N R^{-0.87} \qquad \Lambda(R) = c_s R^{-0.48}. \tag{9.5}$$

with c_N = cm^{-4} (mm $\mathrm{h}^{-1})^{0.87}$ and c_s = 51 cm^{-1} (mm $\mathrm{h}^{-1})^{0.48}$. The size distribution of snow for a given precipitation rate was found to be broader than for rain (Gunn and

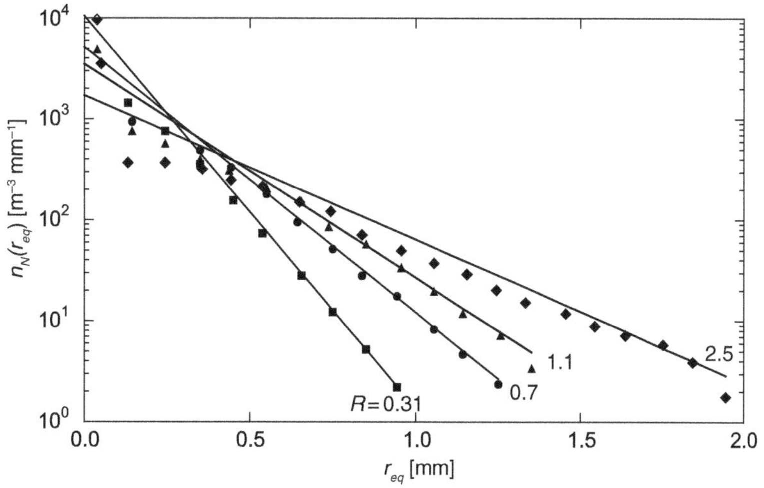

Fig. 9.3 Size distributions of snowflakes $n_N(r_{eq})$ expressed in terms of melted equivalent radii r_{eq} as a function of the precipitation rate R in mm h^{-1}. Figure adapted from Gunn and Marshall (1958) with permission of the American Meteorological Society.

Marshall, 1958). The reason for this is that snowflakes can grow considerably larger than raindrops owing to their lower fall velocities. At the same time, single ice crystals can reach the ground. In addition, the fall velocities of snowflakes are not uniquely dependent on their size but depend also on their density and shape. Owing to their lower fall speeds, snowflakes have less kinetic energy and upon collision burst into pieces less readily. Nevertheless, the fracturing of snowflakes is one mechanism of ice multiplication, as discussed in Section 8.4.1. As for raindrops, this fracturing is likely to be collision induced. While fracturing of raindrops is only a function of droplet size, fracturing of snowflakes may also depend on crystal habit and temperature.

All in all, the size distribution for snow is more variable than for rain, because snowflakes vary much more in density and shape and hence in their fall speeds. Also, it is more difficult to make size-segregated measurements of snowflakes than of raindrops.

9.3 Radar

Microwave radars were developed early in World War II to aid spotting distant ships and airplanes. Nowadays one of their uses is to detect precipitation particles.

"Radar" stands for radio detection and ranging and is an active remote sensing method. Remote sensing means that the instrument is in a location other than that of the objects to be measured. In active remote sensing instruments, an electromagnetic pulse is sent towards the particles or hydrometeors and their scattered signal is measured. In contrast, passive remote sensing refers to instruments that do not send out an electromagnetic pulse but receive radiation emitted from atmospheric gases, particles or hydrometeors. The electromagnetic pulse sent out in radar measurements is intercepted by the objects, as schematically shown in Figure 9.4. Part of the energy is scattered forward and lost to the radar antenna. The radiation that is scattered back to the antenna is focused and recorded by a receiver. Information about the distance of the objects from the radar can be obtained from the run time of the electromagnetic pulse.

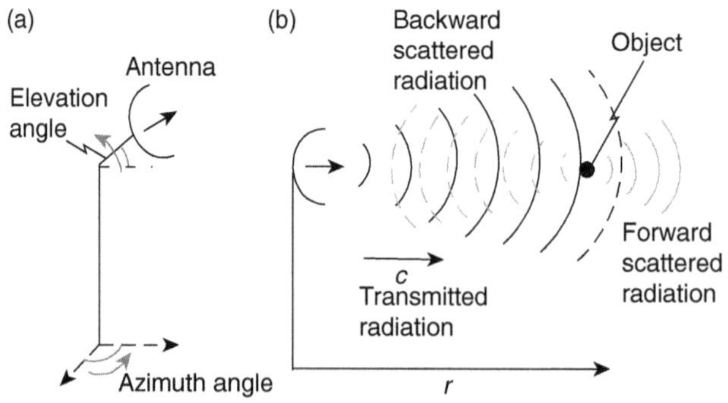

Fig. 9.4 Schematic of a radar device showing (a) the rotation of the radar antenna around an azimuthal angle in the horizontal and an elevation angle in the vertical direction, (b) the transmission of radiation with velocity c, which is then scattered by an object at distance r.

Nowadays, precipitation radars are an essential measurement tool for weather services globally. While rain gauges measure precipitation rates only at individual measurement sites, radars can detect precipitation over large areas and provide advanced warning for hazardous weather. Depending on the wavelength of the weather radar, it is capable of detecting anything from cloud droplets to birds.

The radar signal is proportional to the sixth power of the size r of the object, in the Rayleigh scattering regime. Because of the large size dependence of Rayleigh scattering, cloud droplets and ice crystals (in the size range of μm) cannot be detected with the same wavelength as precipitation particles (in the size range of mm, Table 1.2): if the radar echo (backward scattered radiation) from precipitation particles is sufficiently large, the radar echo from cloud droplets is too weak owing to the r^6 dependence. In fact, this strong dependence of the intensity of the scattered signal on the object size allows the specific detection of precipitation particles, which would otherwise generate a negligible signal compared with the bulk of the cloud droplets. Thus, radars with different wavelengths exist for different purposes. The most common are precipitation or weather radars, emitting electromagnetic pulses with wavelengths between 1 and 10 cm, which will be the focus of this section. These wavelengths are, however, too large to detect cloud droplets. Therefore cloud radars with shorter wavelengths, in the millimeter range, are used to detect cloud droplets, ice crystals and light drizzle.

9.3.1 Scattering regimes

The relationship between the size of the scattering object and the wavelength of the electromagnetic pulse determines the scattering regime. This relationship is commonly expressed by a dimensionless size parameter x given by the ratio of the circumference of a spherical particle and the wavelength:

$$x = \frac{2\pi r}{\lambda}. \tag{9.6}$$

Figure 9.5 shows x as a function of the radius of the object and the wavelength. For the visible part of the electromagnetic spectrum, highlighted in gray, x ranges from much less than 1 to much greater than 1 in going from air molecules, aerosol particles, haze and cloud droplets to raindrops.

Rayleigh scattering takes place when the size of the scattering object is much smaller than the wavelength of the electromagnetic wave. Spherical objects scatter in the Rayleigh regime if their size is smaller than $\sim$ 1/16 of the wavelength (Doviak and Zrnic, 1984) corresponding to $x < 0.4$. This is for instance the case when visible radiation interacts with air molecules. The extinction coefficient (eq. 12.1) in the Rayleigh scattering regime depends inversely on the wavelength to the fourth power (λ^{-4}). This strong wavelength dependence causes visible radiation to be split into its spectral colors when it scatters from air molecules. Rayleigh scattering is thus responsible for the blue sky during the day and for the red sky at dawn and dusk. The parameter x is also in the Rayleigh regime when cloud droplets scatter the signal sent out from cloud radars with millimeter wavelengths and

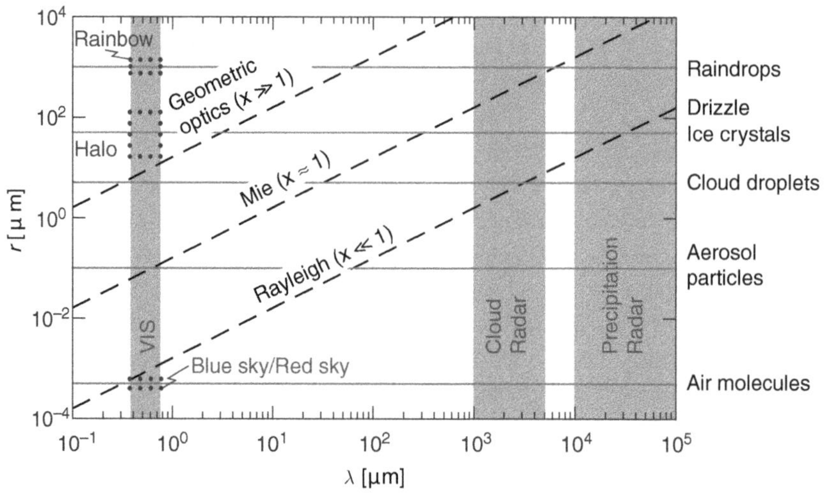

Fig. 9.5 Scattering regimes of spherical particles based on the non-dimensional size parameter x (eq. 9.6) as a function of wavelength λ and particle radius r. On the right-hand side, examples of atmospheric particles are given. The narrowest vertical band, labeled VIS, corresponds to the wavelengths of the visible spectrum.

when raindrops scatter the signal from precipitation radars with centimeter wavelengths. In the Rayleigh regime the scattered radiation is divided evenly between the forward and backward hemispheres of the scattering object. This is important since the radar detects only the backscattered fraction.

If r and λ are on the same order of magnitude, Mie scattering, which favors the forward direction, prevails. This is, for instance, the case when visible radiation interacts with aerosol particles. As the wavelength dependence of the extinction coefficient is much weaker ($\sim \lambda^{-1.3}$) in the Mie regime, all visible wavelengths are scattered similarly; this causes the sky over big cities to be whitish.

If the size of the objects is much larger than the incoming wavelength, scattering processes can be calculated from geometric optics. In this regime, again, scattering takes place predominantly in the forward hemisphere. The most prominent atmospheric examples for geometric optics are rainbows and halos.

9.3.2 Radar reflectivity

Whereas earlier radars provided only a two-dimensional horizontal map as a function of time, nowadays more and more radars provide full three-dimensional information about the precipitation pattern.

The power received at the radar antenna $\overline{P}_R$ averaged over a time interval of approximately 10 ms is given as

$$\overline{P}_R = C_R \frac{Z|K|^2}{r^2}. \tag{9.7}$$

Table 9.1 Typical values of *Z* for various hydrometeors and precipitation types, detected with either a cloud or a precipitation radar. Data based on Houze (2014), Peters and Görsdorf (2010) and Liu *et al.* (2008).

Radar type	Scatterer	*Z* values
Cloud radar	Cloud droplets	−40 to −20 dBZ
Cloud radar	Mixed-phase clouds	−20 to −10 dBZ
Cloud or precip. radar	Drizzle	−20 to 0 dBZ
Precipitation radar	Very light rain or light snow	0 to 10 dBZ
Precipitation radar	Moderate rain and heavier snow	10 to 30 dBZ
Precipitation radar	Melting snow	30 to 45 dBZ
Precipitation radar	Moderate to heavy rain	30 to 60 dBZ
Precipitation radar	Hail	> 60 dBZ

It depends on the radar reflectivity factor Z given in units of mm^6 m^{-3} and decreases quadratically with the distance between the scattering target and the radar r^2; C_R comprises constants and radar parameters: the wavelength, peak power, gain, pulse duration and beamwidth. See for instance Houze (2014). The factor $|K|^2$ is the square of the absolute value of the complex index of refraction, which is different for water and ice. The radar reflectivity factor Z can also be obtained from the sixth moment of the size distribution of the hydrometeors in the volume of air scanned by the radar:

$$Z = 64 \int_0^\infty n_N(r_h) r_h^6 dr_h \approx 64 \sum_i n_N(r_{h,i}) r_{h,i}^6 \Delta r_{h,i} = 64 \sum_i N(r_{h,i}) r_{h,i}^6 \qquad (9.8)$$

where $n_N(r_h)$ is the size distribution of the hydrometeors given in m^{-3} mm^{-1}, $r_{h,i}$ is given in mm^{-1} and $N(r_{h,i}) = n_N(r_{h,i})\Delta r_{h,i}$ is given in m^{-3}. As an example, let us assume a unit volume of air with 15 625 uniform raindrops of 0.5 mm radius. In this case Z is $64 N_{0.5\text{mm}} r_{0.5\text{mm}}^6 = 15\,625$ mm^6 m^{-3}. Because Z is proportional to the sixth moment of the size distribution, the same Z value can be obtained from just one raindrop with $r_h = 2.5$ mm. Note that this 2.5 mm raindrop weighs only 0.065 g whereas the 15 625 raindrops of 0.5 mm radius weigh 8.18 g altogether. Thus, if the radar encounters hydrometeors of different sizes in this scanned volume then the return echo will be dominated by the signal from the largest hydrometeors, which can cause an underestimation of the precipitation amount.

The term r_h^6 in eq. (9.8) causes the values of Z to increase sharply with increasing r_h. Therefore, instead of displaying Z in mm^6 m^{-3}, it is more practical to display it on a logarithmic scale with base 10. The common units for Z are

$$\text{dBZ} = Z[\text{dB}] = 10 \log \left(\frac{Z}{\text{mm}^6\ \text{m}^{-3}} \right).$$

This makes the logarithm dimensionless. Typical values of Z for different hydrometeors are provided in Table 9.1.

In addition to particle size, Z depends also on the phase of the hydrometeors, because $|K|^2$ in eq. (9.7) is different for water and ice. Thus, a better measure of the radar reflectivity is given by the equivalent reflectivity factor, which includes the complex index of reflection K:

$$Z_e = \frac{|K|^2}{0.93} Z, \tag{9.9}$$

with $|K|^2 = 0.197$ for ice particles and 0.93 for liquid particles (Houze, 2014). From eq. (9.9) it follows that, for a given size, liquid hydrometeors have a factor 4.7 higher Z_e values than ice particles, i.e. a reflectivity that is 6.7 dBZ higher in logarithmic units. This explains why in Table 9.1 values between 0 to 10 dBZ refer to light snow but only very light rain and values between 10 and 30 dBZ to heavier snow but only moderate rain.

Note that the radar reflectivity also needs to be corrected for the shape and density of the frozen hydrometeors. Consider a snowfall event with 100 snowflakes m^{-3} having equivalent radii 1.5 mm. According to eqs. (9.9) and (9.8), this would result in $Z_e = 42$ dBZ, which according to Table 9.1 corresponds to melting snow or rain. Therefore correction factors were developed that take the shape and density of the frozen hydrometeors in account (Löffler-Mang and Blahak, 2001). The dependence of Z on the sixth power of the radius of the hydrometeor (eq. 9.8) can be reduced by several orders of magnitude, depending on the type of snow and ice. If the snowflakes are, for example, assumed to be rimed dendrites of 1.5 mm then their reflectivity would need to be reduced by 30 dBZ. This gives a Z_e value of 12 dBZ, which is in agreement with the values reported for light to heavier snow in Table 9.1.

9.3.3 Relation of radar reflectivity to precipitation rate

On the basis of long-standing comparisons between radar reflectivities and rain gauge measurements, empirical relationships between radar reflectivity and rainfall rate are used at many weather services. For example, the Swiss weather service MeteoSwiss (MS) uses the following relationship:

$$Z_{MS} = c_{MS} R^{1.5}, \tag{9.10}$$

where R in mm h^{-1} denotes the rainfall rate in the case of liquid precipitation or the equivalent rainfall rate in the case of solid precipitation and $c_{MS} = 300$ $mm^{4.5}$ $h^{1.5}$ m^{-3}.

However, as illustrated in the small comparison between the 0.5 mm and 2.5 mm raindrops above, neither this empirical relationship nor the exponential distributions for rain or snow (eqs. 9.2 and 9.4) hold for an individual precipitation event, but only when an average over many events is taken. Alternatively Z can be related to R by assuming an average Marshall–Palmer (MP) raindrop size distribution. Inserting eq. (9.2) into eq. (9.8) and using eq. (9.3) to replace Λ yields

$$Z_{MP} = N_0 \frac{6!}{\Lambda^7} = N_0 \frac{6!}{(41)^7} R^{1.47} = c_{MP} R^{1.47}, \tag{9.11}$$

with $c_{MP} = 296$ $mm^{4.53}$ $h^{1.47}$ m^{-3}. This equation is in very good agreement with the empirical formula used by MeteoSwiss. An example of the different values of Z obtained using the empirical relationship (eq. 9.10) or the Marshall–Palmer distribution (eq. 9.11) for various rainfall rates is given in Table 9.2.

Table 9.2 Radar reflectivity Z values as a function of rainfall rate R obtained using the empirical-approach Z_{MS} (eq. 9.10) and the Marshall–Palmer distribution Z_{MP}.

R [mm h^{-1}]	0.1	1	10	100
Z_{MS} [mm^6 m^{-3}]	9.5	300	9500	300 000
dBZ_{MS}	10	25	40	55
dBZ_{MP}	10	25	39	54

9.3.4 Radar images

Depending on how the radar antenna is pointed, different types of radar images are produced. Plan position indicator (PPI) radar images refer to horizontal scans with the radar in the center pointing at a certain elevation angle (typically $< 3°$). The elevation angle to the vertical is used to avoid obstacles such as mountains and to avoid exposing humans to the emitted energy. From PPI scans a horizontal map of radar reflectivity is produced. The more hydrometeors are encountered by the transmitted electromagnetic radiation from the radar, the more the radar signal becomes attenuated. Thus, radar may not be able to detect multiple precipitating clouds if the transmitted radiation is attenuated by closer clouds. Although the scattered signal per unit volume decreases with increasing penetration depth due to attenuation of the radar intensity, the reflectivity remains the same because it is a property of the precipitating cloud (eqs. 9.8 and 9.9). In addition, owing to the quadratic distance law, the intensity of the emitted pulse that is received per unit area decreases with increasing distance from the radar. An attenuated radar signal may cause problems in an area of vigorous thunderstorms because the radar cannot detect what is behind them. Also, owing to the elevation angle of the PPI scan, the signals in the distance stem from higher altitudes; thus precipitation from low clouds in the distance could occur undetected.

Another type of radar image is produced from upward-pointing radars. It is called a height–time indicator (HTI) radar image and provides information on the vertical distribution of hydrometeors and how that distribution changes with time.

Doppler radars offer the possibility of determining the radial velocity of hydrometeors along the beam path of the radar in addition to measuring their reflectivity and distance from the radar. Depending on the scan, either the horizontal or vertical velocity is measured. The Doppler velocity measured by a vertically pointing radar gives the net velocity, i.e. the particle's fall velocity and the vertical wind speed. Thus, in order to estimate the vertical wind speed, the fall velocities of the hydrometeors need to be known.

9.4 Types of precipitation

To a first approximation, precipitation can be classified into continuous large-scale or stratiform precipitation and localized, i.e. small-scale, convective or showery precipitation. Large-scale precipitation is associated with the large-scale ascent of air masses produced by frontal or orographic lifting or by large-scale horizontal convergence. It is mainly produced in nimbostratus (Ns) but also falls from dissipating cumulus and orographic clouds. Stratiform precipitation can last for several hours. In contrast, convective precipitation is associated with cumulus-scale convection in unstable air and is mostly produced in cumulonimbus (Cb) or cumulus congestus (Cu con) (Table 1.1). Whereas the pattern of stratiform precipitation evolves relatively slowly in time, that of convective precipitation changes rapidly and is short lived.

The spatial extent, intensity and lifetime of a precipitation system are mainly controlled by vertical air motions, which in turn are related to the various types of atmospheric instabilities (Chapter 3). The time scale of precipitation events ranges from minutes, for showers from a single thunderstorm, to a few days for stratiform precipitation originating from a low pressure system.

In atmospheric science it is common to distinguish between three spatial scales: the local or microscale, with horizontal dimensions less than 2 km; the regional or mesoscale, between 2 and 2000 km; and the synoptic, global or macroscale, with weather systems larger than 2000 km (Figure 9.6). All these scales can be subdivided. For instance, the mesoscale is subdivided into meso-α (200–2000 km), meso-β (20–200 km) and meso-γ (2–20 km) scales (Orlanski, 1975). Thunderstorms are typical meso-γ-scale phenomena, sea breezes occur on the meso-β scale while fronts and mesoscale convective systems, including squall lines, occur on the meso-α scale (Chapter 10). Tropical cyclones also

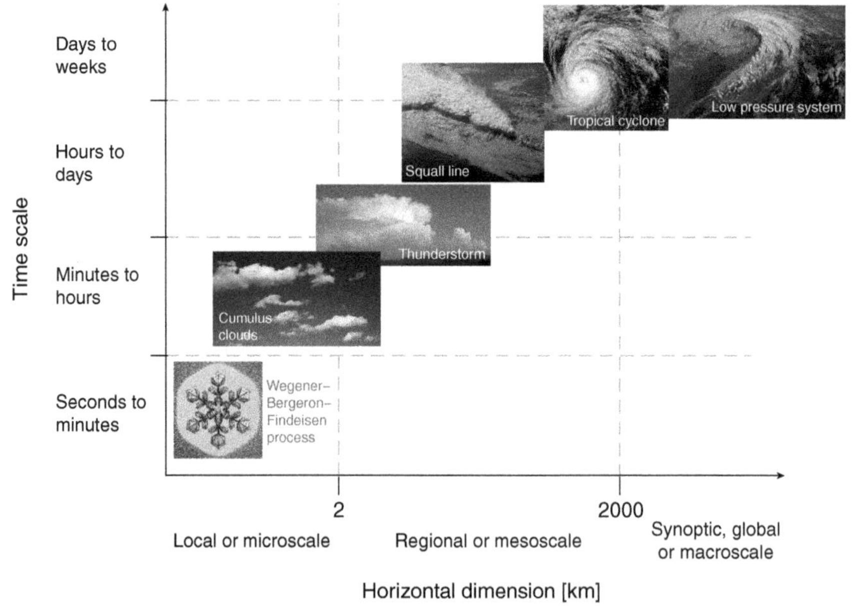

Fig. 9.6 Examples of atmospheric processes that occur at different temporal and spatial scales. The scale separations were defined by Orlanski (1975). Image sources from left to right: Kenneth Libbrecht, Fabian Mahrt (two images), NASA (two images), NOAA image gallery.

belong to the meso-α scale. However, they can extend into the lower end of the synoptic scale, which also hosts low and high pressure systems.

Classifying precipitation as stratiform or convective is problematic because the distinction between them is not always sharp. Widespread precipitation also shows small-scale structure, the most intense precipitation peaks being only a few km in extent. Conversely, convective precipitation can extend over a large area if it stems from a multicell, supercell or mesoscale convective system, as will be discussed in Chapter 10. Thus the definition has been refined, so that convective precipitation is regarded as having locally intense regions which range horizontally from 1 to 10 km and are separated from one another by areas free of precipitation.

One approach to distinguishing stratiform precipitation from convective precipitation is to compare the updraft velocity w inside the cloud with the terminal velocity of snowflakes ($v_{T,snow} \sim$ 1–2 m s^{-1}, Figure 8.17). In all clouds, the vertical air motion w must be large enough to maintain supersaturation. In stratiform clouds the vertical air motions are smaller than the terminal velocity of snow: $w < v_{T,snow}$ (Table 1.3). In contrast, in convective clouds the vertical air motions are typically larger than the terminal velocity of snow: $w > v_{T,snow}$. Distinguishing stratiform and convective clouds in this way was developed for mid-latitudes, where in most cases precipitation is generated via the ice phase (Houze, 2014). The criterion $w < v_{T,snow}$ for stratiform precipitation also holds for warm clouds (i.e. clouds with no ice particles, Chapter 7) because cloud droplets have negligible fall velocities and raindrops have higher fall velocities than snowflakes.

In stratiform clouds the time available for the growth of ice particles sedimenting from cloud top is at least 1–3 hours, which is the time it takes an ice particle falling at 1–2 m s^{-1} minus the updraft velocity to descend through a nimbostratus cloud with a vertical extent of > 7 km (Table 1.3). During this time ice particles can grow considerably by deposition. When falling ice particles descend to a height within 2.5 km of the 0 °C level, aggregation and riming can occur. Aggregation increases in importance within 1 km of the 0 °C level owing to the quasi-liquid layer that forms on the exterior of the snowflakes. It causes the snowflakes to stick together and to form large irregularly shaped snowflakes. This process is called sintering (Section 8.3.2). In addition, ice particles may also grow by riming as long as supercooled cloud droplets are present. This requires w to be strong enough to maintain a relative humidity > 100%, so that the Wegener–Bergeron–Findeisen process (Figure 2.12) does not evaporate all cloud droplets in the vicinity of the ice particles (Figure 8.19).

9.4.1 Stratiform precipitation

The two different types of precipitation (stratiform versus convective) can also be distinguished using radar images, as will be discussed in the following.

9.4.1.1 Plan position indicator (PPI) radar image of stratiform precipitation

Figure 9.7 shows an example of a Doppler radar PPI image of the precipitation and wind velocity associated with the winter storm Lothar. Lothar hit northern France, southern Germany, Austria, Liechtenstein and Switzerland on December 26, 1999 and caused

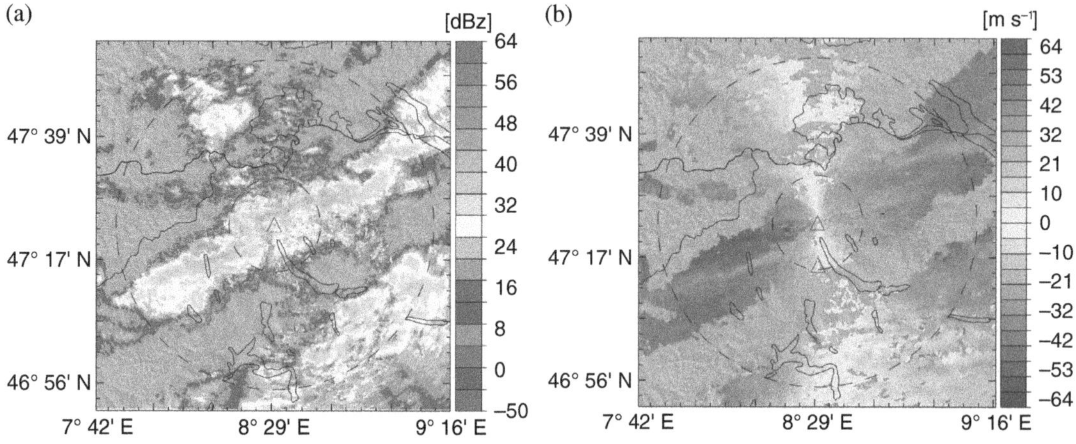

Fig. 9.7 Plan position indicator images of (a) radar reflectivity and (b) Doppler velocity, scanned by the ETH radar in Zurich (triangle in the center) on December 26, 1999. The radar antenna was installed at 600 m above sea level and the scan was performed with a 1.5° elevation angle (Figure 9.4). The broken circles mark PPI heights of 1 km (smaller circle) and 2 km above sea level (larger circle). The brown colors indicate velocities directed away from the radar and bluish colors those directed towards the radar. Figure adapted from Wüest (2001). A black and white version of this figure will appear in some formats. For the color version, please refer to the plate section.

considerable damage. The radar image (Figure 9.7a) shows two rainbands with values above 40 dBZ, indicating heavy rain (Table 9.1). As this precipitation was rather widespread horizontally, it could be characterized as stratiform. The horizontal velocities of the hydrometeors (Figure 9.7b) accompanying the precipitation event show wind speeds up to 50 m s^{-1} relative to the radar beam. The rainbands belong to a cold front that was embedded in a westerly flow as can be inferred from the negative values west of the radar indicating flow towards the radar and positive values to the east of the radar indicating flow away from it. The high radar reflectivity together with high wind speeds suggest the presence of convective rainbands, which could have been caused by slantwise displacement and symmetric instability (Section 3.6).

9.4.1.2 Radar bright band

Stratiform precipitation in mixed-phase clouds is characterized by a layer of high reflectivity, which gives rise to a so-called radar bright band in the signal of a vertically pointing radar. A schematic illustration of the radar bright band associated with stratiform precipitation is shown in Figure 9.8. In a stratiform cloud, ice crystals grow slowly by vapor deposition in the upper part of the cloud, i.e. the layer lying between points 1 and 2 in Figure 9.8. Since this growth is slow, Z_e increases only slightly with decreasing height. In the layer between points 2 and 3, the ice crystals grow more rapidly by aggregation as they approach the 0 °C isotherm. This growth by aggregation sharply increases Z_e because it is proportional to r_h^6 (eq. 9.8).

The layer between points 3 and 5 in Figure 9.8 denotes the radar bright band, in which Z_e can exceed by 10 dBZ_e the values it has in the rain below it or in the snow above it.

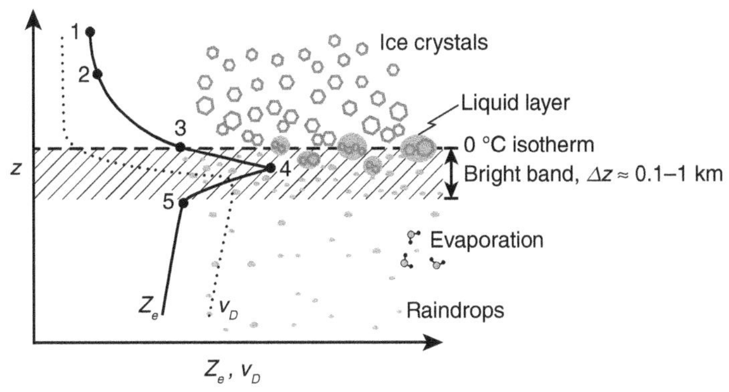

Fig. 9.8 Schematic of a vertical profile of radar data for stratiform precipitation. The solid curve indicates the equivalent reflectivity factor Z_e in dBZ_e and the dotted curve indicates the Doppler velocity v_D for a vertically pointing radar. The microphysical processes occurring in the layers between points 1–5 are explained in the text. The gray disks denote ice crystals that have acquired a liquid layer. Figure adapted from Houze (2014) with permission of Elsevier.

The increase in Z_e between points 3 and 4 is caused by melting accounting for a 4.7 dBZ_e increase (eq. 9.9) along with continuing aggregation; Z_e is highest when snowflakes have acquired a liquid layer without having shrunk in size as indicated by gray disks enclosing aggregated ice particles. Below point 4, continuous melting reduces the size of the snowflakes. The melting snowflakes become more compact and finally collapse or break apart, which reduces the size of the particles and hence reduces Z_e. At the same time the terminal velocity increases from 1–3 m s^{-1} for snowflakes to 5–10 m s^{-1} for raindrops as seen in the sharp increase in the v_D profile. The larger fall velocity of raindrops compared with snowflakes reduces their concentration in space. Together with the ongoing melting and break-up of raindrops, this explains the reduction in reflectivity between points 4 and 5. The layer below point 5 is characterized by rain. If the relative humidity below cloud drops to values below 100%, the raindrops will partly evaporate. This reduces v_D and Z_e.

9.4.1.3 Height–time indicator radar image of stratiform precipitation

An example of a radar bright band associated with a stratiform precipitation event on November 7, 2002 in an HTI scan from a vertically pointing radar is shown in Figure 9.9 (Baschek, 2005). In the reflectivity image of Figure 9.9a the radar bright band is visible as the narrow horizontal band of varying strength located at approximately 700 m above ground, except at the beginning and the end of the measurement where it is at somewhat lower altitudes. Since the radar bright band is a typical indicator of stratiform precipitation the whole event shown in Figure 9.9 could be classified as stratiform. An increasing number and size of regions with negative Doppler velocity can be seen towards the end of this precipitation event (14:30–15:50), being indicative of embedded convective cells. These localized cells are further manifested in the higher reflectivities above the radar bright band towards the end of the precipitation event, indicating enhanced snow production.

A schematic of the microphysical processes leading to stratiform precipitation is shown in Figure 9.10 (Houze, 1981). Note that the cloud boundary and the radar echo boundary

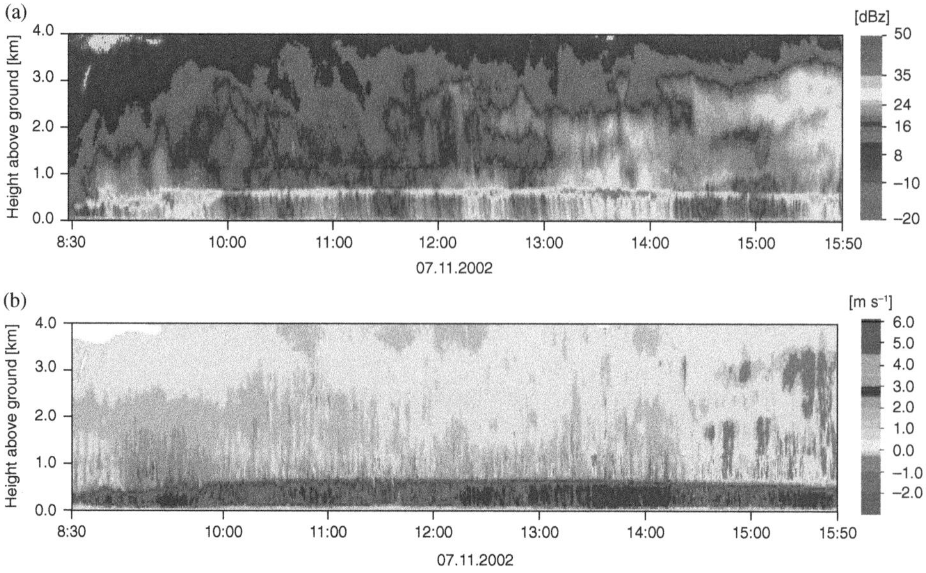

Fig. 9.9 Example of an HTI scan, produced by a vertically pointing radar of a stratiform precipitation event on November 7, 2002. (a) Radar reflectivities and (b) net vertical velocities, where negative values indicate net upward motion. Figure adapted from Baschek (2005). A black and white version of this figure will appear in some formats. For the color version, please refer to the plate section.

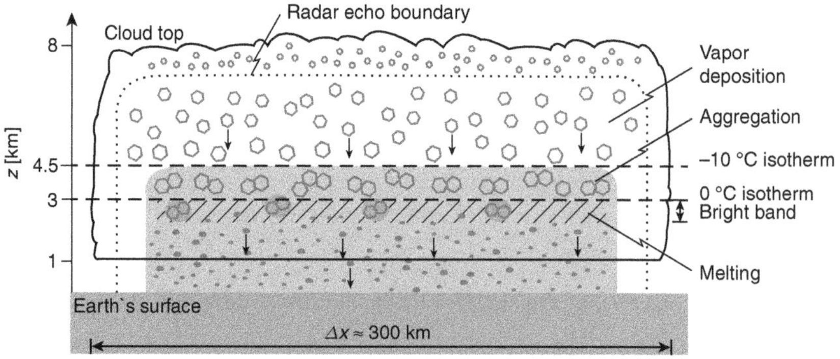

Fig. 9.10 Schematic of the microphysical processes and radar echoes in stratiform precipitation. The bold solid line marks the cloud boundary and the dotted line the radar echo boundary. The white region inside the radar echo boundary corresponds to the lowest reflectivities, the gray region to intermediate reflectivities and the hatched region to the highest reflectivities, associated with the radar bright band. Figure adapted from Houze (1981) with permission of Wiley.

do not coincide; ice crystals in the upper levels of the cloud are too small to be detected by the radar. However, while the cloud has a well-defined base at around 1 km altitude, the radar echo boundary extends down to the surface because precipitation is detected also below cloud base.

In general, precipitation from mid-latitude stratiform clouds develops mainly via the ice phase (Mülmenstädt *et al.*, 2015). Stratiform clouds have a relatively low liquid water content, so that collision–coalescence between cloud droplets is likely to be ineffective. At −12 °C the Wegener–Bergeron–Findeisen process is very effective in causing glaciation of the cloud. Once ice crystals are formed they grow further by vapor deposition. As soon as these ice crystals are large enough to start settling, they serve as embryos for precipitation development owing to growth by accretion at lower levels (Figure 9.10). Recall that accretion refers to the collision and permanent union of two hydrometeors of different sizes and/or phases. Aggregation is the process for ice crystals that is analogous to the collision–coalescence of cloud droplets and it proceeds most rapidly lower in the cloud, at temperatures between −10 °C and 0 °C, i.e. above the radar bright band. Aggregation does not add mass to precipitation but, rather, concentrates the condensate into large particles. Since Z is proportional to r^6, aggregation leads to an increase in the radar echo. The melting layer is 100 m to 1 km thick (Houze, 2014). Below the bright band, the melting aggregates turn into rapidly falling raindrops (Figure. 9.10).

9.4.2 Convective precipitation

Convective precipitation results from cumulus congestus (Cu con) or cumulonimbus (Cb). Isolated convective clouds occur over scales of a few kilometers. Larger convective systems are complexes of this basic structure and will be discussed in Chapter 10. Inside Cu con and Cb the updraft velocities are on the order of 10 m s^{-1} (Table 1.3 and as shown in Figure 9.11b) and thus exceed those of falling ice crystals ($w > v_{T,snow}$). This keeps cloud droplets, ice crystals, snowflakes and small raindrops floating or carried upward in the updraft. Only large raindrops and rimed structures (graupel and hail), for which the gravitational force is large enough, will fall against the updraft. Convective precipitation prevails in the tropics, during summer in mid-latitudes and over warm oceans. It dominates over stratiform precipitation in the global annual mean (Tremblay, 2005).

The precipitation process is much more efficient in convective clouds because its updraft, and hence its liquid water content, is higher than in stratiform clouds (Table 1.3). Therefore the collision–coalescence process is effective and dominates the hydrometeor growth mechanism in the lower levels of the cloud. In the tropics the collision–coalescence process can cause cloud droplets to grow rapidly until they are large enough to overcome the high updraft velocities and to leave the cloud as raindrops. In mid-latitudes, ice-phase processes are normally involved in precipitation formation. Thus, in mid-latitude, convective clouds, the growth of hydrometeors takes place also by riming, resulting in graupel and hail (Chapter 10). Because of the faster precipitation process in convective clouds, the typical time between the occurrence of a cumulus cloud and precipitation can be as short as 20–30 minutes (Section 7.2).

Convective clouds have a radar echo distinctively different from stratiform clouds, as can be seen from a comparison of Figure 9.12 (Houze, 1981) and Figure 9.10. The radar echo of convective clouds evolves with time but is rather constant with altitude. In contrast, the radar echo of stratiform clouds changes with altitude but is rather constant in time.

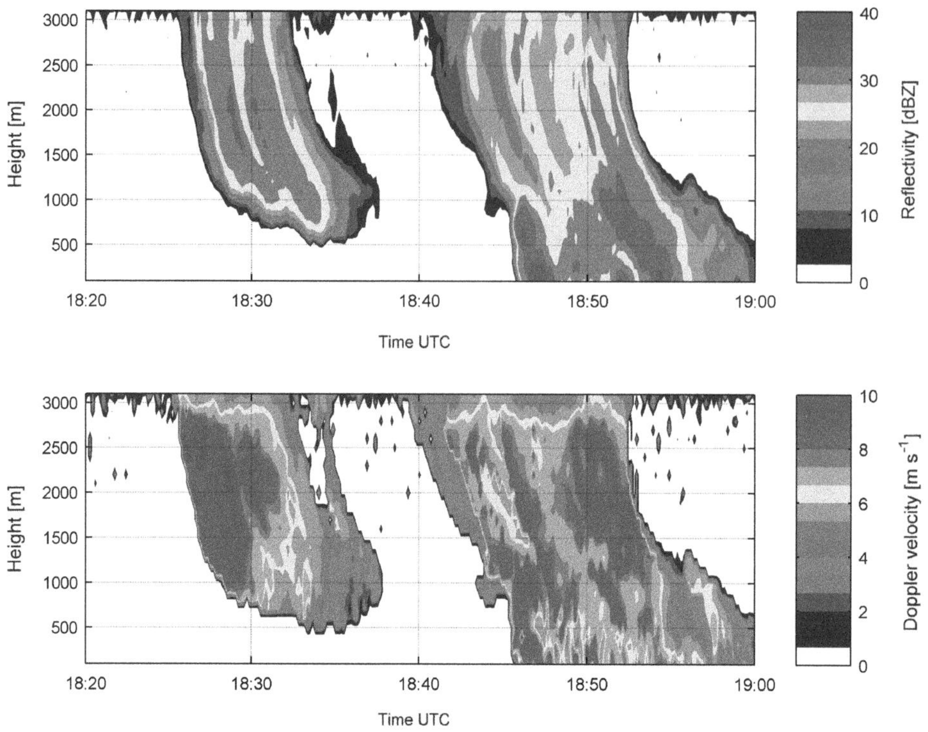

Fig. 9.11 Height–time indicator radar image of convective cloud showers as observed in Zurich, Switzerland on August 13, 2015: (a) reflectivity and (b) Doppler velocity. Data by courtesy of Pascal Graf. A black and white version of this figure will appear in some formats. For the color version, please refer to the plate section.

Convective clouds are characterized by narrow updrafts that are typically a few kilometers or less in width. Convective showers can be as short as 10–15 minutes and occur in pulses, as shown in Figure 9.11. These well-defined vertical cores of maximum reflectivity, so-called convective cells, that consist of large hydrometeors (raindrops, graupel and hail) are typical of convective clouds. Hence, the radar reflectivity does not change noticeably when crossing the 0 °C isotherm (the thin solid line in Figure 9.12) but remains rather constant all the way to the surface. The reasons for the absence of a radar bright band in convective clouds are as follows. First, the 0 °C isotherm is not a straight line but is shifted to higher levels in updrafts owing to the positively buoyant air there and to lower levels in downdrafts, owing to latent heat consumption by melting hydrometeors. Second, convective clouds generally consist of rimed particles (graupel and hail), which have densities and hence reflectivities between those of pure water and pure ice. Third, hailstones produced by a thunderstorm can be much larger than all other precipitation-sized particles.

The white hydrometeors in Figure 9.12 show growing precipitation particles being carried upward by strong updrafts during the developing stage of a cumulus cloud until time t_4. Thereafter, the precipitation particles fall relative to the updraft and reach the ground at time t_6 when the cloud has reached the mature stage. In the dissipating stages of convective clouds (after t_6), the strong upward motion ceases and no longer carries precipitation

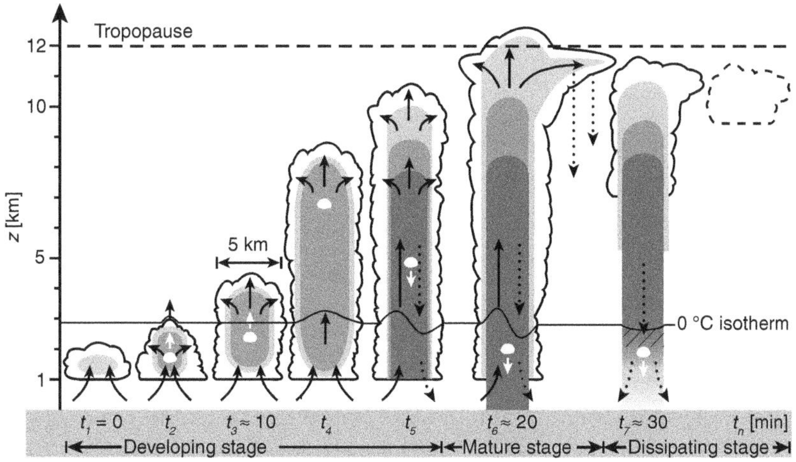

Fig. 9.12 Schematic of the evolution of the radar echo in convective precipitation as a function of time. The bold solid outlines mark the cloud boundaries. The gray regions mark the radar echoes, whereby darker gray shades refer to higher radar reflectivities. The solid and dotted arrows denote updraft and downdraft regions, respectively. The timeline in minutes gives a rough indication of the times associated with the single-cloud stages. The broken outline at t_n indicates a dissipating cloud. The white disks denote the net direction of hydrometeor movement. Figure adapted from Houze (1981) with permission of Wiley.

particles upward or suspends them aloft. The fallout of the particles aloft due to the dying updrafts subsequently causes the core of maximum reflectivity to weaken. The precipitation then takes on a stratiform character, including the radar bright band as seen at t_7 in Figure 9.12. At time t_n only a dissipating cloud remains (the shape with the broken outline).

9.5 Synoptic and mesoscale structure of precipitation

Even though low pressure systems are an example of a synoptic-scale phenomenon (Figure 9.6), they can exhibit mesoscale substructures such as rainbands with embedded mesoscale cells of heavy (convective) precipitation. These substructures are not necessarily resolved by the network of observational stations on which weather forecasts were traditionally based, because the station density is not high enough. In fact the mesoscale structure of large-scale precipitation with embedded convection reveals the interplay of dynamics and microphysics. This could only be observed with the help of radars. Nowadays radars have become part of the observational network, so that more information about the mesoscale structure of precipitation has become available.

Heavy rain, high winds and snow storms are other examples of mesoscale weather. Heavy rain does not have a precise definition because its definition depends on the context. Larger-scale flooding is preceded by continuous precipitation over hours to days. For instance the extreme flood in Central Europe in May and June 2013, which progressed down the rivers Elbe and Danube with high-water flooding along their banks, was a result

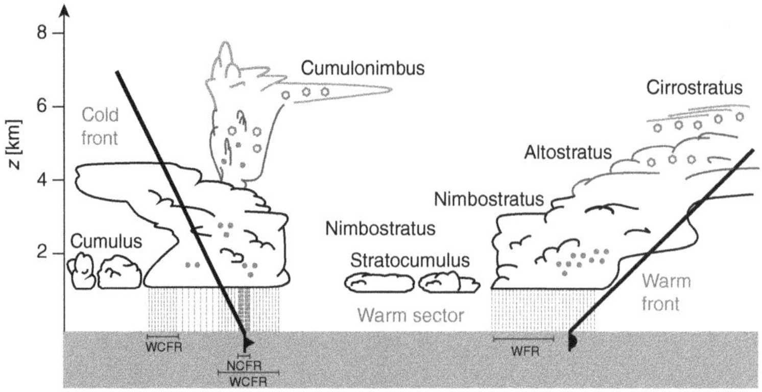

Fig. 9.13 Schematic cross section of an extratropical cyclone (see line AB in the plan view of Figure 9.14b) with typical cloud types (Table 1.1) and rainbands. The solid half disk denotes the surface warm front and the solid triangle the surface cold front. WCFR and NCFR denote the wide and narrow cold frontal rainbands and WFR the warm frontal rainband.

of several days of heavy rain. Another example is the Colorado floods in September 2013, which led to numerous other flash floods.

Over a short time scale (minutes to hours) heavy precipitation will be convective. For extreme events even a single strong convective storm of one hour might be enough to cause flash floods in very small and steep catchments, such as the southern Alps or in the Rocky Mountains.

Heavy rain from convective clouds is often observed to be organized in lines or bands. These rainbands can be distinguished, among others, into wide and narrow cold frontal rainbands (WCFR and NCFR) and warm frontal rainbands (WFR), as shown in Figure 9.13. Its timing and location relative to the surface fronts of a cyclone depend on the airflow, stability, moisture supply and orographic influences. Convective precipitation can also develop in unstable air without frontal lifting and can become organized in bands or lines unrelated to the synoptic pattern.

9.5.1 Norwegian cyclone model

Traditionally, precipitation associated with extratropical cyclones was related to the clouds at the warm and cold fronts, with stratiform precipitation falling from nimbostratus (Figure 1.4c) at the warm front and convective precipitation falling from cumulonimbus (Figure 1.4b) at the cold front, as schematically shown in Figure 9.13. This concept of the cyclogenesis, fronts and associated cloud types is based on the so-called Norwegian cyclone model developed at the Bergen School of Meteorology (Bjerknes and Solberg, 1922). The central idea of the Norwegian cyclone model is that a cyclone can be viewed as a perturbation on an existing frontal boundary. It was mainly developed for frontal structures near Earth's surface and hence does not capture their complete vertical behavior. This model is based on the synoptic pattern and on the concept that warm and cold air flows encounter each other at the polar front, where the warm subtropical air mass pushes polewards and the cold Arctic air mass pushes equatorwards, as shown in Figure 9.14a.

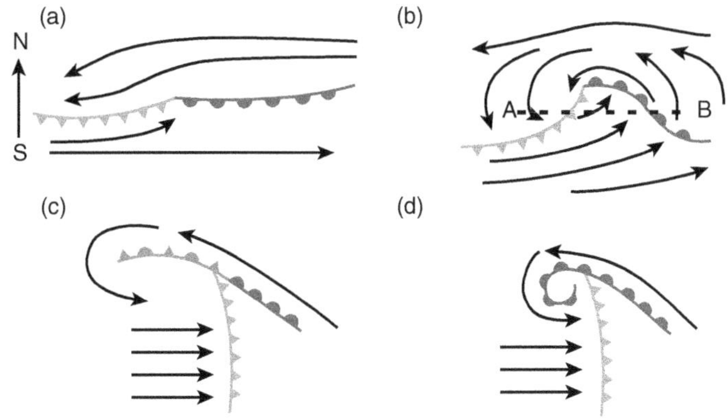

Fig. 9.14 Schematic of the surface fronts and associated flow of the different cyclogenesis phases: (a) and (b), the early phases, (c) the occluded phase, (d) the T-bone frontal structure. The half disks denote the surface warm fronts and the triangles the surface cold fronts.

This causes the deformation of the polar front in form of a wave. Ahead of the wave crest the warm air gains ground and behind it the cold air gains ground (Figure 9.14b). If this wave disturbance develops further, it leads to a pressure drop in the center and causes well-developed warm and cold fronts.

Whereas in the Norwegian cyclone model all cyclones end in an occlusion, new studies suggest that only about one third of all cyclones develop an occlusion as illustrated in Figure 9.14c. Instead, there are other possible ways in which an incipient disturbance can lead to cyclogenesis. Approximately one third of all cyclones develop a T-bone frontal structure, as shown in Figure 9.14d (Shapiro and Keyser, 1990). This structure is characterized by a warm front that is bent backwards and a cold front that is perpendicular to the warm front. The remaining cyclones show frontal structures that differ from the classical T-bone and occlusion structures, mainly because they develop only either a warm or a cold front, or because they have a strongly barotropic structure with very weak surface fronts. This highlights the large variability of frontal structures during cyclogenesis.

9.5.2 Conveyor belt approach

An alternative way to describe the (synoptic) pattern of extratropical cyclones and the associated frontal structures and precipitation patterns is provided by the concept of conveyor belts (e.g. Browning, 1985), shown schematically in Figure 9.15. Key features of the conveyor belt concept are the warm and cold conveyor belts (Carlson, 1980) that give rise to warm- and cold-frontal rainbands and a dry intrusion, as shown in Figure 9.15.

The air movements along conveyor belts are relative to the movement of the low pressure system. As long as no condensation sets in, the potential temperature θ of the air is conserved and movement along the warm conveyor belt takes place on isentropic surfaces, i.e. on surfaces of constant θ (Chapter 2). The isentropic surfaces slope upward towards the poles owing to the latitudinal temperature gradient (Figure 3.9), causing the air in the warm

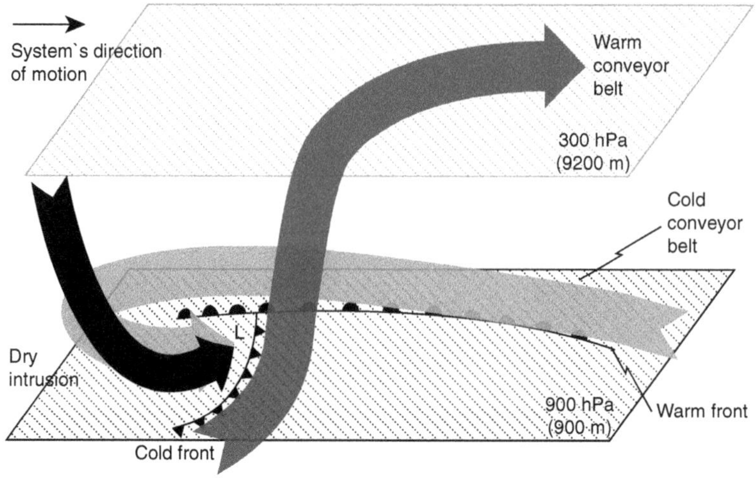

Fig. 9.15 Schematic of a katafront extratropical cyclone with a warm conveyor belt (WCB), shown as the dark gray arrow, a cold conveyor belt, shown as the dark light gray arrow, and a dry intrusion, shown as the black arrow. The altitudes correspond to the average ascent of an air parcel in a WCB. Indicated heights are based on the WCB climatology by Madonna *et al.* (2014).

conveyor belt to rise polewards. Once condensation sets in the latent heat release causes the warm air mass to rise on surfaces of constant equivalent potential temperature θ_e, causing the flow to become cross-isentropic. At the frontal surface, the isentropic surfaces are closer together than elsewhere. As the air mass rises partly on isentropic θ surfaces and partly on θ_e surfaces, the frontal surface is sometimes defined in terms of θ surfaces and other times in terms of θ_e surfaces.

The warm conveyor belt (WCB) is responsible for most precipitation from extratropical cyclones. It originates typically within the boundary layer ($\sim$ 900 hPa) and experiences the largest ascent within the cyclone, of more than 600 hPa. This ascent is caused by large-scale lifting ahead of or across the cold front. Air within the WCB originates partly from the southerly low-level jet ahead of the surface cold front and flows along the length of the cold front, where it ascends to higher levels. Up to this point the conveyor belt model is consistent with the traditional view in which warm air moves on top of the resident cold air at the warm front.

The new aspect of the conveyor belt model is that at higher altitudes the WCB either turns ahead of the warm front and flows parallel to it (Figure 9.15) or crosses the cold front and flows in a direction of the motion opposite to the low pressure system. If the WCB slopes downstream during ascent, relative to the motion of the cold front, the front is called a katafront. In katafronts the main region of ascent, cloudiness and precipitation occurs ahead of the cold front (i.e. it is prefrontal). Precipitation, indicated by the dark gray areas in Figure 9.16a, is found in the warm sector and ahead of the warm front.

The reason for this curved precipitation pattern is the dry air intrusion. In the case of a katafront the dry air, indicated by the light gray broken arrow in Figure 9.16a, crosses the cold front from behind, overrunning the WCB and at the same time restricting it to

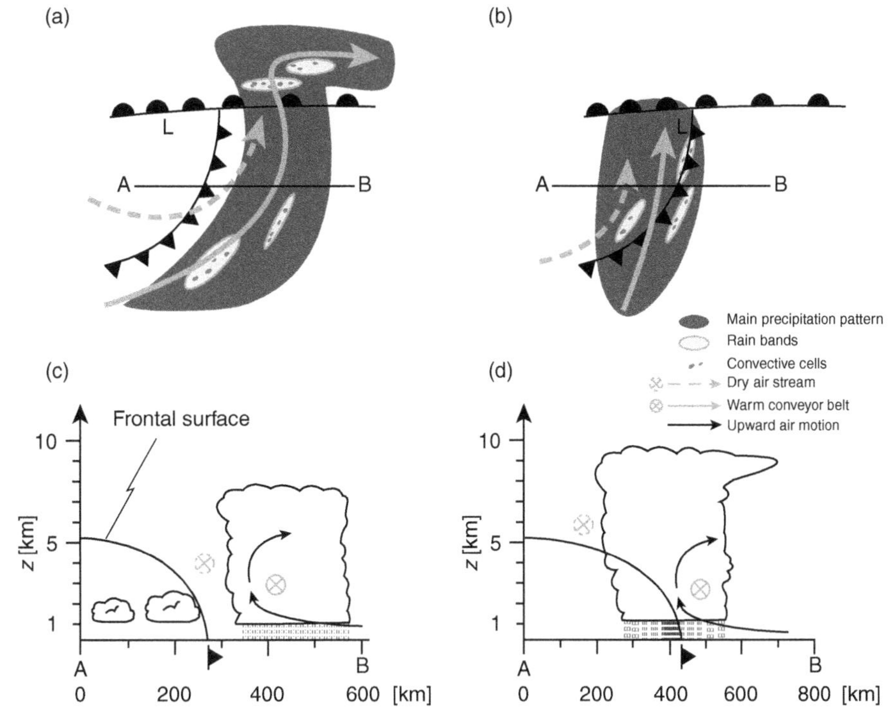

Fig. 9.16 Schematic depictions of warm conveyor belts (pale gray solid arrows) as viewed from above (a) in a katafront and (b) in an anafront. Vertical cross sections through the cold front and the WCB along the lines AB in (a) and (b) are shown in (c) for the katafront and in (d) for the anafront.

areas in the warm section ahead of the cold front. The dry air mass originates from the upper troposphere and has a lower equivalent potential temperature than the air masses of the WCB. Hence air is unstable near the frontal cloud band, where dry cold air overruns the warm moist air of the WCB. This results in the formation of deep convective clouds directly ahead of the warm front (shown by the presence of rainbands in Figure 9.16a), and in stratiform precipitation in the warm section, as shown in the vertical cross sections in Figure 9.16c.

If during ascent the WCB slopes upstream relative to the motion of the cold front, it is called an anafront. In anafronts the main areas of cloudiness and precipitation appear along and behind the cold front, as indicated by the shaded area in Figure 9.16b. Key to understanding this pattern is again the motion of the dry air stream, which in the case of an anafront is nearly parallel to the ascending air of the WCB and allows the WCB to reach behind the cold front during its ascent. The separation of the two air streams can be seen in sharp cloud edges at the rear of the frontal cloud layer (Browning, 1985) and is illustrated in the cross section of the anafront in Figure 9.16d.

In contrast with the WCB, the cold conveyor belt (CCB) is a coherent air stream in the lower troposphere between 700 hPa and the surface flowing on the cold side of the warm front, in a direction opposite to the movement of the cyclone. Cold conveyor belt trajectories frequently turn cyclonically around the center of the storm, as shown in Figure 9.15. A CCB is often associated with strong low-level winds and can trigger moisture transport along the bent-back part of the cold front, also referred to as the deformation zone, and cause low-level wind jets (Schemm and Wernli, 2014).

Because the conveyor belt concept is used to describe the low pressure system and synoptic-scale phenomenon, it cannot resolve the mesoscale structure of embedded convection and rainbands (Figures 9.13 and 9.16). This embedded convection can result from, for instance, symmetric instability (Figure 3.8). Rainbands are also associated with deep convective clouds, shallow convection in the planetary boundary level or they can originate at mid-levels in low-pressure systems. In both katafront and anafronts, these rainbands cause most precipitation associated with frontal systems.

9.5.3 Orographic precipitation

Orographic precipitation is another example of a mesoscale phenomenon. It encompasses all precipitation events where topographic lifting comes into play. Orographic precipitation can have a stratiform or convective character depending on the synoptic situation. If air is forced to rise over a mountain range in a stable stratified atmosphere and reaches its lifting condensation level (Section 2.4) then stratiform clouds will form on the upslope side (the

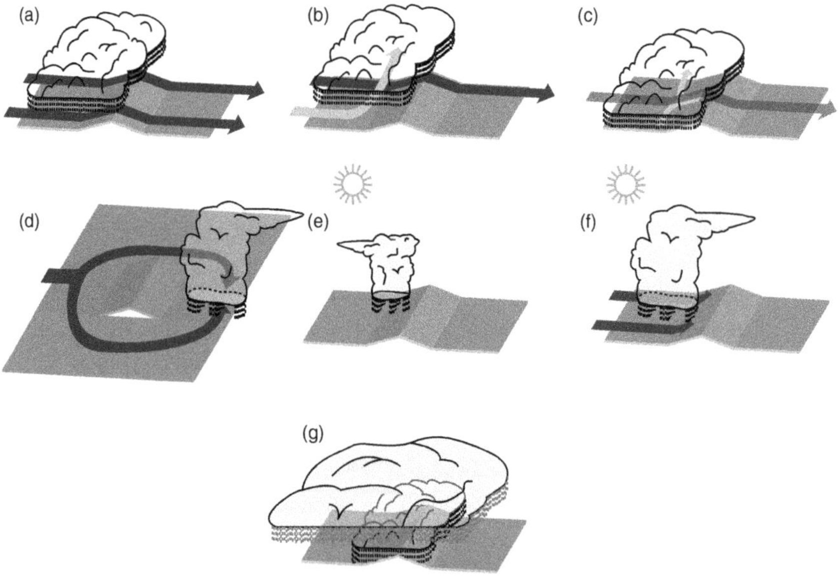

Fig. 9.17 Schematic illustrations of different mechanisms of orographic precipitation. The dark gray arrows indicate the direction of the flow. (a) Stable upslope ascent, (b) partial blocking of an impinging air mass (light gray arrow), (c) down-valley flow induced by evaporative cooling on the windward side (light gray arrows), (d) lee-side convergence, (e) convection triggered by solar heating, (f) convection due to mechanical lifting above the level of free convection and (g) seeder-feeder mechanism. Figure adapted and reprinted from Roe (2005) with permission of Annual Review of Earth and Planetary Sciences.

windward side or luv), as shown in Figure 9.17a (Roe, 2005). When the air descends on the downslope side (lee), the clouds will evaporate again. Depending on the cloud thickness and the height of the mountain range, stratiform precipitation may fall on the windward side.

In Figure 9.17b the flow is partially blocked, meaning that the atmosphere is either too stable or the flow is not strong enough for the lower-level flow to ascend over the mountain. It can either stagnate or gets diverted around the mountain, the latter being the case in Figure 9.17b. A similar case of blocking is shown in Figure 9.17c, where evaporation of the precipitation below cloud base has cooled the air so that a strong down-valley flow results on the windward side, which forces the air to ascend to higher levels over the mountain crest.

Convective clouds may be formed when the air diverges around the mountain and convergence on the lee side causes rising motion (Figure 9.17d). On sunny days, with weak synoptic flow, the Sun warms the hillslope; this can lead to instabilities and trigger convection and cumulus cloud formation, as shown in Figure 9.17e.

If the Sun heats the sun-facing slopes and causes the air to rise, this air will be replaced by air from the valley. This mechanical lifting above the level of free convection can also lead to cumulus clouds, as shown in Figure 9.17f. The up-valley wind is part of the mountain–valley wind system and reverses into a down-valley wind at night. This wind system is comparable to the land–sea breeze (Section 9.6) as both are caused by differential heating (mountain–valley versus land–sea respectively). Depending on the instability associated with the convective clouds and their liquid water content, convective precipitation may fall.

Another mechanism by which orography can cause precipitation is the so-called seeder–feeder mechanism. In the example shown in Figure 9.17g, the "seeder" cloud is at high levels and is undisturbed by the orography below. The cap cloud at low levels that is caused by orographic lifting by the mountain range would not have precipitated by itself because it is too shallow. However, it provides the "food" for the hydrometeors falling from the seeder cloud aloft and thus enhances precipitation from the seeder cloud.

9.6 Precipitation in the present and future climate

Precipitating clouds are an important component of the hydrological cycle, as shown in Figure 9.18. Water changes its phase from gaseous to liquid and/or solid when moving through the hydrological cycle. While water resides for thousands of years in the oceans and in the continental ice shelves, on average it only lasts for 9–11 days in the troposphere (Pruppacher and Jaenicke, 1995). Atmospheric water vapor can be transported over large distances both horizontally and vertically. In fact, one third of the precipitation over land originates from evaporation over the oceans and not from local evaporation from the corresponding land masses. Conversely, the excess precipitation over land enters the oceans as run-off. The global annual mean precipitation rate amounts to 2.7 mm d^{-1}, divided

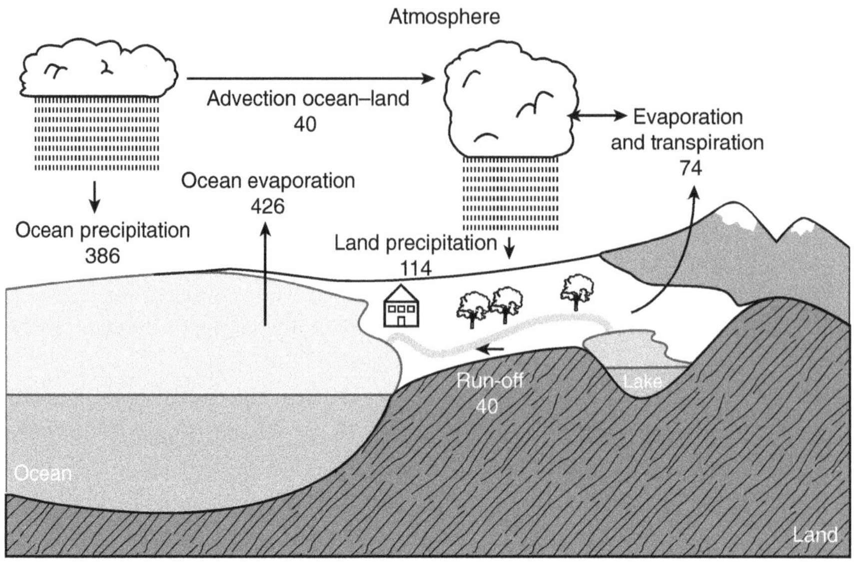

Fig. 9.18 Schematic surface view, with a vertical cross section, of various fluxes and reservoirs within the atmospheric branch of the hydrological cycle, with their annual average magnitudes based on data of Trenberth *et al.* (2011) and references therein. The water vapor reservoirs are given in 10^{15} kg and the fluxes in 10^{15} kg yr^{-1}.

approximately into 2.1 mm d^{-1} over the fractional area of the globe covered with oceans (0.7) and 0.6 mm d^{-1} over that of land (0.3).

The geographical distribution of precipitation for northern hemisphere winter (December–February) and summer (June–August) is shown in Figure 9.19a (Solomon *et al.*, 2007). Not surprisingly, it bears some resemblance to the geographical distribution of total cloud cover as discussed in Chapter 1 and shown in Figure 1.1. The largest amount of precipitation falls in the tropics, in the Intertropical Convergence Zone (ITCZ) and South Pacific Convergence Zone (SPCZ), the same areas that are characterized by a high amount of cloud cover. This precipitation is mostly of a convective nature, falling from isolated thunderstorms or bigger precipitation complexes such as tropical depressions, tropical storms and tropical cyclones (Chapter 10). When one compares the amount of precipitation in December, January and February (DJF) with that of the period June, July and August (JJA), the seasonal cycle of the ITCZ becomes apparent. Its average position is at around 5 °N, because convection is more vigorous over land than over oceans and the amount of land mass is larger in the northern hemisphere than in the southern hemisphere. Because of the smaller heat capacity of land masses as compared with oceans, heating by the Sun warms the land masses more readily. During the day, this causes the air over land to become unstable and to rise. If it is sufficiently moist, cumulus clouds will form. The more unstable and moist the atmosphere is, the deeper the cumulus clouds will be and the more they will precipitate. The rising air over land is replaced by the adjacent air, often coming from the nearby oceans, which thereby provide a source of moisture. This effect can often be observed in mid-latitudes during summer and is referred to as sea breeze. The

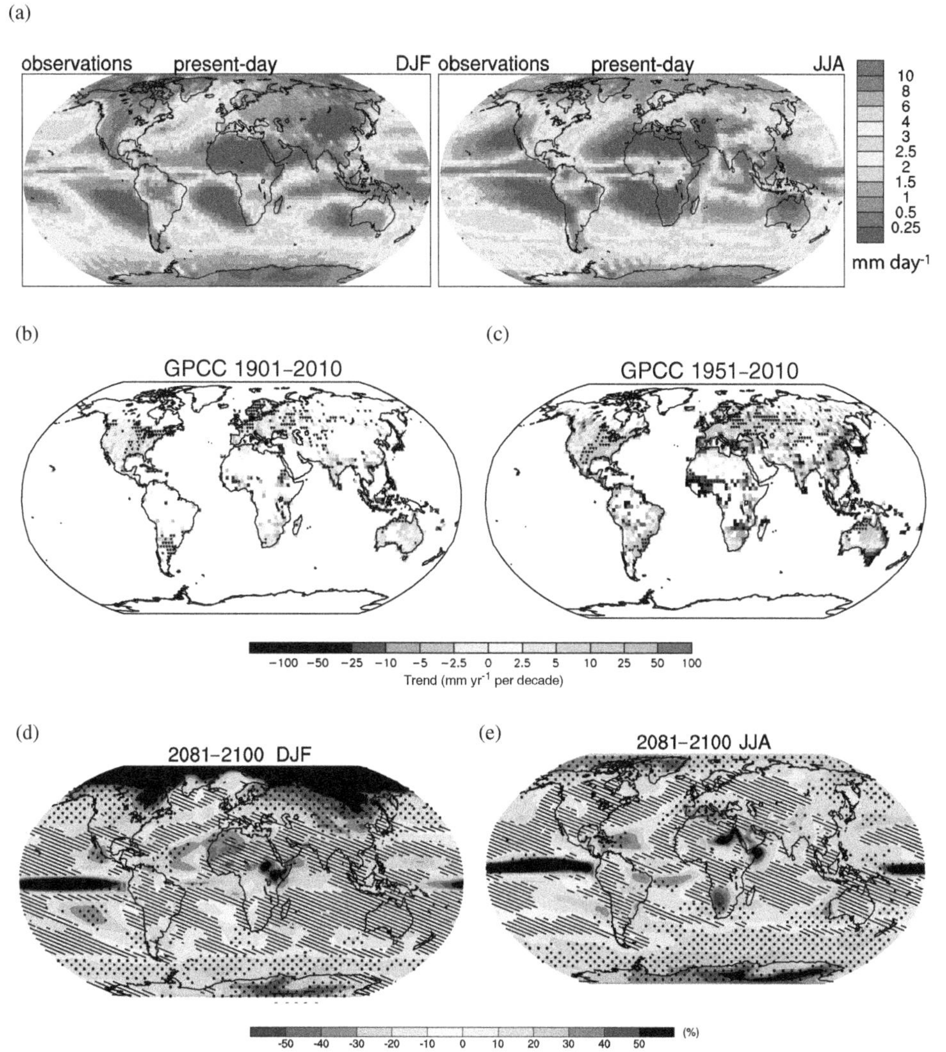

Fig. 9.19 (a) Seasonal mean precipitation rates over the period 1979–1993 in DJF and in JJA. Trends in annual precipitation over land (b) 1901–2010 and (c) 1951–2010. The trends were calculated for grid boxes with $>$ 70% complete records and for $>$ 20% data availability in the first and last deciles of the period. The white areas in (b) and (c) indicate incomplete or missing data and the black plus signs indicate grid boxes where trends are significant. (d), (e) projected percentage changes of seasonal mean precipitation for the period 2081–2100 relative to the reference period 1986–2005 following the RCP8.5 forcing scenario for DJF and JJA. The hatching indicates regions where the multimodel mean change is less than one standard deviation from the natural internal variability in 20-year means. The stippling indicates regions where the multimodel mean change is greater than two standard deviations from the natural internal variability in 20-year means and where $>$ 90% of the models agree on the sign of change. Data sources: (a), Figure TS.30 of IPCC AR4 (Solomon *et al.*, 2007); (b), (c), Figure 2.22 of IPCC AR5 (Hartmann *et al.*, 2013); (d), (e), Figure 12.22 of IPCC AR5 (Collins *et al.*, 2013). A black and white version of this figure will appear in some formats. For the color version, please refer to the plate section.

circulation reverses to a land breeze during night time, when the land cools faster than the ocean.

The same mechanism is responsible for the monsoon circulations on a larger scale and an annual instead of a daily basis. The Indian summer monsoon is the most prominent example. Heating of the Indian subcontinent is so pronounced during JJA that a permanent low pressure system develops, causing the ITCZ to extend by up to 30° N. The Indian summer monsoon is associated with precipitation rates in excess of 10 mm d^{-1} over the Indian subcontinent. In DJF, when the ITCZ is south of the equator, deep convective clouds rise and precipitate over the Indian ocean instead. During that period, a high pressure system develops over the Indian subcontinent associated with the northeast trade winds. It is responsible for the negligible amounts of precipitation over the Indian subcontinent during DJF. The southward movement of the ITCZ is also seen in the location of precipitation in Africa, Central and South America in Figure 9.19a: western Africa south of the Sahara and Central America receive most of their precipitation in JJA whereas southern central Africa and Brazil receive most of their precipitation in DJF.

As discussed above, precipitation in the mid-latitudes is mainly associated with frontal systems. Regions of high annual precipitation are found along the storm tracks shown in Figure 1.2. Convective showers and thunderstorms contribute to precipitation in mid-latitudes, especially in the respective summers. Steep orography causes orographic precipitation, as evident at the coasts of Patagonia, the west coast of North America in DJF and to a lesser degree also in Norway.

The change in annual precipitation over land for the two periods 1901–2010 and 1951–2010 compiled globally and for individual regions by the Fifth Assessment Report (AR5) of the Intergovernmental Panel on Climate Change (IPCC) (Hartmann *et al.*, 2013) and obtained from the Global Precipitation Climatology Centre (GPCC) data sets is shown in Figures 9.19b, c. As precipitation can be very localized in space and time, trends in precipitation are less smooth than for instance those in temperature. Even so, one can see a trend towards increases in precipitation in mid-latitudes and decreases in precipitation in the subtropics, i.e. in those regions that are already rather dry in the present climate. The trends are more pronounced and significant since 1951. This coincides with the period during which the temperature increase has been largest (Stocker *et al.*, 2013). Because the data coverage over the oceans is very sparse, precipitation trends are shown only over land.

In order to make projections about changes in precipitation patterns and magnitudes in the future, one needs to follow a particular climate change scenario. Climate change scenarios take several factors into account, such as demographic development, socio-economic development and technological changes and innovations. Based on these factors, time series (pathways) of future emissions and/or concentrations of anthropogenic greenhouse gases and aerosol particles are projected. These projections are then used within climate models to simulate changes in climate, for example in temperature, precipitation or cloud cover based on the anthropogenic forcing. As the future development of these factors is very uncertain, different scenarios of anthropogenic forcings have to be interpreted as different possibilities of the resulting climate. In the fifth IPCC report, scenarios are defined in terms of representative concentration pathways (RCPs). Each RCP is characterized by a total anthropogenic radiative forcing (RF) in Wm^{-2} in the year 2100 relative to the RF in

1750, on the basis of the emission scenario. The magnitude of the RF corresponding to an RCP is indicated by the number following RCP, for instance RCP8.5 leads to a positive RF of 8.5 Wm^{-2} in the year 2100; it defines a business-as-usual scenario in which the emissions of anthropogenic greenhouse gases continue to increase at the current level. These RCPs are used in climate models of different complexity to project possible future climates.

A possible distribution of precipitation at the end of the twenty-first century following the RCP8.5 forcing scenario is shown in Figures 9.19d,e. The distribution for the other scenarios can be found in Collins *et al.* (2013). They are not shown here because the changes in precipitation are less certain for a smaller RCP, and the areas where at least 90% of the models agree on the sign of the precipitation change are smaller. The data in Figures 9.19d, e give a multimodel mean over 39 coupled atmosphere–ocean general circulation models (AOGCM). These global climate models are currently the most comprehensive in their representation of the climate.

The projections of precipitation differ between different AOGCMs. To account for this variability, robust changes in precipitation are indicated by the stippling in Figures 9.19d, e. These are defined in such a way that the multimodel mean change needs to exceed two standard deviations of natural internal variability in 20-year means and at least 90% of the models need to agree on the sign of change. Robust changes include increased precipitation in high latitudes and decreases in parts of the subtropics. This marks a continuation of the trend already seen in the twentieth century (Figures 9.19b, c). Note that in DJF northern Europe is expected to experience increases in precipitation, while the decrease in precipitation in southern Europe is not significant. In JJA the opposite is true, with southern Europe expecting robust decreases in precipitation. In other words, the Mediterranean climate is expanding polewards for the RCP8.5 scenario. In JJA the change in precipitation is not significant in northern Europe (Scandinavia) because in this region some models predict an increase in precipitation and others a decrease.

In summary, the global mean precipitation will increase in a warmer climate. This increase in precipitation has a distinct geographic pattern; there will be regions in which precipitation increases, does not change at all or even decreases. The contrast between dry and wet regions and dry and wet seasons is likely to increase in most parts of the world in a warmer climate (Collins *et al.*, 2013).

9.7 Exercises

Single/multiple choice exercises

1. Which of the following is/are true?
   (a) Raindrop size can be described best by the log-normal Marshall–Palmer distribution.
   (b) The intercept of the raindrop size distribution depends almost linearly on the precipitation rate.

(c) The slope of the snowflake size distribution increases by approximately 50% with a doubling of the snowfall rate.
(d) The Marshall–Palmer distribution holds for equivalent radii of precipitating particles smaller than 0.5 mm.
(e) Raindrop break-up alters the droplet size distribution.

2. After hydrometeors are formed, they can grow by different microphysical processes until they reach precipitation size. The intensity of the precipitation is described by means of a precipitation rate. Which of the following is/are true?
   (a) Snowfall rates are usually higher than rainfall rates due to the larger particle size of snow.
   (b) The average annual precipitation rate in Europe is around 6000 mm yr^{-1}.
   (c) The rainfall rate depends on the raindrop volume distribution and on the fall velocities of the hydrometeors.
   (d) According to the Marshall–Palmer approximation, the raindrop size distribution has more larger raindrops for increasing rainfall rates.
3. Which particles fall into the Rayleigh regime at the wavelength of the maximum spectral radiance of Earth?
   (a) Raindrops
   (b) Ice particles
   (c) Cloud droplets
   (d) Aerosol particles
   (e) Air molecules
4. Atmospheric particles can act as scatterers when interacting with electromagnetic radiation. Which of the following is/are true?
   (a) The intensity of Rayleigh scattering is proportional to λ^4.
   (b) The angular distribution of Rayleigh scattering favors backward scattering, making Rayleigh scattering important for radars.
   (c) Mie scattering prevails if r and λ are of the same order of magnitude and the intensity is proportional to $\lambda^{-1/3}$.
   (d) In the Rayleigh regime a certain scattering efficiency of a given material corresponds to one particle size only.
5. Why does a cloud radar return negative values, for example -30 to -10 dBZ in mixed-phase clouds?
   (a) A cloud radar can only measure precipitating particles and thus returns wrong values for stationary cloud sheets.
   (b) Owing to the small sizes of cloud particles, the reflectivity is very weak and its logarithm becomes negative.
   (c) The chosen radar wavelength is too long for measuring cloud particles and the signal received originates from other interfering objects such as insects.
   (d) Mixed-phase clouds and ice clouds occur at higher levels where the electromagnetic pulse is reduced to a weak signal.
6. In the atmosphere, snowflakes are in general larger than raindrops. Which of the following is/are true? The fall velocity of snowflakes …
   (a) … is higher than that of raindrops owing to the larger particle size.

(b) ... increases linearly with increasing size.
(c) ... with a maximum dimension of 1 cm is on average 5 m s^{-1}.
(d) ... depends on their size, density and shape.

7. Which description(s) of a radar observation match(es) a typical stratiform precipitation pattern?
   (a) There is no significant change over time but different features can be distinguished with height.
   (b) Continuous development of patterns with time and a major growth region of precipitation particles close to the 0 °C isotherm.
   (c) There is pronounced increase in radar reflectivity close to the 0 °C isotherm.
   (d) The cloud base and cloud top can be recognized readily by a change in reflectivity.
8. Precipitation can be classified into stratiform and convective precipitation. Which of the following is/are true?
   (a) Stratiform precipitation is associated with frontal and orographic lifting.
   (b) A conditionally unstable atmosphere is a prerequisite for the formation of convective precipitation.
   (c) Stratiform precipitation is long lived compared with convective precipitation.
   (d) Radar images can be used to classify the type of precipitation.
9. Which of the following is/are true?
   (a) Ice processes play no role in convective precipitation, whereas stratiform precipitation forms via the ice phase.
   (b) Convective precipitation is associated with higher updraft velocities.
   (c) The radar echo of stratiform precipitation evolves with time, whereas that of convective precipitation evolves with altitude.
   (d) Convection is important in heavy storms with high amounts of precipitation.
   (e) Both the horizontal and temporal extents are typically larger for stratiform clouds.
10. Radars have become important for weather services since they can be used to detect precipitation particles. Which of the following is/are true?
   (a) Radars are passive remote sensing instruments as the instrument has no direct contact with the detected particles.
   (b) The intensity of the backscattered radar signal is almost equal for cloud droplets and raindrops, as they are both liquid, but is different for ice crystals.
   (c) A radar reflectivity of 50 dBZ unambiguously indicates a large number of 1 mm raindrops.
   (d) The radar reflectivity increases linearly with hydrometeor size.
   (e) The radar reflectivity takes the different hydrometeor sizes in a unit volume into account.
11. Why is the equivalent radar reflectivity highest in the bright band?
   (a) Because the equivalent radar reflectivity is proportional to $|K|^2$ and $|K|^2$ is larger for water than for ice.
   (b) Because raindrops may further grow below the cloud base if the relative humidity is sufficiently large.
   (c) Because raindrops generally have larger fall velocities than snowflakes.
   (d) Because break-up decreases the size of raindrops.

Table 9.3 Radius (r_h) and number concentration ($n_N(r_h)$) of hydrometeors.

r_h [mm]	$n_N(r_h)$ [m^{-3} mm^{-1}]
0.2–0.4	200
0.4–0.6	150
0.6–0.8	100
0.8–1.0	50
1.0–1.2	35
1.2–1.4	31
1.4–1.6	20
1.6–1.8	15
1.8–2.0	10
2.0–2.2	15
2.2–2.4	2

12. Imagine another planet where the precipitation consists of ethanol instead of water, but other atmospheric conditions (e.g. air density, temperature) are equal. Which of the following is/are true? You may need to look up some properties of ethanol to answer the question.
    (a) The maximum raindrop size is not expected to be different.
    (b) A cloud drop with diameter 5 μm (at the same altitude) has a larger terminal velocity in the case of ethanol.
    (c) There will be fewer cirrus clouds in such an atmosphere.

Long exercises

1. In Table 9.3 the size and the number concentration of particles larger than 0.2 mm below a certain cloud are given.
    (a) Calculate the radar reflectivity in units of dBZ. For simplicity, you can take the mean radius in each size bin (e.g. 0.3 mm for the 0.2–0.4 mm size bin). What precipitation type is associated with that cloud?
    (b) Estimate the rainfall rate for that cloud using the empirical formula from MeteoSwiss (eq. 9.10).
    (c) What problems arise when estimating precipitation amount from radar reflectivity? Name and explain two.
2. Figure 9.20 shows a radar image of a height–time indicator (HTI) scan.
    (a) The two panels show the Doppler velocity in m s^{-1} and the reflectivity in dBz. Using the information (scale bars) provided, identify which plot shows reflectivity and which shows Doppler velocity.
    (b) What additional information could you gain from the plot showing the Doppler velocity?
    (c) What type of precipitation is shown in the figure? Use the temporal evolution and the vertical structure of the precipitation event shown for your reasoning.

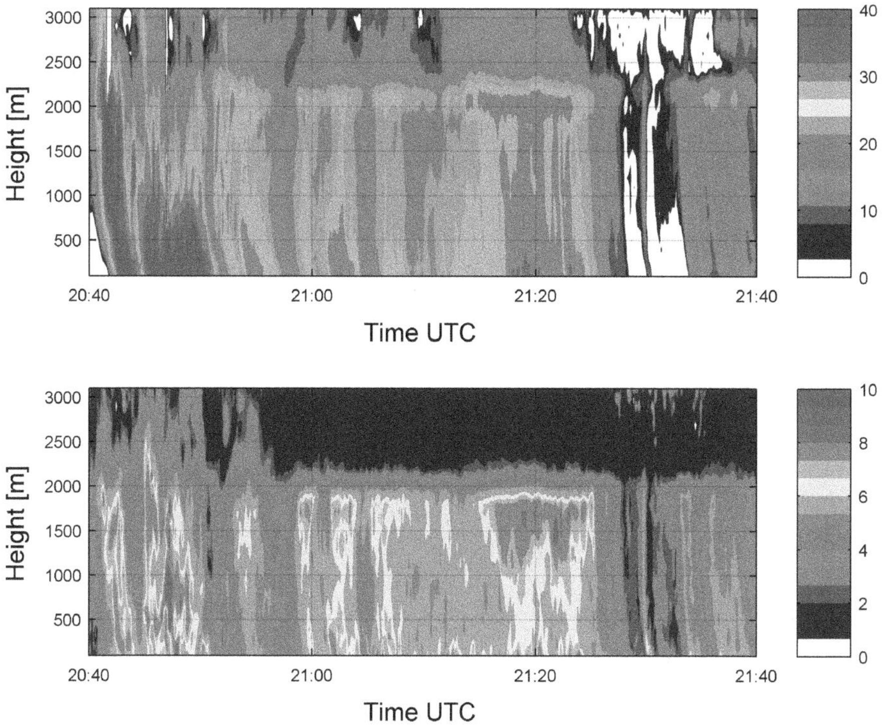

Fig. 9.20 Height–time indicator (HTI) radar images as observed in Zurich, Switzerland on June 21, 2015. Data by courtesy of Pascal Graf. A black and white version of this figure will appear in some formats. For the color version, please refer to the plate section.

(d) Localize the so called “bright band” in the radar images. Can it be identified in both parameter spaces?

(e) Focusing on the period between roughly 20:40–20:50 UTC the reflectivity shows a tendency to increase towards the surface. What could be reasons for this increased radar reflectivity? *Hint: Recall how a HTI scan works and use the slight bending/kink seen in the reflectivity values for your reasoning.*

3. Stratiform versus convective precipitation:

   (a) Explain briefly how stratiform precipitation forms in mid-latitudes.

   (b) For the case of convective precipitation why do only large raindrops and rimed structures such as graupel and hail reach the ground?

   (c) Why does it sometimes take only 20–30 minutes from the formation of a convective cloud until it precipitates?

4. Precipitation rates:

   (a) Plot the Marshall–Palmer distribution for both rain and snow from $r_h = 0.5$ to $r_h = 1.5$ mm for two different rainfall and snowfall rates ($R = 0.2$ and $R = 2$ mm h^{-1}). Describe and interpret the results.

   (b) The highest precipitation rate in Switzerland that has ever been measured (for 1 h) occurred in Locarno-Monti on the August 28, 1997 and reached the value 91.2 mm per hour. Assuming that the Marshall–Palmer distribution holds for this precipitation event, how many raindrops were there with radii between 2 and 4 mm per cubic meter of air?

5. Weather proverbs were developed and used for many centuries before weather forecast models were available for more accurate predictions. They are based on observations and deliver a certain probability for a weather phenomenon to occur. Some weather proverbs simply arose from coincidence, but others reflect common weather patterns. Interpret the following weather proverbs scientifically. Keep in mind that some are only valid for specific locations since they are based on observation. It might be helpful to reread Chapter 1 to explain some of the proverbs.
   (a) "When clouds appear like rocks and towers, Earth's refreshed by frequent showers."
   (b) "A ring around the Sun or Moon means rain or snow coming soon."
   (c) "Dew on the grass, rain won't come to pass."
   (d) "Red sky at night, sailor's delight. Red sky in the morning, sailor take warning."
   (e) "Mackerel sky (altocumulus clouds) and mare's tails (cirrus clouds) make tall ships carry low sails."
   (f) "The more cloud types show, the greater the chance of rain or snow."
   (g) "The sudden storm lasts not three hours. The sharper the blast, the sooner 'tis past."
6. Why is it more difficult to make projections about changes in global precipitation than about changes in global temperature in the future climate?

10 Storms and cloud dynamics

Convective precipitation originates from cumulus clouds, namely cumulus congestus (Cu con) and cumulonimbus (Cb); (Table 1.1, Figure 1.4b and Chapter 9). The prerequisite for precipitation from cumulus clouds is that the hydrometeors grow to precipitation size in a short time. This is most efficiently achieved in Cb, which by definition consists of water and ice and has the largest vertical extent and highest vertical velocities (Table 1.3). Precipitation can also fall from Cu con, but its precipitation is typically less heavy and in the form of short showers, whereas Cb has the potential to turn into a thunderstorm, owing to the coexistence of water and ice, which is a prerequisite for the formation of lightning and thunder (Section 10.2). The various types of thunderstorms can generally be grouped into three categories as follows.

1. Ordinary, isolated or pulse thunderstorms, which consist of only a single convective cell and last approximately 40 minutes to one hour.
2. Multicell thunderstorms, which consist of a number of cells in different stages of development and last for several hours.
3. Supercell thunderstorms, which also last several hours but, in contrast with multicell thunderstorms, consist of only one self-maintaining cell.

Supercell storms are the rarest type of thunderstorm, but the most violent. Whether a given thunderstorm turns out to be a single cell (Section 10.1), multicell or supercell (Section 10.3), depends on both the vertical shear of the horizontal wind (hereafter just referred to as the wind shear) and the static stability of the environment. The largest convective storms are mesoscale convective systems (Section 10.4). They include a large variety of mesoscale systems, such as squall lines, mesoscale convective complexes and tropical cyclones, which we will discuss in terms of their formation and transition into extratropical cyclones (Section 10.5). This chapter ends with a discussion of the observed changes in tropical and extratropical cyclones in the last decades and projected future changes.

10.1 Isolated thunderstorms and hail

An ordinary or isolated thunderstorm is also called a pulse storm because of its short duration (Table 1.3) and because it may generate another ordinary thunderstorm as it weakens. Thus this type of thunderstorm appears to occur in pulses. It undergoes a characteristic life cycle and is often accompanied by graupel and hail.

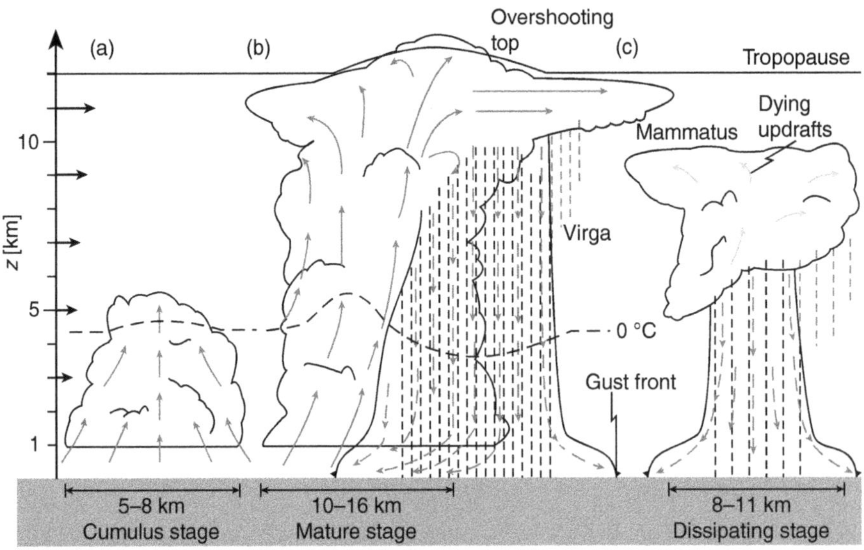

Fig. 10.1 Cross section of the three stages in the life cycle of an isolated single-cell thunderstorm: (a) cumulus stage, (b) mature stage and (c) dissipating stage. The gray arrows denote the air motion and the black arrows along the vertical axis represent the strength of the horizontal winds. The thin solid lines in (b) and (c) enclosing vertical broken lines represent the boundary of the precipitation-cooled air. Adapted from Doswell (1985).

10.1.1 Life cycle of an ordinary thunderstorm

A typical life cycle of an ordinary thunderstorm can be characterized by three stages (Figure 10.1) (Emanuel, 1994). The early (cumulus) stage is characterized by vigorous updrafts of positively buoyant air parcels that form in a statically unstable or conditionally unstable atmosphere. These updrafts, associated with high supersaturations and a large water vapor supply, set the stage for the rapid growth of hydrometeors due to collision–coalescence (Section 7.2) and ice microphysical processes (Section 8.3). The temperature is higher inside the updraft than in the surrounding environment because of the latent heat released by condensation and freezing. In Figure 10.1 this is shown for the 0 °C isotherm, which is at a higher altitude inside the cloud than outside it.

An example of radiosonde profiles on a day in which thunderstorms were observed, in North Carolina on April 18, 2015, is shown in Figure 10.2. The atmosphere was close to saturated neutral (Figure 3.2) with no well-defined inversion on top of the boundary layer. The lifting condensation level (LCL, Figure 4.3) was at 940 hPa, the level of free convection (LFC) at 750 hPa, and the level of neutral buoyancy (LNB) was at 280 hPa. Because the atmosphere was close to saturated neutral, it had a convectively available potential energy (CAPE) value of only 215 J kg^{-1}, corresponding to weak convection (Table 4.1). The wind shear in the lowest 6 km, as can be deduced from the wind barbs at the right-hand side of Figure 10.2, was only 9 m s^{-1}. Thus, this sounding is typical of weak single cell convection.

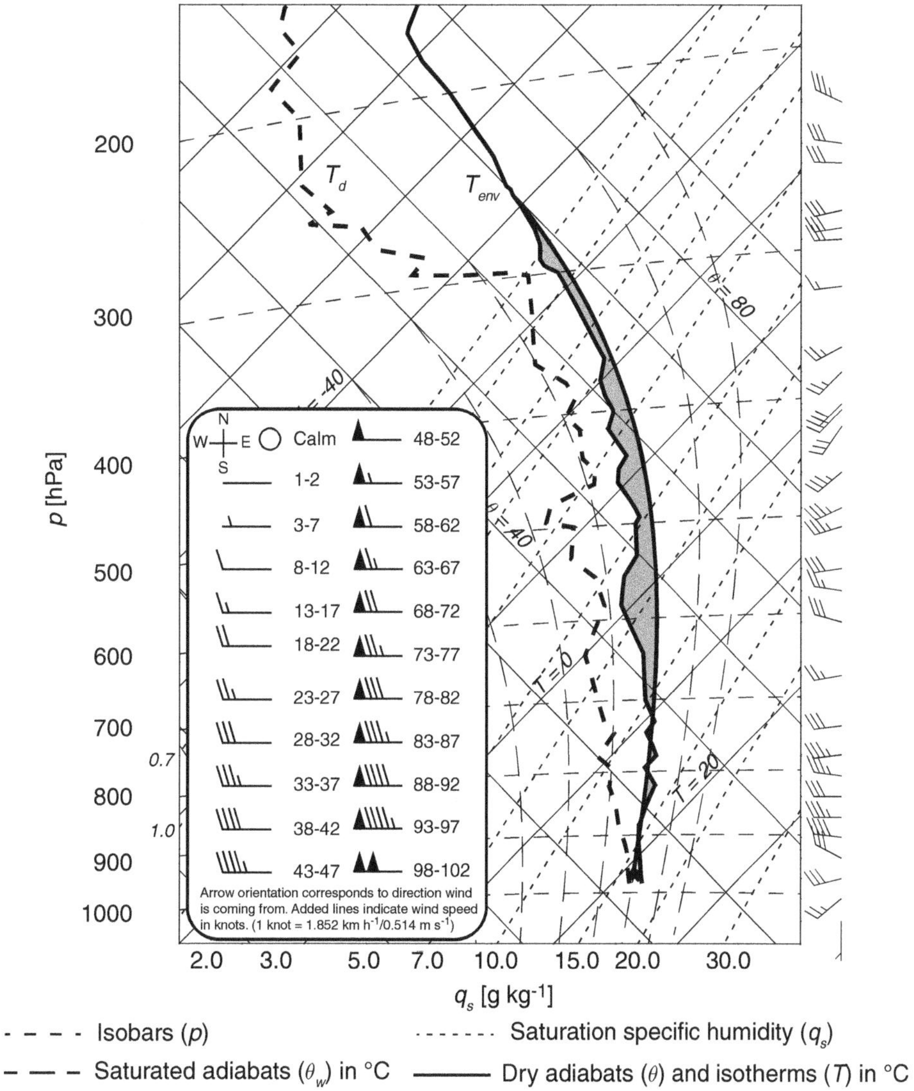

Fig. 10.2 Profiles of T_{env} and T_d and wind barbs in a tephigram in an environment where ordinary thunderstorms formed. The sounding was taken in Greensboro, North Carolina, on April 18, 2015 at 7 am local time. The CAPE value corresponds to the gray area. Data were obtained from the National Climate Data Center in the United States. The wind barbs at the right-hand side of the tephigram indicate wind direction and speed, as shown in the insert.

The mature stage is governed by active updrafts within the cloud and downdrafts associated with a heavy downpour of precipitation (Figure 10.1b). The downdrafts can either coincide with the updrafts or be slightly displaced horizontally. A slight horizontal displacement between the updrafts and the downdrafts is indicative of some weak wind shear in the environment. In the mature phase of the thunderstorm, the updrafts are most vigorous and the 0 °C isotherm is displaced even higher in the updraft region. The updraft speeds are highest in the middle of the cloud, which is not influenced by entrainment

(Section 4.2.4.3). Here they extend slightly above the level of neutral buoyancy (LNB, Section 4.2), as indicated by the overshooting top in the middle of the cloud top in Figure 10.1b. Elsewhere, the updraft basically terminates when the LNB is reached at the temperature inversion of the tropopause, indicated in the figure at 12 km altitude. Since the air above the LNB is stably stratified it acts like a lid, which causes the rising, overshooting, air entering the stratosphere to experience a downward restoring force. The overshooting top is characteristic for all convective clouds.

Overshooting air is sometimes visible in the form of a pileus, or cap cloud, which is a small horizontal lenticularis cloud (Section 1.2.3 and Figure 3.1) that forms on top of the convective cloud. A pileus cloud is indicative of strong updrafts inside the cloud and is often a harbinger of severe weather. It also marks the transition of the cumulus cloud into a cumulonimbus with its characteristic horizontal anvil consisting of ice crystals. The anvil forms because new air is lifted from below and at the same time is capped by the inversion, so that the air at the upper levels has no other option than to spread out horizontally. The anvil is usually much thicker than the pileus cloud. While the anvil is an integral part of a cumulonimbus cloud, the pileus cloud appears as an isolated cloud on top of the cumulus cloud and is only associated with an overshooting top.

In the mature stage, precipitation falls from the majority of the cloud and extends all the way to the surface (Figure 10.1b). Where the relative humidity is too low underneath the anvil, virga can be observed (Figures 1.8c and 7.13). Also, mammatus clouds are sometimes visible. Falling precipitation causes a downward drag as indicated by the broken arrows in Figure 10.1. The 0 °C isotherm extends to lower altitudes inside the downdrafts because of latent heat consumption due to the melting of hydrometeors. Depending on the RH, further cooling of the air within the downdraft region is caused by evaporation below cloud base. The boundary between the cold air and the warmer air at Earth's surface is called a gust front (Figure 10.1b). The gust front is accompanied by locally strong wind gusts and is an example of a density current: the air behind the gust front is cold and dense and pushes the warmer air out of its way as it propagates downstream. An example of a density current in everyday life is spilled water flowing horizontally on a surface because of its higher density than air.

Once the downdrafts reach Earth's surface, the cold air spreads out horizontally underneath the thunderstorm and starts to cut off the supply of warm air. This terminates the mature stage and causes the thunderstorm to turn into its dissipating stage, where only downdrafts can exist (Figure 10.1c). The thunderstorm still precipitates in the dissipating stage, but the intensity of precipitation is weaker than during the mature stage. The anvil with its virga now extends to lower altitudes because of the pre-moistened air from previously sublimated ice crystals and evaporated cloud droplets. In the dissipating stage, the cloud and the precipitation take on a stratiform character. Given that Cb anvils spread out in the upper troposphere, these clouds are also referred to as anvil cirrus.

10.1.2 Hail

Thunderstorms can produce hail. A hailstone is defined as a rimed particle that is larger than 2.5 mm in radius (Table 1.2). Hailstones often exhibit a layered internal structure

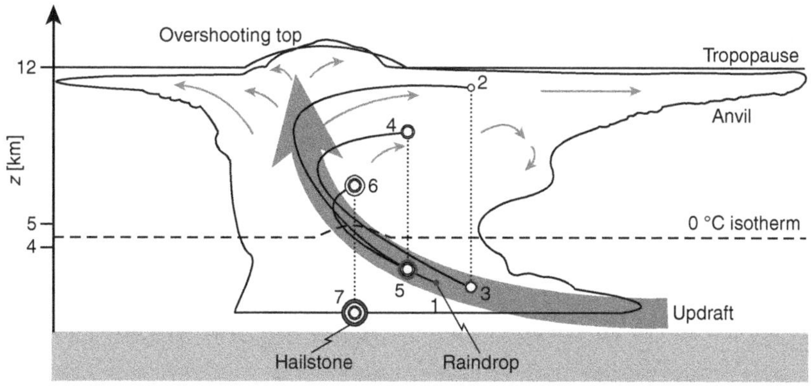

Fig. 10.3 Schematic of hailstone formation. Black circles denote wet growth and white circles dry growth. Upward directed motions of the hailstone are denoted by solid lines and downward motions by dotted lines.

that corresponds to alternating layers of clear, solid and spongy ice. Spongy ice is an indicator of air inclusions. The density of a hail particle can exceed 800 kg m^{-3} if the particle is dominated by solid ice. The largest recorded hailstone fell in a thunderstorm in South Dakota, USA, on July 23, 2010. It had a diameter of 30.32 cm and weighed 0.88 kg. The heaviest recorded hailstone was found in Bangladesh in April 1986 with 1.02 kg (http://wmo.asu.edu/#global).

The typical growth and life cycle of a hailstone is shown schematically in Figure 10.3. Normally, a raindrop first forms (labeled 1 in Figure 10.3). Owing to the strong updrafts associated with thunderstorms in which hail is formed, the raindrop is lifted above the 0 °C level, where it freezes and forms the core of the hailstone (point 2 in Figure 10.3). It grows further by accretion with cloud droplets and ice crystals in the updraft until it becomes sufficiently heavy that it falls down. On its way down (between points 2 and 3 in Figure 10.3) it grows by the further accretion of ice particles and supercooled cloud droplets and raindrops.

Cloud droplets or raindrops colliding with the hailstone can form a water sphere around the frozen core, even if they are slightly supercooled. This so-called wet growth of the hailstone leads to clear, solid, ice layers. Wet growth occurs as long as the temperature of the hailstone remains at or above the melting point, owing to the release of the heat of fusion by the freezing process. How much supercooling a hailstone can tolerate and still form a water sphere depends on its size and density and on the size and number concentration of the particles with which it collides. In laboratory studies, a water sphere was observed down to −10 °C (Garcia-Garcia and List, 1992).

In contrast, if accretion of the hailstone with supercooled water droplets occurs while the temperature of the hailstone remains below the freezing point, its growth is called dry growth. In this case freezing is fast enough that the collected droplets do not have time to spread over the surface of the hailstone. As a result, dry growth produces ice layers with many air cavities and bubbles.

In order to develop a layered structure the growth of a hailstone has to alternate between wet and dry growth. This alteration is typically achieved if the hailstone is carried upward multiple times and thereby crosses the 0 °C isotherm repeatedly: the second time the hailstone enters the updraft is point 3 in Figure 10.3. In the updraft the liquid shell freezes. Again, at some point the hailstone becomes too heavy to be carried further upward and falls down again. Between points 4 and 5 it grows again by accretion of ice particles and cloud droplets. This cycle is repeated until the hailstone no longer enters an updraft or is too heavy to be carried upwards by the updraft and falls down to the surface (point 7 in Figure 10.3), possibly colliding with further raindrops on its way.

10.2 Lightning and thunder

Besides hail, thunderstorms are accompanied by lightning and thunder. In general, cloud-to-cloud lightning, intra-cloud lightning and cloud-to-ground lightning can be distinguished. Lightning requires the simultaneous presence of an ice and a liquid phase inside a cloud and also that the cloud top is colder than −15 to −20 °C. Both criteria are met in cumulonimbus. Lightning is a transfer of charge during which the temperature can reach 30 000 K in a narrow channel (Section 10.2.3). It is accompanied by a pressure enhancement of one to two orders of magnitude, resulting in a sound wave (thunder).

10.2.1 Global electrical circuit

For the occurrence of thunder and lightning a separation of charges is required. A typical charge distribution in a mature thunderstorm and its role in the global electrical circuit are shown in Figure 10.4. In the global electrical circuit, electrified thunderstorm clouds are the generators. Earth's atmosphere has a downward-directed fair weather current whose upper extent is bounded by the electrosphere, which carries a net positive charge. The electrosphere is a layer of the atmosphere that begins a few tens of kilometers above Earth's surface and extends upwards up to the ionosphere. In the electrosphere, the electrical conductivity is so high that the layer essentially carries a constant electric potential. The electrical conductivity increases with altitude owing to the increasing ionization by cosmic rays and it peaks in the ionosphere, which extends from approximately 85 km to 600 km above Earth's surface. The ionosphere consists of electrons and electrically charged atoms and molecules. The electron density increases from 10^4 cm^{-3} at 85 km to 10^6 cm^{-3} in the middle of the ionosphere (at around 300 km). Above 300 km the electron density decreases again owing to the decrease in air density with increasing altitude. The source of the positive charge of the electrosphere is positive charge that leaks from the upper levels in thunderstorms (see below) as indicated by the ionic current in Figure 10.4.

The base of the fair weather current is bounded by Earth's surface, which carries a net negative charge, which is mainly maintained by lightning. According to convention, the global electrical current is shown in Figure 10.4 for the propagation direction of the positive charges. Negative charges, in form of free electrons, flow in the opposite direction. In this circuit, positive charge flows from the tops of electrified thunderstorm clouds to the electrosphere from where the fair weather current, 1500 A (Pruppacher and Klett, 1997),

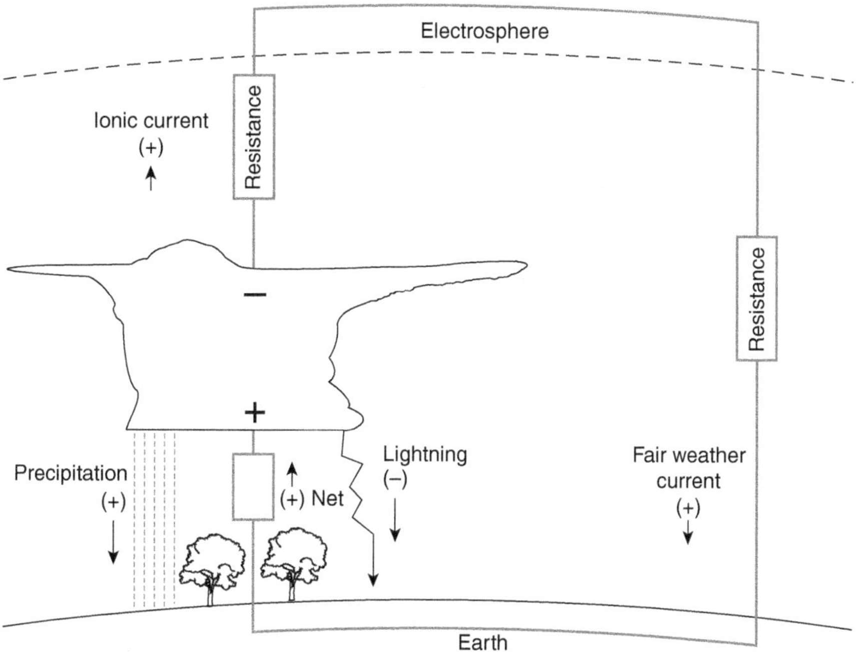

Fig. 10.4 Schematic of the main global electrical circuit (not to scale). The positive and negative signs in parentheses indicate the sign of the transported charges.

continuously leaks positive charge to Earth's surface. In the absence of lightning it would take only around seven minutes for Earth and the atmosphere to be free of charge as a result of charge transport by the fair weather current.

Precipitation particles are polarized by the electric field of the fair weather current as well as by the electric field beneath thunderstorms in such a way that they tend to preferentially collect positive ions as they fall to the ground (Wallace and Hobbs, 2006). Therefore they transfer positive charges to Earth's surface. Thus, lightning needs to supply Earth's surface with negative charges of 1500 A in order to balance the fair weather current and the positive charges carried by falling precipitation particles. Assuming a current of 1 A per thunderstorm cell (Pruppacher and Klett, 1997) implies that globally 1500 thunderstorm cells are active at any one time.

10.2.2 Charge separation within clouds

For lightning to occur, the charges inside the cloud need to be separated. The typical assumption is that before lightning occurs, a Cb cloud has a tripole charge structure, with positive charges near the cloud top and the cloud base. The negatively charged layer in the middle of the cloud is called the main charging zone and usually occurs between the -10 °C and -20 °C isotherms (Figure 10.5).

Different hypotheses exist on how charge separation occurs and the exact conditions are still a matter of debate. The thermoelectric theory (Pruppacher and Klett, 1997) was one of

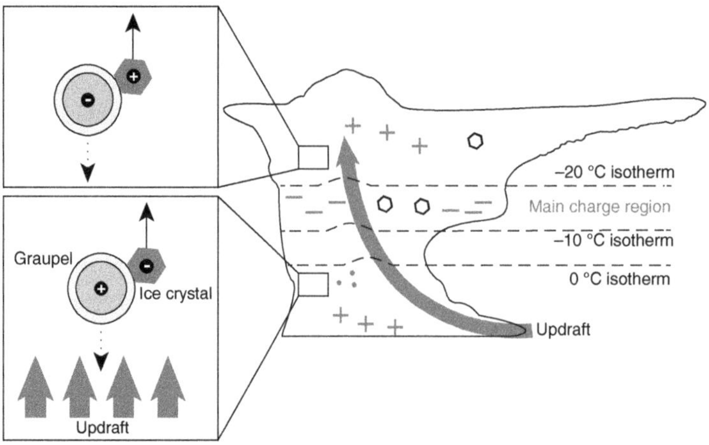

Fig. 10.5 Schematic view of charge generation due to the collision of an ice crystal with a graupel particle and the subsequent charge separation due to the differing terminal velocities of oppositely charged hydrometeors.

the first mechanisms proposed to account for the charge separation observed in laboratory experiments. It invokes the generation of charge during collisions between graupel particles and ice crystals at different temperatures inside the Cb, as shown in Figure 10.5. The separation of charge then occurs, owing to the different terminal velocities of oppositely charged hydrometeors. A review of possible charge separation mechanisms can be found in Saunders (2008).

The thermoelectric theory is used mainly to explain the negatively charged layer in the middle part of the Cb and the positively charged region in its upper part (Figure 10.4). The surfaces of the graupel particles or hailstones are warmer than the surrounding ice crystals because they collide with liquid droplets, which freeze upon contact and thereby release latent heat. In ice the H^+ ions are more mobile than the OH^- ions, so that the H^+ ions can move from the warm hailstone or graupel particle to the cold colliding ice crystal. This causes the graupel particle to become negatively charged. The thermoelectric theory is limited to temperatures below -10 °C, where aggregation is unlikely and ice crystals bounce off graupel after collision. As the small positively charged ice crystals are carried upward with the updraft, a charge separation between the layer of negative charges in the middle of the cloud and the positively charged region in the upper layers of the cloud is created.

The weakly positively charged region near cloud base (Figure 10.4) is regarded as the first step for the initiation of ground flashes. One explanation for the presence of this weakly positively charged region is again riming but with warm ice crystals accreting on the cold graupel particles. Hence, the H^+ ions move from the warm ice crystals to the cold graupel particles. Alternative explanations for the positively charged region near cloud base are that it is due to the melting of ice particles or to mixed-phase processes (Wallace and Hobbs, 2006).

Newer theories relate the charge generation to differences in the diffusional growth rates of graupel particles and ice crystals. They require the collision of ice crystals with graupel or hail particles in the presence of cloud droplets (Saunders, 2008). It is currently not clear whether the diffusional growth rate theory can explain the tripolar structure in most cases. Therefore this structure in thunderstorms may not be explained by a single charge separation mechanism, but different mechanisms could be important in different regions of the cloud.

10.2.3 Ground flashes

Once the electric field caused by the separated charges is stronger than the so-called breakdown field strength (> 150–400 kV m^{-1}, according to Stolzenburg and Marshall (2009)), lightning is likely to occur in order to equilibrate the charge. The electric field is now strong enough to locally ionize the air within a small channel associated with a strong increase in its conductivity. Charge transport then preferentially occurs through this channel, visible as lightning.

Here we will focus on cloud-to-ground lightning flashes or short ground flashes. These account for only 25% of all lightning flashes. Cloud-to-ground lightnings are more common in mid-latitudes where the freezing level and hence the main charging zone (Figure 10.5) is at lower altitudes than in the tropics, where only 10% of all lightning flashes are ground flashes. Each lightning flash consists of several strokes, which are needed to equalize the charges. On average this takes three to four strokes per lightning flash.

Lightning strokes transport electrons from the cloud to the ground while the return strokes carry positive charges from objects at Earth's surface to the cloud base, as illustrated in Figure 10.6 (Uman, 1987) and highlighted in the photograph shown in Figure 10.7. Ground flashes that transport negative charges to the ground originate from the lower, negatively charged part of the cloud. This discharge is called the stepped leader. It moves downward towards Earth in discrete steps of 1 μs (~ 50 m) (Figure 10.6c–e). The cascade nature of this process results from the electric field intensity at the tip of the stepped leader decreasing as some electrons are captured by coagulation with positive ions, atoms and neutral molecules from the environment. These electrons then need to be replenished from higher levels.

A typical process chain leading to ground flashes proceeds as follows (Wallace and Hobbs, 2006). The stepped leader is initiated by a local discharge between a small pocket of positive charges at the base of the Cb, which is at around 1 km above ground over land (Table 1.1), and the lower part of the negatively charged region (Figure 10.6b). As the negatively charged stepped leader approaches the ground, it induces positive charges on high objects, such as tall buildings or trees (Figure 10.6c). In their vicinity the electric field strength increases locally. When the stepped leader is within 10–100 m of the ground, a discharge moves up from the ground to meet it (Figure 10.6d). When this discharge connects with the stepped leader, the breakthrough field strength is exceeded and a large number of electrons flows to the ground (Figure 10.6e).

Afterwards, a highly luminous lightning stroke propagates upward from the ground to the cloud following the path of the stepped leader (Figure 10.6f), called the return stroke. The return stroke with its bright luminous channel is the stroke that is visible by eye. Since a lightning stroke only lasts a few milliseconds, the time is too short for the air to expand. Instead the pressure increases almost instantaneously to values of 10–100 times

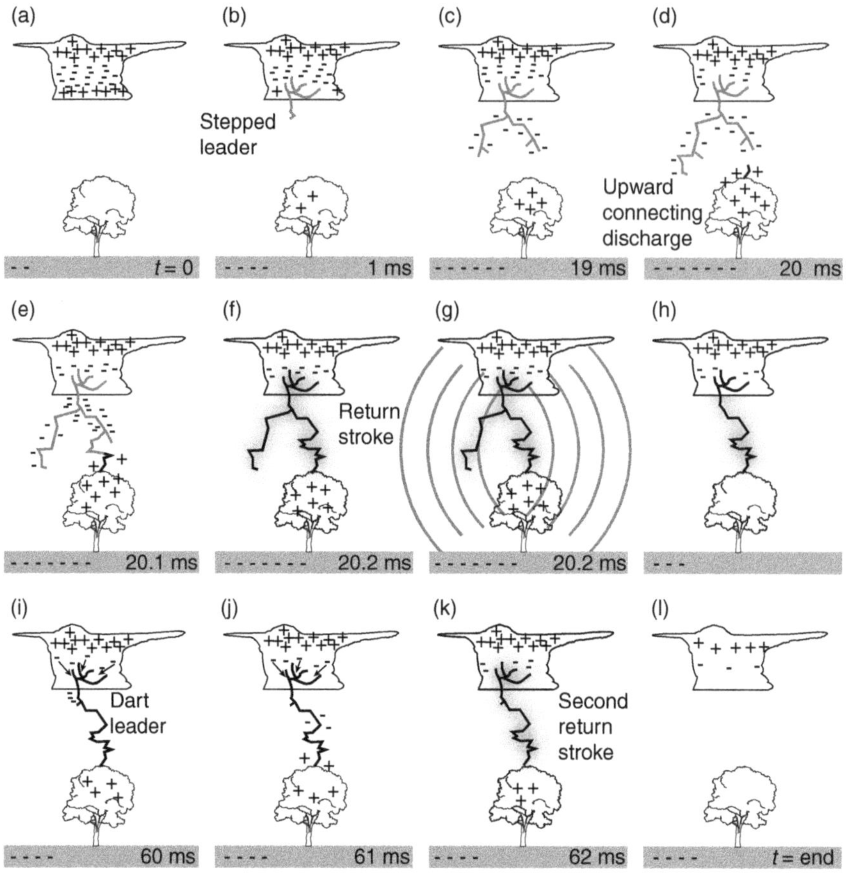

Fig. 10.6 Schematics (not to scale) illustrating processes leading to a ground flash and their associated time steps: (a) cloud charge distribution, (b) preliminary breakdown, (c, d) stepped leader, (e) attachment process, (f, h) first return stroke, (g) thunder, (i, j) dart leaders of subsequent strokes, (k) second return stroke and (l) situation after the lightning strokes have finished. The visible strokes (f, k) are shaded. Figure adapted from Uman (1987) with permission from Elsevier.

the atmospheric pressure. This simultaneously generates a cylindrical shock wave (Figure 10.6g), which propagates outward as a sound wave and is heard as thunder.

Following the first stroke, subsequent strokes can occur along the same main channel as long as the electric field remains sufficiently high. Additional electrons are supplied by the more distant regions of the negatively charged area of the cloud. A negatively charged leader, called the dart leader, then moves downward along the same path and carries further electrons onto the ground (Figures 10.6c, j). The dart leader is followed by another visible return stroke to the cloud (Figure 10.6k). This process continues until the charges are mainly equalized (Figure 10.6l). Then the charge separation process (Section 10.2.2) needs to start again, and a new electric field must be established, before the next lightning sequence can occur.

The lightning flash of the first stroke reaches up to 40 kA, which heats up the channel by ohmic dissipation, i.e. energy loss caused by electrical resistance, to above 30 000 K.

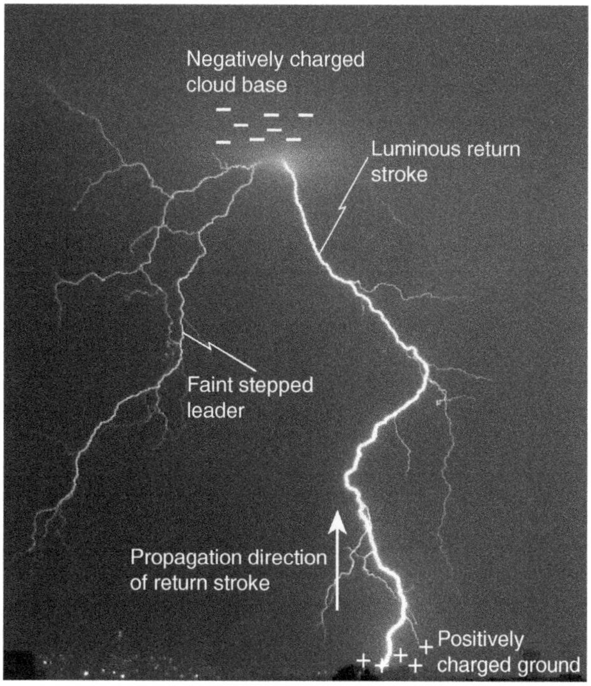

Fig. 10.7 Photograph of a ground flash, in which some of the aspects of Figure 10.6 are highlighted. Photograph taken by Milos Vujovic.

Hence, a lightning stroke is an example of a plasma present in Earth's atmosphere. This plasma, like the plasma in a neon lamp, ionizes the gas and makes the lightning stroke visible as recombination takes place.

The maximum updraft velocity in a convective cloud is a good indicator of the total lightning activity. An analysis of radar data from different thunderstorms in Colorado, Kansas and northern Alabama, USA by Deierling and Petersen (2008) found a correlation of ≥ 0.8 for the lightning rate and maximum updraft velocity or total updraft volume in the main charging zone ($T < -5$ °C). Hence, a knowledge of the updraft velocities in convective clouds can help to predict their lightning activity, which in turn is a good proxy for precipitation intensity.

10.3 Multicell and supercell storms

Thunderstorms can also occur in larger clusters and be part of a bigger system. There are two kinds of larger storms: the multicell storm, which consists of multiple cells, that can produce small to moderate size hail and weak tornadoes and the supercell storm, which is a large single-cell storm that can produce large hail and weak to violent tornadoes. A key to the longevity and strength of both supercell and multicell storms is the wind shear (Figure 10.8 and Table 10.1). This change in wind speed and, for supercells, also in direction with

Table 10.1 Wind shear between 0 and 6 km above ground [m s^{-1}], the associated storm type and its characteristics (Markowski and Richardson, 2010).

Shear	Storm type	Characteristics
$0-10$ m s^{-1}	Single-cell	Short-lived, gust front may initiate new cells
$10-20$ m s^{-1}	Multicell	Gust front repeatedly generates new cells and determines the propagation of the multicell
> 20 m s^{-1}	Supercell	Quasi-steady updrafts; propagation governed by pressure gradients caused by turning shear

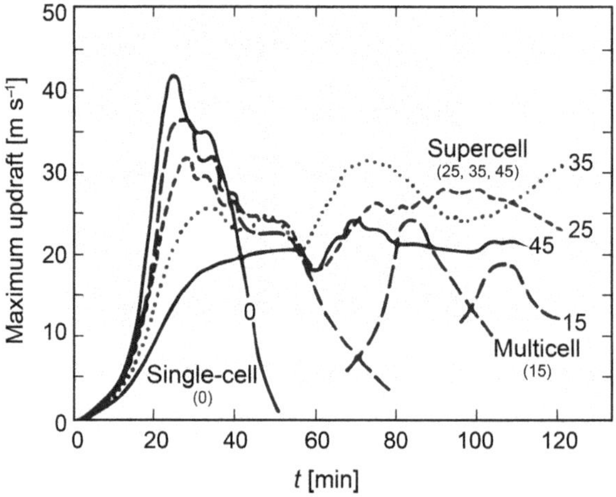

Fig. 10.8 Lifetime of single-cell, multicell and supercell thunderstorms in terms of the maximum updraft velocity. The lines are labeled in m s^{-1} according to the vertical shear of the horizontal wind between the surface and 10 km height. Figure adapted from Weisman and Klemp (1982) and reproduced by courtesy of the American Meteorological Society.

height ensures that the updraft and downdraft regions are horizontally displaced. Therefore the precipitation is separated from the updraft and cannot shut off the updraft.

Figure 10.8 shows simulations based on a three-dimensional cloud model (Weisman and Klemp, 1982) of the maximum updraft velocity as a function of time for different values of the wind shear. In the case of no wind shear the maximum updraft speed, 40 m s^{-1}, is reached after 25 minutes and the storm decays after 50 minutes. This is in good agreement with observations of ordinary single-cell thunderstorms. In single-cell thunderstorms the gust front may initiate new cells, which seems to occur randomly in space and time.

Multicell storms typically have wind shears of around $10-20$ m s^{-1} in the lowest 6 km (Table 10.1). This enables the gust front to remain within the storm system and to trigger new cells repeatedly. As shown in Figure 10.8, simulations for a 15 m s^{-1} shear show a first maximum in vertical velocity after 25 minutes, with new maxima appearing after 85 and 110 minutes.

Supercell storms require the largest wind shear, of more than 20 m s^{-1}, in the lowest 6 km (Table 10.1). In addition to the prerequisite of large vertical shear, it is important that

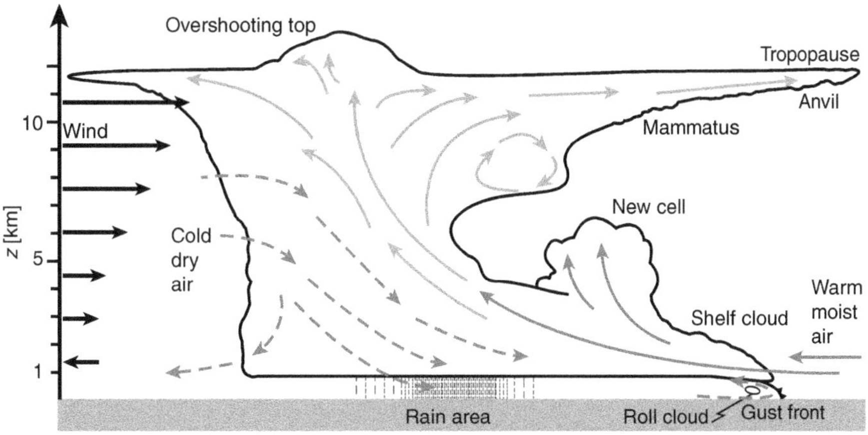

Fig. 10.9 Schematic view of the motion of the warm moist air (solid curved arrows) and the cold dry air (broken curved arrows) in an intense multicell thunderstorm. The storm is moving from left to right. Figure adapted from Ahrens (2009).

the wind speed turns anticyclonically with height (Markowski and Richardson, 2010); this is also referred to as turning shear. As shown in Figure 10.8, simulations with 25 m s^{-1} shear or more show no decrease in maximum updraft speed below 15 m s^{-1}.

10.3.1 Multicell storms

Multicell storms are larger (several tens of kilometers in horizontal dimensions) and last longer than single-cell storms, but they occur less frequently. They can turn into a mesoscale convective system (Section 10.4).

A schematic view of the air motion and other features associated with an intense multicell thunderstorm is shown in Figure 10.9. The name arises because a multicell thunderstorm consists of a number of cells, each in a different stage of development. The individual cells can be characterized to be in an early, a mature or a dissipating stage (Figure 10.1). In Figure 10.9 a small new cell is appearing ahead (downstream) of the mature cell. The inflowing warm moist air in both cells is shown by the curved solid arrows. Like the ordinary thunderstorm in Figure 10.1, a multicell storm also has an overshooting top that protrudes into the stratosphere with a pronounced anvil. Mammatus clouds (Figure 1.8c) can often be seen to hang below the anvil.

In contrast with an ordinary thunderstorm, the updraft is tilted because of the wind shear, indicated by the black arrows along the vertical axis in Figure 10.9. Underneath the multicell storm, the cold air spreads out horizontally. However, it does not cut off the inflow of warm moist air as in the case of an ordinary thunderstorm. Instead, the cold and dense air behind the gust front helps to lift the warm moist air and allows a new cell to form ahead of the main updraft. If the lifted air is sufficiently moist, a shelf cloud (Figure 10.9) forms in the lifted air ahead of the gust front. It is a visual manifestation that the approaching storm will be severe.

10.3.2 Vorticity

The self-organization of thunderstorms in more complex (and dangerous) convective systems is related to the generation of vertical vorticity inside a storm, i.e. of horizontal motions around a vertical rotation axis. A strong vertical vorticity in the cloud acts to lower the pressure, and this in turn, maintains the updrafts. Regions with high values of vertical vorticity are often observed in clouds that are part of mesoscale systems, such as supercell thunderstorms.

In order to understand the reinforcement between vorticity and updrafts we now analyze the sources of vorticity. Vorticity is a quantity measuring the spin of a local portion of fluid. It can be defined at every point of the domain as the curl of the velocity field:

$$\vec{\omega} = \nabla \times \vec{v}. \tag{10.1}$$

Its three components along the x-, y- and z- directions are given by

$$\eta = \frac{\partial w}{\partial y} - \frac{\partial v}{\partial z}, \tag{10.2}$$

$$\xi = \frac{\partial u}{\partial z} - \frac{\partial w}{\partial x}, \tag{10.3}$$

$$\zeta = \frac{\partial v}{\partial x} - \frac{\partial u}{\partial y}, \tag{10.4}$$

where u, v and w denote the x-, y- and z- components of the velocity $\vec{v}$, respectively. Equations (10.2)–(10.4) show that vorticity is generated by wind shear, in both the horizontal and vertical direction. Synoptic-scale atmospheric motions are essentially two-dimensional, because the vertical scale on which they take place ($\sim$ 10 km) is much smaller than the horizontal scale ($\sim$ 1000 km). Neglecting vorticity components other than the vertical is therefore justified on the synoptic scale.

Horizontal vorticity is important for phenomena occurring on the mesoscale, such as thunderstorms, which are characterized by much smaller horizontal extents ($\sim$ 10–100 km, Figure 9.6). Here the height of the troposphere can be comparable with their horizontal extent. The primary source of vorticity on the mesoscale is the wind shear ($\partial u/\partial z$, $\partial v/\partial z$), which creates horizontal vorticity, as shown in Figure 10.10a.

Hence, on the mesoscale it is easier for the atmospheric flow to generate horizontal rather than vertical vorticity. To understand the source of vertical vorticity ζ in mesoscale systems, we start from the equation describing the temporal evolution of ζ for a fluid parcel in an incompressible flow in the horizontal direction:

$$\frac{d\zeta}{dt} = \underbrace{\zeta \frac{\partial w}{\partial z}}_{\text{stretching}} + \underbrace{\xi \frac{\partial w}{\partial y} + \eta \frac{\partial w}{\partial x}}_{\text{tilting}} + \underbrace{F_\zeta}_{\text{mixing}} . \tag{10.5}$$

Equation (10.5) is known as the vorticity equation in the z-direction. It can be obtained by taking the curl of the Navier–Stokes equation (3.1). While the Navier–Stokes equation describes the conservation of linear momentum, the vorticity equation depicts the conservation of angular momentum. The assumption of an incompressible flow implies that

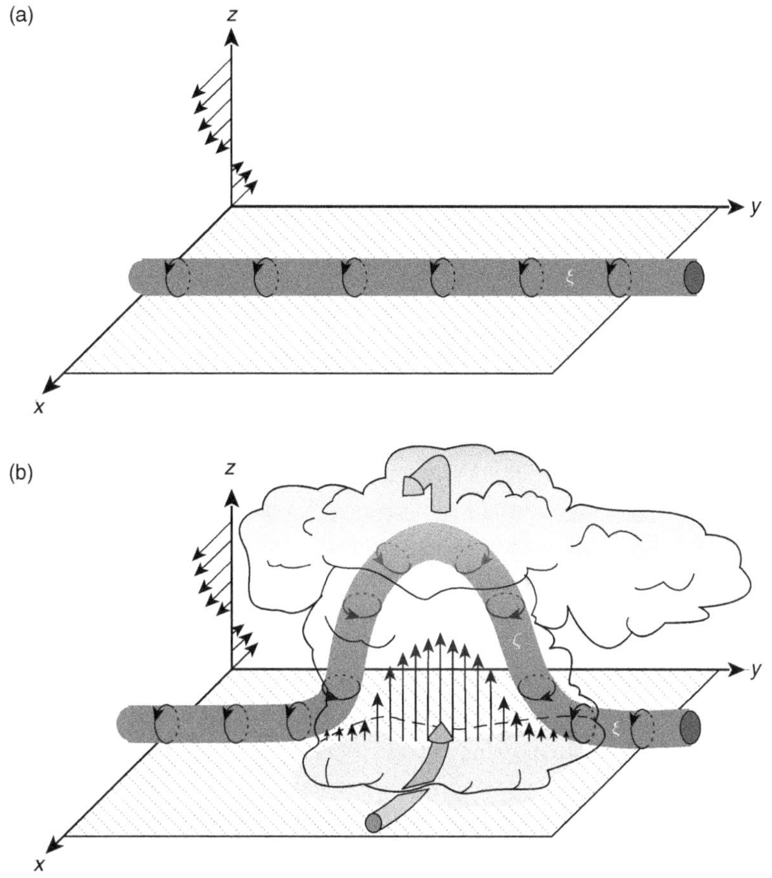

Fig. 10.10 Conversion of horizontal vorticity ξ into vertical vorticity ζ in a thunderstorm in the presence of vertical wind shear $\partial u/\partial z$ indicated by the horizontal arrows. (a) The creation of ξ by $\partial u/\partial z$, (b) The creation of ζ by the superposition of an updraft of a convective cloud on a horizontal vorticity tube. The large mid-gray arrows show the in- and outflow, the vertical arrows denote the updrafts. The arrow loops around the gray tubes indicate the vorticity. Figure adapted from Klemp (1987).

the density remains constant within a parcel of fluid that moves horizontally with the flow; it does not require the fluid itself to be incompressible. This assumption is often applied in the atmosphere because it eliminates changes in air density with time and simplifies the continuity equation to

$$\nabla \cdot \vec{v} = \frac{\partial u}{\partial x} + \frac{\partial v}{\partial y} + \frac{\partial w}{\partial z} = 0. \tag{10.6}$$

The first term on the right-hand side of eq. (10.5) can be linked to the horizontal convergence of the horizontal wind via eq. (10.6). It is a direct consequence of the continuity equation that horizontal convergence leads to vertical motion. Neglecting the tilting and mixing terms in eq. (10.5) and inserting eq. (10.6) into eq. (10.5), we obtain

$$\frac{d\zeta}{dt} = -\zeta \left(\frac{\partial u}{\partial x} + \frac{\partial v}{\partial y} \right). \tag{10.7}$$

This equation means that convergence $(\partial u/\partial x + \partial v/\partial y) < 0$ in the presence of vorticity increases the vertical vorticity, i.e. the fluid spins faster. This effect can be understood using the principle of angular momentum conservation: an ice skater spins faster if she contracts her arms towards her body, because her moment of inertia decreases.

The tilting terms in eq. (10.5) are related to the generation of vertical vorticity ζ by the tilting of horizontal vortex tubes, as illustrated in Figure 10.10b. When the updraft of the convective cloud is superimposed on the vortex tube, the tube is deformed upward in such a way that vorticity develops around the vertical axis, i.e. horizontal vorticity is converted into vertical vorticity ζ. After the creation of some initial vertical vorticity ζ the stretching term in eq. (10.5) becomes important in the updraft region, which is characterized by convergence and ascending motions. The mixing term in eq. (10.5) is usually small and therefore can be neglected.

In the mature stage of the thunderstorm (Figures 10.1b and 10.11b), precipitation creates downdrafts. At altitudes above ≈ 1 km, the air in the downdrafts is warmed by adiabatic compression. Near the surface this downward motion is slowed, and cooling by evaporation of the precipitation proceeds with no compensating adiabatic compression. The negative temperature anomaly at the surface produces a large horizontal buoyancy gradient across the gust front (Section 10.1.1), which ultimately leads to the generation of horizontal vorticity.

To illustrate this, let us assume that a horizontal gradient in the buoyancy force F_B (eq. 3.14) is generated only in the x-direction, as shown in Figure 10.11b. This gradient in F_B then creates horizontal vorticity in the form of two vortices that rotate opposite to each other around the y-axis, as shown in Figure 10.11c. Mathematically this is described using the ξ-component of the vorticity equation in the absence of the stretching and tilting terms:

$$\frac{d\xi}{dt} = \frac{\partial \xi}{\partial t} + u\frac{\partial \xi}{\partial x} + w\frac{\partial \xi}{\partial z} = -\frac{\partial F_B}{\partial x}. \tag{10.8}$$

The horizontal vorticity generated in this way is a second source of horizontal vorticity that can also be converted into vertical vorticity if the gust front remains within the system. The development of strong thunderstorms can thus be summarized as follows. They usually occur in areas with strong vertical wind shear, which implies the generation of horizontal vorticity (eqs. 10.2, 10.3 and Figure 10.10a). Vertical vorticity is first created inside the system by the tilting term in eq. (10.5), where the presence of a local updraft below the cloud leads to horizontal variations of vertical velocity w such that $\partial w/\partial x$ and $\partial w/\partial y$ are non-zero. The storm's gust front also supplies horizontal vorticity at low levels (Figure 10.11c). If the gust front remains within the system, as it does in multi- and supercell thunderstorms, this horizontal vorticity is also available for conversion into vertical vorticity by tilting in the updraft region. The combination of these two effects (horizontal vorticity generation by vertical shear of the environmental wind and by a horizontal gradient in F_B) can lead to very high values of vertical vorticity inside a thunderstorm. These, in turn, reinforce the strong downdrafts that generate large gradients in F_B, increasing the horizontal vorticity and hence establishing a positive feedback loop that ensures the longevity of multi- and supercell thunderstorms. In the case of supercell thunderstorms, this feedback loop can create a mesocyclone (Section 10.3.3).

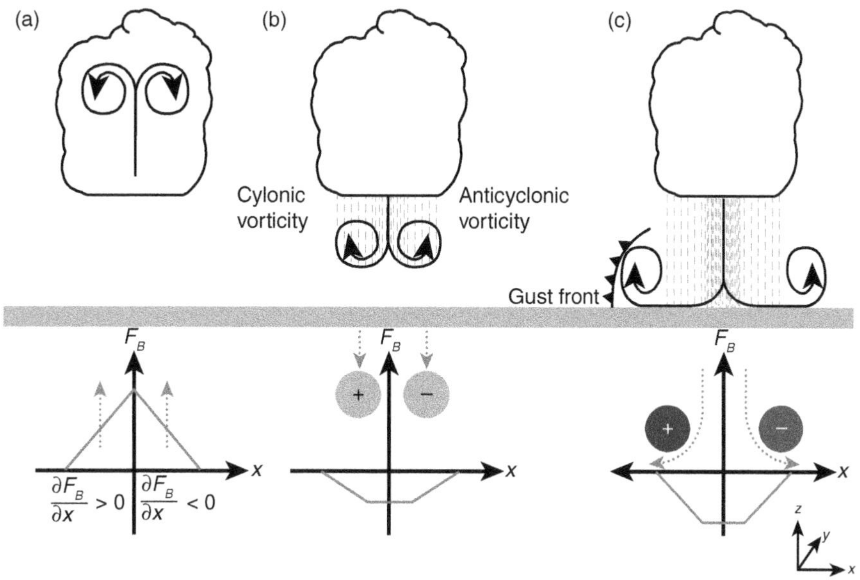

Fig. 10.11 The generation of horizontal vorticity ξ in the y-direction by a gradient in the horizontal buoyancy force F_B in the x-direction: (a) cumulus stage with positive buoyancy; (b), (c), mature stages with negative F_B gradients at the surface creating a pair of vortices of different signs and intensities (shaded disks, the darker disks correspond to greater intensity) according to eq. (10.8).

The horizontal vorticity generated behind the gust front can sometimes be visible if an "arcus" cloud forms above and ahead the gust front, as shown in Figure 10.12. The downdrafts that cause gust fronts are classified according to their horizontal extent. If this is less than 10 km and their diverging wind speeds exceed 10 m s^{-1}, they are known as downbursts (Markowski and Richardson, 2010). Downbursts are divided into microbursts, which last between 1 and 5 minutes in an area of less than 4 km, and macrobursts, which cover an area between 4 and 10 km and last between 5 and 30 minutes. Downbursts are visible if the warm air lifted by the gust front was sufficiently moist.

10.3.3 Supercell storms

Supercell storms are more violent and destructive than multicell thunderstorms. They occur less frequently than multicell storms because they need an even stronger wind shear in the lowest 6 km than multicell storms (Table 10.1), and the shear must turn anticyclonically with altitude (Markowski and Richardson, 2010). A situation where an anticyclonic-turning shear can be found near the surface is, for instance, when low-level air comes off the Gulf of Mexico while the air at mid levels comes from the west or north. Such a setting guarantees a large contrast between warm and moist air masses near the surface and cold and dry air masses aloft and is the reason why supercell storms and tornadoes occur most frequently in the so-called "Tornado Alley", located in the midwest of the United States. In addition to this dynamical prerequisite, the formation of supercell storms also requires ample low-level moisture

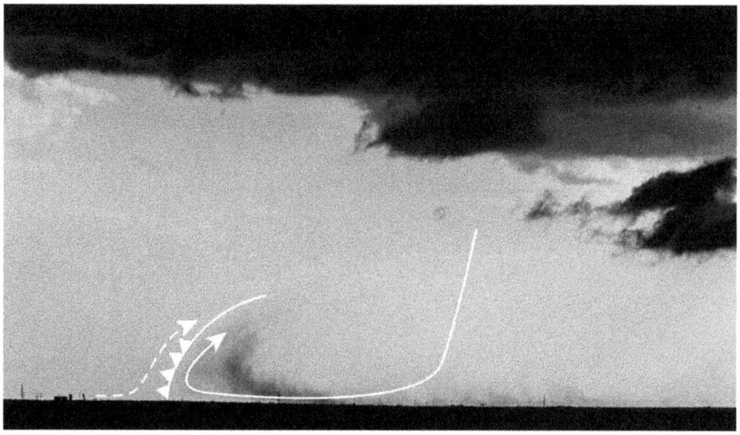

Fig. 10.12 Photograph of a dry downburst (a microburst), in which the associated gust front is observable by the raised dust and is marked with white triangles. The downburst was formed analogously to the wet downburst shown in Figure 10.11. Photograph taken by Zachary Hargrove.

and a sufficiently high CAPE value, i.e. adequate instability for vigorous convection.

Supercell storms have a horizontal size comparable with multicell thunderstorms but consist of a single giant updraft–downdraft pair with strong vertical updrafts of 10 to 40 m s^{-1} (Figure 10.8). Supercell storms are characterized by a mesocyclone, a deep updraft that is persistently rotating around a vertical axis, and are often isolated from other thunderstorms. They usually occur in mid-latitudes and can be found in the warm sector of a low pressure system generally propagating in a northeasterly direction. Supercell storms can occur with or without the formation of a tornado. The opposite is also found: tornadoes can form without a mesocyclone. These are called non-supercell tornadoes, or landspouts, and are typically smaller and weaker than supercell tornadoes. In some vigorous cases the top of the supercell storm may extend into the stratosphere and the width of the storm may exceed 40 kilometers (Ahrens, 2009). Supercell storms often produce hail and can transition to a tornadic phase as schematically shown in Figure 10.13.

As in a multicell storm, the updraft in a supercell storm is tilted because of the wind shear, the areas of precipitation are separated from the updraft and it often produces mammatus clouds (Figure 1.8c) hanging below the anvil. It is characterized by two marked downdrafts, the forward flank downdraft with gust front ahead of the storm and the rear flank downdraft behind the storm. Behind the forward flank gust front the first light rain forms, followed by heavy rain and hail as the center of the forward flank downdraft is approached. A line provided by cumulus clouds builds the flanking line, which forms as surface air is lifted into the storm along the rear flank gust front.

Most precipitation particles in the updrafts are advected downwind into the northeastern part of the supercell, owing to the strong wind shear. This results in a high radar-reflectivity in that region (Figure 10.14). In the forward flank downdraft the precipitation particles fall to the surface, causing negative buoyancy due to their weight (Section 4.2.4) and to melting and evaporative cooling (Markowski, 2002).

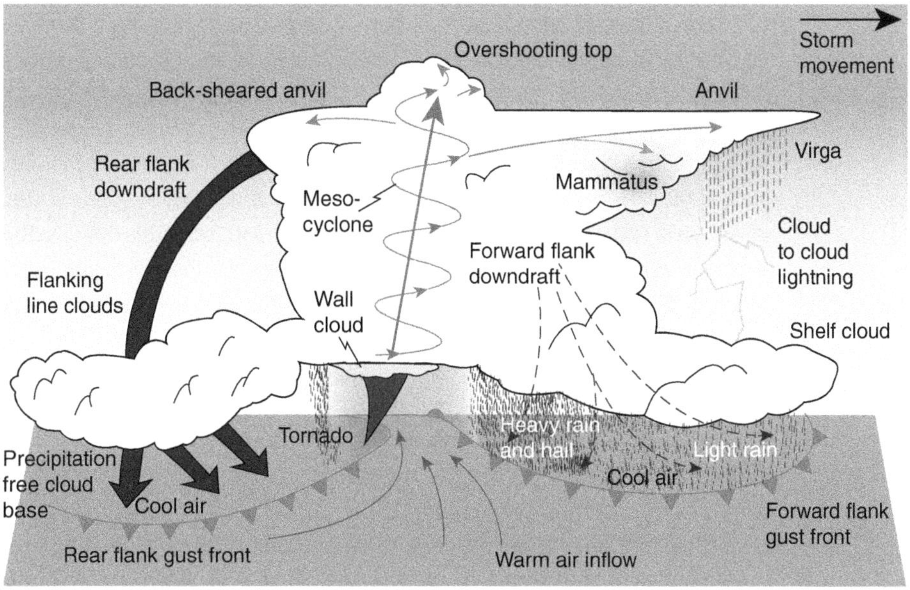

Fig. 10.13 Schematic of the main features, including the updraft and downdrafts and surface air flow, associated with a tornado-breeding supercell thunderstorm. A black and white version of this figure will appear in some formats. For the color version, please refer to the plate section.

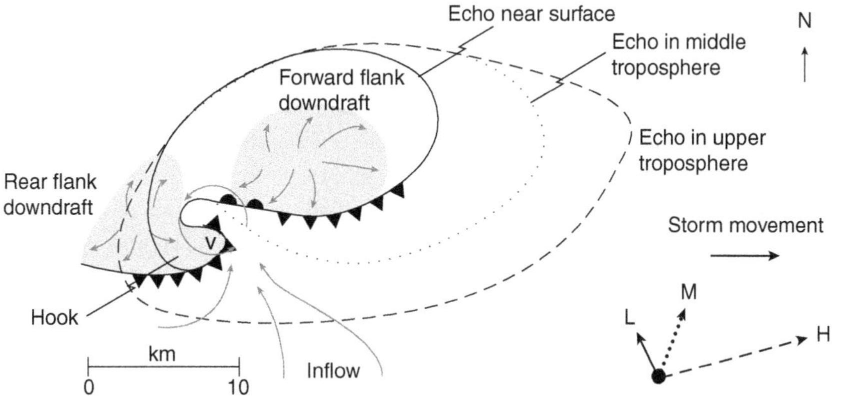

Fig. 10.14 Schematic diagram showing horizontal sections of the precipitation radar echoes at three different levels of a supercell storm. Environmental winds relative to the storm at low, medium and high levels of the troposphere are represented by the three vectors low (L), middle (M) and high (H). The tornado is represented by a V (vault).

The rear flank downdraft is forced downward by mid-level environmental winds that cause the air to warm by adiabatic compression. If dry air is injected into the supercell, the rear flank downdraft is characterized by a slot of dry air with a rain-free base where the temperature increases dry adiabatically. If the air contains precipitation particles, the negative buoyancy in the rear flank downdraft is driven by their loading, by hail melting and by evaporative cooling below cloud base. Normally the cloud base in the area of the rear flank downdraft is higher than elsewhere owing to the dry air. Both the rear flank downdraft and the forward flank downdraft are thought to be important for producing (spawning) tornadoes, as will be discussed below.

Figure 10.14 shows that the inflow at the surface is almost perpendicular to the main wind direction, i.e. the direction of the storm motion. The convergence of the warm moist air and the cooler air in the supercell leads to cold and warm fronts at the surface. Unlike at the cold and warm fronts in an extratropical low pressure system, the cool air in the supercell approaches from both behind and ahead of the warm air. Since the horizontal wind turns anticyclonically with altitude, a good portion of the outflow at upper levels is almost parallel to the main wind direction, as illustrated in Figure 10.14. Supercell storms are characterized by an extensive overhanging echo ahead of the storm which slopes downward to the ground towards its rear flank (the echo in the middle troposphere). Below they have a region of low reflectivity, called the vault or bounded weak-echo region (BWER), which penetrates into the heart of the storm beneath its highest top (the echo in the upper troposphere), as shown in Figure 10.17. The vault is partly surrounded by an intense hook-shaped echo at low levels (the echo near the surface) (Browning, 1964).

The anticyclonic turning of the horizontal wind $\vec{v}_h$ from low to high levels is a prerequisite for the asymmetry of the supercell. It can mathematically be expressed in terms of the helicity H, which is a measure of the transfer of horizontal vorticity $\vec{\omega}_h$ from the environment to an air parcel in a convective cloud. It is defined as follows:

$$H = \int \vec{v}_h \cdot \vec{\omega}_h dz = \int \vec{v}_h \cdot \nabla \times \vec{v}_h dz. \quad (10.9)$$

Equation (10.9) is integrated over the depth of the inflow layer (around 1–3 km) of the convective cloud (Markowski and Richardson, 2010). The result is zero when the horizontal wind is normal to the horizontal vorticity vector. The higher the helicity, the higher the chance of supercell storms and tornadoes.

The helicity causes the majority of the ascending air to descend in the forward flank downdraft and thus establishes the giant updraft–downdraft circulation. Unless the environmental winds have a turning shear, the updraft and downdraft in a thunderstorm cannot become intertwined and neither a mesocyclone nor a tornado can form. The gust front remaining within the system and the turning wind shear, which causes the updraft and downdrafts to be horizontally displaced, together promote the longevity of the supercell.

10.3.4 Tornadoes

Transition to the tornadic phase occurs relatively suddenly (on the order of 10 minutes) after a supercell has slowly evolved over several hours. Tornado-genesis is favored when the wind shear within the lowest 1 km is high, the lifting condensation level is low and the humidity in the planetary boundary layer is high.

A high humidity and a low lifting condensation level (LCL) ensure that the downdraft air is not cooled too much by evaporation. This is especially important for air in the rear flank downdraft, because it intertwines with the warm inflow air and gets lifted again in the mesocyclone (Figure 10.13). A high value of the low-level wind shear ensures that a positive feedback loop can be established between the horizontal vorticity generation behind the gust front (eq. 10.8) and the vertical vorticity generation (eq. 10.5).

When a tornado develops, it forms near the leading edge of the hook echo (shown as V in Figure 10.14). Supercells need strong, storm-scale, rotation to develop tornadoes. Part of this rotation is produced by the turning shear of the environmental wind. In addition, cool air from high altitudes is forced into the rear flank downdraft of the supercell, where it accelerates downward and forms a gust front at the rear flank. This, in addition to creating horizontal vorticity, causes extra lifting of the inflowing warm buoyant air. The warm rising air and the rear flank downdraft start to spiral around each other, as can be seen in the hook-shaped radar echo of the rear flank downdraft (Figure 10.14).

This spiraling marks the formation of a mesocyclone, i.e. a rotating updraft (Figure 10.13). If the rear flank gust front approaches the forward flank gust front, the updraft is weakened in lower levels but remains strong aloft, causing a vertical stretching and intensification of the mesocyclone. Mesocyclones typically have diameters between 3 and 80 km (the meso-γ to meso-β scale; Section 9.4). The formation of a mesocyclone distinguishes a supercell storm from a multicell storm and explains why tornadoes in multicell storms are usually much weaker. The wall cloud underneath the mesocyclone in Figure 10.13 indicates the area of strongest updrafts. Here condensation occurs at altitudes lower than that of the ambient cloud. Together with the wall cloud, the mesocyclone extends to low levels.

A tornado is associated with a strong pressure minimum and very high wind speeds. It is defined as a violently rotating column of air, underneath a cumuliform cloud, which is in contact with the ground. Often (but not always) it is visible as a funnel cloud, as shown in Figure 10.15. A funnel cloud is a cone-shaped or needle-like protuberance hanging downwards from the cloud base. The vortex of the tornado typically has a diameter on the order of 100 meters. Thus, a tornado is the most intense of all circulations on the microscale (Figure 9.6). It usually rotates cyclonically and can reach wind speeds up to 135 m s^{-1} (490 km h^{-1}). The rotational wind speed of a tornado can be calculated from the cyclostrophic wind balance. It is a special case of the gradient wind balance (eq. 3.44) when the horizontal extent of the vortex is so small that the Coriolis force can be neglected. The cyclostrophic wind hence results from the force balance between the pressure gradient force and the centrifugal force:

$$\frac{1}{\rho}\frac{\partial p}{\partial r} = \frac{v^2}{r}. \tag{10.10}$$

An example of two radiosonde profiles in Kansas, USA on May 18, 2013 of an environment where a supercell that spawned tornadoes is shown in Figure 10.16. The radiosonde profiles were taken 10 and 4 hours prior to the formation of a tornado. At 7 am local time, the LCL was at 880 hPa and the humidity in the planetary boundary layer (PBL) was well mixed, with a mean specific humidity of 14.8 g kg^{-1}. The boundary layer was confined by a strong inversion of 7 K at its top with a convective inhibition (CIN) value 400 J kg^{-1}. Above the inversion the atmosphere was conditionally unstable. The level of free convection (LFC) was at 630 hPa and the level of neutral buoyancy (LNB) was at 190 hPa, resulting in a CAPE value of 2600 J kg^{-1}. The wind turned anticyclonically from

Fig. 10.15 Photograph of a tornado, taken by Thomas Winesett.

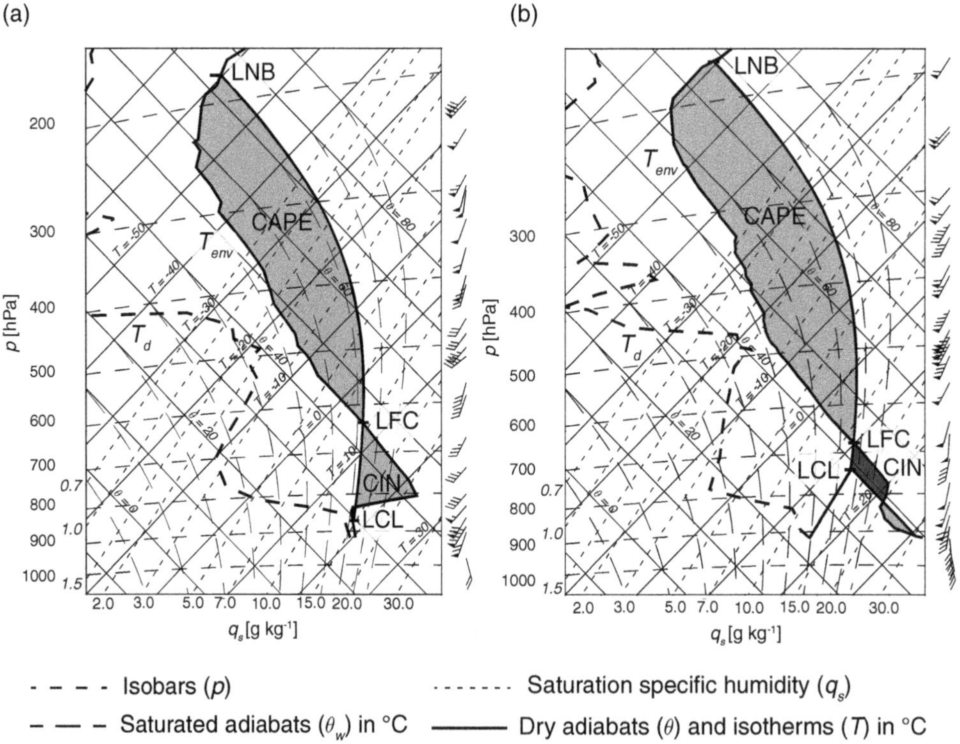

Fig. 10.16 Profiles of T_{env} and T_d and wind barbs in tephigrams in an environment that produced a tornadic supercell storm in Kansas, USA. The soundings were taken at Dodge City, Kansas on May 18, 2013 at (a) 7 am and (b) 1 pm local time. Data were obtained from the National Climate Data Center in the United States.

south-southeast at the surface, with warm and humid air coming from the Gulf of Mexico, to southwest at 1 km above the surface (Figure 10.16a).

This sounding was characterized by extreme instability above the capping inversion. In such a case very rapid thunderstorm development can be expected if the cap is weakened or the air underneath is sufficiently heated to overcome convective inhibition (CIN). Such a sounding is also referred to as a "loaded gun" sounding.

Owing to the solar heating between 7 am and 1 pm local time, the temperature at the surface (790 m) increased from 20.4 to 32 °C and T_d decreased from 19.6 to 17 °C. Hence, the CIN was reduced to 200 J kg^{-1}, the LCL was raised to 730 hPa and the LFC to 670 hPa and the value CAPE increased slightly to 2700 J kg^{-1} (Figure 10.16b), corresponding to typical CAPE values found for tornadoes (Table 4.1). The wind shear in the lowest 6 km was 20 m s^{-1}, which, according to Table 10.1, is borderline for supercell development. Thus only a little extra heating was necessary in order to overcome CIN and for explosive thunderstorms to develop. On that day, the extra heating caused the formation of a supercell at 6 pm that spawned tornadoes between 7 and 8 pm.

A radar image of the supercell storm that developed in these conditions at 7:15 pm on May 18, 2013 can be seen in Figure 10.17. It shows the highest radar reflectivity, with more than 70 dBZ, indicative of hail (Table 9.1), underneath the mesocyclone. The overshooting top reached altitudes up to 18 km and thus extended into the stratosphere. The area of low reflectivity in the lower troposphere between the forward flank downdraft and the rear flank downdraft denotes the bounded weak-echo region.

To date, the largest tornado outbreak occurred between April 25 and April 28, 2011 with 355 recorded tornadoes, four of them being in the most destructive category (EF5 on the Enhanced Fujita Scale; http://www.spc.noaa.gov/efscale/). For comparison, on average less than one EF5 tornado is observed per year. This tornado outbreak was accompanied by supercells with CAPE values of 3000–4000 J kg^{-1} and a mesoscale convective complex (see below). Prior to this event the record number of tornadoes had been counted during the historic 1974 tornado outbreak, where between April 3 and April 4 148 tornadoes were recorded across 13 states of the USA.

10.4 Mesoscale convective systems

Mesoscale convective systems (MCSs) are the largest convective storms. They occur on the meso-α scale (Section 9.4) and are even larger and longer lasting than individual multicell or supercell storms. An MCS is a cloud system consisting of an ensemble of thunderstorms that produces a contiguous precipitation area on the order of 100 km in horizontal dimension in at least one direction.

Mesoscale convective systems account for a large portion of precipitation in the tropics and during the warmer time of the year in mid-latitudes. Characteristic of an MCS is the coexistence of convective and stratiform precipitation. While the convective regions are characterized by intense vertically extending cores and high radar reflectivities (from large hydrometeors), the stratiform regions are of the uniform texture associated with lower

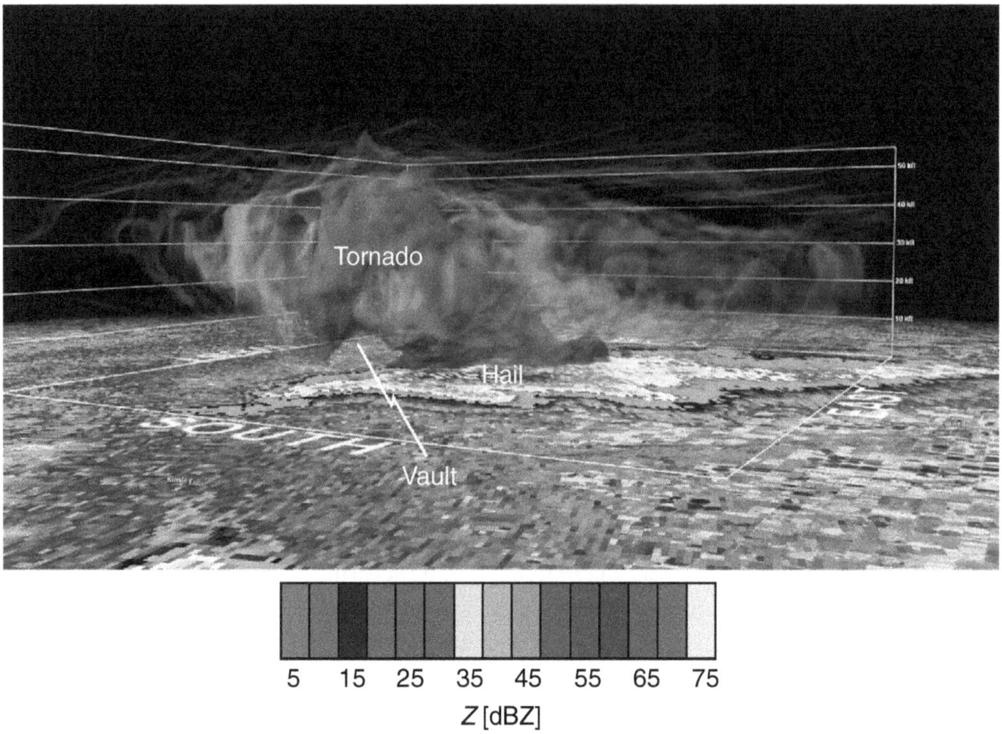

Fig. 10.17 3D radar image of a tornadic supercell over Kansas on May 18, 2013 at 7:15 pm local time. The radar data were obtained from the National Climate Data Center in the United States. The picture was produced with GR2Analyst and kindly provided by Thomas Winesett. A black and white version of this figure will appear in some formats. For the color version, please refer to the plate section.

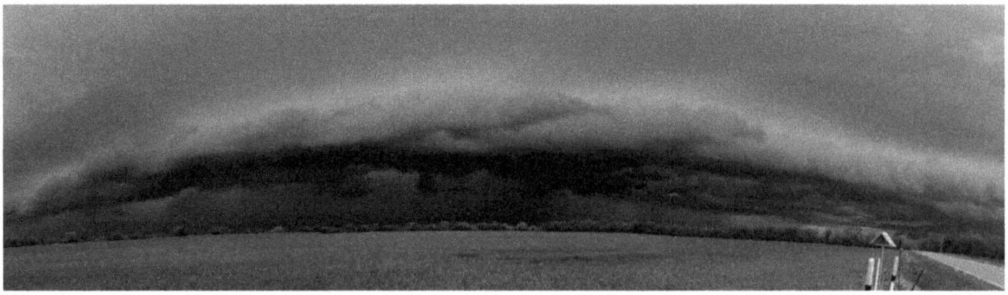

Fig. 10.18 Photograph of a squall line by Brian Johnson.

reflectivities and lighter precipitation. This stratiform precipitation forms as a result of the dissipation of older convective cells and the sloped ascent of a mesoscale layer of air. The name "mesoscale convective systems" indicates that they develop mesoscale circulations as they mature.

The convective elements within MCSs exhibit varying degrees and types of organization, depending on the large-scale environment in which they form. A distinct form of mesoscale organization comprises squall lines composed of groups of cells arranged in a line behind a long continuous gust front, as shown in Figure 10.18. The gust front can coincide with a

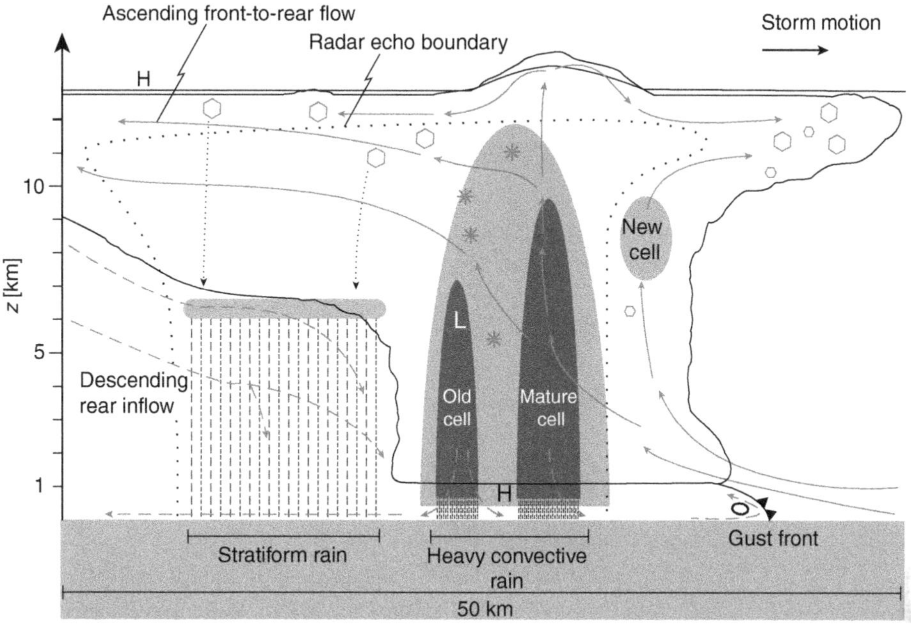

Fig. 10.19 Conceptual model of the kinematic, microphysical and radar-echo structure of a squall line, viewed in a vertical cross section oriented perpendicularly to the squall line and parallel to its motion. Intermediate and strong radar reflectivities are indicated by medium and dark shading, respectively, and the radar-echo boundary by the dotted line. The small horizontal gray area denotes the radar bright band. The points labeled H and L indicate centers of positive and negative pressure perturbations, respectively. The dotted arrows indicate the range of ice particle trajectories, the solid arrows indicate the flow inside the squall line and the broken arrows the descending rear inflow. Figure adapted from Houze (1993) with permission of the American Meteorological Society.

cold front of a low pressure system, but squall lines can also occur ahead of a cold front. A cross section through an idealized squall line which is moving towards the right is shown in Figure 10.19 (Houze, 1993). The air generally ascends from the front to the rear through the squall line, as indicated by the solid arrows. In the direction of the storm motion, the gust front approaches first (at the surface). As warm air is lifted above it, an updraft with a new convective cell is formed. The new cell is followed by a mature cell and an old cell, which are characterized by the strong radar echoes associated with heavy convective showers. The ascent of the mature convective cell is terminated once it overshoots into the stratosphere. Then the air spreads out horizontally, mainly downwind of the storm motion but partly also upwind of it. In fact, this mechanism is the same in every Cb that has a noticeable horizontal anvil.

The anvil that develops upwind is called the stratiform anvil because of the stratiform character of the precipitation that forms underneath. It is fed by large ice crystals formed inside the mature convective cells as indicated by the range of ice particle trajectories in the stratiform region shown by the dotted arrows. First, cloud droplets grow by collision–coalescence in the updraft of the convective region. Upon freezing they continue to grow by accretion and riming below 0 °C. The ice crystals spread out from the convective region into the stratiform anvil. Here the ice crystals continue to grow by deposition, which is the dominant growth mechanism at low temperatures. Growth by aggregation and riming (Sections 8.3.2 and 8.3.3) becomes important at temperatures > -10 °C in the vicinity of the melting layer. This layer constitutes the radar bright band and is indicated in the figure by the horizontal gray shaded area of intermediate radar reflectivity in the stratiform region. It is displaced from the convective anvil because it requires uniform horizontal structures inside the cloud. Below the bright band, precipitation leaves the MCS. It is less intense than the precipitation originating directly from the convective towers. It is, however, more intense than typical stratiform rain as a result of the seeder–feeder mechanism (Section 9.5.3) triggered by the convective core. Below the stratiform anvil, cold environmental air descends along isentropic surfaces owing to mass continuity. This ensures the supply of cold air for the gust front.

In addition to heavy rain, a squall line frequently produces lightning, strong straight-line winds and possibly tornadoes or waterspouts. A waterspout in its common form is an example of a non-supercell tornado and, as the name suggests, it forms over water. Tornadic waterspouts have also been reported but less frequently. Waterspouts are typically less violent than tornadoes.

The strong straight-line winds associated with squall lines are often referred to as *derechos* (John and Hirt, 1987; Hinrichs, 1888). Wind speeds in *derechos* can exceed 160 km h^{-1} and cause considerable damage.

A mesoscale convective complex (MCC) is an example of an MCS with well-defined characteristics that allow its identification in infrared satellite images. Its size is measured in terms of its cloud top temperature. It has cloud top temperatures below -32 °C over an area of 100000 km^2 or cloud top temperatures below -52 °C over an area of 50000 km^2 for at least six hours. It is a fairly round system with an eccentricity (the ratio of the minor and the major axis) of at least 0.7 at its maximum extent.

10.5 Tropical cyclones

Tropical cyclones (TCs) are among the most devastating natural phenomena. They are also important because they redistribute heat excess in the tropical oceans to higher latitudes and are therefore a key component of the global circulation (Section 12.2); they are known by different names, depending on where they arise (see below). They range in diameter from 100 to 4000 km and belong mainly to the meso-α scale but can extend into the lower end of the synoptic scale (Figure 9.6). The largest TCs occur in the northwest Pacific Ocean. The largest Atlantic TC on record was Hurricane Sandy, with a diameter of 1800 km, which hit the eastern United States on October 29, 2012.

10.5.1 General characteristics

As the name suggests, TCs develop in the tropics. A TC is characterized by a cloudless eye in its center, as illustrated in Figure 10.20. The eye ranges between 8 and 200 km in diameter. It is surrounded by the eyewall, in which the most vigorous convection with thunderstorm activity occurs. However, convection is not restricted to the eyewall but also

Table 10.2 Saffir–Simpson scale of TC severity (http://www.nhc.noaa.gov/aboutsshws.shtml). The categories are based solely on the maximum wind speed sustained over 1 minute at 10 meters. The storm surge and minimum pressure values are for reference only.

TC	Maximum wind speeds [m s^{-1}]	Storm surge [m]	Minimum pressure [hPa]
1	33–42	1.0–1.7	980–994
2	43–49	1.8–2.6	965–979
3	50–58	2.7–3.8	945–964
4	59–69	3.9–5.6	920–944
5	≥ 70	≥ 5.7	≤ 920

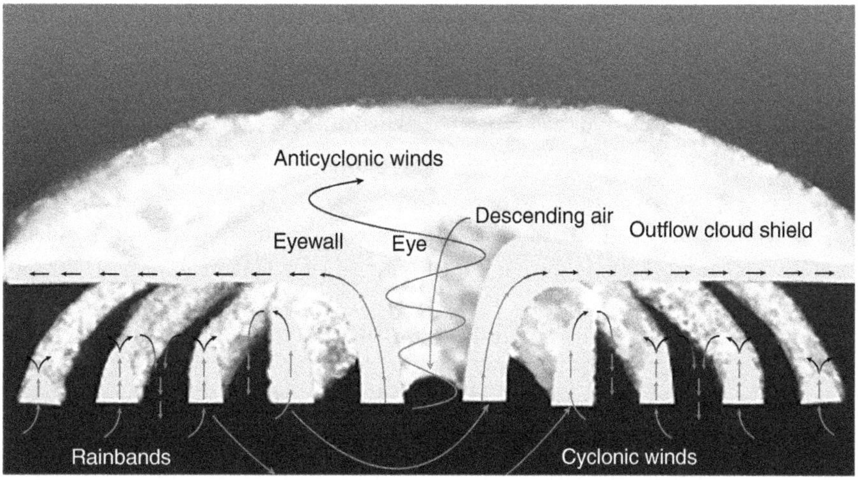

Fig. 10.20 Schematic of the three-dimensional structure of a tropical cyclone. The solid gray arrows at the surface denote the primary circulation; in the eye the downward arrow indicates the subsidence and the upward circulating arrow indicates the change in rotation axis from the surface to the top. The short gray arrows denote the rising and sinking air and the short black arrows the outflow at cloud top.

exists further away from the center of the TC, where it occurs in bands that are separated by areas of descending motions. These rainbands are caused by symmetric instability (Figure 3.8). The highest wind speeds near the surface are found underneath the eyewall. The maximum sustained wind speed over 1 minute at 10 meters above ground is used to categorize the intensity of tropical cyclones according to the Saffir–Simpson scale, which has five categories (Table 10.2). Tropical cyclones in category 5 are the most severe, with wind speeds exceeding 250 km h^{-1}. In the case of typhoon Haiyan, which hit the Philippines on November 7, 2013, one-minute-sustained wind speeds of 315 km h^{-1} were measured.

As mentioned above, tropical cyclones have different names depending on the ocean basin in which they form. Tropical cyclones formed in the north Atlantic, the northeast Pacific east of the dateline and the south Pacific east of 160° E are called hurricanes. Typhoons are those TCs that form in the northwest Pacific west of the dateline. Severe tropical cyclones form in the southwest Pacific west of 160° E and in the southeast Indian

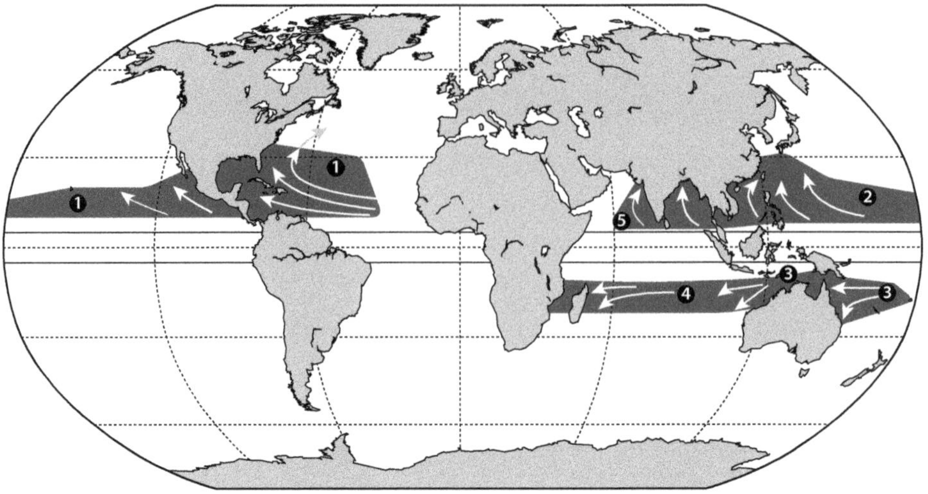

Fig. 10.21 Schematic of the occurrence of tropical cyclones. The numbers refer to the different regions and the associated names: (1) hurricanes, (2) typhoons, (3) severe tropical cyclones, (4) tropical cyclones and (5) severe cyclonic storms.

Ocean east of 90° E. Severe cyclonic storms form in the north Indian Ocean and tropical cyclones form in the southwest Indian Ocean (Figure 10.21).

10.5.2 Prerequisites for tropical cyclone formation

In general, there are six prerequisites for a TC to form. These can be divided into three thermodynamic criteria (1–3) and three dynamic criteria (4–6).

First, the sea surface temperature (SST) needs to be sufficiently high because evaporation of water from the ocean surface provides the energy for a TC. It was found that 98% of all tropical cyclones between 1981 and 2008 formed when the SST was higher than 26.5 °C (Tory and Dare, 2015). As the strong winds associated with a TC stir up colder water from lower levels in the ocean, not only the surface water but also the water below, to a depth of at least 50 m, needs to be sufficiently warm. A shallower warm water depth would imply that cold water could be stirred up through wave action, which would cut off the energy supply of the TC.

The second thermodynamic criterion states that the atmosphere has to be conditionally unstable, to facilitate deep convection and to liberate the heat stored in the ocean (Emanuel, 1986). This is fulfilled if the temperature in the atmosphere quickly decreases with height. There is, however, some debate about how important this criterion is, because TCs usually have low CAPE values (Table 4.1).

Third, the mid-troposphere needs to be sufficiently moist because otherwise the entrainment of dry air could prevent convective clouds from rising all the way to the tropopause.

The word "cyclone" suggests that a TC is a rotating system. It can only form some degrees away from the equator as the Coriolis force is required to initiate rotation of the moving air. Note that the Coriolis force is negligible for the rotation of a tornado (Section 10.3.4) because of its smaller horizontal extent and much shorter lifetime. Most TCs form at least 5° away from the equator, which defines the fourth criterion.

Fifth, it has been found that a TC arises from an initial disturbance, such as a tropical easterly wave or an upper-level trough. A tropical easterly wave is a north–south-oriented elongated area of high vorticity and relatively low pressure that travels from east to west in the tropics. It has its origin in the Intertropical Convergence Zone (Figure 1.2). The most prominent example is provided by African easterly waves, which originate over North Africa from instabilities in the African easterly jet (AEJ). The AEJ forms as a result of the strong temperature and moisture gradient between the hot and dry Sahara in the north and the cooler Atlantic ocean to the south. To remain in thermal wind balance (Section 3.5.2), the atmosphere develops a strong wind shear. This causes the low-level winds to reverse their direction and flow towards the east as part of the African monsoon circulation. The AEJ itself flows towards the west. Instabilities within this jet form the seed circulation for a tropical easterly wave, from which thunderstorms develop. In order for a TC to form, these thunderstorms must cluster together and start to rotate.

Last but not least, the wind shear must not be too strong because otherwise it would tear off the tops of the convective clouds. Such erosion of the overcasting clouds would allow more infrared radiation to escape to space, which would cool the core and weaken the storm. Vertical wind shear tilts the TC, causing heat and moisture to be removed from the center of the TC. Also, vertical wind shear enables dry low-energy air from the environmental mid troposphere to enter the storm, further weakening it. All together, it becomes a less efficient heat engine (Section 10.5.5).

10.5.3 Circulation within a tropical cyclone

The circulation within a TC can be divided into a dynamical and a thermodynamical part, where the primary circulation is the dynamical part and the secondary circulation is the thermodynamic part. The primary circulation comprises horizontal spiraling winds, i.e. the general rotation of the TC, as shown in Figure 10.22a. It can be described by the gradient wind balance (eq. 3.44), i.e. the balance between the Coriolis, pressure gradient and centrifugal forces (Section 3.5). As shown in Figure 10.22b, the pressure is lowest in the eye, whereas the tangential wind speed has its maximum underneath the eyewall.

The primary circulation can be visualized simply when stirring a cup of tea made with tea leaves, even though the horizontal forces at play are different. If one stopped stirring, the tea leaves would accumulate in the center. The same would happen to the air in a TC in the absence of the secondary circulation because friction due to the ocean waves causes the air to converge in the center.

The secondary circulation describes the vertical circulation. It is also called the "in-up-and-out circulation" and is shown in Figure 10.22b. Air convergence at lower levels ("in") is associated with rising air masses in the eyewall ("up"), leading to thunderstorm clouds and the release of latent heat. The rising of the warm, positively buoyant, air stops at the tropopause, where it spreads out anticyclonically ("out").

10.5.4 Differences between tropical and extratropical cyclones

The differences between tropical and extratropical cyclones are visualized in Figure 10.23. An extratropical cyclone is caused by baroclinic instability (Section 3.5.2), i.e. by

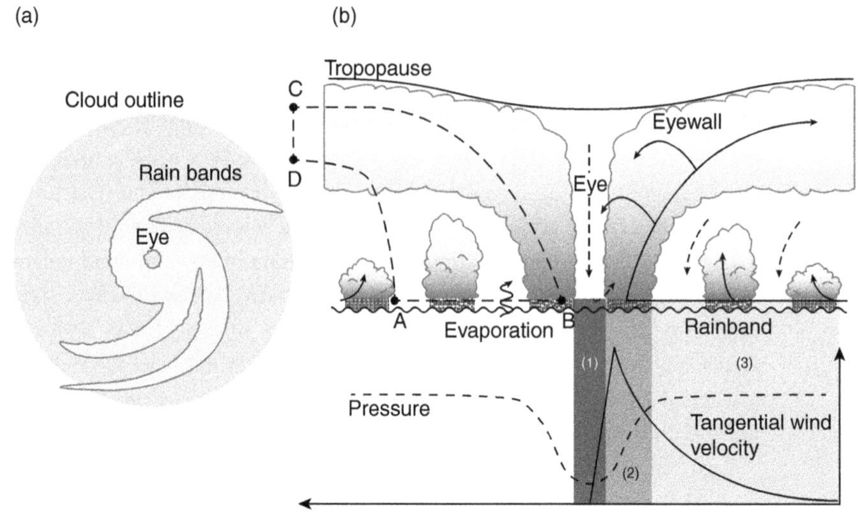

Fig. 10.22 Schematic of (a) the primary circulation viewed from above and (b) the secondary circulation in a cross section of a TC. The different regions in a TC are indicated by numbers: (1) refers to the eye, (2) to the eyewall and (3) to the outer region. The letters A–D denote the different stations in the Carnot process (Figure 10.24). The lower part of (b) shows qualitatively the pressure and tangential wind distribution in a TC.

instability due to intersecting isobars and isotherms. In contrast, a TC receives its energy from the evaporation of warm ocean water and the associated latent heat release in the atmosphere.

Both tropical and extratropical cyclones rotate cyclonically near the surface. However, one distinct difference between the two cyclone types is that a TC has a warm core, whereas an extratropical cyclone is cold-cored. A warm core causes the cyclonic circulation to weaken with height, according to the hypsometric equation (3.10), until the pressure gradient vanishes in the mid-troposphere (Figure 10.23e). At higher levels the circulation reverses, so that the outflow at upper levels is anticyclonic. In contrast, an extratropical cyclone has a low pressure anomaly (the pressure is lower than the environment) throughout its vertical extent and a cold core in higher levels (Figure 10.23f). Therefore the air rotates cyclonically at every altitude. In fact, because of the cold core the pressure anomaly increases with height, again according to the hypsometric equation (3.10) and this causes the highest wind speeds to be located in the upper troposphere.

A tropical cyclone is symmetric which allows the eye to be visible. In contrast, an extratropical cyclone tilts backwards with height as can be seen by the isobars in Figure 10.23f. In addition, the Coriolis parameter is of much larger importance in extratropical cyclones.

10.5.5 Tropical cyclone as a heat engine

A TC can be understood as a Carnot heat engine, discussed in Section 2.2.4. In a TC the energy required for isothermal expansion of the Carnot process (from A to B in Figures 10.22 and 10.24) is provided in the form of latent heat deriving from the evaporation of

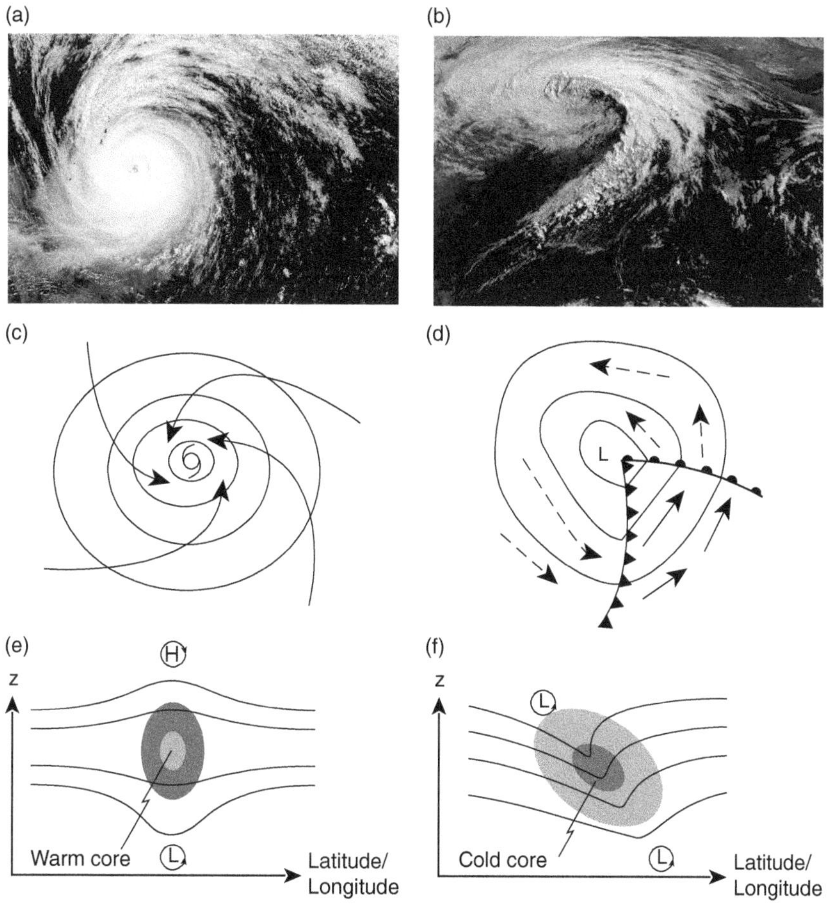

Fig. 10.23 Schematics of the three-dimensional structures of a TC (a), (c), (e) and an extratropical cyclone (b), (d), (f). The top panels are satellite pictures. The middle panels are schematic views at the surface. The solid arrows denote the flow of warm air masses and the broken arrows denote cold air masses. The solid lines are isobars. The bottom panels show cross sections of latitude or longitude versus height along a line going through the center of the cyclones. Image credits: (a) NOAA Goddard MODIS Rapid Response Team and (b) NOAA image gallery.

water from the ocean. Because the energy of the evaporated water is stored in the form of water vapor, the temperature remains approximately constant at 300 K during its inflow. It expands owing to the decrease in pressure towards the TC center, which is from 1000 hPa to 900 hPa in Figure 10.24. Moist adiabatic expansion occurs when the air is lifted and the water vapor condenses in the eyewall. This ascent closely follows a constant equivalent-potential-temperature surface. During the ascent the temperature decreases from 300 K near the surface to about 200 K at the tropopause, where the ascent terminates. Just outside the edge of the TC at point C, the air sinks because of clear-sky radiative cooling. As this radiative cooling is balanced by adiabatic warming, the temperature remains approximately constant until point D. Between D and A the air warms due to dry adiabatic compression and the cycle is closed. Whereas in an idealized Carnot cycle the adiabatic compression

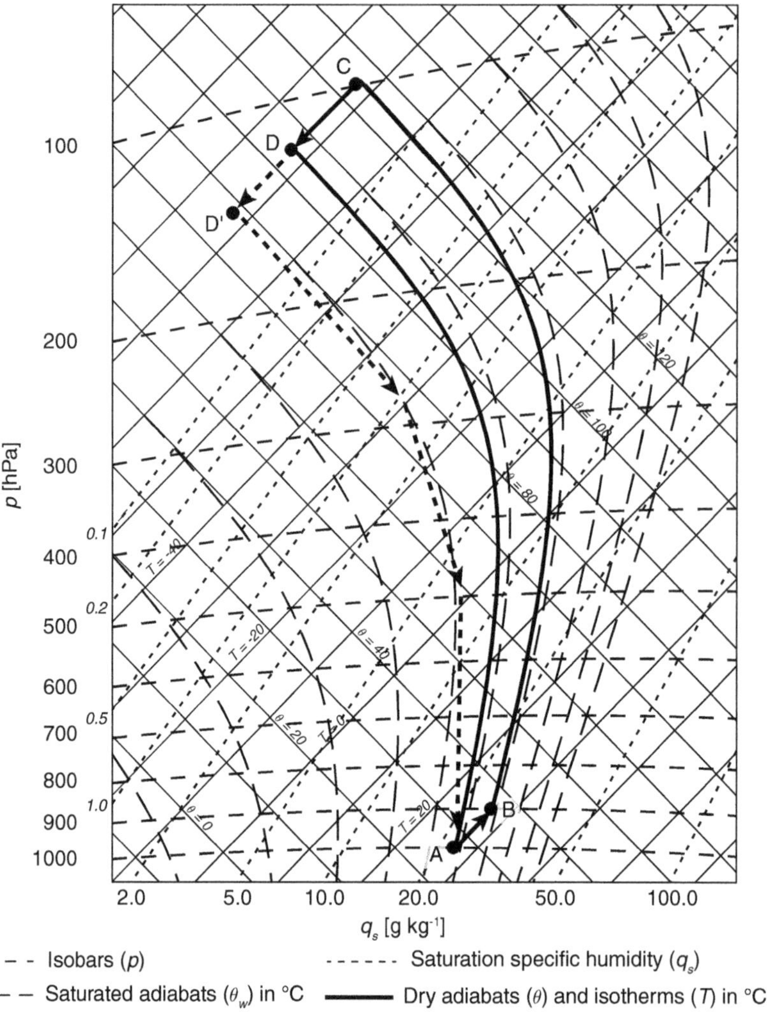

Fig. 10.24 Carnot cycle of a major tropical cyclone in a tephigram with points A–D as labeled in Figure 10.22.

also needs to follow an equivalent-potential-temperature surface back down again, in reality the descending air will follow a path between a dry and moist adiabat, as shown by the broken arrows from D′ to A (in Figure 10.24).

The efficiency of a Carnot cycle in which thermal energy is converted into mechanical energy can be obtained from the difference between the temperature at which thermal energy is put into the system, T_{in}, and the temperature at which energy leaves the system, T_{out}, divided by the input temperature. In our case T_{in} equals the sea surface temperature (SST, approximately 300 K) and T_{out} is the temperature at the tropopause, T_{tropo} (approximately 200 K), which results in an efficiency η of approximately one third:

$$\eta = \frac{T_{in} - T_{out}}{T_{in}} \approx \frac{SST - T_{tropo}}{SST} = \frac{300 - 200}{300} = \frac{1}{3}. \tag{10.11}$$

This means that one third of the heat energy in a TC is converted into kinetic energy in the form of high wind speeds, which may cause major damage when the TC makes landfall.

10.5.6 Decay of tropical cyclones

The main source of energy for TCs is the latent heat delivered by the warm ocean due to evaporation. If the heat source is cut off, the TC will die very soon as for instance happens when TCs make landfall. If landfall occurs over a swamp, the TC can still pick up sufficient energy to retard its dissipation to some extent (Emanuel, 2005); TCs also weaken and eventually die off if they move over colder ocean waters, where again they are deprived of their main energy source.

Upon landfall, TCs cause serious damage with many casualties. On average, TCs claim 80% of their victims when they first strike shore. Most damage is associated with storm surges, i.e. the increase in sea level, which can flood coastal communities. The storm surge is worst when it coincides with high tides.

The wind speeds are highest on the right quadrant with respect to the motion of the TC due to the superposition of translational and rotational winds. They decay exponentially when TCs hit land, losing half their landfall value in about 7 h, 75% in 15 h and 90% in 24 h. This decay is enhanced when they hit mountainous terrain because of its higher friction. Rainfall generally diminishes after landfall, but not as fast as wind speed because the TC contains a lot of water vapor and clouds when it moves over land.

About 50% of tropical cyclones undergo extratropical transition; this occurs when they move polewards and turn into extratropical cyclones. The transition into extratropical cyclones normally occurs at 30°–40° latitude. During the extratropical transition the cyclone loses its symmetric warm core and becomes asymmetric with a cold core; this is the typical characteristic of an extratropical low pressure system and implies that the cyclone center tilts backwards with height (Figure 10.23f). The cyclone connects to the upper-level flow and gains asymmetry by developing fronts in which the warm air and polar air move against each other.

Hurricane Sandy underwent extratropical transition on October 29, 2012, when making landfall in New Jersey and hitting New York City. Sandy first gained strength by intensifying the warm core. It started to become slightly asymmetric early on while still increasing in strength. Upon landfall the central pressure quickly increased and the warm core weakened. The polar jet stream extended rather far to the south, enabling Sandy to connect with a trough of Arctic air and to develop fronts and a cold-core structure.

After an extratropical transition, the main energy source is no longer the release of latent heat in the deep convective clouds of the eyewall but the horizontal temperature contrast, which causes baroclinic instability. A cyclone can increase in size and can also re-strengthen owing to baroclinic instability after it has undergone extratropical transition.

10.6 Cyclones and climate change

Whether, and in what ways, tropical and extratropical cyclones have changed over time is a research subject of growing interest. Their change with time is not easy to determine because the capabilities for observing cyclones over the oceans were rather limited before the usage of satellite data. While it is clear that higher SSTs provide more energy for TCs in a warmer climate, evaluation of the changes in all six prerequisites for TC formation (Section 10.5) is necessary to understand and project the changes in TC frequency and intensity. Globally no significant trend in TC frequency could be observed in recent decades. On a regional scale only the increase in intensity of the most severe TCs in the north Atlantic since 1983 appears statistically robust (Kossin *et al.*, 2007) but here also no robust trend in the annual number of TCs could be identified over the past century (Hartmann *et al.*, 2013).

Future projections of TC frequency and intensity hinge on the quality of global climate models (GCMs) and on uncertainties in emission scenarios. While the horizontal resolution of GCMs has increased in recent decades, the grids of current GCMs are still too coarse to resolve TCs properly (Bretl *et al.*, 2015). As summarized by the IPCC (Stocker *et al.*, 2013) it is likely (occurs with a probability of 66%–100%) that the global number of TCs will not increase in the twenty-first century but might actually decrease. At the same time, it is likely that the maximum wind speed and rain rates of TCs will increase. Probably the effect of climate change on TCs will not be the same in all ocean basins but, for the time being, region-specific projections have low confidence. As such, it is more likely than not (50%–100% probability of the outcome) that the frequency of the most intense TCs will increase in some basins.

Concerning extratropical cyclones, an upward trend in the frequency and intensity has been observed during the last 60 years in the north Atlantic in winter (Hartmann *et al.*, 2013). On the global scale, however, the confidence in large-scale trends in extratropical storms in the last 100 years is low. Evidence is also not sufficient to detect robust trends in small-scale severe weather events such as hail or thunderstorms (Stocker *et al.*, 2013).

The expected future trends in the frequency of extratropical cyclones are also rather uncertain but, in contrast with TCs, here the IPCC states that the global number is unlikely to decrease by more than a few percent. As for TCs, it is projected that extratropical cyclones will produce more precipitation, most noticeably in winter in the Arctic, northern Europe, North America and the mid-to-high latitudes of the southern hemisphere (Stocker *et al.*, 2013).

10.7 Exercises

Single/multiple choice exercises

1. An isolated thunderstorm is generally characterized by three stages. Which of the following is/are true?
   (a) Gust fronts are responsible for the termination of the major stage.
   (b) The anvil is characteristic for all three stages.
   (c) The horizontal spread-out of the anvil can be explained by the low air density in the stratosphere.

(d) The 0 °C isotherm in the major stage varies horizontally because of latent heat release and consumption.
(e) Precipitation in the major stage is mainly stratiform.

2. Hailstones can undergo wet or dry growth. Which of the following is/are true?
   (a) Wet growth takes place at positive and dry growth at subzero temperatures (°C) in the cloud.
   (b) Ice layers formed by wet growth have a larger density than layers formed by dry growth.
   (c) Wet growth defines accretion of a hailstone with cloud droplets and dry growth accretion with ice crystals.
   (d) The thickness of the hailstone layers depends (among other factors) on the vertical velocity in the cloud.
3. Which criteria need to be fulfilled for lightning to occur?
   (a) The presence of both ice and liquid phases.
   (b) A cloud top colder than −15 °C.
   (c) The separation of charges between the electrosphere and Earth's surface.
   (d) A weak fair weather current.
   (e) The presence of a negatively charged region in the middle of the thunderstorm cloud.
4. The generation of vertical vorticity inside a cloud is important in terms of the self-organization of thunderstorms. Which of the following is/are true?
   (a) The sources of vertical vorticity are stretching, tilting and mixing, which are usually all of equal importance.
   (b) Horizontal vorticity is mainly created by vertical shear of the horizontal wind.
   (c) Vertical vorticity is generated by the tilting of the rotation axis from horizontal to vertical.
   (d) Downdrafts reduce the generation of horizontal vorticity as they cut off the updrafts.
   (e) For a unidirectional wind shear, the horizontal vortex tube is perpendicular to the direction of the vertical shear of the horizontal wind.
5. Assume you observe a cloud developing an anvil with a fuzzy outline on a nice summer's day. Which of the following is/are true?
   (a) The observed cloud is a cumulonimbus.
   (b) The anvil indicates the presence of ice in the cloud.
   (c) Lightning, hail, and heavy rain will occur from this cloud at a later time.
   (d) The cloud could belong to a single-cell, multicell, or supercell storm.
6. Multicell storms consist of multiple cells. Which of the following is/are true?
   (a) Multicell storms have a longer lifetime and a larger spatial extent than single-cell storms.
   (b) Multicell storms are less frequent than supercell storms.
   (c) Updrafts and downdrafts are horizontally displaced in multicell storms.
   (d) In multicell storms, all cells are in the same development stage.

7. Multicell and supercell storms can both cause severe damage. Which of the following is/are true?
   (a) Supercell storms are generally more severe than multicell storms.
   (b) The vertical displacement of updrafts and downdrafts is key to explaining the strength of both storm types.
   (c) Gust fronts are important for the formation of new cells in a multicell storm.
   (d) A multicell storm can transition into a supercell storm if the vertical wind shear weakens.
8. A tornado is a vortex of wind that is in contact with both the ground and the cloud base. Which of the following is/are true?
   (a) Tornadoes are produced only by supercell storms.
   (b) Waterspouts are less common and less severe than supercell tornadoes.
   (c) The only term of the Navier–Stokes equation that is negligible for the calculation of the wind speed of a tornado is friction.
   (d) Tornadoes rotate cyclonically.
9. Some prerequisites must be fulfilled for the formation of tropical cyclones. Which of the following is/are true?
   (a) Tropical cyclones are most prevalent in the West Pacific Warm Pool because of its high static stability.
   (b) A clustering of thunderstorms can lead to the formation of a tropical cyclone.
   (c) The higher the vertical shear of the horizontal wind, the stronger the tropical cyclone.
   (d) The formation of tropical cyclones can be triggered by African easterly waves.
10. Tropical cyclones can be described by a primary and secondary circulation. Which of the following is/are true?
   (a) The secondary circulation is also referred to as in-up-and-down circulation.
   (b) The primary circulation of a tropical cyclone is in geostrophic wind balance.
   (c) The secondary circulation of a tropical cyclone can be described by a Carnot cycle.
   (d) Rainbands are indicative of the cyclonic rotation of a tropical cyclone.
11. Tropical cyclones are one of the most destructive weather phenomena. Which of the following is/are true?
   (a) According to the Saffir–Simpson scale, the intensity of tropical cyclones is categorized into five categories by using maximum wind speeds, storm surge and minimum pressure.
   (b) The largest damage associated with tropical cyclones is caused by high wind speeds, of up to 260 km h^{-1}.
   (c) A conditionally unstable atmosphere is a prerequisite for the formation of tropical cyclones.
   (d) All tropical cyclones undergo extratropical transition when they cross the 20° latitude line.
12. Tropical and extratropical cyclones differ in some properties. Which of the following is/are true?
   (a) Tropical cyclones form at the equator, whereas extratropical cyclones occur at latitudes of around 40°.

(b) Both tropical and extratropical cyclones gain their energy mainly from latent heat release in the atmosphere.
(c) The Coriolis force can be neglected for tropical cyclones but not for extratropical cyclones.
(d) Only extratropical cyclones are associated with fronts.

Long exercises

1. The occurrence of thunderstorms is inhomogeneous in time and space. Where and when do you expect most thunderstorms to occur?
   (a) At which time of the day?
   (b) In which latitudes?
   (c) Over land or over sea?
2. Three pictures (Picture10-2a.png, Picture10-2b.png and Picture10-2c.png) that show damage after the passage of severe storms can be downloaded from: www.cambridge.org/clouds. Distinguish in the different pictures whether a downburst or tornado caused the damage.
3. Imagine you have three weather stations which measure wind velocity. One is located (with position coordinates in meters) at point A (0, 0, 10), one at point B (100, 100, 10) and one at point C (100, 0, 10). One day, you measure wind velocities Y (u, v, w) (in m s^{-1}) at the same time at the different measurement stations X: A (1, 2, 2), B (5, 7, 2), C (8, 10, 2). You can assume that the partial derivatives are constant in space on this scale. On the basis of these measurements, what vertical vorticity do you expect to occur 5 seconds later?
4. Horizontal vorticity may be generated by horizontal buoyancy gradients. This is observed in gust fronts and in downbursts. The temperature distribution inside an axially symmetric downburst may be represented as $T = T_{env} - T_0\, e^{-ax^2}$, with $a = 2 \times 10^{-6}$ m^{-2}, $T_{env} = 293$ K and $T_0 = 5$ K.
   (a) Plot the temperature inside the downburst from -3000 to 3000 m.
   (b) Assuming that the generation of vorticity is driven only by horizontal gradients of buoyancy

$$T_B = g\frac{T - T_{env}}{T_{env}},$$

   derive an expression for the generation of horizontal vorticity $\partial\xi/\partial t$ as a function of x and compute its maximum value.
5. Tropical versus extratropical cyclones: both evolve intense precipitation and high wind speeds, but they are fundamentally different.
   (a) Describe four differences between tropical cyclones and mid-latitude extratropical cyclones.
   (b) Under what conditions can a tropical cyclone transform into an extratropical cyclone? Is the opposite process possible too?

6. Tropical cyclones can be qualitatively described by the gradient wind balance:

$$\frac{v^2}{r} + fv = \frac{1}{\rho}\frac{\partial p}{\partial r}.$$

The first term on the left-hand side originates from the centrifugal acceleration, so v is the velocity along the cyclonically rotating air parcel.

   (a) Assume that the angular velocity of the cyclone $\omega = v/r$ is constant; at $r = 10$ km, you measure v to be 60 m s^{-1}. The cyclone has a radius 15 km. What is the minimum surface pressure in the center of the cyclone if the ambient pressure (and the pressure at the edge of the cyclone) is 1000 hPa? Assume $f = 2.5 \times 10^{-5}$ s^{-1} and $\rho = 1.2$ kg m^{-3}.
   (b) Assume again that ω is constant, what angular velocity do you need for a tornado with a much smaller outer radius of 50 m to achieve the same absolute pressure difference? To what v does this angular velocity correspond at a distance of 40 m from the core? Is your result realistic?
   (c) Which term in the equation of the gradient wind balance can you neglect in the case of the tornado?

7. The ocean is the main energy source for tropical cyclones. In this exercise, we will consider a simplistic model to calculate the energy that can be taken up from the ocean by a TC. Assume a sea surface temperature $SST = 28.35$ °C, a tropopause temperature $T_{tropo} = -73$ °C, a surface pressure 1000 hPa and an air density $\rho_{env} = 1.15$ kg m^{-3}. Further, assume a perfectly circular TC with a radius of 200 km, where the wind speed, $u = 15$ m s^{-1}, is constant throughout the TC.
   (a) Determine the efficiency with which the heat taken up from the ocean can be transformed into mechanical energy by the TC (acting as a Carnot cycle).
   (b) The latent heat flux (LHF) at the ocean–air interface can be calculated as (Singh *et al.*, 2005):

$$\text{LHF} = \rho_{env} L_v C_E u (q_s - q_v), \qquad (10.12)$$

   where ρ_{env} is the air density, L_v the latent heat of evaporation, C_E the dimensionless bulk transfer coefficient for water vapor, u the wind speed at a height of 10 m and q_s and q_v the saturation specific humidity and the specific humidity, respectively. Assuming $C_E = 1.104 \times 10^{-3}$ and $q_v = 23.6$ g kg^{-1}, calculate the LHF, i.e. the maximum heat energy that can be released from the ocean and taken up by the TC. (*Hint: q_s is calculated from the SST, assuming saturation at the ocean surface.*)
   (c) What is the maximum energy that can be taken up by a TC of the given size?
   (d) Determine the maximum energy available that can be converted into mechanical energy.
   (e) Comment on the assumption that the wind speed is constant throughout the TC and draw a schematic of how the wind speed changes from the eye of the TC to its outer boundary.
   (f) So far you have calculated the available energy for the TC on the basis of the given temperatures only. Are there other parameters that affect the energy available for TC formation? If yes, what are they and how do they provide energy?

11 Global energy budget

The Sun is the source of energy for all life on Earth. The temporal and spatial variations in the solar radiation received on Earth drive the general circulation of the atmosphere, with all its wind systems and the hydrological cycle. In this chapter we discuss the energy fluxes in the global long-term annual mean.

In equilibrium there is no net gain or loss of energy at the top of the atmosphere (TOA), within the atmosphere or at Earth's surface, i.e. the energy budget is closed everywhere. Therefore it can be studied separately at the TOA (Section 11.1), within the atmosphere (Section 11.2) and at the surface (Section 11.3). Clouds have a large effect on Earth's radiative budget, as will be demonstrated. Furtheremore greenhouse gases (GHGs) strongly influence Earth's radiative budget. The presence of GHGs leads to a warming known as the greenhouse effect. The impact of clouds is more complex and depends strongly on the cloud top height and the associated effective temperature. Their influence on the solar and terrestrial radiation can be isolated and described in terms of the cloud radiative effects, as will be discussed in Section 11.4.

11.1 Energy balance at the top of the atmosphere

The global long-term annual mean energy budget is shown in Figure 11.1. It is divided into shortwave or solar radiation, with wavelengths between 0.2 and 4 μm, and longwave or terrestrial radiation, which refers to wavelengths beyond 4 μm. Whereas solar radiation can be transmitted, absorbed or scattered in the atmosphere, terrestrial radiation is absorbed and re-emitted. Solar radiation can be subdivided further into ultraviolet radiation ($\lambda < 0.38$ μm), visible radiation (0.38 μm $< \lambda <$ 0.75 μm) and near infrared radiation ($\lambda > 0.75$ μm). Terrestrial radiation consists entirely of infrared radiation (0.75 μm $< \lambda <$ 1 mm).

The amount of radiation emitted by a body depends on its temperature, according to the Stefan–Boltzmann law. For the longwave radiation emitted by Earth it is given by

$$F_{LW} = \epsilon \sigma T^4. \quad (11.1)$$

Here ϵ is the emissivity, σ the Stefan–Boltzmann constant (5.67×10^{-8} W m^{-2} K^{-4}) and T the temperature. Hence, the total radiation emitted over the whole surface area by the Sun is

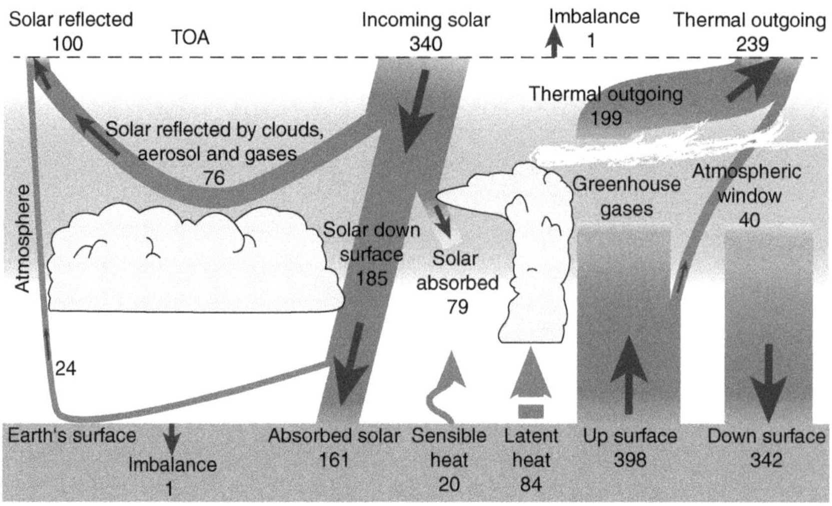

Fig. 11.1 Earth's annual global mean energy budget. The numbers give the individual energy fluxes in W m^{-2} and are based on values of the IPCC AR5 (Hartmann *et al.*, 2013). The imbalances at the TOA and at the surface amount to 0.6 W m^{-2} and have been rounded to 1 W m^{-2} in order to close the budget.

$$F_{Sun} = 4\pi R_{Sun}^2 \sigma T_{Sun}^4, \tag{11.2}$$

where $R_{Sun} = 6.98 \times 10^8$ m is the radius of the Sun and $T_{Sun} = 5769.6$ K is its temperature. The total radiation F_{Sun} amounts to 3.85×10^{26} W. This energy, when distributed homogeneously over a sphere with a radius that equals the distance between the Sun and Earth, $r_{Sun-Earth} = 1.5 \times 10^{11}$ m, yields the solar constant S_0:

$$S_0 = \frac{F_{Sun}}{4\pi r_{Sun-Earth}^2} = 1360 \text{ W m}^{-2}. \tag{11.3}$$

At any one time, a disk with cross-sectional area of πr_{Earth}^2 is sunlit. Because this amounts to one quarter of Earth's surface area, Earth's surface receives only 1/4 of S_0 in the global annual mean, as illustrated schematically in Figure 11.2.

From the average 340 W m^{-2} solar radiation received at the TOA, 100 W m^{-2} are reflected back to space, as shown in Figure 11.1. The ratio of the reflected and the received radiation is called the planetary albedo α_p and amounts to 100/340 = 29.4%. Thus only 24 W m^{-2} is reflected by Earth's surface. The majority of the reflected radiation (76 W m^{-2}) is due to reflection within the atmosphere, mainly by clouds (47 W m^{-2}, Section 11.4). The rest is due to scattering by air molecules and aerosol particles.

The difference between the incoming and reflected radiation amounts to 240 W m^{-2} and is absorbed by the surface and the atmosphere. By far the largest amount of the absorbed energy is re-emitted as longwave radiation. Earth can be treated as a black body, i.e. an emissivity of unity can be assumed. The outgoing longwave radiation is emitted from the whole surface area of Earth. Because it will again be absorbed and re-emitted by

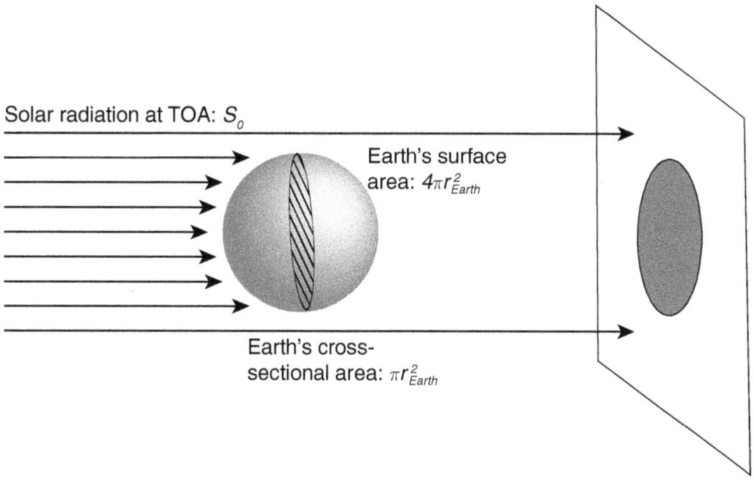

Fig. 11.2 Schematic of the solar radiation intercepted by Earth at the TOA. The projected area is πr^2_{Earth}.

air molecules (Section 11.2), the radiation leaving the atmosphere originates from higher levels in the atmosphere.

In equilibrium the net shortwave radiation F_{SW} (the incoming minus the reflected radiation) is balanced by the outgoing longwave radiation F_{LW}:

$$F_{SW} = \frac{S_o}{4}(1 - \alpha_p) = \epsilon\sigma T^4 = F_{LW}. \tag{11.4}$$

However, owing to the increases in greenhouse gas concentrations, the Earth–atmosphere system is not in equilibrium. There is a 0.6 W m^{-2} imbalance between the net shortwave radiation at the TOA of 240 W m^{-2} and the outgoing radiation at the TOA of 239 W m^{-2} which amount to 1 W m^{-2} in the rounded values shown in Figure 11.1. Earth has thus not yet reached the final temperature that is needed to emit 240 W m^{-2} into space, so Earth will continue to warm until the radiation is in a new equilibrium. This can only be reached if the radiative forcings do not continue to change over time and requires that the concentrations of greenhouse gases and aerosol particles stabilize.

The outgoing longwave radiation, 239 W m^{-2}, corresponds to an effective temperature T_e of 255 K according to the Stefan–Boltzmann law (eq. 11.1), assuming an emissivity of unity. This is 34 K colder than the average temperature of Earth's surface T_s in the present climate. In the absence of an atmosphere T_e would equal T_s. Of the 34 K difference between T_s and T_e, 33 K is caused by the so-called natural greenhouse effect. It is dominated by the natural greenhouse gases, i.e. water vapor, carbon dioxide and methane. The natural greenhouse effect has to be clearly distinguished from the anthropogenic greenhouse effect due to anthropogenic emissions since pre-industrial times. This effect caused an increase in T_s of 0.85 K between 1880 and 2012 (Stocker *et al.*, 2013).

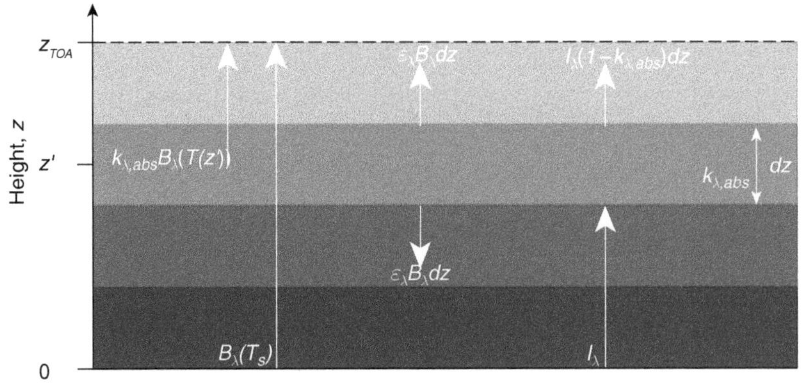

Fig. 11.3 Schematic layer model of the Earth–atmosphere system illustrating longwave radiative transfer through an absorbing medium, as described in eq. (11.5).

11.2 Energy balance in the atmosphere

The altitude at which Earth effectively emits radiation to space, i.e. where the atmospheric temperature is 255 K, is in the mid-troposphere at around 5–5.5 km, assuming an average atmospheric lapse rate 6.5 K km^{-1}. Simply speaking, Earth's atmosphere is too dense below this height, i.e. photons emitted by air molecules at lower altitudes are absorbed and re-emitted by other air molecules. Hence, only a small part of the longwave radiation emitted at Earth's surface reaches the TOA directly, and is emitted to space. This can be understood by looking at a simple layer model for the atmosphere shown schematically in Figure 11.3. The radiative transfer equation for longwave radiation is given by the so-called Schwarzschild equation describing the transmitted radiation through an absorbing medium:

$$\frac{dI_\lambda}{dz} = -k_{\lambda,abs}I_\lambda + k_{\lambda,abs}B_\lambda(T), \tag{11.5}$$

where I_λ is the wavelength-dependent intensity of radiation in W m^{-2}, $k_{\lambda,abs}$ is the wavelength-dependent absorption coefficient for greenhouses gases, aerosol particles and cloud hydrometeors in m^{-1} and B_λ is the wavelength-dependent black body source function (the Planck function).

For each layer the emission of electromagnetic radiation equals the absorption in order for it to be in radiative equilibrium, i.e. $\epsilon_\lambda = k_{\lambda,abs}$. Each layer emits both upwards and downwards. When radiation is emitted by a layer at a height z', it is partly absorbed and re-emitted by each layer above z' before it reaches the TOA. Hence, the outgoing longwave radiation at the TOA is found by integrating eq. (11.5) over all layers dz in the atmosphere, taking into account the radiation emitted from the surface and all layers beneath z' that is not attenuated by absorption. The downward-emitted longwave radiation can be calculated accordingly from its atmospheric emission and absorption in all layers.

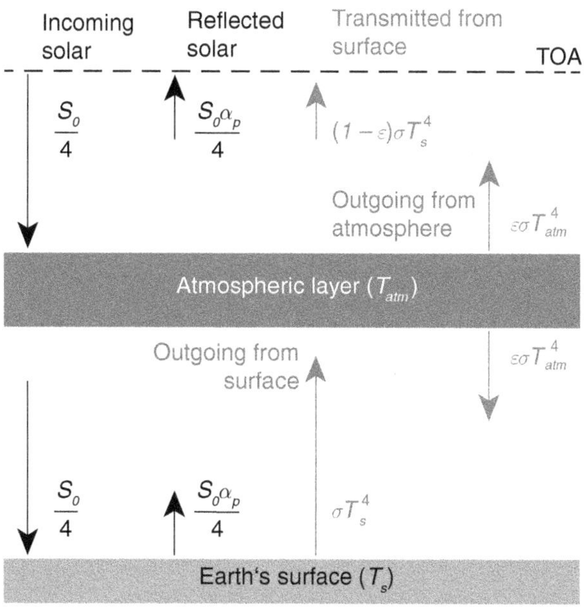

Fig. 11.4 Simple model of the greenhouse effect if the atmosphere is taken to be a single isothermal layer (eq. 11.6).

The atmosphere receives 199 W m^{-2}, as shown in Figure 11.1. This is sum of the 84 W m^{-2} latent heat (Section 11.3), 20 W m^{-2} sensible heat, 79 W m^{-2} (240 W m^{-2} – 161 W m^{-2} in Figure 11.1) absorbed shortwave radiation and 16 W m^{-2} absorbed longwave radiation (the 398 W m^{-2} emitted from the surface minus the 40 W m^{-2} directly emitted to space and minus the 342 W m^{-2} emitted back to the surface). The energy balance within the atmosphere is closed by the 199 W m^{-2} of longwave radiation emitted to space from the atmosphere and clouds.

A simple model of the greenhouse effect is shown in Figure 11.4. It assumes that the incoming solar radiation at the TOA is balanced by the radiation emitted at the surface and from within the atmosphere, according to

$$\frac{1}{4}S_0(1-\alpha_p) = (1-\epsilon)\sigma T_s^4 + \epsilon\sigma T_{atm}^4, \tag{11.6}$$

where T_{atm} is the temperature of the atmosphere viewed as a single isothermal layer. The model further assumes that the atmosphere is completely transparent for solar radiation, Earth's surface has emissivity unity and temperature T_s and that the emissivity of the atmosphere is between zero and unity depending on the greenhouse gas concentration. Non-radiative energy components such as latent heat are not taken into account. The surface temperature T_s is given by

$$T_s = \sqrt[4]{\frac{S_0(1-\alpha_p)}{4(1-\epsilon/2)\sigma}}. \tag{11.7}$$

Inserting the numbers from the Earth's energy budget provided above and T_s = 289.5 K yields emissivity $\epsilon = 0.79$. This implies that almost 80% of the energy emitted to space in eq. (11.6) originates from the atmosphere and only $\sim$ 20% from Earth's surface. Within the atmosphere the energy budget is given as

$$\epsilon\sigma T_s^4 = 2\epsilon\sigma T_{atm}^4. \tag{11.8}$$

Inserting the above values, an atmospheric temperature T_{atm} of 243.4 K is obtained. While this is 12 K lower than the effective temperature with which the Earth–atmosphere system emits to space, it nicely illustrates the greenhouse effect in a simplified manner.

11.3 Energy balance at the surface

The surface energy balance is given as

$$c_p \frac{\partial T_s}{\partial t} = F_{SW}^{\downarrow} - F_{LW}^{\uparrow} + F_{LW}^{\downarrow} - \text{LH} - \text{SH}, \tag{11.9}$$

where LH and SH are the latent and sensible heat fluxes.

Of the net solar radiation, 240 W m^{-2}, absorbed by the Earth–atmosphere system, the surface absorbs a contribution $F_{SW}^{\downarrow} = 161$ W m^{-2}. Without an atmosphere, Earth's surface would need to re-emit this 161 W m^{-2} in a steady state. Assuming black body radiation ($\epsilon = 1$) and a global mean surface temperature of 289.5 K yields 398 W m^{-2} of emitted longwave radiation $F_{LW}^{\uparrow}$ compared with the received 161 W m^{-2}. This discrepancy is again due to the greenhouse effect of the atmosphere. As a consequence of the presence of greenhouse gases (natural and anthropogenic) and clouds, only 40 W m^{-2} can directly escape to space (Kiehl and Trenberth, 1997). This 40 W m^{-2} escapes through the infrared atmospheric window between wavelengths of 8 and 14 μm where the transmissivity of the atmosphere is very high. The remaining 358 W m^{-2} is absorbed by greenhouse gases and clouds and re-emitted according to the temperature of these, again following the Stefan–Boltzmann law (eq. 11.1).

A so-called back radiation $F_{LW}^{\downarrow} = 342$ W m^{-2} is emitted from the atmosphere back to the surface. Hence, the net longwave radiation leaving Earth's surface amounts to only 56 W m^{-2} (40 W m^{-2} + 358 W m^{-2} − 342 W m^{-2}). Note that the fact that the atmosphere emits only 16 W m^{-2} to space but 342 W m^{-2} towards Earth's surface is due to the vertical density gradient in the atmosphere. This means that 105 W m^{-2} (161 W m^{-2} − 56 W m^{-2}) needs to leave the surface by other means than radiation in order to approximately balance the energy budget at Earth's surface.

In the atmosphere, heat is transported not only by radiation but also by moving air masses and can be stored in the form of water vapor evaporated from Earth's surface. Sensible heat can directly be felt and measured in terms of temperature. The sensible heat flux SH depends on the temperature gradient between the surface and the overlying air as well as wind speed. The sensible heat flux transports a global annual mean of 20 W m^{-2} into the atmosphere. The latent heat flux LH comprises the evaporation of water from the oceans and from land, evapotranspiration from plants and the sublimation of ice and snow. The latent heat flux transports 84 W m^{-2} into the atmosphere. The latent heat is stored in

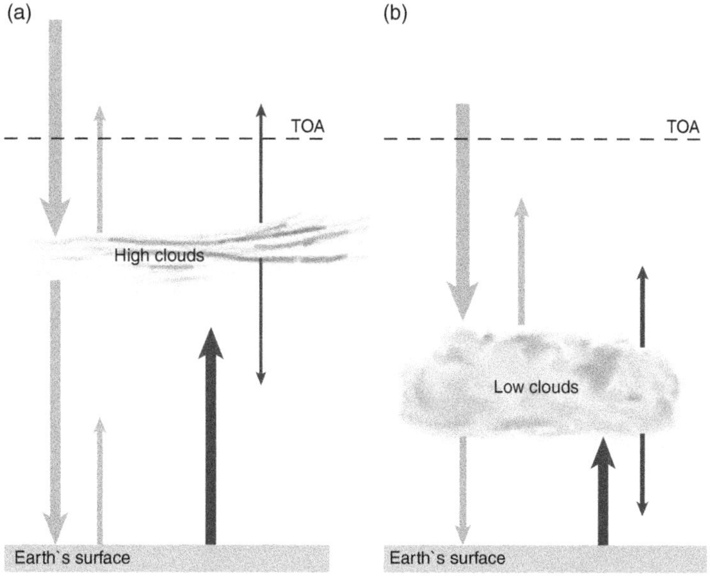

Fig. 11.5 Schematic of the radiative effect of (a) high-level clouds and (b) low-level clouds, where the shortwave radiative fluxes are shown in light gray and the longwave radiative fluxes in dark gray.

the gas phase as a result of the evaporation or sublimation process. It is released back to the atmosphere when water vapor condenses onto droplets or deposits onto ice crystals. The vertical transport of heat by means of water vapor is due to the fact that latent heat consumption and latent heat release occur at different altitudes.

In equilibrium the surface energy balance would be closed and no heating of Earth's surface would occur. As shown in Figure 11.1, the surface has the same 0.6 W m^{-2} imbalance as the TOA radiation budget. It causes Earth to warm until a new equilibrium is reached.

11.4 Cloud radiative effects

The reflection of solar radiation by clouds, which reduces the amount of solar radiation reaching the surface, is also called the albedo effect of clouds. This has nothing to do with the cloud albedo effect of aerosol particles, which will be discussed in Section 12.1. There the term "cloud albedo effect" refers to the effect aerosol particles have on the cloud albedo. Here it refers to the effect that the cloud albedo has on the radiation budget of Earth. Besides their albedo effect, clouds also have a greenhouse effect. As they absorb part of the longwave radiation emitted from Earth and the lower atmosphere and re-emit some of this absorbed radiation back to lower levels in the atmosphere, clouds reduce the amount of longwave radiation effectively emitted to space. In doing so, clouds warm the Earth–atmosphere system. Whether the albedo effect or the greenhouse effect dominates for a given cloud depends on the altitude and thickness of the cloud, on its microphysical properties and on the surface albedo, as shown in Figure 11.5.

As discussed in Section 11.2, the atmosphere is dense in the lower part of the troposphere, so that most of the radiation emitted from there is absorbed by other air molecules or particles before it is emitted to space. Therefore the cloud greenhouse effect is only important for mid-level clouds, cirrus clouds, nimbostratus and deep convective clouds (Table 1.1). Of these, cirrus clouds and deep convective clouds have the largest greenhouse effect because of their high altitudes, which correspond to low cloud-top temperatures. As a consequence, they emit the least amount of longwave radiation to space and thus warm the Earth–atmosphere system.

Cirrus clouds are semi-transparent to shortwave radiation. They thus have a small albedo effect and their greenhouse effect dominates. Deep convective clouds, however, have a large albedo effect. In these clouds the albedo effect and the greenhouse effect almost balance each other.

Low-level clouds also reflect large amounts of shortwave radiation. Their albedo effect dominates over their greenhouse effect and is particularly large if they lie over a dark surface with a low albedo, as for example the ocean or tropical forests. Only when they lie over a very reflective surface such as snow or ice can their greenhouse effect dominate over their albedo effect. As these low-level clouds are located at altitudes where the air density is rather high, the radiation emitted by them in the upward direction will be absorbed and re-emitted by overlying air molecules and will not be directly emitted to space. Thus the presence of low-level clouds is not noticeable to the outgoing longwave radiation and their greenhouse effect is small.

An exception are Arctic stratus clouds, which are very important for the Arctic climate. They have a net heating effect despite their low altitude. These clouds warm the surface because at the high Arctic latitudes the solar zenith angle is so large that they reflect less solar radiation than they emit towards the surface.

A second effect contributing to the net greenhouse effect of Arctic low clouds is that bright surfaces such as fresh snow and ice have an enhanced reflectivity also in the infrared. Therefore, longwave radiation emitted from clouds to these surfaces is not entirely absorbed there but largely reflected. The reflected part may then be partly absorbed and re-emitted a second time by the cloud. Such multiple cycles of reflection, absorption and re-emission can enhance the greenhouse effect of a low-lying cloud compared with the case of a single reflection. When the surface albedo exceeds ~ 0.7, as for fresh snow, multiple reflections between a cloud and the surface can cause the atmosphere to absorb more radiation than in the clear-sky case (Kiehl, 1992).

The so-called cloud radiative effect (CRE) is defined as the difference in net radiation at the TOA between a cloudy atmosphere and a cloud-free atmosphere (Ramanathan *et al.*, 1989). It can be divided into a shortwave (SCRE) and a longwave (LCRE) component:

$$\text{SCRE} = F_{SW} - F_{SW}^{cs}, \tag{11.10}$$

$$\text{LCRE} = F_{LW}^{cs} - F_{LW}, \tag{11.11}$$

where cs denotes clear sky, F_{SW} the net shortwave radiation (the incoming shortwave radiation minus the reflected shortwave radiation) and F_{LW} the emitted longwave radiation. All radiative fluxes in the above equations are defined as positive regardless of their direction. The SCRE component is also referred to as the albedo effect of clouds and the LCRE component as the greenhouse effect of clouds.

In the global and annual mean, the SCRE amounts to −47.3 W m^{-2} and the LCRE to 26.2 W m^{-2} (Boucher *et al.*, 2013). Thus, the net effect of clouds on the TOA radiation balances CRE = −21.1 W m^{-2}. Hence clouds have a net cooling effect on the Earth–atmosphere system. To put this number into perspective, it is roughly 5–6 times as large as the positive radiative forcing (RF) associated with the doubling of carbon dioxide relative to pre-industrial conditions (Section 12.3), which amounts to 3.7 W m^{-2} (Myhre *et al.*, 1998; Solomon *et al.*, 2007).

The geographical distributions of the SCRE, LCRE and CRE as obtained from the CERES satellite data (Wielicki *et al.*, 1996) are shown in Figure 11.6 (Boucher *et al.*, 2013). It can be seen that the SCRE is largest along the northern hemisphere storm tracks, in the southern ocean and in the tropics, for example in the Intertropical Convergence Zone (ITCZ), the South Pacific Convergence zone (SPCZ), China, Central America and equatorial Africa, where convective activity with optically thick clouds dominates (Figure 1.2). In addition the stratus decks off the west coasts of southern Africa and North and South America are clearly visible, as these are bright clouds over the much darker ocean surface. These stratus decks disappear in the LCRE.

Low-level clouds with a large vertical extent, such as nimbostratus and cumulonimbus, have an effect on both longwave and shortwave radiation. For tropical cumulonimbus, the SCRE and LCRE are comparable. This near cancelation causes the net cloud radiative effect (Figure 11.6c) to be between −10 and 10 W m^{-2} over large areas in the tropics. For instance, the West Pacific Warm Pool, which is the area of high sea surface temperatures close to Indonesia, is characterized by the largest LCRE, of more than 70 W m^{-2}, and a large SCRE of ∼ −90 W m^{-2}, leading to a CRE of only ∼ −20 W m^{-2}. In mid-latitudes, where the tropopause is lower than in the tropics, the cloud tops are also lower. Consequently the cloud top temperatures are higher, causing the LCRE to be smaller in mid-latitudes than in the tropics. Therefore the LCRE cannot balance the SCRE in mid-latitudes, and a negative net CRE of several tens of W m^{-2}, found in storm tracks (Figure 1.2) over the southern ocean and stratus decks (Figure 1.1), clearly dominates the LCRE.

Positive values of the net CRE exceeding 10 W m^{-2} are restricted to regions with a high albedo: ice-covered regions and deserts. In ice-covered regions the combined albedo of the cloud and the snow- or ice-covered surface can be smaller than that of the surface alone, resulting in a positive SCRE. As discussed above this starts to become important if the surface albedo exceeds approximately 0.7. Bear in mind, though, that satellite retrievals have difficulties in distinguishing clouds from snow and ice and thus have the largest uncertainties over these snow- and ice-covered areas.

Since cloud cover, cloud type, cloud altitude or cloud microphysical properties can change in a warmer climate, clouds constitute a climate feedback in response to changing environmental conditions (Section 12.3). To compensate the 3.7 W m^{-2} RF due to the doubling of carbon dioxide, the SCRE, for example, would need to increase by 8% (e.g., from −47.3 W m^{-2} to −55 W m^{-2}) in a warmer climate. If, on the contrary, the SCRE were to decrease by the same amount, the warming would be twice as large as from the doubling of CO_2 alone.

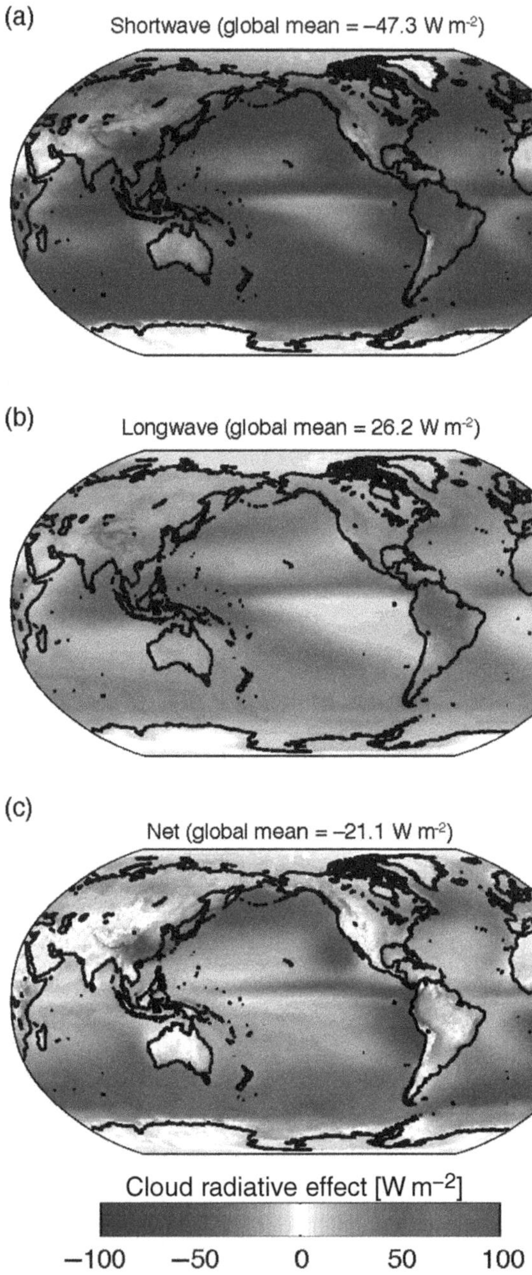

Fig. 11.6 Annual mean (a) shortwave (SCRE), (b) longwave (LCRE) and (c) net cloud radiative effect (CRE) averaged over the years 2001 to 2011, as obtained from the CERES satellite data. Data source: Figure 7.7 from IPCC AR5 (Boucher *et al.*, 2013). A black and white version of this figure will appear in some formats. For the color version, please refer to the plate section.

11.5 Exercises

1. Consider the global energy budget:
   (a) Assuming that $R_{Sun} = 6.98 \times 10^8$ m and that $T_{Sun} = 5769.56$ K, calculate the total radiation emitted by the Sun.
   (b) From your result in (a) calculate the solar constant, assuming a distance between Sun and Earth of 149.6×10^9 m. How would your result change if this distance was doubled?
   (c) Calculate the mean temperature of Earth without the natural greenhouse effect of the atmosphere. Assume Earth to radiate as a black body and a planetary albedo of $\alpha_p = 0.3$.
2. Corti and Peter (2009) developed a simple model to calculate cloud radiative forcing. With their model, the (clear-sky) longwave net flux F_{LW}^{CS} can be calculated as follows:

$$F_{LW}^{CS} \approx \beta T_s^{k*},$$
$$F_{LW} \approx (1 - \epsilon_*)\beta T_s^{k*} + \epsilon_* \beta T_{cld}^{k*}.$$

The shortwave cloud radiative effect SCRE is given by

$$\text{SCRE} \approx -I \times 0.73 \times (1 - \alpha)\frac{R_{c,1} - \alpha R_{c,2}}{1 - \alpha R_{c,2}}.$$

To calculate the SCRE and LCRE, the following equations are needed:

$$\epsilon_* \approx 1 - e^{-0.75\tau}$$
$$R_{c,1} \approx \frac{\tau/\xi}{\gamma_* + \tau/\xi}$$
$$R_{c,2} \approx \frac{2\tau}{\gamma_* + 2\tau}$$
$$I = S_0 \cdot \xi = S_0 \cos z,$$

where S_0 is the solar constant (1360 W m^{-2}), z is the solar zenith angle, τ is the cloud optical depth, α is the surface albedo, T_s is the surface temperature and T_{cld} is the cloud top temperature. Assume that $\beta = 1.607 \times 10^{-4}$ W m^{-2} K^{-1}, $k_* = 2.528$ and $\gamma_* = 7.7$.
   (a) Calculate both the shortwave and longwave cloud radiative effects of a tropical cloud around noon with the following properties: the cloud optical depth is 20, the solar zenith angle is 10°, the surface temperature is 300 K and the cloud top temperature is 265 K. The surface albedo is 0.05. Does the cloud have a net cooling or warming effect on the climate?
   (b) Now think of an Arctic stratus cloud. Which input parameters will be considerably different in the Corti and Peter (2009) model? How do you expect the net cloud radiative effect to change?
3. The total irradiance at the surface of a black body can be described by the Stefan–Boltzmann law.

(a) Using the Stefan–Boltzmann law, derive an expression for the sensitivity of the emitted radiation to temperature changes.
(b) Assuming an average temperature of 300 K for the tropics and of 250 K for the poles, calculate the sensitivities for these two regions.
(c) A doubling of CO_2 relative to pre-industrial conditions would lead to a radiative forcing of 3.7 W m^{-2}. What would be the temperature response needed to equilibrate this radiative forcing (i) in the tropics and (ii) at the poles, respectively?
(a) Which region is more sensitive? Why?

12 Impact of aerosol particles and clouds on climate

Aerosol particles affect the climate by scattering solar radiation and by absorbing solar and terrestrial radiation, as discussed in Chapter 11. In addition they modify cloud properties by acting as CCN (Chapter 6) and INPs (Chapter 8). These so-called aerosol–cloud interactions play a key role in the anthropogenic radiative forcing of the climate system (Boucher *et al.*, 2013). They continue to remain the most uncertain of all forcing agents and are still associated with a low level of scientific understanding. The impact of aerosol particles on the radiative budget at TOA is expressed in terms of the radiative forcing (RF) as discussed in Section 12.1. The aerosol RF is negative and hence partly offsets the greenhouse gas warming.

Clouds have a large effect on Earth's radiative budget (Chapter 11). Vice versa, clear-sky radiative cooling and solar heating destabilize the atmosphere and thus drive convection and cloud formation. Also, the emission of longwave radiation from a cloud top destabilizes the air above it, allowing the cloud to grow vertically, while the absorption of solar radiation within the cloud can cause its dissipation. The climate impact of clouds depends on the altitude and the geographical locations where they form. These two aspects will be discussed in Section 12.2 together with the change in global cloud cover over the last 40–60 years.

The increase in greenhouse gases causes Earth's temperature to increase. In addition to the radiative forcing due to greenhouse gases, feedbacks also operate in the climate system in such a way that the water vapor mixing ratio increases or snow and ice melt. These changes in turn cause the temperature to increase even more; thus, they are positive feedbacks. How they work, and the role clouds play in a warmer climate, is the topic of Section 12.3.

Given that aerosol particles partly offset greenhouse gas warming, it has been suggested that aerosol particles should be deliberately injected into the atmosphere to cool the climate. This so-called climate engineering involving aerosol particles and clouds is discussed in Section 12.4.

12.1 Aerosol radiative forcing

Aerosol particles affect the climate directly by scattering and absorbing radiation as well as by modifying cloud properties, as shown in Figure 12.1 as quantified in terms of the RF in W m^{-2}. This is defined as an external perturbation of the climate system, examples

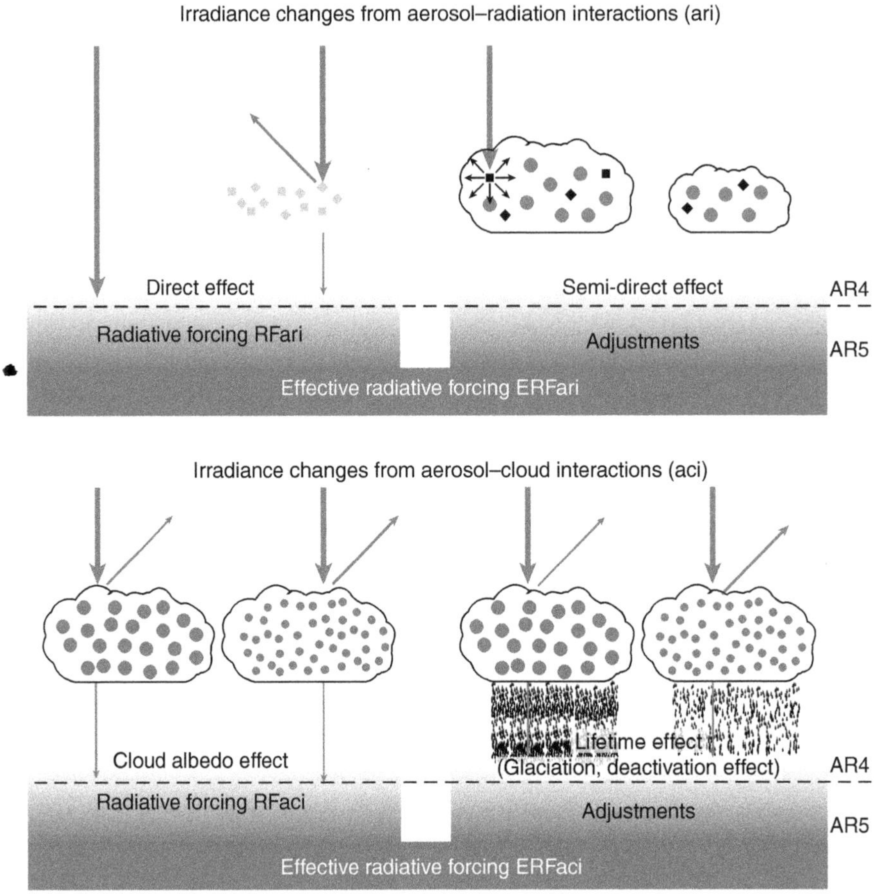

Fig. 12.1 Schematic of radiative forcing (RF) and effective radiative forcing (ERF) due to aerosol–radiation interactions (ari) and to aerosol–cloud interactions (aci). The terminology used here may be compared with the earlier terminology used in IPCC AR4 (Forster *et al.*, 2007; Denman *et al.*, 2007). The thick arrows depict the incoming solar radiation and the thin arrows the reflected and transmitted solar radiation. The gray and black squares in the upper figure denote scattering and absorption aerosol particles and the gray disks in the upper and lower figures denote larger and smaller cloud droplets, whereby smaller gray disks refer to smaller cloud droplets. Modified from Figure 7.3 of IPCC AR5 (Boucher *et al.*, 2013).

of which are volcanic eruptions, changes in solar activity and emissions of greenhouse gases and aerosol particles. The radiative forcing is calculated as the difference in the net radiative flux either at the TOA or at the tropopause between two radiative transfer simulations that differ only by the amount of forcing agent under consideration. A positive RF leads to a warming and a negative RF to a cooling of the Earth–atmosphere system.

As shown in Figure 12.1, the RF for aerosols can be divided into that due to aerosol–radiation interactions (RFari), formerly called the direct aerosol effect, and the RF due to aerosol–cloud interactions (RFaci), also referred to as the indirect aerosol effect(s), the cloud albedo effect or the Twomey effect (Twomey, 1974; Denman *et al.*, 2007). Aerosol perturbations will cause not only an RF but also changes in the vertical structure of the atmosphere in terms of temperature, relative humidity and precipitation formation. All adjustments in the cloud microphysical structure, temperature and relative humidity that accompany RFari and RFaci occur on time scales of minutes to days and so are much faster than the time scale of global climate change. Therefore these fast adjustments are included, in addition to RF, in the so-called effective radiative forcing (ERF) shown in Figure 12.1 (Boucher *et al.*, 2013). The sum of all adjustments related to anthropogenic aerosol particles either cause the ERFaci+ari to be larger than RFaci+ari alone or act against RFaci+ari and so reduce (buffer) the RFaci+ari (Stevens and Feingold, 2009). The different aspects of Figure 12.1 are discussed in detail in the following sections.

12.1.1 Radiative forcing due to aerosol–radiation interactions

Radiative forcing due to aerosol–radiation interactions is proportional to the amount of aerosol particles in the atmosphere and to their ability to extinguish radiation. The extinction of radiation is the sum of the scattering and absorption of radiation, as discussed in Section 5.2. It is proportional to the aerosol mass mixing ratio and is a function of particle size and wavelength. Extinction is usually defined in terms of an extinction coefficient specific to each aerosol type (eq. 12.2); see below. Here we are interested in the mass extinction efficiency (MEE), which is the extinction coefficient divided by the air density and is given in m^2 kg^{-1}.

The scattering and absorption of solar radiation by aerosol particles reduces the amount of solar radiation absorbed by the surface, especially over dark regions such as oceans. However, absorption of longwave radiation by aerosol particles enhances the greenhouse effect. Aerosol radiative properties can vary greatly, depending on their scattering and absorbing properties. These depend on their shapes, sizes (larger aerosols absorb more and scatter less), chemical compositions and mixing states. The shortwave RFari is typically negative, since aerosol scattering prevails over absorption for most aerosol types.

Larger aerosols (in the micrometer size range, e.g. mineral dust) have a significant positive RF in the longwave mainly due to their absorptive properties, while their longwave scattering is typically of minor importance. For instance, sulfate aerosol particles exhibit a net negative RF as long as they are smaller than 2 μm and a net positive RF at larger wavelengths (Lacis *et al.*, 1992). In the global annual mean the positive longwave RF of aerosol particles is smaller than the negative shortwave RF, so that the aerosol RF is negative.

The scattering of solar radiation by aerosol particles is dominated by Mie scattering (Figure 9.5). The scattering efficiency is largest for aerosol particles that have diameters comparable with the wavelengths of the electromagnetic radiation with which they interact. For the scattering of visible light, this corresponds to accumulation mode aerosol particles (Table 5.1). In addition some aerosol particles, such as soot, also absorb solar radiation. The median values of the MEEs from 20 global climate models are 8.9 m^2 g^{-1} for black carbon, 8.5 m^2 g^{-1} for sulfate, 5.7 m^2 g^{-1} for particulate organic matter, 3 m^2 g^{-1} for sea salt aerosols and 0.95 m^2 g^{-1} for mineral dust (Kinne *et al.*, 2006). That the MEEs for sulfate and carbonaceous aerosols are larger than those for sea salt and mineral dust is due to their generally smaller sizes, with a larger fraction of accumulation mode particles.

The vertically integrated amount of solar radiation extinguished by aerosol particles is described in terms of the aerosol optical depth (AOD). The degree of extinction of radiation traveling a distance ds through the atmosphere depends on the amount of matter (gas molecules, aerosol particles and cloud hydrometeors) in the light path s and the wavelength λ. The change in the radiation intensity due to extinction, dI_λ, can be expressed by Beer's law:

$$\frac{dI_\lambda}{ds} = -k_{\lambda,ext} I_\lambda, \tag{12.1}$$

where I_λ is the wavelength-dependent intensity of radiation in W m^{-2} and $k_{\lambda,ext} = k_{\lambda,abs} + k_{\lambda,scat}$ denotes the wavelength-dependent extinction coefficient in m^{-1}, being the sum of the absorption and scattering coefficients. Beer's law is similar to Schwarzschild's equation (11.5) in the absence of a black body source function. Note that, in contrast with eq. (11.5), where only the absorption coefficient $k_{\lambda,abs}$ needed to be considered, here the full extinction coefficient $k_{\lambda,ext}$ is necessary.

The amount of light extinguished at a height z depends on the intensity of radiation at z, the density of gases, aerosol particles and hydrometeors and the effectiveness of these in extinguishing light. Neglecting extinction by gases and performing the calculation in the absence of clouds, we can obtain the wavelength-dependent extinction coefficient by integrating over the scattering efficiency times the cross-sectional area of the aerosol number size distribution $n_N(r)$:

$$k_{\lambda,ext} = \int_0^\infty \pi r^2 n_N(r) Q_{\lambda,ext}(r) dr, \tag{12.2}$$

where $Q_{\lambda,ext}$ denotes the wavelength dependent dimensionless scattering efficiency. Considering monochromatic light, we can simplify eq. (12.1) by neglecting the dependence on λ. The net extinction due to aerosol particles of a light beam traveling vertically downwards from TOA to a height z is obtained by integrating eq. (12.1), with $ds = dz$, between z and TOA:

$$\frac{I(z)}{I_0} = \exp\left[-\int_z^{\mathrm{TOA}} k_{ext} dz'\right] = \exp(-\tau_{AP}). \tag{12.3}$$

Here, I_0 is the intensity of radiation at the TOA, i.e. the solar constant S_0 (eq. 11.3) and τ_{AP} is the AOD. By definition τ_{AP} is zero at the TOA and increases downward. If $\tau_{AP} = 1$, the solar radiation is reduced to $1/e = 36.8\%$ by the presence of aerosol particles. Note that the use of the optical depth assumes that the path of the incident solar radiation is perpendicular to Earth's surface, i.e. that the solar zenith angle is 0 °.

The AOD aggregates the radiative influence of the various aerosol species into one quantity. The geographical distribution of the observed annual mean AOD as obtained from the MODIS satellite at a wavelength of 550 nm in cloud-free regions, averaged over the time period March 2000 to July 2015, is shown in Figure 12.2. The wavelength 550 nm was chosen because it lies in the middle of the visible wavelength range. The AOD for land can best be retrieved over dark surfaces with a homogeneous surface albedo, because the

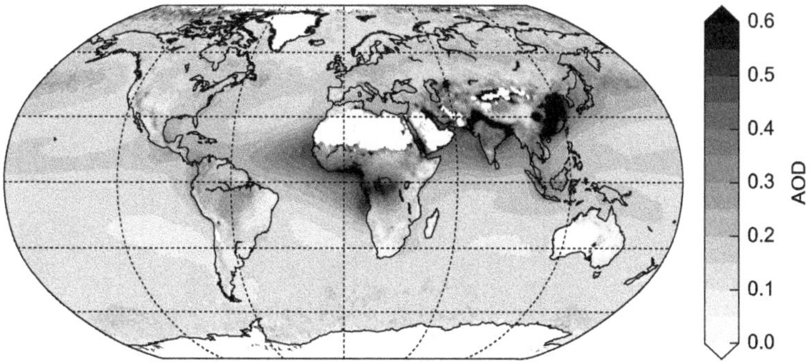

Fig. 12.2 Annual mean aerosol optical depth at wavelength 550 nm as determined from the MODIS satellite retrieval from March 2000 to July 2015.

reflected radiation from the surface and the extinction by gases need to be subtracted from the reflected radiation in order to isolate the AOD.

Therefore in areas with a bright surface albedo over land, such as the Sahara, there are no data in the level-1 data set provided by the Atmosphere Archive and Distribution System (LAADS). Along the same line, MODIS cannot distinguish between reflection from surface ice and from clouds. This explains the lack of data over Greenland and Antarctica. Maxima in AOD are visible downwind of the Sahara desert, off the west coast of Africa. Besides, biomass burning causes high AOD values in central Africa, the Amazon and southeast Asia. Fossil fuel combustion adds to the high AOD values in east Asia and India. Other sources of high AOD values include wildfires and volcanic eruptions.

The annual global mean AOD at wavelength 550 nm from observations and various global climate models (GCMs) is given in Figure 12.3 (Kinne *et al.*, 2006). The annual global mean AOD is estimated from surface remote sensing sites (Ae in Figure 12.3) to amount to 0.135 and when derived from satellite data (S* in Figure 12.3) to amount to 0.15. While only the total AOD can be estimated from observations, GCMs predict the contributions of individual aerosol types. Most GCMs include sea salt (SS), mineral dust (DU), particulate organic matter (POM), black carbon (BC) and sulfate (SU). Their simulated AOD is mostly lower than that obtained from observations (Figure 12.3). More strikingly, the contributions of the different aerosol types differ vastly between the 18 GCMs. While mineral dust has the largest aerosol burden, its global contribution to AOD is only 27% in the multimodel mean (Kinne *et al.*, 2006). Thus it is as high as the contributions from sulfate and sea salt, which have burdens that are only 10% and 40%, respectively, relative to that of mineral dust (Table 5.3). The over-proportionally large contribution of sulfate aerosol particles to the AOD is caused by their three and nine times larger MEE values than those of sea salt and mineral dust, respectively. The remaining contribution to the AOD stems from POM and BC.

Black carbon also contributes to ERFari. Despite its small AOD, of only 0.04, BC modifies the static stability of the atmosphere (Section 3.2) owing to its large absorption cross section, which results in local heating. This heating can affect cloud formation or the vertical extent of a cloud. The absorption of solar radiation by BC inside cloud droplets

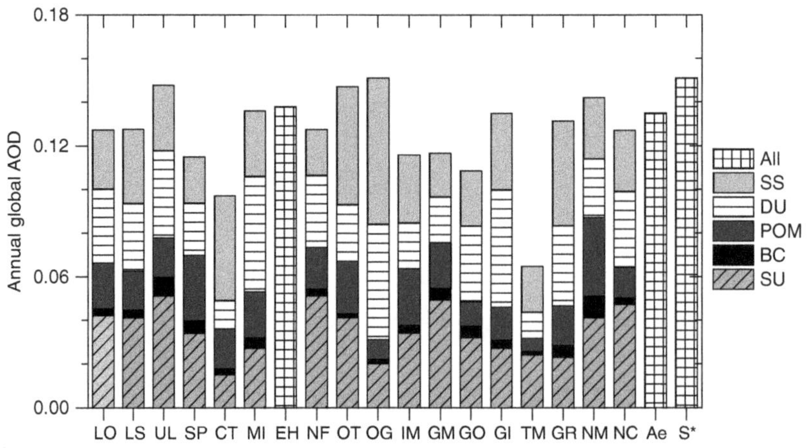

Fig. 12.3 Simulated contributions of the five aerosol types, sea salt particles (SS), mineral dust particles (DU), particulate organic matter (POM), black carbon particles (BC) and sulfate particles (SU) to the annual global aerosol optical depth (AOD) at a wavelength of 550 nm by different global models, shown on the *x*-axis. The bars with small squares refer to AOD observations from ground-based AERONET data (Ae) and satellite-composite data from the MODIS instrument over oceans and the MISR instrument over land (S*). Figure adapted from Kinne *et al.* (2006), in which the climate models are listed.

may also cause these droplets to evaporate. This in turn causes the cloud cover to decrease, allowing more solar radiation to be absorbed in the Earth–atmosphere system. Because of its absorption and the associated changes in static stability, the increase in BC due to anthropogenic activity has been hypothesized to have caused shifts in precipitation patterns in India and China (Menon *et al.*, 2002).

The reduction of solar radiation at the surface by anthropogenic aerosol particles between the 1950s and the 1980s has been termed global dimming. With the introduction of clean air acts in some countries, anthropogenic aerosol emissions have been declining in Europe and North America since then and the dimming trend has been reversed into a global brightening trend between the 1980s and 2000s (Wild, 2012).

12.1.2 Radiative forcing due to aerosol–cloud interactions

As discussed in Chapters 6 and 8, aerosol particles can act as CCN and INPs and therefore can affect cloud properties and the radiation balance. An increase in the number concentration of CCN and/or INPs will increase the number concentration of cloud droplets and/or ice crystals, respectively. If the liquid (ice) water content remains constant, this increase in CCN (INPs) will cause the cloud droplets (ice crystals) to be smaller. The larger number concentration of smaller cloud hydrometeors increases the total surface area of the cloud droplets (ice crystals). As the fraction of reflected solar radiation is proportional to the total cross-sectional area of hydrometeors per projected area of a cloud, a polluted cloud has a higher cloud albedo. It thus reflects a larger fraction of solar radiation back to space and causes a negative RFaci (Figure 12.1).

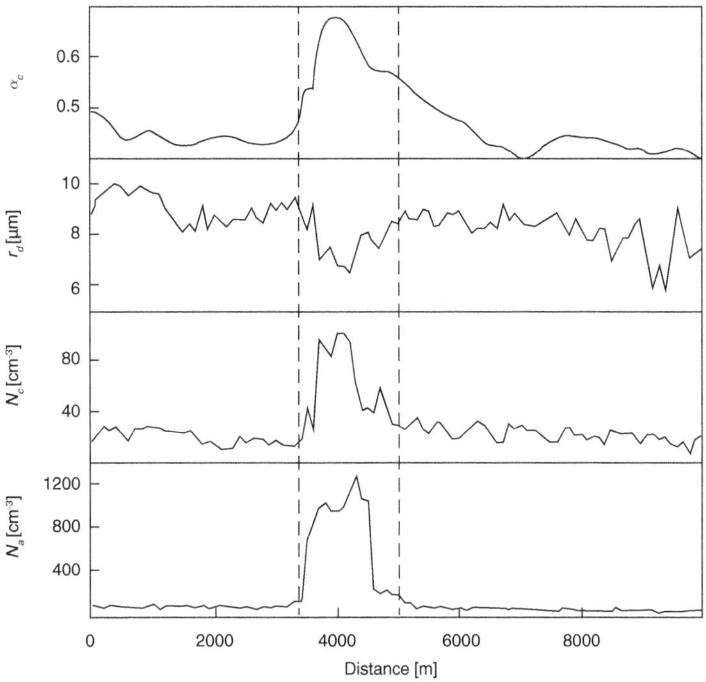

Fig. 12.4 Aircraft measurements of cloud albedo α_C, cloud droplet radius r_d, cloud droplet number concentration N_C, and accumulation-mode (0.1–1 μm) aerosol number concentrations N_a across a ship track off the coast of California. The ship track is situated between the broken vertical lines. The measurements were taken about 30 min after emission from the ship: the albedo values were obtained from radiation measurements 150 m above the cloud layer and all other data were obtained from flights 20–50 m below the cloud top. Figure schematically adapted from Durkee *et al.* (2000).

Next to the effect on the RFaci, an increase in aerosol particles will lead to adjustments in the cloud microphysical properties. One adjustment that predominantly influences boundary layer clouds is the so-called "cloud lifetime effect" (Albrecht, 1989). If a polluted cloud consists of more but smaller cloud droplets, the formation of precipitation will be slowed down because of the lower collision efficiency of the smaller droplets (Section 7.2). This may prolong the cloud's lifetime because a polluted cloud does not precipitate quickly. On the contrary, in polluted clouds the smaller droplets can evaporate faster. Changes in the INPs and the number and size of cloud droplets can also cause changes in ice nucleation and thus in the frequency of occurrence of glaciation of supercooled clouds or the vertical extent of clouds (Lohmann and Feichter, 2005).

Evidence for RFaci can be seen in satellite pictures of so-called "ship tracks". An example is shown in Figure 1.4f, where the white lines indicate recent ship routes embedded in a background stratus cloud deck. The changes in cloud microphysics in a ship track off the coast of California compared with the adjacent background stratus cloud are shown in Figure 12.4 (Durkee *et al.*, 2000). The figure shows aircraft measurements taken on a flight through a ship track. Within the ship track, indicated by the broken vertical lines, a tenfold increase in the accumulation-mode aerosol concentration was observed. This caused an increase in the cloud droplet number concentration from 20 cm^{-3} to 90 cm^{-3} and a decrease in the cloud droplet radius by approximately 2 μm. As discussed above, a larger concentration of smaller droplets increases the total projected area of the cloud droplets and thus the cloud albedo α_c, from below 0.5 to almost 0.7.

In clean, pristine, conditions, where only a few CCN are present, the cloud consists of fewer cloud droplets with a rather broad size distribution having at least some large droplets, typical of the marine cumuli shown in Figure 1.10b. These large droplets cause a faster initiation of the collision–coalescence process (Section 7.2) and form drizzle. If the drizzle does not evaporate below cloud base and reaches the surface, it removes the aerosol particles within the drizzle drops by wet scavenging (Section 5.4.2). Additionally, ambient aerosol particles below cloud base are removed from the atmosphere when colliding with falling drizzle drops.

In contrast, a polluted cloud that forms on many and generally smaller CCN typically has a narrower cloud droplet size distribution, typical of the continental cumuli shown in Figure 1.10c, because more and smaller cloud droplets have to compete for the available water vapor and so grow more slowly. These smaller cloud droplets have a smaller collision efficiency and therefore the drizzle formation rate is retarded compared with that for a clean cloud discussed above. A reduction in drizzle is associated with a smaller wet scavenging rate of aerosol particles, causing them to accumulate in the boundary layer.

The marine boundary layer thus has two stable states, a clean state that tends to remain clean and a polluted state that tends to remain polluted (Baker and Charlson, 1990). Evidence of these two states of the marine boundary layer is seen in open versus closed cells. Open cells refer to rings of clouds in which air rises at the cell boundaries and subsides in the cell interior. Clouds in these open cells produce drizzle, removing a fraction of the aerosol particles, and thus remain clean (Stevens *et al.*, 2005). Closed cells, however, have a circulation opposite to that in open cells, with air rising in the middle and subsiding at the cell boundaries. They are characterized by the absence of drizzle. Rosenfeld *et al.* (2006) hypothesized that open cells could transition to the closed-cell cloud regime by anthropogenic emissions from ships that prevent drizzle formation. A detailed modeling study by Feingold *et al.* (2015) supports this hypothesis in the case where the aerosol intrusions are large, reach the cloud layer before the heaviest drizzle occurs, in the early morning hours, and the free troposphere is cloud-free.

As it takes longer for precipitation-sized droplets to form in polluted clouds, it has been hypothesized that polluted clouds live longer and hence reflect more solar radiation back to space than clean clouds. However, a detailed model study, which traced individual air parcels having different aerosol concentrations, from very clean to very polluted, did not find an increase in lifetime for polluted clouds. The smaller cloud droplets in polluted clouds evaporate faster, which enhances entrainment (Section 4.2.4.3), causes further evaporation and reduces the vertical extent and ultimately the lifetime of the cloud. With this faster evaporation, the lifetime of polluted clouds was found to be comparable to that of clean clouds (Jiang *et al.*, 2006).

Sometimes a build-up of cloud liquid water is observed in ship tracks and other polluted clouds owing to the reduction in precipitation efficiency in accordance with a cloud lifetime effect. If the air above the boundary layer is, however, dry, then the enhanced cloud top cooling and the associated enhanced cloud-top entrainment in polluted clouds causes faster evaporation of cloud droplets, which subsequently leads to a reduced liquid water content compared with that of a clean cloud (Ackerman *et al.*, 2004). In other words, the studies by Ackerman *et al.* (2004) and Jiang *et al.* (2006) question the existence and importance of a cloud lifetime effect.

Another example of fast adjustments that contribute to the ERFaci are aerosol effects on mixed-phase clouds. If anthropogenic activity leads to an increase in INPs, this would result in the rapid glaciation of a supercooled liquid water cloud owing to the Wegener–Bergeron–Findeisen process; this is the glaciation effect in Figure 12.1. As long as supercooled cloud droplets are present, the cloud is at or close to water saturation and so is supersaturated with respect to ice. Thus the ice crystals can quickly reach precipitation size and potentially turn a non-precipitating cloud into a precipitating cloud. The glaciation effect was found to lead to a reduction in cloud cover, resulting in a decrease in reflected shortwave radiation, a response opposite to that of adding CCN (Lohmann, 2002).

At the same time, anthropogenic activity leads to an increase of secondary aerosol particles (Table 5.3). If sulfate, nitrate or organics condense on INPs, such a coating with soluble material can cover their active sites (Section 8.1.3.2). Being coated with soluble material reduces their ability to act as INPs for contact nucleation, and instead converts them into INPs acting in the immersion mode (Figure 8.4b). Given that immersion freezing was found to be initiated at lower temperatures than contact freezing for all substances except soot (Ladino Moreno *et al.*, 2013), this so-called deactivation effect (Figure 12.1) causes more clouds to remain supercooled. Hence, precipitation production is less efficient and more shortwave radiation is reflected to space (Hoose *et al.*, 2008; Storelvmo *et al.*, 2008). Whether the glaciation or the deactivation mechanism dominates, how large they are and hence to what degree these effects contribute to ERFaci is still a matter of debate.

12.1.3 Comparison of anthropogenic forcings

The RF of the different radiative forcing agents between 1750 (pre-industrial times) and 2011 is summarized in Figure 12.5 (Stocker *et al.*, 2013). Since 1750, well-mixed greenhouse gases (WMGHGs) have caused a positive radiative forcing of 2.83 (2.54 to 3.12) W m^{-2} (the numbers in brackets denote the 5%–95% confidence interval). The WMGHGs are the long-lived greenhouse gases, with lifetimes long enough to be relatively homogeneously mixed in the troposphere. They include carbon dioxide (CO_2), methane (CH_4), nitrous oxide (N_2O) and halocarbons.

Carbon dioxide is the main contributor to the radiative forcing of WMGHGs with 1.82 (1.63 to 2.01) W m^{-2}, followed by CH_4 with 0.48 (0.43 to 0.53) W m^{-2}, halocarbons with 0.36 (0.32 to 0.4) W m^{-2} and N_2O with 0.17 (0.14 to 0.20) W m^{-2}. Owing to their homogeneous distribution in the troposphere, only a few observational sites are necessary to measure their concentration. Therefore the behavior of WMGHGs is well understood and the RF of WMGHGs is associated with a very high confidence level.

The RF of ozone is two-sided. Halocarbons reaching the stratosphere destroy ozone, causing a small negative radiative forcing of -0.05 (-0.15 to $+0.05$) W m^{-2}. This forcing

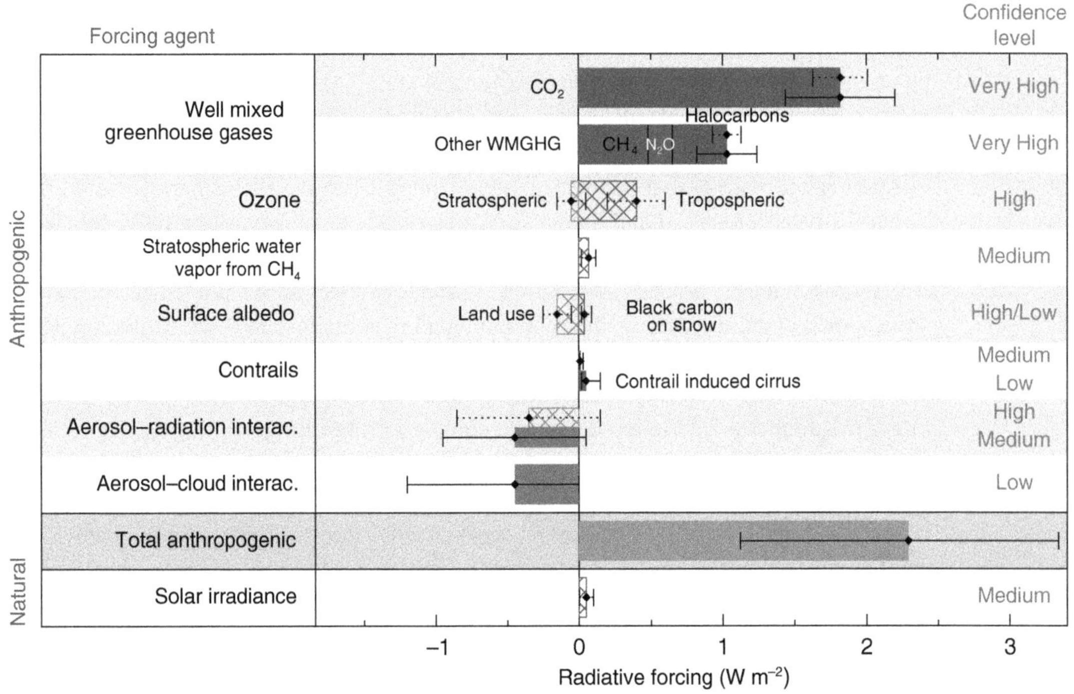

Fig. 12.5 Global mean RF estimates in 2011 relative to 1750, along with uncertainties, for the different forcing agents. The best estimates of the RFs are shown as black diamonds with corresponding uncertainty intervals. The solid bars are for the ERF and the hatched bars are for the RFs. Volcanic forcing is not included as its episodic nature makes it difficult to compare with other forcing mechanisms. Figure adapted from Figure TS.6 of IPCC AR5 (Stocker *et al.*, 2013).

is negative because a decrease in stratospheric ozone leads to less absorbed ultraviolet radiation and to a decrease in absorbed and emitted infrared radiation. In contrast, tropospheric ozone has increased. It is formed mainly during summer by the photo-chemical oxidation of its precursor gases (nitrogen oxides, carbon monoxide and hydrocarbons). Another contribution of the positive RF of tropospheric ozone is the decrease in stratospheric ozone, which allows more ultraviolet radiation to be absorbed by tropospheric ozone. Altogether, the positive RF of tropospheric ozone amounts to 0.4 (0.20 to 0.60) W m^{-2}.

While water vapor is the largest natural greenhouse gas, human activity increases water vapor concentrations only locally, for instance near irrigated fields. In addition water vapor only has an average lifetime on the order of 10 days in the troposphere (Section 2.5), which is much shorter than that of WMGHGs. But water vapor concentrations respond to the warming of the atmosphere and amplify climate change, as will be discussed in Section 12.3. In the stratosphere, however, water vapor is produced by the oxidation of methane with the hydroxyl radical. Here its lifetime is on the order of years, resulting in a small positive RF of 0.07 (0.02 to 0.12) W m^{-2}.

As can be seen from Figure 12.5, aerosol particles are associated with the largest negative RF but also with a much lower level of scientific understanding. As anthropogenic aerosol particles cause changes both in ERFari and ERFaci, which are hard to disentangle, the fifth IPCC report provided only a combined estimate of ERFaci+ari as their most robust estimate. It amounts to −0.9 W m^{-2} with an uncertainty range of −1.9 to −0.1 W m^{-2} (Boucher *et al.*, 2013). This estimate was obtained as an expert judgment of the respective literature. It was guided by estimates from GCM simulations and estimates involving satellite data. The ERFari value was estimated from GCMs as −0.45 W m^{-2} with a range between −0.95 and +0.05 W m^{-2} (Myhre *et al.*, 2013). Here the 95% range extends to slightly positive values because some studies attribute a very large positive RF to black carbon. In order to provide an estimate of ERFaci only and to be consistent with both the ERFaci+ari and the ERFari estimates, additivity of ERFaci and ERFari had to be assumed. This yields an ERFaci value of −0.45 W m^{-2} (Figure 12.5). However, estimates of ERFaci from GCMs are much more negative and are inconsistent with this estimate. Thus the three possibilities are that ERFaci and ERFari cannot be added linearly, that GCMs systematically overestimate ERFaci or that ERFaci+ari is more negative. As discussed in Boucher *et al.* (2013), there is a tendency for GCMs to overestimate ERFaci, but whether this is the only reason for this inconsistency is still an open issue.

The remaining anthropogenic radiative forcings, including land use changes, the presence of black carbon on snow that decreases the surface albedo, contrails and contrail-induced cirrus, are only of minor importance. Thus, the net anthropogenic forcing is dominated by positive forcing from the long-lived greenhouse gases and negative forcing due to aerosol particles; it amounted to 2.3 W m^{-2} in 2011, with an uncertainty ranging from 1.1 to 3.3 W m^{-2}. Compared with the anthropogenic forcings, the natural forcing caused by changes in solar irradiance is negligible.

In summary, the negative RF of aerosol particles has partly offset the positive RF of greenhouse gases. Over the period 1951–2010 the warming due to the RF of greenhouse gases has been assessed in different studies to amount to 0.9 °C. This has partly been offset by a 0.25 °C cooling, mainly due to the negative RF of aerosol particles, resulting in a net warming of 0.65 °C over this period (Stocker *et al.*, 2013).

12.2 Clouds and climate

12.2.1 Clouds at different altitude levels

As discussed in Chapter 1, clouds can be categorized into different altitude levels. This categorization is important for their radiative properties, as discussed in Section 11.4. An overview of the cloud cover at different levels of the atmosphere, based on satellite data (Mace *et al.*, 2009; Chepfer *et al.*, 2010), is given in Figure 12.6, where averages over boreal winter (December–February, DJF) and boreal summer (June–August, JJA) are shown. The satellite data originate from the CloudSat and CALIPSO (Cloud-Aerosol Lidar and Infrared Pathfinder Satellite Observation) data sets for the period 2006–2011 (Boucher *et al.*, 2013). These satellites are equipped with a cloud radar and a cloud-aerosol lidar, which allow them to provide information about the vertical structure of clouds in terms of their frequency of occurrence and liquid and ice water contents. However, in regions with

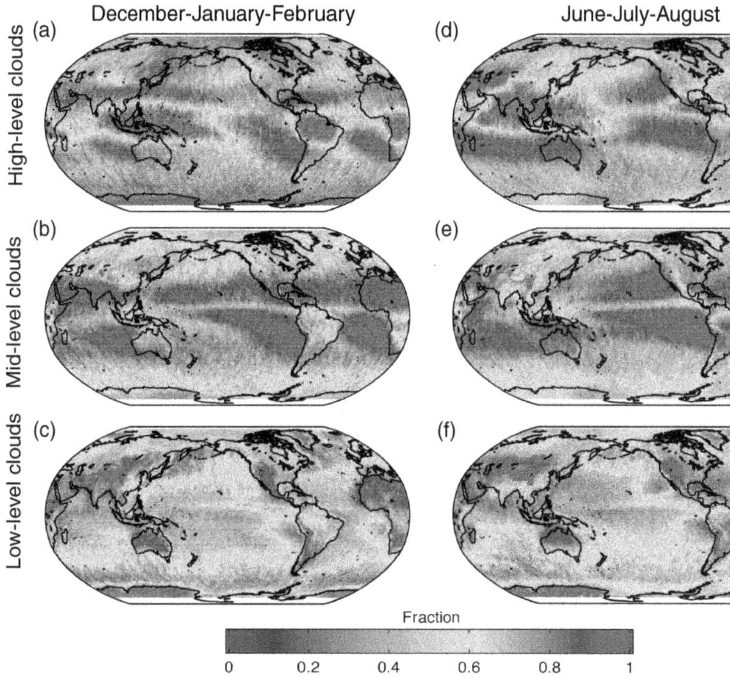

Fig. 12.6 (a)–(c) DJF and (d)–(f) JJA averaged high-level, mid-level and low-level cloud coverage based on data from CloudSat and CALIPSO for the period 2006–2011 (Mace *et al.*, 2009). For low-level clouds, only the CALIPSO data set for 2007–2010 (Chepfer *et al.*, 2010) is used at locations where it indicates a larger fractional cloud cover than the CloudSat data set, because the latter removes some clouds with tops at altitudes below 750 m. Figure adapted from Figure 7.6 of IPCC AR5 (Boucher *et al.*, 2013). A black and white version of this figure will appear in some formats. For the color version, please refer to the plate section.

high clouds, low-level clouds are underestimated by satellite data (Chepfer *et al.*, 2008), owing to the shadowing by overlying clouds.

Note that, from the satellite perspective, low clouds with vertical extents, which are considered as a separate class in the WMO cloud type definition (Table 1.1), are included in the high cloud category because of their high cloud tops. The areas with the largest fraction of high-level clouds are found in the tropics. More precisely, regions associated with deep convection, as shown in Figures 12.6a, d encompass the West Pacific Warm Pool, the Amazon and central Africa. Also visible are the Intertropical Convergence Zone (ITCZ) and the South Pacific Convergence Zone (SPCZ), as shown in Figure 1.2. The ITCZ forms where the air masses of the south and north trade winds converge. As a consequence, the ITCZ is on average at 5° S in DJF but moves to, on average, 15° N in JJA owing to the larger area of land masses in the northern hemisphere, which warm faster than the oceans. The largest area of high cloud cover in JJA is associated with the deep convective clouds and torrential rainfalls that characterize the Indian summer monsoon. The northward movement of the high cloud coverage over Central America and Africa owing to the shift in the ITCZ is not as pronounced as over the

Indian subcontinent. In DJF, mid-latitude high-level clouds are associated with extratropical cyclones (Figure 9.13) in the southern ocean, in the north Pacific and in the north Atlantic along the extratropical storm tracks (Figure 1.2) while in JJA mid-latitude high-level clouds in the northern hemisphere are mainly associated with summer-time convection.

Mid-level clouds (altostratus and altocumulus) are found least frequently of all clouds because the mid-troposphere is on average rather dry (Figure 12.6b, e). Usually deep convective clouds are accompanied by clouds in their vicinity whose tops reach only into the mid-troposphere, giving rise to the maxima in mid-level clouds seen in the ITCZ and SPCZ. At mid and high latitudes, mid-level clouds form in connection with extratropical cyclones. Unlike deep convective clouds, which are predominantly caused by conditional instability (Section 3.3), mid-level clouds can also form by shear instability. In the atmosphere, shear instability is commonly caused by a vertical gradient in the horizontal wind. Therefore, it is a common cause of mid-level clouds in the extratropics.

Low-level clouds form under the subsiding branch of the tropical Hadley circulation cells (Section 12.2.2) and over cold ocean currents such as the California Current off the coast of California, the Humboldt Current off the west coast of South America, the Benguela Current off the west coast of southern Africa, the Canary Current off the west coast of northern Africa or the West Australian Current. Their vertical extent is limited by the temperature inversion at the top of the boundary layer. In addition, low-level clouds also occur in the storm tracks of both hemispheres, where their coverage can exceed 80% (Figure 12.6c, f). They are found in the warm sector of extratropical cyclones and can also occur behind cold fronts (Figure 9.13). The Arctic ocean has a high coverage of stratus clouds during summer, when the sea-ice amount is small and the Arctic ocean serves as a moisture source for cloud formation (Kay and Gettelman, 2009).

12.2.2 Cloud regimes

There are preferred geographical locations for some cloud types. The reason is that their formation is associated with typical synoptic conditions. It has thus been found useful to define cloud regimes in which self-similar cloud scenes are grouped together, an example being the different marine stratocumulus regions shown in Figure 1.2. Other methods to distinguish cloud regimes are according to cloud top properties retrieved from satellite data (the cloud top pressure and optical depth) or according to large-scale subsidence and/or lower tropospheric static stability, as the cloud regimes found in areas with little instability are different from those in areas with large instability. Examples of cloud regimes that form in subsidence regions are polar stratus clouds, closed cells over the cold ocean currents or associated with cold air outbreaks, and marine stratocumulus. Cloud regimes that require a more unstable environment are marine shallow cumulus and open cells. For instance, deep convective clouds such as those found in the ITCZ and SPCZ (Figure 1.2) mark a cloud regime characterized by the highest instability. Within a given dynamically defined regime, further geographic refinements are possible. The stratus cloud regime can be distinguished

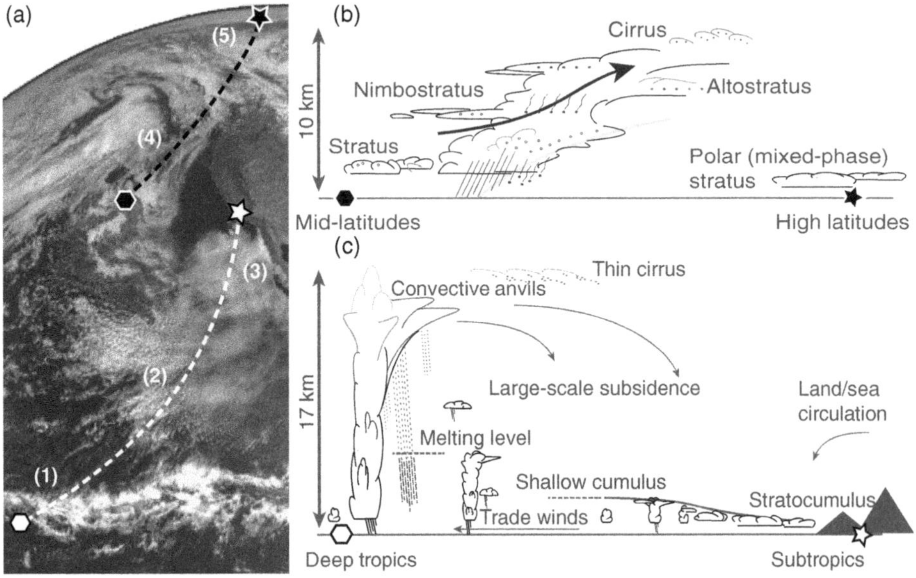

Fig. 12.7 (a) A visible-wavelength geostationary satellite image over the northern west Pacific. The numbers denote some important cloud regimes: (1) organized tropical convection, (2) trade wind cumulus, (3) marine stratocumulus, (4) stratiform clouds of a warm front and (5) polar stratus. (b) A schematic of the different cloud types associated with extratropical cyclones along the broken black line from mid to high latitudes in (a). (c) A schematic of the different cloud types found along the broken white line in (a) going from the subtropics off the coast of California to the tropics. Figure adapted from Figure 7.4 of IPCC AR5 (Boucher *et al.*, 2013).

into warm-phase low- and mid-latitude stratus decks, which cool the climate, and polar mixed-phase stratus decks, which warm the climate.

Figure 12.7a shows a geostationary satellite image on which some important cloud regimes can be identified. Going from south to north these are (1) organized tropical convection, (2) trade wind cumulus, (3) marine stratocumulus, (4) stratiform clouds of a warm front and (5) polar stratus. The typical cloud types associated with a warm front of an extratropical cyclone are shown in Figure 12.7b. An approaching warm front is visible by its shield of cirrus clouds. These cirrus clouds turn into altostratus and then into an extensive region of nimbostratus associated with frontal uplift. Behind the warm front, low-level stratus forms in the warm sector. Along the cold front, convective clouds are found because of the stronger cold frontal lifting (Figure 9.13).

Figure 12.7c shows schematically the Hadley circulation and the typical cloud types associated with it. The term "Hadley circulation" describes a thermally driven tropical circulation cell resulting from a latitudinal energy distribution. In the tropics more shortwave radiation is absorbed than longwave radiation is emitted back to space. In contrast, the polar regions receive less shortwave radiation than the longwave radiation they emit to space. This results in an energy surplus in the tropics and an energy deficit in the polar regions. Thus energy needs to be transported from the tropics to the polar regions. Part of this poleward energy transport occurs via ocean currents and the other part occurs within the atmosphere. The atmospheric energy transport of the global circulation starts with the Hadley cell, which is the largest overturning circulation cell in the atmosphere. It is associated with different cloud regimes and cloud types. Heating from the Sun causes the air in the ITCZ to rise and to form precipitating deep convective clouds with extensive cirrus anvils. They mark the ascending branch of the Hadley cell in the subtropics (Figure 12.7a, cloud regime (1)). Owing to their low cloud top temperature, they emit the least amount of longwave radiation and appear white in the satellite image.

The deep convective clouds in the ITCZ can occur in isolation or can form mesoscale convective complexes, which are an example of mesoscale convective systems (Section 10.4). The outflow from the deep convective anvils towards the subtropics is marked by thin cirrus clouds. The Hadley circulation extends approximately up to 30° latitude in the annual mean. Here, owing to the Coriolis force, the winds at high altitudes no longer have a poleward component but only flow eastward. At these latitudes the air masses from the Hadley cell converge with those from the mid-latitude Ferrel cell, which is another overturning circulation in the atmosphere. Owing to mass continuity the air needs to sink here, causing the subtropics to be characterized by large-scale subsidence. Adiabatic compression in the subsiding, mainly cloud-free, air warms this air and leads to a marked temperature inversion at the top of the boundary layer over the cold oceans.

In the shallow boundary layer beneath the inversion, evaporation from the ocean leads to a high relative humidity, so stratus and stratocumulus clouds with cloud bases of only 400–600 m above the ocean can form (Eastman and Warren, 2013). In Figure 12.7a they are visible as Californian stratus and stratocumulus towards the northern end of the broken white line (cloud regime (3)). The Hadley circulation is closed by the trade winds near the surface, which advect the subsided air masses equatorwards. As the air gets closer to the equator, it encounters warmer ocean temperatures; thus the boundary layer depth increases and the stratus clouds break up first into stratocumulus and then into shallow cumulus, as shown in Figure 12.7a (cloud regime (2)) and Figure 12.7c. Close to the ITCZ the rising air becomes so positively buoyant that cumulus clouds with higher tops, such as Cu med, Cu con and Cb, form.

12.2.3 Trends in cloud cover

While reliable observations of temperature reach back into the eighteenth century, reliable observations of cloud cover obtained from surface observations only date back to the 1950s. The extended cloud report archive (Hahn and Warren, 2009; Eastman and Warren, 2013) contains cloud observations from ships since 1952 and from land stations since 1971. Certain data extend further back, especially over land. However, prior to the dates stated above, data were taken differently. In order to have a homogeneous time series, they are not included in the cloud report archive.

As shown in Figure 12.8, the total cloud cover over land decreased by 4% per century between 1971 and 2009 on the basis of ground measurements (Eastman and Warren, 2013) and that over ocean increased by a similar amount. The trend over land is caused by decreases in all stratiform clouds (stratus, nimbostratus, altostratus and cirrus) whereas the convective cloud types (cumulus, stratocumulus and altocumulus) showed slight increases

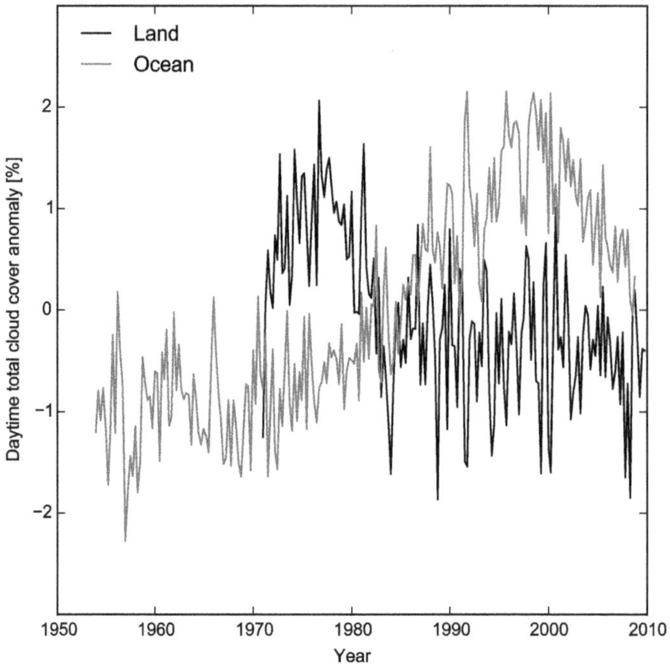

Fig. 12.8 Time series of global cloud cover anomalies over land and ocean. Both time series are based on averages of seasonal anomalies from the long-term mean, allocated to individual 10° grid boxes weighted by land or ocean area and relative box size. Figure adapted from Eastman and Warren (2013) with permission of the American Meteorological Society.

over this period. The long-term variations in oceanic cloud cover are in contradiction with satellite-data-based studies, which found only small trends in total cloud cover from 1979 to 2001 (Wylie *et al.*, 2005); therefore their trend has to be taken with caution. Ground-based observations of total cloud cover are biased towards low-level clouds because they may obscure mid- and high-level clouds. Satellite-based observations, however, are most sensitive to high clouds, which might explain the contrary cloud cover trends over the oceans.

12.3 Climate feedbacks

After having considered forcing agents and the associated RF, we will now discuss how the climate system responds to RF. If the Earth–atmosphere system is perturbed by a radiative forcing ΔF, for instance a doubling of CO_2, the radiation balance at the TOA will be affected as follows:

$$\Delta F = \Delta R + \Delta H + \lambda \Delta T_s. \tag{12.4}$$

Here ΔR represents the net radiative imbalance at the TOA, ΔH is the heat taken up by the ocean, ΔT_s is the net change in global mean surface temperature and λ is the climate

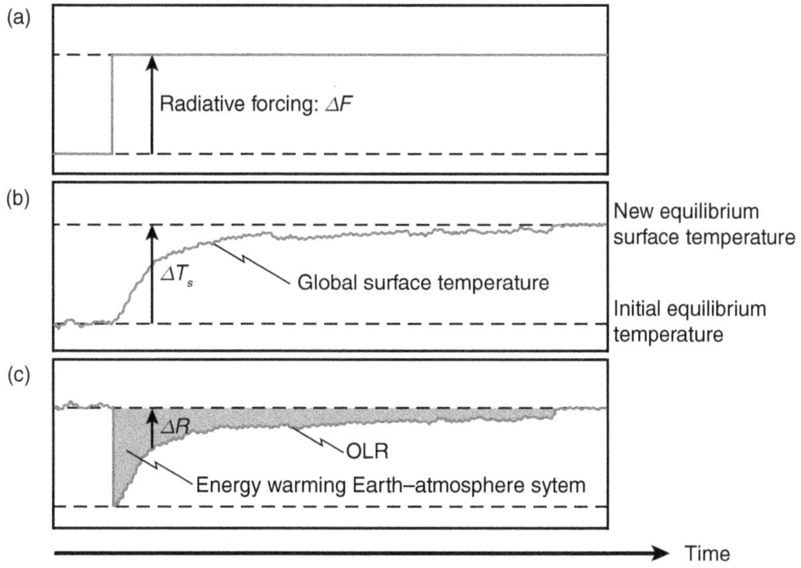

Fig. 12.9 (a) Schematic of an instantaneous radiative forcing ΔF, (b) response of the Earth–atmosphere system in terms of the global mean surface temperature T_S and (c) the net radiative imbalance ΔR and outgoing longwave radiation (OLR). Figure adapted from Murphy *et al.* (2009).

sensitivity parameter. All these quantities change over time except for ΔF, which in our example is caused by a step change from $1 \times CO_2$ to $2 \times CO_2$ as shown in Figure 12.9a. After this step change to $2 \times CO_2$ is imposed, the radiation at the TOA is no longer balanced, as shown in Figure 12.9c, where the outgoing longwave radiation at the TOA decreases step-wise at the same time as CO_2 doubles. Initially this imbalance, ΔR, equals ΔF. If ΔF and thus ΔR are positive then over time the land and ocean will warm. According to the Stefan–Boltzmann law (eq. 11.1) the land and the ocean will emit more radiation in response to their higher temperatures. In the new equilibrium, land and ocean emit as much longwave radiation as needed to balance ΔF at the TOA. Hence, the radiative imbalance ΔR at the TOA will be zero once the new equilibrium is reached (Figure 12.9c). Also, the ocean does not take up any more heat at this point and therefore $\Delta H = 0$ in the new equilibrium. Consequently, once the new equilibrium is reached, eq. (12.4) reduces to

$$\Delta F = \lambda \Delta T_s. \tag{12.5}$$

12.3.1 Planck feedback

The radiative forcing ΔF for $2 \times CO_2$ was obtained from different radiative transfer models by Myhre *et al.* (1998). From these calculations a simplified expression has been derived:

$$\Delta F = 5.35\,\mathrm{W\,m^{-2}} \ln \frac{[CO_2]_t}{[CO_2]_0}, \tag{12.6}$$

where the square brackets indicate a concentration. The logarithmic dependence of the radiative forcing on the CO_2 concentration is a result of the high optical depth of the latter. For the other greenhouse gases, with small optical depth, the dependence is linear. For t at the time of CO_2 doubling, the expression within the logarithm is equal to 2, yielding

$$\Delta F = 5.35 \text{ W m}^{-2} \ln 2 = 3.7 \text{ W m}^{-2}. \tag{12.7}$$

In climate simulations the CO_2 concentration is typically doubled from the pre-industrial value of 280 ppm in 1750 (Myhre *et al.*, 2013) to 560 ppm. Splitting ΔF into its longwave (LW) and shortwave (SW) components yields

$$\lambda = \frac{\Delta F_{LW} - \Delta F_{SW}}{\Delta T_s}. \tag{12.8}$$

Without feedbacks that affect the temperature, the change in F_{LW} due to $2 \times CO_2$ in the new equilibrium with $\Delta R = \Delta H = 0$ is given by the derivative of the Stefan–Boltzmann law (eq. 11.1) with respect to temperature:

$$\frac{\Delta F_{LW}}{\Delta T_s} \approx \frac{dF_{LW}}{dT_s} = 4\epsilon\sigma T_s^3 = \frac{4F_{LW}}{T_s}. \tag{12.9}$$

The value of the climate sensitivity parameter calculated using eq. (12.9) is called the Planck feedback because the changes in temperature are described by the Stefan–Boltzmann law, which is the Planck law integrated over all wavelengths. (This calculation ignores Earth's atmosphere because it calculates dF_{LW}/dT_s from $4\epsilon\sigma T_s^3$.) Doing this results in a null climate sensitivity parameter λ_0, or theoretical Planck feedback, as follows:

$$\lambda_0 = \frac{\Delta F_{LW}}{\Delta T_s} = \frac{4F_{LW}}{T_s} = \frac{4 \times 398 \text{ W m}^{-2}}{289.5 \text{ K}} = 5.5 \text{ W m}^{-2} \text{ K}^{-1}, \tag{12.10}$$

where the numbers for the emitted longwave radiation and the surface temperature have been taken from Figure 11.1. However, this value is far too high because in fact the outgoing longwave radiation at the TOA only amounts to 239 W m^{-2} (Figure 11.1). Therefore the Planck feedback needs to be estimated from models. Using the multimodel GCM estimate of λ_0, 3.2 W m^{-2} K^{-1} (Soden and Held, 2006), and rearranging eq. (12.5) yields

$$\Delta T_s = \frac{\Delta F}{\lambda_0} = \frac{3.7 \text{ W m}^{-2}}{3.2 \text{ W m}^{-2} \text{ K}^{-1}} = 1.2 \text{ K}. \tag{12.11}$$

To put it differently, a simple doubling of CO_2 leads to an increase in temperature of 1.2 K (assuming a climate sensitivity of 3.2 W m^{-2} K^{-1}).

12.3.2 Water vapor, lapse rate and ice-albedo feedback

In addition to this Planck feedback several other feedbacks occur in response to an increase in temperature, as shown in Figure 12.10. Therefore these additional feedbacks are often referred to as temperature feedbacks. As the temperature increases, the saturation vapor pressure increases according to the Clausius–Clapeyron equation (2.59). The relative humidity has been found to remain more or less constant when the temperature changes (Soden *et al.*, 2002), which implies that the specific humidity increases in a warmer climate. Because water vapor itself is a greenhouse gas, this feedback is positive. More

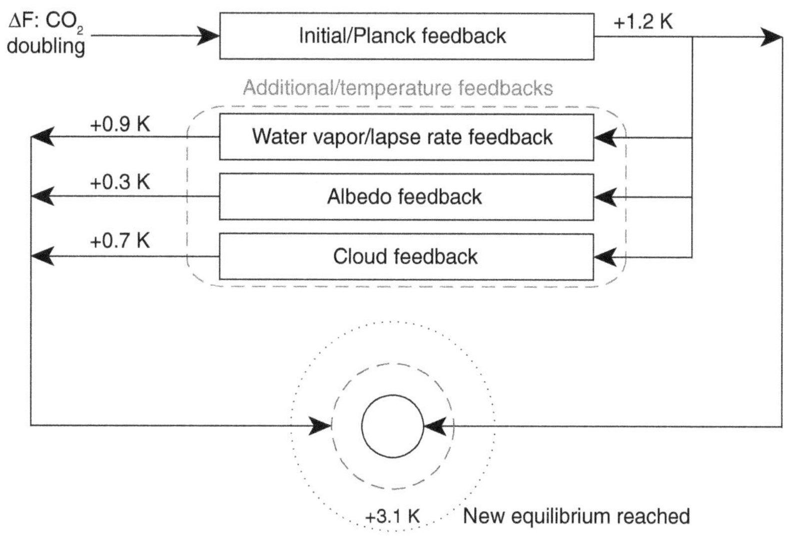

Fig. 12.10 Components of the feedbacks associated with a doubling of CO_2, and the respective temperature changes required to reach this new equilibrium. The numbers are taken from Dufresne and Bony (2008).

water vapor at higher temperatures implies, on the contrary, that the vertical temperature gradient is reduced because the moist adiabatic lapse rate is smaller for higher temperatures than for lower temperatures. In other words, because convection is more vigorous in a warmer climate and convection moistens and warms the mid-troposphere due to latent heat release, the vertical temperature gradient, i.e. the lapse rate, is smaller in a warmer climate. Since the emission of longwave radiation to space predominantly stems from the mid-troposphere, a warmer troposphere implies that more longwave radiation is emitted to space, which corresponds to a negative feedback. Since the positive water vapor feedback and the negative lapse-rate feedback are both linked to water vapor, they are often considered together. The average temperature increase from 12 global climate models for a doubling of CO_2 due to the combined water vapor and lapse rate feedback amounts to $+0.9\pm0.2$ K (Dufresne and Bony, 2008).

An increase in temperature leads to the melting of snow and ice surfaces. This causes bright surfaces to be replaced by darker surfaces, which absorb more solar radiation. Unlike the Planck feedback and the combined water vapor and lapse rate feedback, this positive surface-albedo feedback is not of global extent but is limited to areas presently covered by snow and ice. Therefore it only gives rise to a 0.3 ± 0.1 K temperature increase in the global annual mean for a doubling of CO_2 (Dufresne and Bony, 2008). The temperature increase is strongly amplified in polar and mountainous regions.

12.3.3 Cloud feedback

Of the feedbacks shown in Figure 12.10, the cloud feedback has the largest spread between different GCMs and thus is the most uncertain (Vial *et al.*, 2013; Boucher *et al.*, 2013). It results in a temperature increase of 0.7 ± 0.5 K (Dufresne and Bony, 2008). The reasons for this spread and uncertainty are twofold. (i) In a warmer climate, clouds may form at higher altitudes and may have different water contents, phases and microphysical properties. (ii) Because clouds can be smaller than the grid box of a GCM they need to be parameterized, that is, described in terms of large-scale (grid mean) variables. Therefore GCMs have problems in simulating certain cloud types and cloud microphysical properties.

This large uncertainty contributes to the spread in the estimated change of the global mean surface temperature, between 1.5 and 4.5 K for 2 × CO_2 (Collins *et al.*, 2013). If we add the above-mentioned averaged-model predicted temperature increases due to cloud feedback (0.7±0.5 K), the combined water vapor and lapse rate feedback (0.9±0.2 K) and the albedo feedback (0.3±0.1 K), and take the average temperature increase from these GCMs of 3.1±0.7 K for 2 × CO_2 (Dufresne and Bony, 2008), the initial warming of 1.2 K from the Planck feedback is increased by 150% (Figure 12.10).

The importance of the changes in low-level clouds for global warming is highlighted in Figure 12.11, which shows the response, i.e. the calculated total warming (with all the feedbacks discussed above included), of 16 global climate models at the time of 2 × CO_2. The temperature increase predicted by individual GCMs ranges between 1.8 and 5 K.

The right-hand panels show the changes in the amount of low-level clouds at 2 × CO_2 for two GCMs that fall at either end of the projected warming range. In the GCM with the lower temperature increase (lower panel) the coverage of low-level clouds increases locally by more than 4%, causing a negative cloud feedback in these regions. Hence the temperature increase calculated by this GCM for a CO_2 doubling is modest (1.8 K). In contrast, the coverage of low clouds decreases in the other GCM shown in the upper panel. In this GCM the change in low-level clouds is a positive cloud feedback, causing the GCM to predict a warming of 4.5 K at the time of CO_2 doubling. All in all, the inter-model differences in cloud feedback to global warming, especially those due to changes in boundary-layer clouds, give rise to one of the largest uncertainties for estimates of the increase in global mean surface temperature (Boucher *et al.*, 2013).

Eastman *et al.* (2011) evaluated anomalies in daytime low-level cloudiness as a function of anomalies in sea surface temperature (SST) from ship-based observations in regions dominated by marine stratus and stratocumulus clouds. They found that the cumulus cloud amount increases with increasing SST while the stratus and stratocumulus cloud amount decreases with increasing SST. This is a direct consequence of the fact that static stability decreases with increasing SST, favoring the formation of cumulus clouds over stratocumulus and stratus clouds. In four of the six investigated regions, the coverage of stratus and stratocumulus clouds decreased more than that of cumulus clouds increased. This points to a positive cloud feedback for increasing SST. In conclusion, their findings suggest that those GCMs that simulate decreasing low-level clouds in response to a doubling of CO_2 (such as in the upper panel in Figure 12.11) agree better with observations.

The change in cirrus cloud fraction with increasing temperature is shown in Figure 12.12a. Data were obtained from regressing the monthly mean anomalies of the cirrus cloud fraction from CALIPSO satellite data versus the monthly mean anomalies of surface temperature (Zhou *et al.*, 2014). The surface temperature anomalies were taken from the European Centre for Medium-Range Weather Forecasts (ECMWF) reanalysis (ERA-interim) project (Dee *et al.*, 2011). Figure 12.12a indicates that cirrus clouds will shift to

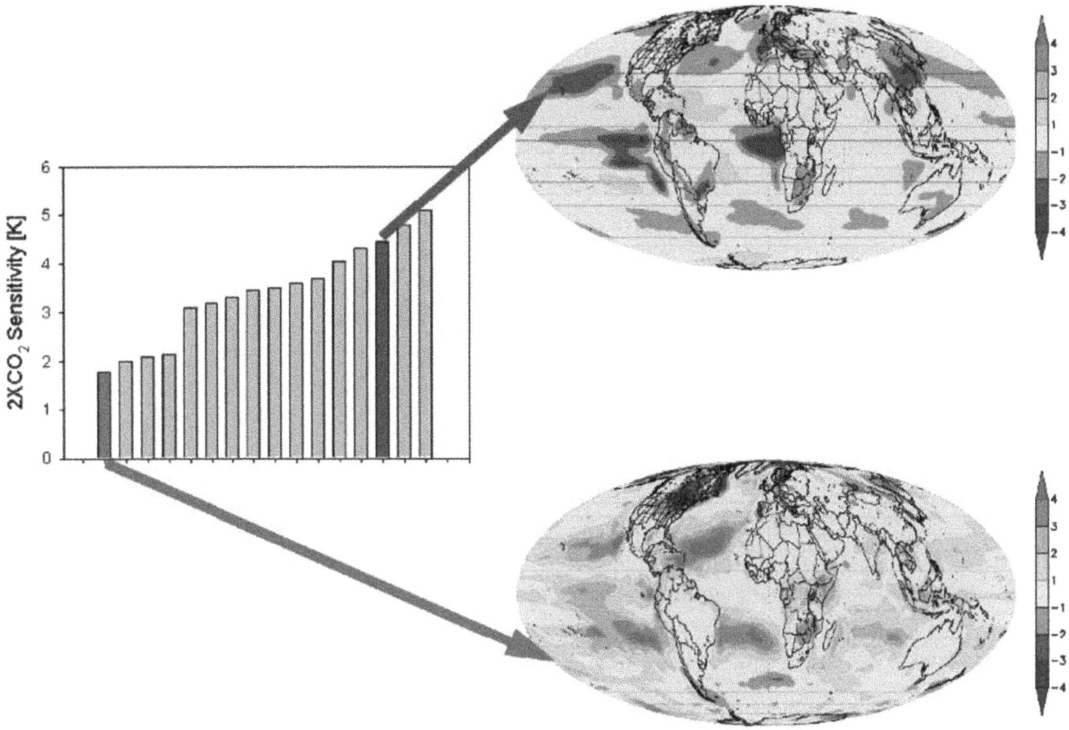

Fig. 12.11 Left: global mean temperature change of 16 GCMs as a response to a doubling of CO_2. Right: the change in low cloud amount in % K^{-1} from two individual GCMs. Figure reprinted with permission from Brian Soden.

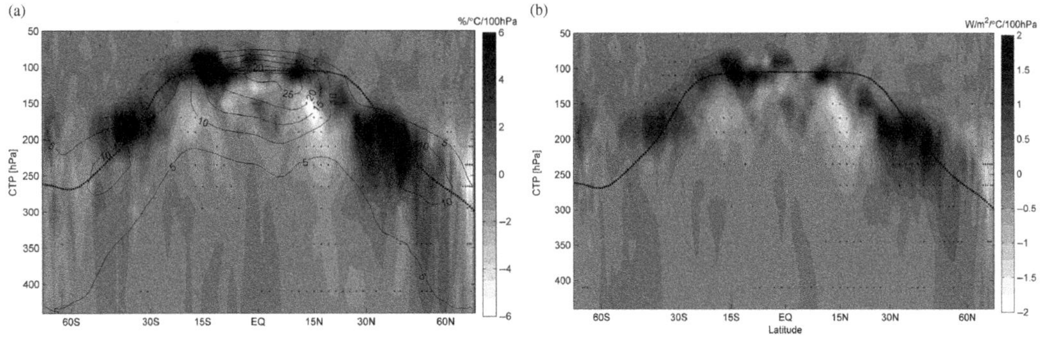

Fig. 12.12 (a) Zonal mean response of the cirrus cloud fraction to inter-annual surface warming, calculated by regressing the monthly mean anomalies of cloud fraction from CALIPSO satellite data against the monthly mean anomalies of global mean surface temperature (from ERA-interim). The contours give the six-year mean cirrus cloud fraction in % per °C, and per 100 hPa and the bold dotted line denotes the ERA interim climatological tropopause pressure. (b) Cirrus feedback as a function of latitude and cloud top pressure (CTP) in W m^{-2} per °C and per 100 hPa. The dots denote pixels where the linear regression slope is statistically distinguishable from zero. Figure adapted from Zhou *et al.* (2014) with permission from the American Geophysical Union. A black and white version of this figure will appear in some formats. For the color version, please refer to the plate section.

higher altitudes in response to a global surface warming. This can most clearly be seen in the tropics (30° N – 30° S) where the cirrus cloud fraction generally decreases by between 250 hPa and 130 hPa (white areas) and increases above these altitudes (dark gray shading). This change in cirrus cloud fraction amounts to up to 6% per K and per 100 hPa. The mid-latitudes of both hemispheres are characterized by an increase in cirrus cloud fraction within the upper troposphere, showing that the amount of cirrus cloud will increase with increasing surface temperature. From this fractional change in cirrus cloud cover the cirrus cloud feedback has been estimated and is shown in Figure 12.12b. A positive cloud feedback (dark gray shading) can be seen in areas where the cirrus cloud cover increases in Figure 12.12a and where the occurrence of cirrus clouds was shifted to higher altitudes. The higher the cirrus clouds are, the colder are their cloud top temperatures and, according to the Stefan–Boltzmann law (eq. 11.1), the less longwave radiation is emitted to space. Thus, their longwave cloud radiative effect (LCRE, Section 11.4) increases. The cirrus cloud feedback is negative at lower altitudes, where the cirrus cloud cover decreases, as seen in Figure 12.12a. In the global annual mean the cirrus cloud feedback is positive, at 0.2 ± 0.2 W m^{-2} K^{-1} (Zhou *et al.*, 2014), and thus would be a substantial fraction of the cloud feedback if confirmed by other studies.

A summary of robust cloud responses to greenhouse gas warming is shown in Figure 12.13 (Boucher *et al.*, 2013). Robust cloud responses are defined as those responses that have been simulated by most models and possess some kind of independent support or understanding. The reduction in low-level cloud discussed above is still uncertain, as indicated by the pale gray font color. Robust positive feedback contributions are expected from the rising of the cloud tops of clouds which form or extend into the upper troposphere. The rising of high-level clouds implies that the LCRE (Section 11.4) increases. This decreases the amount of energy emitted to space and therefore leads to a positive cloud feedback.

In a warmer climate the Hadley cell is expected to broaden, which would mean that the storm tracks migrate polewards. This causes low-level clouds to prevail in areas that experience less solar radiation, which decreases their SCRE and thus also constitutes a positive cloud feedback. So far no robust mechanisms have been identified that contribute to a significant negative cloud feedback. All in all the cloud feedback is positive.

12.4 Climate engineering involving aerosol particles and clouds

As discussed in Section 12.1, aerosol particles partly offset the greenhouse gas warming by aerosol–radiation and aerosol–cloud interactions. On this basis, different methods have been suggested of deliberately altering the climate system in order to alleviate the impacts of climate change. This is called climate engineering. While climate engineering consists of methods to remove CO_2 and methods to change the radiation balance, here only the methods affecting the radiation balance by aerosol particle injections are discussed. These methods are referred to as solar radiation management (SRM) and aim at reducing the incoming solar radiation either by injecting aerosol particles into the stratosphere or by enhancing the albedo of marine stratiform clouds, as shown in Figure 12.14. In

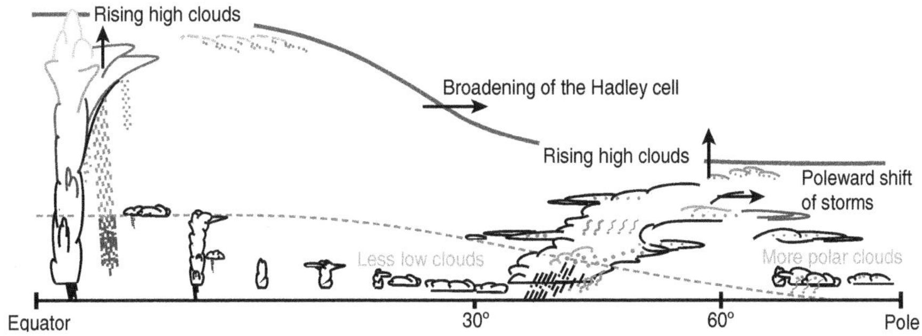

Fig. 12.13 Schematic of the robust cloud responses to greenhouse gas warming. The tropopause and melting level are shown by the thick solid line and thin gray broken line, respectively. The changes anticipated in a warmer climate shown in black indicate a robust positive feedback contribution and those in pale gray a small and/or highly uncertain feedback contribution. Figure modified from Figure 7.11 of IPCC AR5 (Boucher *et al.*, 2013).

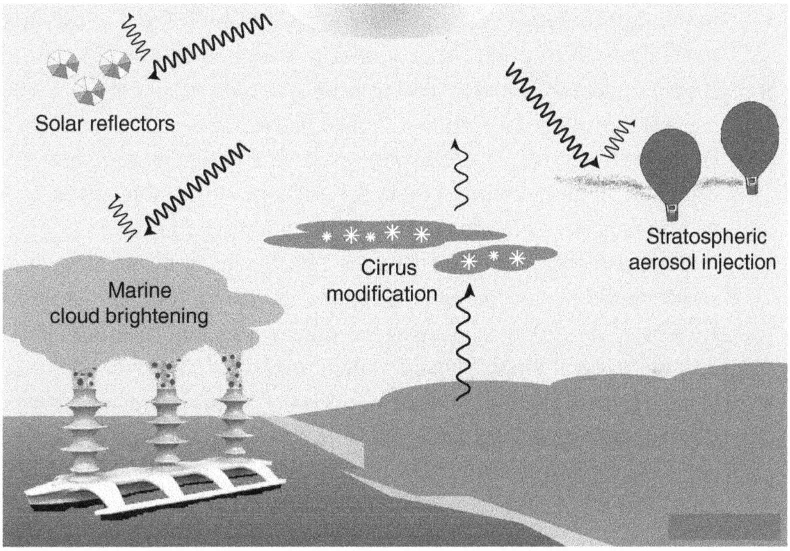

Fig. 12.14 Schematic of the climate engineering methods discussed in Section 12.4.

addition, a method to change cirrus cloud properties by injecting INPs is discussed. This method would affect both the shortwave and the longwave radiation, as explained below in Section 12.4.3.

In this section only the physical background for climate engineering involving aerosol particles is explained, as an application of the aerosol–cloud interactions discussed so far. The topic of climate engineering itself is highly debated not only in science but also in politics. It is important to note that this chapter is not recommending the employment of climate engineering to offset greenhouse gas warming. Rather, we discuss suggested climate engineering methods in order to investigate the possible outcomes of these approaches, with their uncertainties and risks. For the side effects of these methods and their technical implementation the reader is referred to the assessment of aerosol climate engineering in the fifth assessment report of the IPCC (Boucher *et al.*, 2013).

12.4.1 Stratospheric aerosol injections

Major volcanic eruptions during which aerosol particles are injected into the stratosphere lead to a substantial cooling of the global mean temperature. These eruptions can thus be inferred from the global mean temperature record. For instance the Mt. Pinatubo eruption in June 1991 caused an input of 9.25 ± 2 megatonnes (Mt) of sulfur (S) into the stratosphere (Guo *et al.*, 2004). This caused the global mean lower tropospheric temperature to drop by 0.5 °C as a result of the increased planetary albedo due to the presence of sulfate aerosol particles (Soden *et al.*, 2002), which have a lifetime of a couple of years in the stratosphere (Figure 5.17).

Knowledge of the cooling effect of major volcanic eruptions gave birth to the idea of engineering the climate with stratospheric aerosol injections (Figure 12.14) and with that to mimic permanent volcanic eruptions. Crutzen (2006) was the first to propose an SRM method in which 5 Mt S per year are continuously injected into the stratosphere. This should offset the radiative forcing from $2 \times CO_2$ of 3.7 W m^{-2}. Newer studies, however, suggest that at least 10 Mt S would be needed annually to maintain an RF of -4 W m^{-2} (Heckendorn *et al.*, 2009; Niemeier *et al.*, 2011). Ten megatonnes of S corresponds to only 8% of all sulfur emissions (Table 5.3), thus no severe side effects in the troposphere, such as acid rain, would be expected from this additional sulfur.

Since the SRM study of Crutzen (2006), many GCM simulations have investigated stratospheric sulfate aerosol injections, targeting different aspects of it. For instance, Robock *et al.* (2008) compared the effectiveness of injecting either 2.5 or 5 Mt S yr^{-1} into the tropical stratosphere or alternatively 1.5 Mt S yr^{-1} into the Arctic stratosphere to offset the temperature increase following a business-as-usual scenario (Section 9.6) in coupled atmosphere–ocean GCM simulations. The strongest decrease in temperature was observed for the scenario of injecting 5 Mt S yr^{-1} into the tropical stratosphere. In this case the global average surface-air temperature dropped to values last reached before 1920. The 1.5 Mt S yr^{-1} Arctic injection was designed to see whether it would be possible just to cool the Arctic and thus partly cancel the amplification in Arctic warming. In this scenario the global average surface-air temperature dropped only by about 0.2 K.

The study by Robock *et al.* (2008) further examined the effect of an abrupt end of SRM with stratospheric aerosols. In this case, the temperatures in all three SRM scenarios increased very quickly. Within a decade after the ending of SRM the temperature difference without and with SRM was reduced from 1.3 K to 0.2 K for the 5 Mt S yr^{-1} tropical injection. Besides, both tropical and Arctic stratospheric aerosol injections were found to alter the Asian and African precipitation patterns.

To test SRM methods in different GCMs, an SRM model intercomparison project (GeoMIP) was initiated in which the solar constant was reduced by successively larger decrements (Kravitz *et al.*, 2011, 2015). This method would correspond to employing successively more solar reflectors in space (Figure 12.14), or it can be regarded as a simplified way to study the effect of steadily increasing stratospheric sulfate

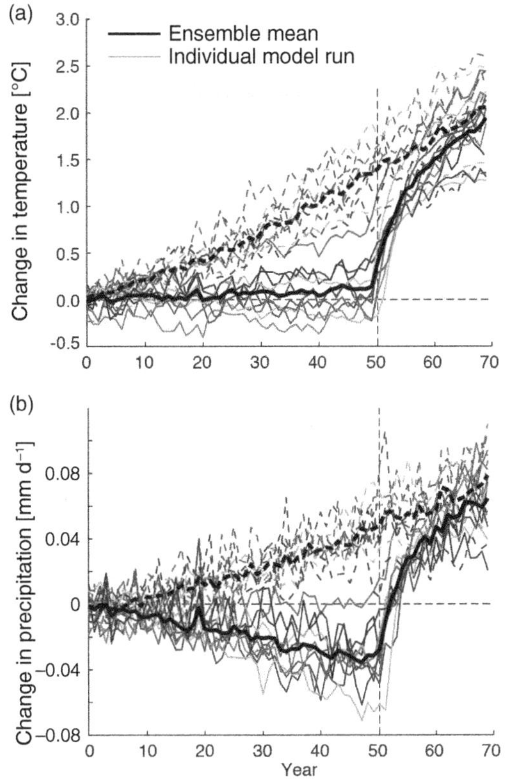

Fig. 12.15 Time series of the change of (a) the globally averaged surface temperature and (b) the precipitation relative to each model's 1 × CO_2 reference simulation. The solid lines correspond to simulations using solar radiation management (SRM) to balance the temperature increase from a 1% per year CO_2 increase until year 50 after which SRM is stopped. The broken lines correspond to simulations with a 1% per year CO_2 increase and no SRM. Figure adapted from Figure 7.24 of IPCC AR5 (Boucher *et al.*, 2013).

aerosol injections in GCM simulations that are conducted without an interactive aerosol model.

Figure 12.15 shows simulations of the change in global mean temperature and precipitation for a period of 70 years, in which the CO_2 concentration increased by 1% per year starting from pre-industrial CO_2 concentrations (280 ppm). In the SRM simulations the solar constant was successively decreased more for the first 50 years in order to keep the temperature constant despite the 1% CO_2 increase per year. The changes in global mean temperature and precipitation are shown relative to each model's 1 × CO_2 reference simulation, which shows the variations in temperature and precipitation with time due only to natural variability. After 50 years the SRM was stopped. As in the case of the results from Robock *et al.* (2008), reducing the solar constant as an analog to stratospheric aerosol injection works in keeping the surface temperature at any given value. However, if SRM is stopped, it only takes a decade until the projected global mean surface temperature and precipitation values are very similar to those projected without SRM (Jones *et al.*, 2013).

Moreover, these GeoMIP simulations clearly demonstrate that it is not possible simultaneously to restore both temperature and precipitation to their pre-industrial values with SRM methods. This can be understood in terms of the surface energy budget (eq. 11.9). Because the SRM method is applied in such a way that the surface temperature remains constant, $c_p \partial T_s/\partial t$ is zero. In this case a decrease in the downward shortwave radiation flux $F_{SW}^{\downarrow}$ needs to be balanced by changes in the heat fluxes LH + SH and the net longwave radiation flux $F_{LW} = F_{LW}^{\uparrow} - F_{LW}^{\downarrow}$:

$$F_{SW}^{\downarrow} = \mathrm{LH} + \mathrm{SH} + F_{LW}. \qquad (12.12)$$

If Earth were able to balance a change in the incoming solar radiation $F_{SW}^{\downarrow}$ by the same change in the net longwave radiation flux F_{LW} then the heat fluxes would remain unaffected. However, this is not the case. Given that the surface temperature remains constant, the changes in the longwave radiation flux and the sensible heat flux will be smaller than the changes in the latent heat flux. Thus the reduction in solar radiation needs to be largely balanced by a reduction in the latent heat flux, i.e. in evaporation. Since the annual global mean evaporation rate is balanced by the annual global mean precipitation rate, a reduction in solar radiation leads to a decrease in precipitation. In the case of the GeoMIP scenario shown in Figure 12.15b, precipitation is reduced on average by 0.03 mm d^{-1} after 50 years, which is 1% of the present-day global annual mean precipitation. Moreover, it is worth noting that the spatial pattern of the precipitation changes are not uniform (Boucher *et al.*, 2013).

12.4.2 Marine cloud brightening

As discussed in Section 12.1.2 and shown in Figures 1.4f and 12.4, anthropogenic emissions of aerosol particles or their precursor gases into the marine boundary layer can lead to the formation of ship tracks. However, ship tracks are not visible all the time. Estimates of the radiative forcing of shipping emissions from GCMs range between −0.1 and −0.6 W m^{-2} (Lauer *et al.*, 2007; Peters *et al.*, 2012). The RF just from ship tracks is only a minor fraction of this and has been estimated to amount to between −0.0004 and −0.009 W m^{-2} (Schreier *et al.*, 2007). As discussed above, anthropogenic emissions of aerosol particles into the marine boundary layer have been suggested in order to convert an open-cell cloud regime to a closed-cell cloud regime (Rosenfeld *et al.*, 2006) and thus to increase the planetary albedo (Section 12.1.2).

Given that an increase in the amount or reflectivity of marine boundary layer clouds would increase planetary albedo, the idea of systematically brightening marine boundary layer clouds (Figure 12.14) was initially put forward by Latham (1990) as another SRM method. It requires the injection of aerosol particles acting as CCN and has been investigated in various modeling studies since then. However, the modeling studies produced controversial results (Boucher *et al.*, 2013). In simple modeling studies the cloud droplet number concentration was increased by a factor 5 in 75% of the marine boundary layer clouds. Those studies concluded that the resulting cloud brightening would be sufficient to offset a doubling of CO_2. Subsequent studies that employ more complete treatments of aerosol–cloud interactions have found stronger or weaker changes. In particular, a cloud droplet number concentration necessary to offset a doubling of CO_2 cannot be achieved in regions with weak updraft velocities (Pringle *et al.*, 2012). Therefore IPCC AR5 concluded that evidence concerning the effectiveness of cloud brightening methods enhancing the albedo of marine clouds is ambiguous (Boucher *et al.*, 2013). The many uncertainties associated with aerosol–cloud interactions will have to be studied more carefully before more robust conclusions can be drawn.

12.4.3 Cirrus modification

Cirrus modification (Figure 12.14) is the only climate engineering method associated with the injection of aerosol particles that affects both the shortwave and the longwave radiation. It was put forward by Mitchell and Finnegan (2009) and is based on a GCM study by Lohmann *et al.* (2008) evaluating the difference in the TOA radiation balance under the assumption that all cirrus clouds form either by heterogeneous nucleation on INPs or by the homogeneous freezing of solution droplets (Chapter 8). Comparing Figures 12.16a and 12.16b, the immersion-freezing INP concentration of mineral dust aerosols is much lower than the concentration of solution droplets. Therefore, the ice crystal number concentrations are much smaller in cirrus clouds formed heterogeneously than those formed by homogeneous nucleation, shown in Figure 12.16c. Heterogeneous nucleation occurs at lower supersaturations and higher temperatures than homogeneous nucleation, so that the cirrus clouds in the simulation with heterogeneous nucleation form at lower altitudes, as shown in Figure 12.16.

The smaller number concentration of ice crystals in cirrus clouds formed by heterogeneous nucleation can grow to larger sizes and thus can sediment faster. This reduces the ice water content of these cirrus clouds. Combined with the lower number concentrations, this leads to a reduced optical depth and thus less reflected solar radiation of cirrus clouds formed by heterogeneous nucleation than in the case of cirrus clouds formed by homogeneous nucleation. In the global annual mean this causes a difference in SCRE of +2.7 W m^{-2} between the simulations with heterogeneous and homogeneous nucleation. However, this effect is more than offset by the changes in LCRE of these clouds. Because the heterogeneously nucleated cirrus clouds develop at lower altitudes, they have a smaller LCRE (Sections 11.4 and 12.3). In addition their lower optical depth results in a reduced emissivity, which allows more longwave radiation from below to be emitted to space. Thus the difference in the global annual mean LCRE between simulations with heterogeneous and those with homogeneous nucleation is −4.7 W m^{-2}. Overall this leads to a more negative net cloud radiative effect (CRE) of heterogeneously nucleated cirrus clouds, 2 W m^{-2}, as compared with homogeneously nucleated cirrus clouds (Lohmann *et al.*, 2008).

These simulations triggered a proposal by Mitchell and Finnegan (2009) to inject INPs into the upper troposphere, specifically into regions that are prone to cirrus formation. The resulting cirrus clouds would then form heterogeneously, leading to a cooling of the

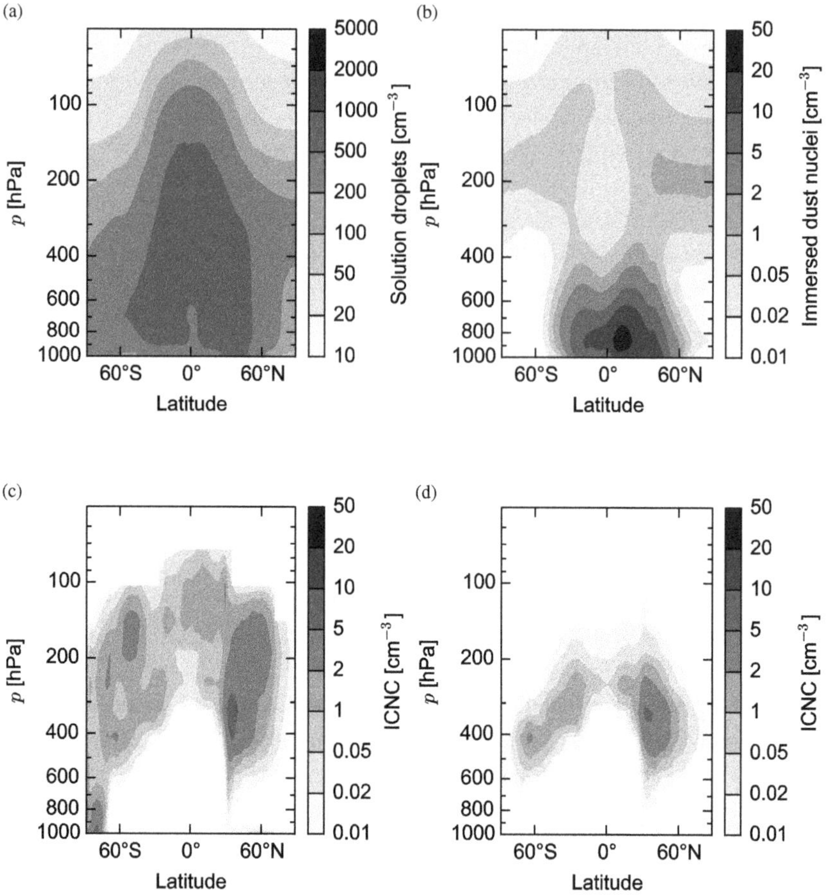

Fig. 12.16 Annual zonal mean latitude versus pressure cross sections of (a) number concentration of solution droplets, (b) number concentration of immersed dust nuclei, (c) ice crystal number concentrations (ICNC) assuming homogeneous nucleation and (d) ICNC assuming heterogeneous nucleation. Note the different color scale in panel (a). Figure adapted from Lohmann *et al.* (2008) and reproduced by permission of IOP Publishing.

atmosphere. A recent modeling study by Storelvmo *et al.* (2013) that explicitly increased INPs confirmed that a negative CRE of 2 W m^{-2} could be achieved given that the cirrus clouds in the present-day climate are mainly formed by homogeneous nucleation and that the method of injecting suitable INPs is technically feasible. However, it should be pointed out that whether this method results in a net cooling or warming of the Earth–atmosphere system depends crucially on the number of injected INPs. Seeding with too many INPs can lead to cirrus clouds that are more opaque than those formed homogeneously. This would change the sign of the effect from a net cooling to a net warming. A similar or even slightly larger global mean reduction in CRE can be achieved if only the winter hemisphere is seeded (Storelvmo and Herger, 2014) because there the negative response in LCRE is most pronounced.

12.4.4 Summary of the climate engineering discussion

The methods discussed above originate from our knowledge about aerosol–cloud interactions, but are still in their infancy (Boucher *et al.*, 2013). While the potential benefits of offsetting the greenhouse gas warming to some degree may appear attractive, climate engineering in terms of SRM comes with side effects, risks and costs. Some side effects of stratospheric aerosol injections have been identified, among them regional decreases in precipitation (Robock *et al.*, 2009) and negative impacts on the tropospheric and stratospheric ozone concentration and ultraviolet radiation (Tilmes *et al.*, 2015). A side effect of any SRM method is a decrease in global mean precipitation. Also, it is questionable which portion of the global warming can be offset by climate engineering methods involving aerosol particles and clouds.

In addition, neither SRM methods nor cirrus modification can compensate ocean acidification due to the increasing CO_2 concentration. These methods aim purely to cure the symptoms but not to reduce the CO_2 concentration. A side effect of these climate engineering methods is that abruptly stopping climate engineering causes the temperature to increase rapidly. This would put stress on ecosystems and species that are sensitive to the rate of climate change. Only if climate engineering involving aerosol particles and clouds is accompanied by aggressive greenhouse gas mitigation might it be possible to avoid transitions across climate thresholds.

12.5 Exercises

Single/multiple choice exercises

1. Aerosol particles alter the amount of solar radiation reaching Earth's surface by scattering and absorption. Which of the following is/are true?
   (a) The mass extinction efficiency is given in $m^2\ kg^{-1}$ and depends, among other factors, on the chemical composition of the aerosol particles, on their size and shape and on their height in the atmosphere.
   (b) The distribution of annual mean AOD shows high contributions from sulfate and sea salt aerosols, owing to the large burden of these aerosol particles in the atmosphere.
   (c) Overall, the warming effect of aerosol particles in the longwave radiative spectrum is outweighed by the scattering and absorption of solar radiation, which results in a cooling effect.
2. Which of the following statements about INPs and the deactivation indirect effect is/are true?
   (a) The deactivation indirect effect is highly important in Arctic cirrus clouds.
   (b) The deactivation indirect effect describes a process where INPs are altered in such a way that they become ineffective.

(c) The concentration of anthropogenic emissions strongly determines the significance of the deactivation indirect effect.
(d) Anthropogenic emissions can speed up the glaciation of clouds by providing more INPs or can reduce the ice nucleation efficiency through the deactivation indirect effect.

3. Imagine two air parcels with the same volume at the same altitude. The first air parcel contains 10 particles per cm^3 and all particles have radius 1 μm. The second air parcel contains 1000 particles per cm^3 and all particles have radius 100 nm. The aerosol particles all have the same chemical composition and are spherical. Which of the air parcels contributes more to the scattering of solar radiation?
   (a) The first air parcel.
   (b) The second air parcel.
   (c) Both have the same contribution.
4. By injecting aerosol particles into the stratosphere a cooling of the global mean temperature can be achieved that is similar to the cooling following a volcanic eruption. This is an example of a solar radiation management (SRM) method. Consider a scenario where a yearly increasing amount of CO_2 is emitted. Which of the following is/are true?
   (a) When using SRM, the global mean temperature can be kept constant despite a CO_2 increase but precipitation patterns will change.
   (b) If SRM is stopped, the global mean temperature rapidly increases to values from simulations without climate engineering but the precipitation does not.
   (c) SRM was found only to be effective when applied in the tropical regions.
   (d) SRM alters the radiation balance, which results in a reduction in the latent heat flux and the sensible heat flux.
   (e) The injection of sulfate aerosol particles reduces the planetary albedo and thus the downward shortwave radiative flux.
   (f) In the annual global mean, the reduced solar radiation due to aerosol injections reduces the precipitation rate.
5. Which of the following is/are true?
   (a) The radiative forcing of climate includes only anthropogenic forcing agents and is dominated by the positive forcing from well-mixed greenhouse gases.
   (b) Aerosol particles have a net negative radiative forcing although some individual aerosol particles make a positive contribution to the radiative forcing.
   (c) The regeneration of the ozone layer due to regulation of the emissions of halocarbons will lead to a positive radiative forcing in the stratosphere.
   (d) The largest uncertainty in the radiative forcing arises from the greenhouse gases owing to their longevity.
6. The null climate sensitivity parameter λ_0 describes the ratio of radiative forcing and equilibrium temperature change solely due to the Planck feedback. An increase in the concentration of a greenhouse gas implies an additional forcing of the climate. Imagine that the CO_2 concentration increases to a value three times that of today. How large is the temperature increase when considering only initial effects but no temperature feedbacks?

(a) 1.6 K
(b) 1.8 K
(c) 2.2 K
(d) 2.9 K
(e) 3.3 K

7. The Hadley cell . . .
   (a) . . . is the largest circulation cell.
   (b) . . . is caused by the surplus of energy in high latitudes.
   (c) . . . has a descending branch, which is marked by cirrus anvils with low cloud top temperatures.
   (d) . . . extends approximately up to 30° latitude in the annual mean, owing to the rotation of the Earth.
8. Cloud regimes . . .
   (a) . . . are often characterized by stability or small-scale subsidence.
   (b) . . . are often distinguished by the cloud base pressure, which can be retrieved from satellite data.
   (c) . . . associated with the highest instability are deep convective clouds.
   (d) . . . are for example marine stratocumulus, trade wind cumulus or polar stratus.
9. Climate engineering is discussed as a technical option to offset greenhouse gas warming. Which of the following statements about climate engineering is/are true?
   (a) Climate engineering involving aerosol particles can maintain a constant global mean surface temperature, but it decreases the global mean precipitation.
   (b) Climate engineering involving cirrus clouds affects both the shortwave and the longwave radiation.
   (c) Increasing the reflectivity of marine stratocumulus clouds is only possible for clean clouds in the southern hemisphere.
   (d) Limited climate engineering in high latitudes could save the Greenland and Antarctic ice sheets without side effects.

Long exercises

1. The aerosol optical depth (AOD) describes the vertically integrated amount of light extinguished due to the presence of aerosol particles.
   (a) Consider two monodisperse aerosol populations, of the same aerosol type, having the same number of aerosol particles but different radii. Which aerosol population has the higher AOD and why? What assumption do you have to make in order to answer this questions?
   (b) What AOD is necessary in order to have a transmission of solar radiation of 8%? Is such an AOD realistic?
   (c) Does doubling of the AOD value calculated above lead to a transmittance of 4%?
2. Climate sensitivity:
   (a) Explain the terms "radiative forcing" and "climate sensitivity".

Table 12.1 Greenhouse gas concentrations.

Greenhouse gas	1750 (pre-industrial)	2011 (present-day)
CO_2	278 ppm	391 ppm
CH_4	722 ppb	1803 ppb
N_2O	270 ppb	324 ppb

(b) A likely range of the equilibrium climate sensitivity is $\lambda = 0.8-2.5$ W m^{-2} K^{-1}. What does that mean for the future temperature change?

(c) Model estimates for climate sensitivity require a long integrating time, i.e. the model must calculate over several thousand years. Why is this necessary?

3. Radiative forcing:

(a) Calculate the direct radiative forcing RF of the greenhouse gases CO_2, CH_4 and N_2O using the following equations:

$$\delta \mathrm{RF}_{CO_2} = \alpha_1 \ln \frac{C}{C_0},$$
$$\delta \mathrm{RF}_{CH_4} = \alpha_2(\sqrt{M} - \sqrt{M_0}) - [f(M,N) - f(M_0,N)],$$
$$\delta \mathrm{RF}_{N_2O} = \alpha_3(\sqrt{N} - \sqrt{N_0}) - [f(M,N) - f(M,N_0)],$$
$$f(M,N) = 0.47 \ln[1 + 2.01 \times 10^{-5} \times (MN)^{0.75} + 5.31 \times 10^{-15} M(MN)^{1.52}].$$

Here C, M and N stand for the present-day, and C_0, M_0 and N_0 for the pre-industrial, concentrations of CO_2, CH_4 and N_2O, respectively. The concentrations, which have to be inserted in ppm for CO_2 and in ppb for CH_4 and N_2O, can be found in Table 12.1. Take $\alpha_1 = 5.35$, $\alpha_2 = 0.036$ and $\alpha_3 = 0.12$. (The equations and concentrations can be found in the supplementary material of Myhre *et al.* (2013).)

(b) Methane is partly responsible for the observed tropospheric ozone increase in recent decades. Furthermore, its reaction with the OH radical forms water vapor, e.g. in the upper troposphere. These effects are not included in the equations given above. Do you expect the actual radiative forcing of methane to be larger or smaller than the radiative forcing calculated in (a)? Give a short explanation.

4. Future temperature predictions are complicated by the fact that the temperature increase caused by enhanced greenhouse gas concentrations has feedbacks that affect the temperature. Explain two important temperature feedbacks.

5. Early climate models often described RFaci by a direct relationship between the aerosol mass and the cloud droplet number concentration.

(a) Use the following equations to estimate the cloud optical thickness of a continental stratus cloud (Boucher and Lohmann, 1995):

$$N_t = 10^{2.24+0.257 \cdot \log(m_{SO_4})},$$

$$r_v = \left(\frac{M_l \rho}{\frac{4}{3} \pi \rho_l N_c} \right)^{1/3},$$

$$r_e = 1.1 r_v,$$

$$\tau = \frac{3}{2} \frac{\text{LWP}}{r_e}.$$

Here N_c is the cloud droplet number concentration in cm^{-3}, m_{SO_4} is the aerosol sulfate mass concentration in $\mu g\ m^{-3}$, r_v is the mean volume cloud droplet radius, M_l is the liquid water content, ρ is the air density, ρ_l is the density of liquid water, r_e is the effective droplet radius, τ is the cloud optical depth and LWP is the liquid water path (i.e. the vertically integrated M_l value for the cloud). The sulfate mass concentration inside the cloud is 1 $\mu g\ m^{-3}$, the liquid water content of the cloud is 0.3 $g\ m^{-3}$ and the liquid water path of the cloud is 150 $g\ m^{-2}$. Assume the pressure and virtual temperature of the cloud to be 800 hPa and 278 K. In the last equation, the liquid water path and the equivalent radius must be inserted in $g\ m^{-2}$ and μm, respectively.

(b) Explain the problems when calculating the N_c values with this approach.

5. Assume that you inject aerosol particles into the stratosphere, resulting in an aerosol number density $N_0 = 10\ cm^{-3}$ at 15 km, which linearly decreases with altitude to 1 cm^{-3} at 25 km. Assume a monodisperse size distribution, so that all particles have radius 0.2 μm. If each particle acts like a mirror without shadowing other particles, what percentage of the incoming solar radiation would be reflected back to space by this aerosol layer?

References

Abbatt, J. P. D., Benz, S., Cziczo, D. J., Kanji, Z. A., Lohmann, U., and Mohler, O. 2006. Solid ammonium sulfate aerosols as ice nuclei: a pathway for cirrus cloud formation. *Science*, **313**, 1770–1773.

Ackerman, A. S., Kirkpatrick, M. P., Stevens, D. E., and Toon, O. B. 2004. The impact of humidity above stratiform clouds on indirect climate forcing. *Nature*, **432**, 1014–1017.

Ahrens, C. D. 2009. *Meteorology Today: An Introduction to Weather, Climate, and the Environment*. 9th edn. Brooks/Cole Cengage Learning.

Albrecht, B. 1989. Aerosols, cloud microphysics, and fractional cloudiness. *Science*, **245**, 1227–1230.

Andreae, M. O., and Rosenfeld, D. 2008. Aerosol–cloud-precipitation interactions. Part 1. The nature and sources of cloud-active aerosols. *Earth Sci. Rev.*, **89**, 13–41.

Andronache, C., Grönholm, T., Laakso, L., Phillips, V., and Venäläinen, A. 2006. Scavenging of ultrafine particles by rainfall at a boreal site: observations and model estimations. *Atmos. Chem. Phys.*, **6**(12), 4739–4754.

Ansmann, A., Tesche, M., Althausen, D., Müller, D., Seifert, P., Freudenthaler, V. *et al.* 2008. Influence of Saharan dust on cloud glaciation in southern Morocco during the Saharan Mineral Dust Experiment. *J. Geophys. Res.*, **113**, doi:10.1029/2004JD005000.

Atkinson, J. D., Murray, B. J., Woodhouse, M. T., Whale, T. F., Baustian, K. J., Carslaw, K. S. *et al.* 2013. The importance of feldspar for ice nucleation by mineral dust in mixed-phase clouds. *Nature*, **498**, 355–358.

Ayala, O., Rosa, B., Wang, L. P., and Grabowski, W. W. 2008. Effects of turbulence on the geometric collision rate of sedimenting droplets. Part 1. Results from direct numerical simulation. *New J. Phys.*, **10**, doi:10.1088/1367–2630/10/7/075015.

Baker, M., and Charlson, R. J. 1990. Bistability of CCN concentrations and thermodynamics in the cloud-topped boundary layer. *Nature*, **345**, 142–145.

Baron, P. A., and Willeke, K. 2001. *Aerosol Measurement: Principles, Techniques, and Applications*. 2nd edn. Wiley.

Baschek, B. 2005. Influence of updrafts and embedded convection on the microphysics of riming. Ph.D. thesis, ETH Zurich.

Beard, K. V. 1976. Terminal velocity and shape of cloud and precipitation drops aloft. *J. Atmos. Sci.*, **33**(5), 851–864.

Beard, K. V., and Pruppacher, H. R. 1971. A wind tunnel investigation of the rate of evaporation of small water drops falling at terminal velocity in air. *J. Atmos. Sci.*, **28**(8), 1455–1464.

Berry, E. X. 1967. Cloud droplet growth by collection. *J. Atmos. Sci.*, **24**(6).

Bjerknes, J., and Solberg, H. 1922. On the life cycle of cyclones and the polar front theory of atmospheric circulation. *Mon. Wea. Rev.*, **50**(9), 468–473.

Blanchard, D. O. 1998. Assessing the vertical distribution of convective available potential energy. *Weather and Forecasting*, **13**(3), 870–877.

Borovikov, A. M., Gaivoronskii, I. I., Zak, E. G., Kostarev, V. V., Mazin, I. P., Minervin, V. E. *et al.* 1963. Cloud physics. In: *Israel Program of Scientific Translations*. Jerusalem.

Boucher, O., and Lohmann, U. 1995. The sulfate-CCN-cloud albedo effect: a sensitivity study with two general circulation models. *Tellus B*, **47**, 281–300.

Boucher, O., Randall, D., Artaxo, P., Bretherton, C., Feingold, G., Forster, P. *et al.* 2013. Clouds and aerosols. Pp. 571–657 in Stocker, T., Qin, D., Plattner, G.-K., Tignor, M., Allen, S. K., Boschung, J. *et al.* (eds.), *Climate Change 2013: The Physical Science Basis. Contribution of Working Group I to the Fifth Assessment Report of the Intergovernmental Panel on Climate Change*. Cambridge University Press.

Boy, M., and Kulmala, M. 2002. Nucleation events in the continental boundary layer: influence of physical and meteorological parameters. *Atmos. Chem. Phys.*, **2**, 1–16.

Brazier-Smith, P. R., Jennings, S. G., and Latham, J. 1973. Raindrop interactions and rainfall rates within clouds. *Q. J. R. Meteorol. Soc.*, **99**(420), 260–272.

Brenguier, J.-L., Pawlowska, H., Schüller, L., Preusker, R., Fischer, J., and Fouquart, Y. 2000. Radiative properties of boundary layer clouds: droplet effective radius versus number concentration. *J. Atmos. Sci.*, **57**, 803–821.

Bretl, S., Reutter, P., Raible, C. C., Ferrachat, S., Poberaj, C. S., Revell, L. E. *et al.* 2015. The influence of absorbed solar radiation by Saharan dust on hurricane genesis. *J. Geophys. Res.*, **120**(5), 1902–1917.

Broekhuizen, K., Kumar, P. P., and Abbatt, J. P. D. 2004. Partially soluble organics as cloud condensation nuclei: role of trace soluble and surface active species. *Geophys. Res. Lett.*, **31**, doi: 10.1029/2003GL018203.

Browning, K. A. 1964. Airflow and precipitation trajectories within severe local storms which travel to the right of the winds. *J. Atmos. Sci.*, **21**(6), 634–639.

Browning, K. A. 1985. Conceptual models of precipitation systems. *Meteorol. Mag.*, **114**(1359), 293–319.

Carlson, T. N. 1980. Air-flow through mid-latitude cyclones and the comma cloud pattern. *Mon. Weat. Rev.*, **108**(10), 1498–1509.

Chepfer, H., Bony, S., Winker, D., Chiriaco, M., Dufresne, J. L., and Seze, G. 2008. Use of CALIPSO lidar observations to evaluate the cloudiness simulated by a climate model. *Geophys. Res. Lett.*, **35**, L15704.

Chepfer, H., Bony, S., Winker, D., Cesana, G., Dufresne, J. L., Minnis, P. *et al.* 2010. The GCM-oriented CALIPSO cloud product (CALIPSO-GOCCP). *J. Geophys. Res.*, **115**, D00H16.

Collins, M., Knutti, R., Arblaster, J., Dufresne, J.-L., Fichefet, T., Friedlingstein, P. *et al.* 2013. Long-term climate change: projections, commitments and irreversibility. Pp. 1029–1136 in Stocker, T., Qin, D., Plattner, G.-K., Tignor, M., Allen, S. K., Boschung, J. *et al.* (eds.), *Climate Change 2013: The Physical Science Basis. Contribution of Working Group I to the Fifth Assessment Report of the Intergovernmental Panel on Climate Change*. Cambridge University Press.

Corti, T., and Peter, T. 2009. A simple model for cloud radiative forcing. *Atmos. Chem. Phys.*, **9**(15), 5751–5758.

Cotton, W. R., Bryan, G. H., and van den Heever, S. C. 2011. *Storm and Cloud Dynamics*. Academic Press.

Croft, B., Lohmann, U., Martin, R. V., Stier, P., Wurzler, S., Feichter, J. *et al.* 2009. Aerosol size-dependent below-cloud scavenging by rain and snow in the ECHAM5-HAM. *Atmos. Chem. Phys.*, **9**, 4653–4675.

Crosier, J., Bower, K. N., Choularton, T. W., Westbrook, C. D., Connolly, P. J., Cui, Z. Q. *et al.* 2011. Observations of ice multiplication in a weakly convective cell embedded in supercooled mid-level stratus. *Atmos. Chem. Phys.*, **11**(1), 257–273.

Crutzen, P. J. 2006. Albedo enhancement by stratospheric sulfur injections: a contribution to resolve a policy dilemma? *Climatic Change*, **77**(3–4), 211–219.

Curry, J. A., and Webster, P. J. 1999. *Thermodynamics of Atmospheres and Oceans*. Academic Press.

Cziczo, D. J., Froyd, K. D., Hoose, C., Jensen, E. J., Diao, M. H., Zondlo, M. A. *et al.* 2013. Clarifying the dominant sources and mechanisms of cirrus cloud formation. *Science*, **340**(6138), 1320–1324.

De Mott, P. J. 1995. Quantitative descriptions of ice formation mechanisms of silver iodide-type aerosols. *Atmos. Res.*, **38**, 63–99.

Dee, D. P., Uppala, S. M., Simmons, A. J., Berrisford, P., Poli, P., Kobayashi, S. *et al.* 2011. The ERA-Interim reanalysis: configuration and performance of the data assimilation system. *Q. J. R. Meteorol. Soc.*, **137**(656), 553–597.

Deierling, W., and Petersen, W. A. 2008. Total lightning activity as an indicator of updraft characteristics. *J. Geophys. Res.*, **113**(D16), D16210.

Denman, K., Brasseur, G., Chidthaisong, A., Ciais, P., Cox, P., Dickinson, R. *et al.* 2007. Couplings between changes in the climate system and biogeochemistry. Pp. 499–588 in Solomon, S., Qin, D., Manning, M., Chen, Z., Marquis, M., Averyt, K. B. *et al.* (eds.), *Climate Change 2007: The Physical Science Basis. Contribution of Working Group I to the Fourth Assessment Report of the Intergovernmental Panel on Climate Change.* Cambridge University Press.

Dentener, F., Kinne, S., Bond, T., Boucher, O., Cofala, J., Generoso, S. *et al.* 2006. Emissions of primary aerosol and precursor gases in the years 2000 and 1750 prescribed data-sets for AeroCom. *Atmos. Chem. Phys.*, **6**, 4321–4344.

Detwiler, A. G., and Vonnegut, B. 1981. Humidity required for ice nucleation from the vapor onto silver-iodide and lead iodide aerosols over the temperature range −6 to −67 degrees C. *J. Appl. Meteorol.*, **20**(9), 1006–1012.

Doswell, Charles A. III. 1985. The operational meteorology of convective weather: storm scale analysis. Report, National Oceanic and Atmospheric Administration (NOAA).

Doviak, R. J., and Zrnic, D. S. 1984. *Doppler Radar and Weather Observations*. Academic Press.

Drobinski, P., Steinacker, R., Richner, H., Baumann-Stanzer, K., Beffrey, G., Benech, B. *et al.* 2007. Föhn in the Rhine Valley during MAP: a review of its multiscale dynamics in complex valley geometry. *Q. J. Royal Meteorol. Soc.*, **133**, 897–916. 625.

Dubovik, O., Holben, B. N., Eck, T. F., Smirnov, A., Kaufman, Y. J., King, M. D. *et al.* 2002. Variability of absorption and optical properties of key aerosol types observed in worldwide locations. *J. Atmos. Sci.*, **59**, 590–608.

Dufresne, J.-L., and Bony, S. 2008. An assessment of the primary sources of spread of global warming estimates from coupled atmosphere–ocean models. *J. Climate*, **21**(19), 5135–5144.

Durant, A. J., and Shaw, R. A. 2005. Evaporation freezing by contact nucleation inside-out. *Geophys. Res. Lett.*, **32**, doi:10.1029/2005GL024175.

Durkee, P. A., Noone, K. J., Ferek, R. J., Johnson, D. W., Taylor, J. P., Garrett, T. J. *et al.* 2000. The impact of ship-produced aerosols on the microstructure and albedo of warm marine stratocumulus clouds: a test of MAST hypothesis 1i and 1ii. *J. Atmos. Sci.*, **57**, 2554–2569.

Eastman, R., and Warren, S. G. 2013. A 39-yr survey of cloud changes from land stations worldwide 1971–2009: long-term trends, relation to aerosols, and expansion of the tropical belt. *J. Climate*, **26**, 1286–1303.

Eastman, R., Warren, S. G., and Hahn, C. J. 2011. Variations in cloud cover and cloud types over the ocean from surface observations, 1954–2008. *J. Climate*, **24**(22), 5914–5934.

Ehn, M., Thornton, J. A., Kleist, E., Sipila, M., Junninen, H., Pullinen, I. *et al.* 2014. A large source of low-volatility secondary organic aerosol. *Nature*, **506**(7489), 476–479.

Emanuel, K. A. 1986. An air sea interaction theory for tropical cyclones: 1. steady-state maintenance. *J. Atmos. Sci.*, **43**, 586–604.

Emanuel, K. A. 1994. *Atmospheric Convection*. Oxford University Press.
Emanuel, K. A. 2005. *Divine Winds*. Cambridge University Press.
Ervens, B., and Feingold, G. 2013. Sensitivities of immersion freezing: reconciling classical nucleation theory and deterministic expressions. *Geophys. Res. Lett.*, **40**(12), 3320–3324.
Evans, L. F. 1965. Requirements of an ice nucleus. *Nature*, **206**(4986), 822.
Federer, B., Waldvogel, A., Schmid, W., Schiesser, H. H., Hampel, F., Schweingruber, M. *et al.* 1986. Main results of Grossversuch-IV. *J. Clim. Appl. Meteorol.*, **25**(7), 917–957.
Feingold, G., Cotton, W. R., Kreidenweis, S. M., and Davis, J. T. 1999. The impact of giant cloud condensation nuclei on drizzle formation in stratocumulus: implications for cloud radiative properties. *J. Atmos. Sci.*, **56**, 4100–4117.
Feingold, G., Koren, I., Yamaguchi, T., and Kazil, J. 2015. On the reversibility of transitions between closed and open cellular convection. *Atmos. Chem. Phys.*, **15**(13), 7351–7367.
Field, P. R., and Heymsfield, A. J. 2015. Importance of snow to global precipitation. *Geophys. Res. Lett.*, **42**, 9512–9520.
Fletcher, N. H. 1962. *Physics of Rain Clouds*. Cambridge University Press.
Forster, P., Ramaswamy, V., Artaxo, P., Berntsen, T., Betts, R., Fahey, D. W. *et al.* 2007. Radiative forcing of climate change. Pp. 129–234 in Solomon, S., Qin, D., Manning, M., Chen, Z., Marquis, M., Averyt, K. B. *et al.* (eds.), *Climate Change 2007: The Physical Science Basis. Contribution of Working Group I to the Fourth Assessment Report of the Intergovernmental Panel on Climate Change*. Cambridge University Press.
Garcia-Garcia, F., and List, R. 1992. Laboratory measurements and parameterizations of supercooled water skin temperatures and bulk properties of gyrating hailstones. *J. Atmos. Sci.*, **49**(22), 2058–2073.
Greenfield, S. M. 1957. Rain scavenging of radioactive particulate matter from the atmosphere. *J. Meteor.*, **14**(2), 115–125.
Gunn, K. L. S., and Marshall, J. S. 1958. The distribution with size of aggregate snowflakes. *J. Meteorol.*, **15**(5), 452–461.
Guo, S., Bluth, G. J. S., Rose, W. I., Watson, I. M., and Prata, A. J. 2004. Re-evaluation of SO_2 release of the 15 June 1991 Pinatubo eruption using ultraviolet and infrared satellite sensors. *Geochem., Geophys., Geosyst.*, **5**(4).
Hahn, C. J., and Warren, S. G. 2009. Extended edited cloud reports from ships and land stations over the globe, 1952–1996 (2009 update). Techical reptort, Carbon Dioxide Information Analysis Center Numerical Data Package NDP-026C.
Hale, B. N., and Plummer, P. L. M. 1974. Molecular model for ice nucleation in a supersaturated vapor. *J. Chem. Phys.*, **61**(10), 4012–4019.
Hall, W. D., and Pruppacher, H. R. 1976. The survival of ice particles falling from cirrus clouds in subsaturated air. *J. Atmos. Sci.*, **33**(10), 1995–2006.
Hallett, J., and Mossop, S. C. 1974. Production of secondary ice particles during riming process. *Nature*, **249**, 26–28.
Hartmann, D. L., Klein Tank, A. M. G., Rusticucci, M., Alexander, L. V., Brönnimann, B., Charabi, Y. *et al.* 2013. Observations: atmosphere and surface. Pp. 159–254 in Stocker, T., Qin, D., Plattner, G.-K., Tignor, M., Allen, S. K., Boschung, J. (eds.), *Climate Change 2013: The Physical Science Basis. Contribution of Working Group I to the Fifth Assessment Report of the Intergovernmental Panel on Climate Change*. Cambridge University Press.
Heckendorn, P., Weisenstein, D., Fueglistaler, S., Luo, B. P., Rozanov, E., Schraner, M. *et al.* 2009. The impact of geoengineering aerosols on stratospheric temperature and ozone. *Env. Res. Lett.*, **4**(4).

Heintzenberg, J., Covert, D. C., and Van Dingenen, R. 2000. Size distribution and chemical composition of marine aerosols: a compilation and review. *Tellus, B*, **52**, 1104–1122.

Henneberger, J., Fugal, J. P., Stetzer, O., and Lohmann, U. 2013. HOLIMO II: a digital holographic instrument for ground-based in situ observations of microphysical properties of mixed-phase clouds. *Atmos. Meas. Tech.*, **6**.

Hess, S. L. 1959. *Introduction to Theoretical Meteorology*. New York: Henry Holt.

Hinrichs, G. 1888. Tornadoes and derechoes. *Amer. Meteorolog. J.*, **5**, 306.

Herich, H., Tritscher, T., Wiacek, A., Gysel, M., Weingartner, E., Lohmann, U. *et al.* 2009. Water uptake of clay and desert dust aerosol particles at sub- and supersaturated water vapor conditions. *Phys. Chem. Chem. Phys.*, **11**, 7804–7809.

Hinds, William C. 1999. *Aerosol Technology: Properties, Behavior, and Measurement of Airborne Particles*. 2nd edn. Wiley.

Holton, J. R. 2004. *An Introduction to Dynamic Meteorology*. Elsevier Academic Press.

Hoose, C., and Möhler, O. 2012. Heterogeneous ice nucleation on atmospheric aerosols: a review of results from laboratory experiments. *Atmos. Chem. Phys.*, **12**, 9817–9854.

Hoose, C., Lohmann, U., Erdin, R., and Tegen, I. 2008. Global influence of dust mineralogical composition on heterogeneous ice nucleation in mixed-phase clouds. *Environ. Res. Lett.*, **3**, doi:10.1088/1748–9326/3/2/025003.

Hoose, C., Kristjansson, J. E., Chen, J. P., and Hazra, A. 2010. A classical-theory-based parameterization of heterogeneous ice nucleation by mineral dust, soot, and biological particles in a global climate model. *J. Atmos. Sci.*, **67**(8), 2483–2503.

Houze, R. A. 1981. Structures of atmospheric precipitation systems – a global survey. *Radio Science*, **16**(5), 671–689.

Houze, Robert A. 1993. *Cloud Dynamics*. Academic Press.

Houze, Robert A. 2014. *Cloud Dynamics*. 2nd edn. Academic Press.

Howard, Luke. 1803. LXIV. On the modifications of clouds, and on the principles of their production, suspension, and destruction; being the substance of an essay read before the Askesian Society in the session 1802–3. *Philosophical Magazine Series 1*, **16**(64), 344–357.

Hoyle, C. R., Pinti, V., Welti, A., Zobrist, B., Marcolli, C., Luo, B. *et al.* 2011. Ice nucleation properties of volcanic ash from Eyjafjallajökull. *Atmos. Chem. Phys.*, **11**(18), 9911–9926.

Ickes, L., Welti, A., Hoose, C., and Lohmann, U. 2015. Classical nucleation theory of homogeneous freezing of water: thermodynamic and kinetic parameters. *Phys. Chem. Chem. Phys.*, **17**(8), 5514–5537.

Intrieri, J. M., and Shupe, M. D. 2004. Characteristics and radiative effects of diamond dust over the western Arctic Ocean region. *J. Clim.*, **17**(15), 2953–2960.

Intrieri, J. M., Shupe, M. D., Uttal, T., and McCarty, B. J. 2002. An annual cycle of Arctic cloud characteristics observed by radar and lidar at SHEBA. *J. Geophys. Res.*, **107**, doi: 10.1029/2000JC000423.

Iribane, J. V., and Godson, W. L. 1981. *Atmospheric Thermodynamics*. Kluwer.

Jaenicke, R. 1988. Aerosol physics and chemistry. Pp. 391–457 in *Landolt-Bornstein, New Ser.*, vol. V/4b. Springer-Verlag.

Jeske, H. 1988. *Special Optical Phenomena Produced by Water Droplets and Ice Crystals in the Atmosphere.* Vol. V, Geophysics 4b. Springer Verlag.

Jiang, H. L., Xue, H. W., Teller, A., Feingold, G., and Levin, Z. 2006. Aerosol effects on the lifetime of shallow cumulus. *Geophys. Res. Lett.*, **33**, doi: 10.1029/2006GL026024. L14806.

Johns, R. H., and Hirt, W. D. 1987. Derechos: widespread convectively induced windstorms. *Weather and Forecasting*, **2**, 32–49.

Johnson, R. H., Rickenbach, T. M., Rutledge, S. A., Ciesielski, P. E., and Schubert, W. H. 1999. Trimodal characteristics of tropical convection. *J. Clim.*, **12**(8), 2397–2418. Part 1.

Jones, A., Haywood, J. M., Alterskjaer, K., Boucher, O., Cole, J. N. S., Curry, C. L. *et al.* 2013. The impact of abrupt suspension of solar radiation management (termination effect) in experiment G2 of the Geoengineering Model Intercomparison Project (GeoMIP). *J. Geophys. Res.*, **118**(17), 9743–9752.

Jorgensen, D. P., and Lemone, M. A. 1989. Vertical velocity characteristics of oceanic convection. *J. Atmos. Sci.*, **46**(5), 621–640.

Joss, J., and Waldvogel, A. 1967. A spectrograph for raindrops with automatic interpretation. *Pure Appl. Geophys.*, **68**(3), 240–246.

Kadoya, K., Matsunaga, N., and Nagashima, A. 1985. Viscosity and thermal-conductivity of dry air in the gaseous-phase. *J. Phys. Chem. Ref. Data*, **14**(4), 947–970.

Kajava, A. V., and Lindow, S. E. 1993. A model of the 3-dimensional structure of ice nucleation proteins. *J. Mol. Biol.*, **232**(3), 709–717.

Kanji, Z. A., and Abbatt, J. P. D. 2006. Laboratory studies of ice formation via deposition mode nucleation onto mineral dust and n-hexane soot samples. *J. Geophys. Res.*, **111**(D16).

Kanji, Z. A., Welti, A., Chou, C., Stetzer, O., and Lohmann, U. 2013. Laboratory studies of immersion and deposition mode ice nucleation of ozone aged mineral dust particles. *Atmos. Chem. Phys. Discuss.*, **13**(4), 8701–8767.

Kay, J. E., and Gettelman, A. 2009. Cloud influence on and response to seasonal Arctic sea ice loss. *J. Geophys. Res.*, **114**, D18204.

Khvorostyanov, V. I., and Curry, J. A. 2014. *Thermodynamics, Kinetics and Microphysics of Clouds*. Cambridge University Press.

Kiehl, J. T. 1992. Atmospheric general circulation modeling. Pp. 319–370 in Trenberth, K. (ed.), *Climate System Modeling*. Cambridge University Press.

Kiehl, J. T., and Trenberth, Kevin E. 1997. Earth's annual global mean energy budget. *Bull. Amer. Meteorol. Soc.*, **78**, 197–208.

Kinne, S., Schulz, M., Textor, C., Guibert, S., Balkanski, Y., Bauer, S. *et al.* 2006. An AeroCom initial assessment – optical properties in aerosol component modules of global models. *Atmos. Chem. Phys.*, **6**, 1815–1834.

Kinzer, Gilbert D., and Gunn, Ross. 1951. The evaporation, temperature and thermal relaxation-time of freely falling waterdrops. *J. Meteorology*, **8**(2), 71–83.

Kirkby, J. *et al.* 2011. Role of sulphuric acid, ammonia and galactic cosmic rays in atmospheric aerosol nucleation. *Nature*, **476**.

Klemp, J. B. 1987. Dynamics of tornadic thunderstorms. *Ann. Rev. Fluid Mech.*, **19**, 369–402.

Köhler, H. 1922. Zur Kondensation des Wasserdampfes in der Atmosphäre. *Geofysiske Publ.*, **2**(6).

Koop, T., Luo, B., Tsias, A., and Peter, T. 2000. Water activity as the determinant for homogeneous ice nucleation in aqueous solutions. *Nature*, **406**, 611–614.

Korolev, A. V., Isaac, G. A., Cober, S. G., Strapp, W., and Hallett, J. 2003. Microphysical characterization of mixed-phase clouds. *Q. J. R. Meteorol. Soc.*, **129**, 39–65.

Kossin, J. P., Knapp, K. R., Vimont, D. J., Murnane, R. J., and Harper, B. A. 2007. A globally consistent reanalysis of hurricane variability and trends. *Geophys. Res. Lett.*, **34**(4).

Krämer, M., Schiller, C., Afchine, A., Bauer, R., Gensch, I., Mangold, A. *et al.* 2009. Ice supersaturations and cirrus cloud crystal numbers. *Atmos. Chem. Phys.*, **9**(11), 3505–3522.

Kravitz, B., Robock, A., Boucher, O., Schmidt, H., Taylor, K. E., Stenchikov, G. *et al.* 2011. The Geoengineering Model Intercomparison Project (GeoMIP). *Atmos. Sci. Lett.*, **12**(2), 162–167.

Kravitz, B., Robock, A., Tilmes, S., Boucher, O., English, J. M., Irvine, P. J. *et al.* 2015. The Geoengineering Model Intercomparison Project Phase 6 (GeoMIP6): simulation design and preliminary results. *Geosci. Model Dev. Discuss.*, **8**(6), 4697–4736.

Kubar, T. L., Hartmann, D. L., and Wood, R. 2009. Understanding the importance of microphysics and macrophysics for warm rain in marine low clouds. Part I: satellite observations. *J. Atmos. Sci.*, **66**(10), 2953–2972.

Lacis, A., Hansen, J., and Sato, M. 1992. Climate forcing by stratospheric aerosols. *Geophys. Res. Lett.*, **19**(15), 1607–1610.

Ladino, L., Stetzer, O., Luond, F., Welti, A., and Lohmann, U. 2011. Contact freezing experiments of kaolinite particles with cloud droplets. *J. Geophys. Res.*, **116**, doi:10.1029/2011JD015727.

Ladino Moreno, L. A., Stetzer, O., and Lohmann, U. 2013. Contact freezing: a review of experimental studies. *Atmos. Chem. Phys.*, **13**, 9745–9769.

Lamarck, J. B. 1802. Sur la forme des nuages. *Annuaire Météorologique pour l'an XIV de l'ère de la République Française*, **No. 3** (Un systême général de météorologie), 149–164.

Lamb, D., and Verlinde, J. 2011. *Physics and Chemistry of Clouds*. Cambridge University Press.

Latham, J. 1990. Control of global warming. *Nature*, **347**(6291), 339–340.

Lau, K. M., and Wu, H. T. 2003. Warm rain processes over tropical oceans and climate implications. *Geophys. Res. Lett.*, **30**, doi: 10.1029/2003GL018567.

Lauer, A., Eyring, V., Hendricks, J., Jockel, P., and Lohmann, U. 2007. Global model simulations of the impact of ocean-going ships on aerosols, clouds, and the radiation budget. *Atmos. Chem. Phys.*, **7**, 5061–5079.

Lehmiller, G. S., Bluestein, H. B., Neiman, P. J., Ralph, F. M., and Feltz, W. F. 2001. Wind structure in a supercell thunderstorm as measured by a UHF wind profiler. *Mon. Wea. Rev.*, **129**(8), 1968–1986.

Lemmon, E. W. 2015. *Thermophysical Properties of Water and Steam. Handbook of Chemistry and Physics*, 96th edn. CRC Press. Book Section 6: Fluid properties.

Li, Y., and Somorjai, G. A. 2007. Surface premelting of ice. *J. Phys. Chem. C*, **111**(27), 9631–9637.

Libbrecht, K. G. 2005. The physics of snow crystals. *Rep. Progr. Phys.*, **68**(4), 855–895.

Liu, Y., Geerts, B., Miller, M., Daum, P., and McGraw, R. 2008. Threshold radar reflectivity for drizzling clouds. *Geophys. Res. Lett.*, **35**(3).

Löffler-Mang, M., and Blahak, U. 2001. Estimation of the equivalent radar reflectivity factor from measured snow size spectra. *J. Appl. Meteorol.*, **40**(4), 843–849.

Lohmann, U. 2002. A glaciation indirect aerosol effect caused by soot aerosols. *Geophys. Res. Lett.*, **29**, doi: 10.1029/2001GL014357.

Lohmann, U., and Diehl, K. 2006. Sensitivity studies of the importance of dust ice nuclei for the indirect aerosol effect on stratiform mixed-phase clouds. *J. Atmos. Sci*, **63**, 968–982.

Lohmann, U., and Feichter, J. 2005. Global indirect aerosol effects: a review. *Atmos. Chem. Phys.*, **5**, 715–737.

Lohmann, U., Spichtinger, P., Jess, S., Peter, T., and Smit, H. 2008. Cirrus cloud formation and ice supersaturated regions in a global climate model. *Env. Res. Lett.*, **3**(045022), doi:10.1088/1748-9326/3/4/045022.

Low, T. B., and List, R. 1982. Collision, coalescence and breakup of raindrops. 1. Experimentally established coalescence efficiencies and fragment size distributions in breakup. *J. Atmos. Sci.*, **39**(7), 1591–1606.

Lüönd, F., Stetzer, O., Welti, A., and Lohmann, U. 2010. Experimental study on the ice nucleation ability of size-selected kaolinite particles in the immersion mode. *J. Geophys. Res.*, **115**, doi: 10.1029/2009jd012959.

Mace, G. G., Zhang, Q. Q., Vaughan, M., Marchand, R., Stephens, G., Trepte, C. *et al.* 2009. A description of hydrometeor layer occurrence statistics derived from the first year of merged Cloudsat and CALIPSO data. *J. Geophys. Res.*, **114**, D00A26.

Madonna, E. 2009. Cloud condensation nuclei measurements of urban aerosols. M.Phil. thesis, ETH Zurich.

Madonna, E., Wernli, H., Joos, H., and Martius, O. 2014. Warm conveyor belts in the ERA-interim dataset (1979–2010). Part I: Climatology and potential vorticity evolution. *J. Climate*, **27**, 3–26.

Magee, N., Moyle, A. M., and Lamb, D. 2006. Experimental determination of the deposition coefficient of small cirrus-like ice crystals near 50 degrees C. *Geophys. Res. Lett.*, **33**, L17813.

Marcolli, C. 2014. Deposition nucleation viewed as homogeneous or immersion freezing in pores and cavities. *Atmos. Chem. Phys.*, **14**, 2071–2104.

Marcolli, C., Gedamke, S., Peter, T., and Zobrist, B. 2007. Efficiency of immersion mode ice nucleation on surrogates of mineral dust. *Atmos. Chem. Phys.*, **7**, 5081–5091.

Markowski, P. M. 2002. Hook echoes and rear-flank downdrafts: a review. *Mon. Weat. Rev.*, **130**(4), 852–876.

Markowski, P., and Richardson, Y. 2010. *Mesoscale Meteorology in Midlatitudes*. John Wiley & Sons.

Marshall, J. S., and Palmer, W. M. 1948. The distribution of raindrops with size. *J. Meteorol.*, **5**(4), 165–166.

Martin, M., Chang, R. Y.-W., Sierau, B., Sjogren, S., Swietlicki, E., Abbatt, J. P. D. *et al.* 2011. Cloud condensation nuclei closure study on summer arctic aerosol. *Atmos. Chem. Phys.*, **11**, 11335–11350.

Mason, B. J. 1971. *The Physics of Clouds*. Clarendon Press.

Menon, S., Hansen, J., Nazarenko, L., and Luo, Y. 2002. Climate effects of black carbon aerosols in China and India. *Science*, **297**, 2250–2253.

Miles, N. L., Verlinde, J., and Clothiaux, E. E. 2000. Cloud droplet size distributions in low-level stratiform clouds. *J. Atmos. Sci.*, **57**(2), 295–311.

Mitchell, D. L. 1996. Use of mass- and area-dimensional power laws for determining precipitation particle terminal velocities. *J. Atmos. Sci.*, **53**(12), 1710–1723.

Mitchell, D. L., and Heymsfield, Andrew J. 2005. Refinements in the treatment of ice particle terminal velocities, highlighting aggregates. *J. Atmos. Sci.*, **62**, 1637–1664.

Mitchell, D. L., and Finnegan, W. 2009. Modification of cirrus clouds to reduce global warming. *Env. Res. Lett.*, **4**(4).

Möhler, O., Field, P. R., Connolly, P., Benz, S., Saathoff, H., Schnaiter, M. *et al.* 2006. Efficiency of the deposition mode ice nucleation on mineral dust particles. *Atmos. Chem. Phys.*, **6**, 3007–3021.

Moller, U., and Schumann, G. 1970. Mechanisms of transport from atmosphere to Earth's surface. *J. Geophys. Res.*, **75**(15), 3013.

Mossop, S. C. 1985. The origin and concentration of ice crystals in clouds. *Bull. Amer. Meteorol. Soc.*, **66**(3), 264–273.

Mülmenstädt, J., Sourdeval, O., Delanoe, J., and Quaas, J. 2015. Frequency of occurrence of rain from liquid-, mixed- and ice-phase clouds derived from A-Train satellite retrievals. *Geophys. Res. Lett.*, **42**.

Murphy, D. M., and Koop, T. 2005. Review of the vapour pressures of ice and supercooled water for atmospheric applications. *Q. J. R. Meteorol. Soc.*, **131**(608), 1539–1565.

Murphy, D. M., Solomon, S., Portmann, R. W., Rosenlof, K. H., Forster, P. M. de F., and Wong, T. 2009. An observationally based energy balance for the Earth since 1950. *J. Geophys. Res.*, **114**, doi:10.1029/2009JD012105. D17107.

Murray, B. J., Broadley, S. L., Wilson, T. W., Atkinson, J. D., and Wills, R. H. 2011. Heterogeneous freezing of water droplets containing kaolinite particles. *Atmos. Chem. Phys.*, **11**, 4191–4207.

Murray, B. J., O'Sullivan, D., Atkinson, J. D., and Webb, M. E. 2012. Ice nucleation by particles immersed in supercooled cloud droplets. *Chem. Soc. Rev.*, **41**, 6519–6554.

Myhre, G., Highwood, E. J., Shine, K. P., and Stordal, F. 1998. New estimates of radiative forcing due to well mixed greenhouse gases. *Geophys. Res. Lett.*, **25**, 2715–2718.

Myhre, G., Samset, B. H., Schulz, M., Balkanski, Y., Bauer, S., Berntsen, T. K. *et al.* 2013. Radiative forcing of the direct aerosol effect from AeroCom Phase II simulations. *Atmos. Chem. Phys.*, **13**(4), 1853–1877.

Nagare, B., Marcolli, C., Stetzer, O., and Lohmann, U. 2015. Estimating collision efficiencies from contact freezing experiments. *Atmos. Chem. Phys. Discuss.*, **15**(8), 12167–12212.

Neubauer, D. 2009. Modellierung des indirekten Strahlungseffekts des Hintergrundaerosols in Österreich. Ph.D. thesis, University of Vienna.

Neubauer, D., Lohmann, U., Hoose, C., and Frontoso, M. G. 2014. Impact of the representation of marine stratocumulus clouds on the anthropogenic aerosol effect. *Atmos. Chem. Phys.*, **14**, 11997–12022.

Niedermeier, D., Shaw, R. A., Hartmann, S., Wex, H., Clauss, T., Voigtländer, J. *et al.* 2011. Heterogeneous ice nucleation: exploring the transition from stochastic to singular freezing behavior. *Atmos. Chem. Phys.*, **11**(16), 8767–8775.

Niemeier, U., Schmidt, H., and Timmreck, C. 2011. The dependency of geoengineered sulfate aerosol on the emission strategy. *Atmos. Sci. Lett.*, **12**, 189–194.

North, G. R., and Erukhimova, T. L. 2009. *Atmospheric Thermodynamics*. Cambridge University Press.

Orlanski, I. 1975. Rational subdivision of scales for atmospheric processes. *Bull. Amer. Meteorol. Soc.*, **56**(5), 527–530.

Ostwald, W. 1897. Studien über die Bildung und Umwandlung fester Körper. 1. Abhandlung: Übersättigung und Überkaltung. *Z. Phys. C.*, **22**, 289–330.

Paramonov, M., Kerminen, V.-M., Gysel, M., Aalto, P. P., Andreae, M. O., Asmi, E. *et al.* 2015. A synthesis of cloud condensation nuclei counter (CCNC) measurements within the EUCAARI network. *Atmos. Chem. Phys.*, **15**, 12211–12229.

Park, S. H., Jung, C. H., Jung, K. R., Lee, B. K., and Lee, K. W. 2005. Wet scrubbing of polydisperse aerosols by freely falling droplets. *J. Aerosol Sci.*, **36**(12), 1444–1458.

Peng, Y., Lohmann, U., Leaitch, R., Banic, C., and Couture, M. 2002. The cloud albedo–cloud droplet effective radius relationship for clean and polluted clouds from RACE and FIRE.ACE. *J. Geophys. Res.*, **107**, doi: 10.029/2000JD000281.

Penner, J. E., Lister, D. H., Griggs, D. J., Dokken, D. J., and McFarland, M. (ed.), 1999. Aviation and the global atmosphere. A special report of working group III of the Intergovernmental Panel on Climate Change. Cambridge University Press.

Peters, G., and Görsdorf, U. 2010. Wolkenradar – Prinzipien und Messungen. *Promet.*, **36**, 144–153.

Peters, K., Stier, P., Quaas, J., and Grassl, H. 2012. Aerosol indirect effects from shipping emissions: sensitivity studies with the global aerosol-climate model ECHAM-HAM. *Atmos. Chem. Phys.*, **12**(13), 5985–6007.

Petters, M. D., and Kreidenweis, S. M. 2007. A single parameter representation of hygroscopic growth and cloud condensation nucleus activity. *Atmos. Chem. Phys.*, **7**, 1961–1971.

Petzold, A., Ogren, J. A., Fiebig, M., Laj, P., Li, S.-M., Baltensperger, U. *et al.* 2013. Recommendations for reporting "black carbon" measurements. *Atmos. Chem. Phys.*, **13**(16), 8365–8379.

Pinti, V., Marcolli, C., Zobrist, B., Hoyle, C. R., and Peter, T. 2012. Ice nucleation efficiency of clay minerals in the immersion mode. *Atmos. Chem. Phys.*, **12**, 5859–5878.

Pokharel, B., Geerts, B., Jing, X., Friedrich, K., Aikins, J., Breed, D. *et al.* 2014. The impact of ground-based glaciogenic seeding on clouds and precipitation over mountains: a multi-sensor case study of shallow precipitating orographic cumuli. *Atmos. Res.*, **147–148**, 162–182.

Pringle, K. J., Carslaw, K. S., Fan, T., Mann, G. W., Hill, A., Stier, P. *et al.* 2012. A multi-model assessment of the impact of sea spray geoengineering on cloud droplet number. *Atmos. Chem. Phys.*, **12**(23), 11647–11663.

Pruppacher, H. R., and Beard, K. V. 1970. A wind tunnel investigation of the internal circulation and shape of water drops falling at terminal velocity in air. *Q. J. R. Meteorol. Soc.*, **96**(408), 247–256.

Pruppacher, H. R., and Jaenicke, R. 1995. The processing of water-vapor and aerosols by atmospheric clouds, a global estimate. *Atmos. Res.*, **38**, 283–295.

Pruppacher, H. R., and Klett, J. D. 1978. *Microphysics of Clouds and Precipitation*. D. Reidel, Hingham, Massachusetts.

Pruppacher, H. R., and Klett, J. D. 1997. *Microphysics of Clouds and Precipitation*. Kluwer Academic.

Pruppacher, H. R., and Rasmussen, R. 1979. A wind tunnel investigation of the rate of evaporation of large water drops falling at terminal velocity in air. *J. Atmos. Sci.*, **36**(7), 1255–1260.

Quante, M. 2004. The role of clouds in the climate system. *J. Phys. Iv*, **121**, 61–86.

Ramanathan, V., Cess, R. D., Harrison, E. F., Minnis, P., Barkstrom, B. R., E., Ahmad, *et al.* 1989. Cloud–radiative forcing and climate: results from the Earth radiation budget experiment. *Science*, **243**, 57–63.

Rangno, A. L. 2002. Clouds classification. Pp. 467–475 in Holton, J. R., Curry, J. A., and Pyle, J. A. (eds.), *Encyclopedia of Atmospheric Sciences*. Academic Press.

Rangno, A. L. 2015. Classification of clouds. Pp. 141–160 in North, G. R., Pyle, J. A., and Zhang, F. (eds.), *Encyclopedia of Atmospheric Sciences*. 2nd edn. Elsevier.

Raschke, E., Ohmura, A., Rossow, W. B., Carlson, B. E., Zhang, Y.-C., Stubenrauch, C. *et al.* 2005. Cloud effects on the radiation budget based on ISCCP data (1991 to 1995). *Int. J. Climatol.*, **25**, 1103–1125.

Rissler, J., Vestin, A., Swietlicki, E., Fisch, G., Zhou, J., Artaxo, P. *et al.* 2006. Size distribution and hygroscopic properties of aerosol particles from dry-season biomass burning in Amazonia. *Atmos. Chem. Phys.*, **6**, 471–491.

Roberts, G. C., and Nenes, A. 2005. A continuous-flow streamwise thermal-gradient CCN chamber for atmospheric measurements. *Aerosol Sci. Technol.*, **39**(3), 206–221.

Roberts, P., and Hallett, J. 1968. A laboratory study of ice nucleating properties of some mineral particulates. *Q. J. R. Meteorol. Soc.*, **94**(399), 25–34.

Robock, A., Oman, L., and Stenchikov, G. L. 2008. Regional climate responses to geoengineering with tropical and Arctic SO_2 injections. *J. Geophys. Res.*, **113**, D16101.

Robock, A., Marquardt, A., Kravitz, B., and Stenchikov, G. 2009. Benefits, risks, and costs of stratospheric geoengineering. *Geophys. Res. Lett.*, **36**, L19703.

Roe, G. H. 2005. Orographic precipitation. *Ann. Rev. Earth Planet. Sci.*, **33**, 645–671.

Rogers, R. R., and Yau, M. K. 1989. *A Short Course in Cloud Physics*. Pergamon.

Rosenberg, R. 2005. Why is ice slippery? *Physics Today*, **58**(12), 50–55.

Rosenfeld, D., Rudich, Y., and Lahav, R. 2001. Desert dust suppressing precipitation: a possible desertification feedback loop. *Proc. Natl. Acad. Sci.*, **98**, 5975–5980.

Rosenfeld, D., Kaufman, Y. J., and Koren, I. 2006. Switching cloud cover and dynamical regimes from open to closed Benard cells in response to the suppression of precipitation by aerosols. *Atmos. Chem. Phys.*, **6**, 2503–2511.

Rossow, W. B., and Schiffer, R. A. 1999. Advances in understanding clouds from ISCCP. *Bull. Amer. Meteorol. Soc.*, **80**, 2261–2287.

Salam, A., Lohmann, U., Crenna, B., Lesins, G., Klages, P., Rogers, D. *et al.* 2006. Ice nucleation studies of mineral dust particles with a new continuous flow diffusion chamber. *Aerosol Sci. Technol.*, **40**, 134–143.

Sant, V., Lohmann, U., and Seifert, A. 2013. Performance of a triclass parameterization of the collision–coalescence process in shallow clouds. *J. Atmos. Sci.*, **70**, 1744–1767.

Sassen, K., DeMott, P. J., Prospero, J. M., and Poellot, M. R. 2003. Saharan dust storms and indirect aerosol effects on clouds: Crystal-face results. *Geophys. Res. Lett.*, **30**, doi: 10.1029/2003GL017371.

Saunders, C. 2008. Charge separation mechanisms in clouds. *Space Sci. Rev.*, **137**(1–4), 335–353.

Schaller, R. C., and Fukuta, N. 1979. Ice nucleation by aerosol-particles – experimental studies using a wedge-shaped ice thermal-diffusion chamber. *J. Atmos. Sci.*, **36**, 1788–1802.

Schemm, S., and Wernli, H. 2014. The linkage between the warm and the cold conveyor belts in an idealized extratropical cyclone. *J. Atmos. Sci.*, **71**, doi: 10.1175/JAS–D–13–0177.1.

Schlamp, R. J., Grover, S. N., Pruppacher, H. R., and Hamielec, A. E. 1976. Numerical investigation of the effect of electric charges and vertical external electric fields on the collision efficiency of cloud drops. *J. Atmos. Sci.*, **33**(9), 1747–1755.

Schreier, M., Mannstein, H., Eyring, V., and Bovensmann, H. 2007. Global ship track distribution and radiative forcing from 1 year of AATSR data. *Geophys. Res. Lett.*, **34**(17). L17814.

Schween, J. H., Kuettner, J., Reinert, D., Reuder, J., and Wirth, V. 2007. Definition of "banner clouds" based on time lapse movies. *Atmos. Chem. Phys.*, **7**(8), 2047–2055.

Seaborn, J. B. 2002. *Mathematics for the Physical Sciences*. Springer.

Sehmel, G. A., and Sutter, S. L. 1974. Particle deposition rates on a water surface as a function of particle diameter and air velocity. *J. Rech. Atmos.*, **8**(3–4), 911–920.

Seifert, P., Ansmann, A., Mattis, I., Wandinger, U., Tesche, M., Engelmann, R. *et al.* 2010. Saharan dust and heterogeneous ice formation: Eleven years of cloud observations at a central European EARLINET site. *J. Geophys. Res.*, **115**, D20201. doi:10.1029/2009JD013222.

Seinfeld, J. H., and Pandis, S. N. 2006. *Atmospheric Chemistry and Physics: From Air Pollution to Climate Change*. 2nd ed. Wiley.

Shapiro, M. A., and Keyser, D. 1990. *Fronts, Jet Streams and the Tropopause*. American Meteorolog. Society, pp. 167–191.

Singh, R., Joshi, P. C., and Kishtawal, C. M. 2005. A new technique for estimation of surface latent heat fluxes using satellite-based observations. *Month. Wea. Rev.*, **133**.

Sipilä, M., Berndt, T., Petäjä, T., Brus, D., Vanhanen, J., Stratmann, F. *et al.* 2010. The role of sulfuric acid in atmospheric nucleation. *Science*, **327**(5970), 1243–1246.

Skeie, R. B., Fuglestvedt, J., Berntsen, T., Lund, M. T., Myhre, G., and Rypdal, K. 2009. Global temperature change from the transport sectors: historical development and future scenarios. *Atmos. Env.*, **43**(39), 6260–6270.

Skrotzki, J., Connolly, P., Schnaiter, M., Saathoff, H., Möhler, O., Wagner, R. *et al.* 2013. The accommodation coefficient of water molecules on ice – cirrus cloud studies at the AIDA simulation chamber. *Atmos. Chem. Phys.*, **13**(8), 4451–4466.

Slinn, W. G. N., Hasse, L., Hicks, B. B., Hogan, A. W., Lal, D., Liss, P. S. *et al.* 1978. Some aspects of transfer of atmospheric trace constituents past air-sea interface. *Atmos. Env.*, **12**(11), 2055–2087.

Soden, B. J., and Held, I. M. 2006. An assessment of climate feedbacks in coupled ocean–atmosphere models. *J. Climate*, **19**(23), 6263.

Soden, B. J., Wetherald, R. T., Stenchikov, G. L., and Robock, A. 2002. Global cooling after the eruption of Mount Pinatubo: a test of climate feedback by water vapor. *Science*, **296**, 727–730.

Solomon, S., Qin, D., Manning, M., Alley, R. B., Berntsen, T., Bindoff, N. L. *et al.* 2007. Technical summary. In: Solomon, S., Qin, D., Manning, M., Chen, Z., Marquis, M., Averyt, K. B. (eds.), *Climate Change 2007: The Physical Science Basis. Contribution of Working Group I to the Fourth Assessment Report of the Intergovernmental Panel on Climate Change*. Cambridge University Press.

Spichtinger, P., and Krämer, M. 2012. Tropical tropopause ice clouds: a dynamic approach to the mystery of low crystal numbers. *Atmos. Chem. Phys. Discuss.*, **12**(10), 28109–28153.

Squires, P. 1958. The microstructure and colloidal stability of warm clouds. *Tellus*, **10**(2), 256–261.

Stevens, B., and Feingold, G. 2009. Untangling aerosol effects on clouds and precipitation in a buffered system. *Nature*, **461**(7264), 607–613.

Stevens, B., Vali, G., Comstock, K., Wood, R., van Zanten, M. C., Austin, P. H. *et al.* 2005. Pockets of open cells and drizzle in marine stratocumulus. *Bull. Amer. Meteorol. Soc.*, **86**, 51–57.

Stevenson, D. S., Dentener, F. J., Schultz, M. G., Ellingsen, K., van Noije, T. P. C., Wild, O. *et al.* 2006. Multimodel ensemble simulations of present-day and near-future tropospheric ozone. *J. Geophys. Res.*, **111**, doi:10.1029/2005JD006338.

Stocker, T. F., Qin, D., Plattner, G.-K., Alexander, L. V., Allen, S. K., Bindoff, N. L. *et al.* 2013. Technical summary. Pp. 33–115 in Stocker, T. F., Qin, D., Plattner, G.-K., Tignor, M., Allen, S. K., Boschung, J. *et al.* (eds.), *Climate Change 2013: The Physical Science Basis. Contribution of Working Group I to the Fifth Assessment Report of the Intergovernmental Panel on Climate Change*. Cambridge University Press.

Stolzenburg, M., and Marshall, T. C. 2009. Electric field and charge structure in lightning-producing clouds. Pp. 57–82 in Betz, H. D., Schumann, U., and Laroche, P. (eds.), *Lightning: Principles, Instruments and Applications*. Springer.

Storelvmo, T., and Herger, N. 2014. Cirrus cloud susceptibility to the injection of ice nuclei in the upper troposphere. *J. Geophys. Res.*, **119**(5), 2375–2389.

Storelvmo, T., Kristjánsson, J.-E., and Lohmann, U. 2008. Aerosol influence on mixed-phase clouds in CAM-Oslo. *J. Atmos. Sci.*, **65**, 3214–3230.

Storelvmo, T., Kristjansson, J. E., Muri, H., Pfeffer, M., Barahona, D., and Nenes, A. 2013. Cirrus cloud seeding has potential to cool climate. *Geophys. Res. Lett.*, **40**, 178–182.

Stubenrauch, C. J., Kinne, S., and GEWEX Cloud Assessment Team. 2009. Assessment of global cloud climatologies. In: *GEWEX Newsletter*, vol. 19. GEWEX.

Textor, C., Schulz, M., Guibert, S., Kinne, S., Balkanski, Y., Bauer, S. *et al.* 2006. Analysis and quantification of the diversities of aerosol life cycles within AeroCom. *Atmos. Chem. Phys.*, **6**(7), 1777–1813.

Textor, C., Schulz, M., Guibert, S., Kinne, S., Balkanski, Y., Bauer, S. *et al.* 2007. The effect of harmonized emissions on aerosol properties in global models – an AeroCom experiment. *Atmos. Chem. Phys.*, **7**(17), 4489–4501.

Tilmes, S., Mills, M. J., Niemeier, U., Schmidt, H., Robock, A., Kravitz, B. *et al.* 2015. A new Geoengineering Model Intercomparison Project (GeoMIP) experiment designed for climate and chemistry models. *Geosci. Model Dev.*, **8**(1), 43–49.

Timmreck, C., Graf, H.-F., and Kirchner, I. 1999. A one and a half year interactive simulation of Mt. Pinatubo aerosol. *J. Geophys. Res.*, **104**, 9337–9360.

Tory, K. J., and Dare, R. A. 2015. Sea surface temperature thresholds for tropical cyclone formation. *Journal of Climate*, **28**(20), 8171–8183.

Tremblay, A. 2005. The stratiform and convective components of surface precipitation. *J. Atmos. Sci.*, **62**(5), 1513–1528.

Trenberth, K. E., Fasullo, J. T., and Mackaro, J. 2011. Atmospheric moisture transports from ocean to land and global energy flows in reanalyses. *J. Climate*, **24**(18), 4907–4924.

Twomey, S. A. 1974. Pollution and the planetary albedo. *Atmos. Env.*, **8**(12), 1251 – 1256.

Uman, M. A. 1987. *The Lightning Discharge*. International Geophysics Series, vol. 39. Academic Press.

Vali, G. 1985. Atmospheric ice nucleation – a review. *J. Rech. Atmos.*, **19**, 105–115.

Vargaftik, N. B., Volkov, B. N., and Voljak, L. D. 1983. International tables of the surface-tension of water. *J. Phys. Chem. Ref. Data*, **12**(3), 817–820.

Vial, J., Dufresne, J.-L., and Bony, S. 2013. On the interpretation of inter-model spread in CMIP5 climate sensitivity estimates. *Clim. Dyn.*, **41**(11/12), 3339–3362.

Voigt, M., and Wirth, V. 2014. Mechanisms of banner cloud formation. *J. Atmos. Sci.*, **70**, 3631–3640.

Wagner, W., and Pruss, A. 2002. The IAPWS formulation 1995 for the thermodynamic properties of ordinary water substance for general and scientific use. *J. Phys. Chem. Ref. Data*, **31**(2), 387–535.

Wallace, J. M., and Hobbs, P. V. 2006. *Atmospheric Science: An introductory Survey*. Academic Press.

Wegener, A. 1911. *Thermodynamik der Atmosphäre*. Barth, Leipzig, Germany.

Weisman, M. L., and Klemp, J. B. 1982. The dependence of numerically simulated convective storms on vertical wind shear and buoyancy. *Mon. Wea. Rev.*, **110**(6), 504–520.

Welti, A., Lüönd, F., Stetzer, O., and Lohmann, U. 2009. Influence of particle size on the ice nucleating ability of mineral dusts. *Atmos. Chem. Phys.*, **9**(18), 6705–6715.

Welti, A., Lüönd, F., Kanji, Z. A., Stetzer, O., and Lohmann, U. 2012. Time dependence of immersion freezing: an experimental study on size selected kaolinite particles. *Atmos. Chem. Phys.*, **12**(20), 9893–9907.

Welti, A., Kanji, Z. A., Lüönd, F., Stetzer, O., and Lohmann, U. 2014. Exploring the mechanisms of ice nucleation on kaolinite: from deposition nucleation to condensation freezing. *J. Atmos. Sci.*, **71**(1), 16–36.

Whitby, K. T., and Sverdrup, G. M. 1980. California aerosols – their physical and chemical characteristics. *Adv. Env. Sci. Technol. (United States)*, 477.

Wiacek, A., Peter, T., and Lohmann, U. 2010. The potential influence of Asian and African mineral dust on ice, mixed-phase and liquid water clouds. *Atmos. Chem. Phys.*, **10**, 8649–8667.

Wielicki, B. A., Barkstrom, B. R., Harrison, E. F., Lee, R. B., Smith, G. L., and Cooper, J. E. 1996. Clouds and the Earth's radiant energy system (CERES): an earth observing system experiment. *Bull. Amer. Meteorol. Soc.*, **77**, 853–868.

Wild, Martin. 2012. Enlightening global dimming and brightening. *Bull. Amer. Meteorol. Soc.*, **93**(1), 27–37.

Wise, M. E., Baustian, K. J., Koop, T., Freedman, M. A., Jensen, E. J., and Tolbert, M. A. 2012. Depositional ice nucleation onto crystalline hydrated NaCl particles: a new mechanism for ice formation in the troposphere. *Atmos. Chem. Phys.*, **12**(2), 1121–1134.

WMO, 1975. International cloud atlas. Volume I: Manual on the observation of clouds and other meteors. WMO vol. 407.

Wood, R. 2012. Stratocumulus clouds. *Mon. Wea. Rev.*, **140**(8), 2373–2423.

Wood, R., Kubar, T. L., and Hartmann, D. L. 2009. Understanding the importance of microphysics and macrophysics for warm rain in marine low clouds. Part II: Heuristic models of rain formation. *J. Atmos. Sci.*, **66**(10), 2973–2990.

Wüest, M. 2001. De-aliasing wind information from doppler radar for operational use. Ph.D. thesis, ETH Zurich.

Wurm, G., and Krauss, O. 2008. Experiments on negative photophoresis and application to the atmosphere. *Atmos. Env.*, **42**(11), 2682–2690.

Wylie, D., Jackson, D. L., Menzel, W. P., and Bates, J. J. 2005. Trends in global cloud cover in two decades of HIRS observations. *J. Climate*, **18**(15), 3021–3031.

Zhou, C., Dessler, A. E., Zelinka, M. D., Yang, P., and Wang, T. 2014. Cirrus feedback on inter-annual climate fluctuations. *Geophys. Res. Lett.*

Zipser, E. J. 2003. Some views on "hot towers" after 50 years of tropical field programs and two years of TRMM data. Pp. 49–58 in Tao, W.-K., and Adler, R. (eds.), *Cloud Systems, Hurricanes and the Tropical Rainfall Measuring Mission TRMM*. American Meteorological Society.

Zobrist, B., Marcolli, C., Koop, T., Luo, B. P., Murphy, D. M., Lohmann, U. *et al.* 2006. Oxalic acid as a heterogeneous ice nucleus in the upper troposphere and its indirect aerosol effect. *Atmos. Chem. Phys.*, **6**, 3115–3129.

Index